Advanced Research on the Sustainable Maritime Transportation

Advanced Research on the Sustainable Maritime Transportation

Editors

Xianhua Wu

Jian Wu

Lang Xu

Basel • Beijing • Wuhan • Barcelona • Belgrade • Novi Sad • Cluj • Manchester

Editors

Xianhua Wu
School of Economics and
Management
Shanghai Maritime
University
Shanghai
China

Jian Wu
School of Economics and
Management
Shanghai Maritime
University
Shanghai
China

Lang Xu
College of Transport and
Communications
Shanghai Maritime
University
Shanghai
China

Editorial Office
MDPI AG
Grosspeteranlage 5
4052 Basel, Switzerland

This is a reprint of articles from the Special Issue published online in the open access journal *Journal of Marine Science and Engineering* (ISSN 2077-1312) (available at: https://www.mdpi.com/journal/jmse/special_issues/C50QNR8457).

For citation purposes, cite each article independently as indicated on the article page online and as indicated below:

Lastname, A.A.; Lastname, B.B. Article Title. *Journal Name* **Year**, *Volume Number*, Page Range.

ISBN 978-3-7258-1839-6 (Hbk)
ISBN 978-3-7258-1840-2 (PDF)
doi.org/10.3390/books978-3-7258-1840-2

Contents

About the Editors

Xianhua Wu

Xianhua Wu is a distinguished professor and Dean of the School of Economics and Management, Shanghai Maritime University, China. His areas of specialty include disaster risk analysis, big data analysis, and applied statistics. He has published seven books and more than 130 academic articles in journals like *Risk Analysis, Ecological Economics, Annals of Operations Research, Transportation Research Part, Science of the Total Environment, Journal of International Trade & Economic Development, Weather, Climate and Society*, and others. As a principal investigator, he currently has seven national research projects and is also a member of the Steering Committee of Economics Education (2018-2022), MOE, China.

Jian Wu

Jian Wu is a distinguished professor of the "Ocean Going Plan" at Shanghai Maritime University and a doctoral supervisor. He is a "Minjiang Scholar" chair professor in Fujian Province and a Zhe Jiang Young Scholar in Zhejiang Province. He is also an IEEE Senior Member and has been included in Stanford University's "Global Top 2% Scientists List." His research focuses on decision science and social network analysis, among other areas. He has received the Emerald Citations of Excellence for 2017 (Outstanding Citations Award) and was nominated for the Eni Award in 2023. His research has been recognized by the IEEE Computational Intelligence Society as a "Research Frontier".

Lang Xu

Lang Xu is an associate professor at the College of Transportation and Communication, Shanghai Maritime University, China. He was a visiting scholar and research assistant at Shanghai Jiao Tong University, Nanyang Technological University, and the University of York, and a part-time researcher at the Decision-Making Consulting Base of the Development Research Center of Shanghai Municipal People's Government and Shanghai Social Sciences Innovation Research Base. Currently, he is engaged in research on sustainable port and shipping operations, port and shipping measurement and statistical analysis, and logistics facility layout optimization. In recent years, he has published more than 70 academic articles in journals like *Marine Pollution Bulletin, Maritime Policy & Management, Computers & Industrial Engineering, Ocean & Coastal Management, International Journal of Logistics Research & Application, Operational Research, Computers & Industrial Engineering, Journal of Cleaner Production, Resources Conservation and Recycling, Complexity, International Journal of Low-carbon Technologies, and RAIRO—Operations Research*. He has also been invited as a Guest Editor for journals including Frontiers in Marine Science, Sustainability, and Environment Science and Pollution Research, and has served as a reviewer for journals like *Production and Operations Management, Transportation Research Part B & E, Marine Pollution Bulletin, Marine Policy, European Journal of Operational Research, International Journal of Production Economics, and Ocean and Coastal Management*. He has led or participated in one National Key R&D Program, three National Natural Science Foundation projects, four Shanghai Science and Technology Innovation Action Plan Soft Science Research Projects, and one Shanghai Philosophy and Social Science Planning Young Scholars Project.

Journal of
Marine Science and Engineering

Editorial

Advanced Research on the Sustainable Maritime Transportation

Xianhua Wu [1,*], Jian Wu [1] and Lang Xu [2]

[1] School of Economics and Management, Shanghai Maritime University, Shanghai 201306, China; jyajian@163.com
[2] College of Transport and Communications, Shanghai Maritime University, Shanghai 201306, China; xulang@shmtu.edu.cn
* Correspondence: ghwu@shmtu.edu.cn

Citation: Wu, X.; Wu, J.; Xu, L. Advanced Research on the Sustainable Maritime Transportation. *J. Mar. Sci. Eng.* **2024**, *12*, 1104. https://doi.org/10.3390/jmse12071104

Received: 13 May 2024
Accepted: 20 May 2024
Published: 29 June 2024

With globalization and environmental sustainability growing in importance, the sustainable development of maritime transportation, as the main mode of international trade, is particularly significant. This Special Issue "Advanced Research on Sustainable Maritime Transportation" contains a total of 15 high-quality research papers, covering research on many aspects from green shipping technologies and maritime safety to sustainable port operations, reflecting the multi-dimensional efforts to transform the industry into a low-carbon, efficient, safe, and environmentally friendly one. Hereafter, the research content of these papers will be reviewed according to different themes.

The first theme is green shipping technology. Focusing on this topic, Lebedevas and Milašius [1] studied the challenges in terms of reliability in the transition to low-carbon fuels for maritime diesel engines. By employing statistical and single-zone mathematical models such as IMPULS and AVL BOOST, the study evaluated the combustion cycle parameters and thermal loads of cylinder–piston assemblies for low-carbon fuels, aiming to optimize energy efficiency and reduce harmful emissions. Zhang et al. [2] studied and developed a new fault-tolerant control strategy suitable for permanent magnet synchronous motor drive systems in electric ships. This strategy is based on sinusoidal pulse width modulation with a hierarchical reconstruction of the carrier configuration, which can improve the fault tolerance of the inverter without requiring additional hardware. Xu et al. [3] studied the impact of emission control areas on orders for ecologically designed ships using a difference-in-difference method and a regional-level ship data set. The results show that this implementation significantly increases the number of orders for eco-designed ships. Zhang et al. [4] compared the effectiveness of five machine learning methods in predicting a ship's carbon intensity index. The study used a variety of models, including artificial neural networks, which had the lowest error among all of them and can provide ship operators with an effective carbon emission management tool. Chen et al. [5] used a multinomial logistic model to analyze the key decision-making factors that influence ship owners when choosing fuel for new ships. The study, based on global new ship order data, revealed that shipowners' choices are significantly affected by their nationality, ship type, and economic factors. Wei et al. [6] assessed the maturity and suitability of alternative fuels against the International Maritime Organization's greenhouse gas emission reduction targets by reviewing material compatibility and storage technologies. The results suggest that multi-product ports may have the potential to serve as multi-fuel hubs, while the remaining ports favor specific fuels. Elkafas et al. [7] explored the feasibility of replacing conventional power systems with alternative clean power systems during short sea journeys. The results show that hydrogen-powered proton exchange membrane fuel cells are the best clean power system to ensure zero-emission journeys. Wang et al. [8] developed a nonlinear optimization model to optimize shipping companies' sailing speed and route deployment strategies under the EU's new emissions trading system policy. The results show that appropriate speed adjustment and ship deployment can effectively reduce operating costs. Eom et al. [9] developed and implemented a digital twin port model to optimize ship scheduling and port operations. Their research shows that this model can significantly improve operational efficiency and reduce carbon emissions. Tu et al. [10] developed a cost–benefit model based on marginal abatement cost to evaluate the economics of adopting different greenhouse gas emission reduction measures in the shipping industry. The study provides shipping companies and maritime authorities with a comprehensive analytical framework for developing decarbonization strategies.

The second theme is maritime security. Focusing on this topic, Yang et al. [11] used bibliometric analysis methods to systematically review the scientific literature on unmanned surface vessels from 2000 to 2023. This study evaluates the new technology development trends, main research hotspots, and future development directions of unmanned surface vessels, providing important guidance and inspiration for the advancement and cooperation of academia and industry in unmanned surface vessel technology. Xu et al. [12] conducted a systematic and bibliometric review of articles related to maritime transportation safety management from 2011 to 2022, comprehensively evaluating the progress of research on safety risk analysis, emergency management, and resilience measurement. The study points out the key research directions and challenges of maritime safety management and provides a direction and foundation for future research on this topic.

The third theme is sustainable port operations. Lebedevas and Čepaitis [13] explored the potential of using the organic Rankine cycle in maritime transport as an energy-saving solution to recover waste heat on ships. The study used numerical simulation methods to evaluate the energy efficiency potential of different heat sources under various engine load conditions. The results show that the organic Rankine cycle can significantly improve the energy efficiency of the main engine. Drazdauskas and Lebedevas [14] studied the performance of the mixed combustion of ammonia and diesel in marine diesel engines. The study optimized combustion cycle parameters by adjusting the injection stage and pressure, aiming to reduce emissions of greenhouse gases, NOX, and unburned NH3 and providing a practical solution for decarbonizing maritime transport. Wang et al. [15] developed a mixed-integer linear programming model aimed at helping shipping companies optimize their operating strategies in response to the EU's new marine fuel policy. The study demonstrates the adaptability of the model under fluctuations in fuel prices, ship costs, and fleet size, providing shipping companies with important guidance to effectively reduce costs and comply with EU carbon emission reduction policies.

In summary, this Special Issue not only reveals the latest technologies and methods in the field of sustainable maritime transportation but also highlights the importance of technological innovation and policy adjustment in achieving sustainability goals in this industry. From the application of low-carbon fuels and technological advancements in electric ships to the digital management of ports, as well as adapting and responding to global environmental policies, these studies collectively build a diversified solution framework that provides scientific guidance and a practical direction for the sustainable development of maritime transportation. These studies exemplify how the industry is gradually moving in a more environmentally friendly, safe, and efficient direction consistent with global sustainable development goals.

Author Contributions: Conceptualization, writing, review and editing: X.W., J.W. and L.X. All authors have read and agreed to the published version of the manuscript.

Funding: National Social Science Foundation of China (22FGLB113).

Conflicts of Interest: The authors declare no conflicts of interest.

References

1. Lebedevas, S.; Milašius, E. Methodological Aspects of Assessing the Thermal Load on Diesel Engine Parts for Operation on Alternative Fuel. *J. Mar. Sci. Eng.* **2024**, *12*, 325. [CrossRef]
2. Zhang, F.; Zhang, Z.; Zhang, Z.; Wang, T.; Han, J.; Amirat, Y. A Fault-Tolerant Control Method Based on Reconfiguration SPWM Signal for Cascaded Multilevel IGBT-Based Propulsion in Electric Ships. *J. Mar. Sci. Eng.* **2024**, *12*, 500. [CrossRef]

3. Xu, L.; Zou, Z.; Liu, L.; Xiao, G. Influence of Emission-Control Areas on the Eco-Shipbuilding Industry: A Perspective of the Synthetic Control Method. *J. Mar. Sci. Eng.* **2024**, *12*, 149. [CrossRef]
4. Zhang, C.; Lu, T.; Wang, Z.; Zeng, X. Research on Carbon Intensity Prediction Method for Ships Based on Sensors and Meteorological Data. *J. Mar. Sci. Eng.* **2023**, *11*, 2249. [CrossRef]
5. Chen, S.; Wang, X.; Zheng, S.; Chen, Y. Exploring Drivers Shaping the Choice of Alternative-Fueled New Vessels. *J. Mar. Sci. Eng.* **2023**, *11*, 1896. [CrossRef]
6. Wei, H.; Müller-Casseres, E.; Belchior, C.; Szklo, A. Evaluating the Readiness of Ships and Ports to Bunker and Use Alternative Fuels: A Case Study from Brazil. *J. Mar. Sci. Eng.* **2023**, *11*, 1856. [CrossRef]
7. Elkafas, A.; Rivarolo, M.; Barberis, S.; Massardo, A. Feasibility Assessment of Alternative Clean Power Systems on board Passenger Short-Distance Ferry. *J. Mar. Sci. Eng.* **2023**, *11*, 1735. [CrossRef]
8. Wang, H.; Liu, Y.; Li, F.; Wang, S. Sustainable Maritime Transportation Operations with Emission Trading. *J. Mar. Sci. Eng.* **2023**, *11*, 1647. [CrossRef]
9. Eom, J.; Yoon, J.; Yeon, J.; Kim, S. Port Digital Twin Development for Decarbonization: A Case Study Using the Pusan Newport International Terminal. *J. Mar. Sci. Eng.* **2023**, *11*, 1777. [CrossRef]
10. Tu, H.; Liu, Z.; Zhang, Y. Study on Cost-Effective Performance of Alternative Fuels and Energy Efficiency Measures for Shipping Decarbonization. *J. Mar. Sci. Eng.* **2024**, *12*, 743. [CrossRef]
11. Yang, P.; Xue, J.; Hu, H. A Bibliometric Analysis and Overall Review of the New Technology and Development of Unmanned Surface Vessels. *J. Mar. Sci. Eng.* **2024**, *12*, 146. [CrossRef]
12. Xu, M.; Ma, X.; Zhao, Y.; Qiao, W. A Systematic Literature Review of Maritime Transportation Safety Management. *J. Mar. Sci. Eng.* **2023**, *11*, 2311. [CrossRef]
13. Lebedevas, S.; Čepaitis, T. Complex Use of the Main Marine Diesel Engine High- and Low-Temperature Waste Heat in the Organic Rankine Cycle. *J. Mar. Sci. Eng.* **2024**, *12*, 521. [CrossRef]
14. Drazdauskas, M.; Lebedevas, S. Optimization of Combustion Cycle Energy Efficiency and Exhaust Gas Emissions of Marine Dual-Fuel Engine by Intensifying Ammonia Injection. *J. Mar. Sci. Eng.* **2024**, *12*, 309. [CrossRef]
15. Wang, H.; Liu, Y.; Wang, S.; Zhen, L. Optimal Ship Deployment and Sailing Speed under Alternative Fuels. *J. Mar. Sci. Eng.* **2023**, *11*, 1809. [CrossRef]

Journal of
*Marine Science
and Engineering*

Article

Study on Cost-Effective Performance of Alternative Fuels and Energy Efficiency Measures for Shipping Decarbonization

Huan Tu [1,2,*], Zheyu Liu [3] and Yufeng Zhang [1,2]

[1] Wuhan Rules and Research Institute, China Classification Society, Wuhan 430022, China; yufengz@ccs.org.cn
[2] Marine Sustainable Energy and Green Technology Laboratory, China Classification Society, Wuhan 430034, China
[3] Naval Architecture and Shipping College, Guangdong Ocean University, Zhanjiang 524088, China; liuzheyu40@gmail.com
* Correspondence: htu@ccs.org.cn

Abstract: Within the context of global initiatives to address climate change, the shipping industry is facing increasingly intensified pressure to decarbonize. The industry is engaging in the exploration and implementation of greenhouse gas (GHG) emission reduction measures, including energy efficiency technologies and alternative fuels, with the objective of accelerating the progression towards greenhouse gas mitigation. The application of various GHG emission reduction measures usually requires different levels of investment costs, and economic feasibility is a key factor influencing policy formulation and investment decisions. In this regard, this paper developed a cost-effective model for energy efficiency measures and alternative fuels based on the marginal abatement cost (MAC) methodology. This model can distinguish the differences between energy efficiency measures and alternative fuels in terms of Tank-to-Wake emissions and Well-to-Wake emissions in the GHG emission evaluation system. By taking typical ship types with significant emission contributions as study cases, i.e., bulk carriers (61–63K DWT), container ships (8000 TEU), product tankers (115K DWT), crude oil tankers (315–320K DWT), and Ro-Ro passenger ferries (3500 DWT), the GHG abatement cost-effective performance of major categories of measures such as operational measures, technical measures, renewable energy sources, and alternative fuels were calculated. According to the MAC results, the marginal abatement cost curves were plotted based on the ranking of energy efficiency measures and alternative fuels, respectively. The impacts of bunker fuel prices and carbon market prices on the cost-effectiveness were analyzed. The research results provided the GHG abatement potential of the integrated application of cost-effective energy efficiency measures, the cost-effectiveness ranking of alternative fuels, and the carbon emission price expected to bridge the price gap between alternative fuels and conventional bunker fuel. The presented methodology and conclusions can be used to assist shipping companies in selecting emission reduction measures, and to support maritime authorities in developing market-based measures.

Keywords: greenhouse gas abatement measures; energy efficiency measures; alternative fuels; marginal abatement cost curves; cost-effectiveness

Citation: Tu, H.; Liu, Z.; Zhang, Y. Study on Cost-Effective Performance of Alternative Fuels and Energy Efficiency Measures for Shipping Decarbonization. *J. Mar. Sci. Eng.* **2024**, *12*, 743. https://doi.org/10.3390/jmse12050743

Academic Editor: Theocharis D. Tsoutsos

Received: 9 April 2024
Revised: 27 April 2024
Accepted: 27 April 2024
Published: 29 April 2024

1. Introduction

In the context of addressing climate change collectively across nations and industries globally, the international shipping industry is accelerating the process of reducing greenhouse gas emissions. In July 2023, the International Maritime Organization (IMO) adopted the 2023 IMO Strategy on the Reduction of GHG Emissions from Ships at the 80th session of the Marine Environment Protection Committee (MEPC 80), setting out a new goal of reaching net-zero GHG emissions by or around 2050 as well as taking up zero or near-zero GHG emission technologies, fuels, and/or energy sources to represent at least 5%, striving for 10%, of the energy used by international shipping by 2030 [1].

In response to the increasingly stringent greenhouse gas emission reduction targets, the industry is actively developing and applying emission reduction measures, including technical energy efficiency measures, operational energy efficiency measures, and low-carbon/zero-carbon alternative fuels [2]. In policy formulation and investment decisions concerning greenhouse gas reduction measures, the emissions reduction potential and the implementation cost are two crucial considerations [3]. The industry seeks to prioritize measures with high emissions reduction potential and low cost implications, but in practice, these two factors are often challenging to balance. In other words, measures with a high emissions reduction potential often involve significant costs, while those with low implementation costs typically offer a limited emissions reduction potential. Therefore, policy makers and investment decision makers are confronted with the challenge of selecting suitable and cost-effective carbon reduction measures [4–6]. To this end, marginal abatement costs (MACs) and marginal abatement cost curves (MACCs) have been widely used to assess the cost-effectiveness of various greenhouse gas emission reduction measures for ships [7–9]. The marginal abatement cost refers to the cost of reducing one additional unit of emission. Based on the relationship between marginal abatement cost and emission reduction potential, MACC can be plotted [10,11]. With MACC to represent the economic feasibility of various emission reduction measures, it is possible to rank all the measures according to their cost-effectiveness and propose priority measures for implementation. The MACC method has become an effective analytical tool providing support for policy formulation and investment decisions [12–14].

1.1. Literature Review

According to the relevant literature on the application of the MACC method in reducing greenhouse gas emissions in the shipping industry, there are typically two main methods used to develop MACC models [15]. In the first method, the MACC models are based on the cumulative assessment of the integrated application of various emissions reduction measures by a fleet, generating a linear cost-effectiveness trend line corresponding to the abatement potential for a specific year. The first method is primarily applied at the macro-analysis level of fleets from companies, countries, or globally. In the second method, the MACC models are based on an individual assessment of various emission reduction measures applied in a specific ship, with the cost-effectiveness and abatement potential of each measure assessed in isolation. Subsequently, the emission reduction measures are ranked from lowest to highest to cost-effectiveness, forming step-form curves that represent the MAC of the abatement measures over their whole lifetime. The second method is primarily applied to the micro-level analysis of individual vessels adopting different emissions reduction measures. A brief literature review extracted from various studies is presented in Table 1, with considerations in terms of method category, application ship types, and emission reduction measures to review the related studies.

Previous research indicates that, since 2009, both industry and academia have carried out a series of studies on marginal abatement costs in the field of greenhouse gas emissions reduction for ships. Viewed from the perspective of the MACC method, research reports from the IMO and related maritime consulting agencies typically use the first MACC method, aiming to investigate the overall abatement potential and cost-effectiveness of the global fleet in implementing greenhouse gas emission reduction measures, and thus assess the global fleet's emission reduction potential and establish rational and feasible emission reduction targets. For shipping companies, maritime authorities, and relevant research scholars, there is a preference to focus on the cost-effectiveness of applying different emissions reduction measures on specific vessel types. This assists in investment decisions and regulatory policy making and hence typically adopt the second MACC method. Viewed from the perspective of the investigated abatement measures, the development trend of research hotspots in emissions reduction measures is closely related to the greenhouse gas emissions reduction pathways and technology development in the shipping industry. In recent years, with the strengthening of emissions reduction targets in shipping, the research focus has gradually shifted towards alternative fuels.

Table 1. Literature review on previous MACC studies of GHG abatement measures for ships.

MACC Method	Country/Organization	Year	Application Ships	Abatement Measures	Ref.
First type: cumulative assessment for cost-effectiveness of integrated application of various abatement measures by a fleet (linear cost-effectiveness trend line)	IMO	2009	Global fleet (14 ship types)	25 energy efficiency measures (10 groups)	[16]
	Netherlands	2009	Global fleet (14 ship types)	29 energy efficiency measures (12 groups)	[17]
	IMO	2014	Global fleet (14 ship types)	22 measures (including LNG and biofuel)	[18]
	IMO	2020	Global fleet (13 ship types)	34 energy efficiency measures (3 groups) + 10 alternative fuels	[19]
	Norway	2011	Global fleet (7 ship types)	25 energy efficiency measures (3 groups) + LNG	[20]
	United States of America	2012	Global fleet (14 ship types)	12 energy efficiency measures (6 operational + 6 technical)	[21]
	Germany	2012	Global fleet (14 ship types)	22 energy efficiency measures (15 groups)	[22]
	China	2019	Global fleet (tankers, containers, and bulk carriers)	14 energy efficiency measures (5 optional + 9 technical)	[8]
Second type: individual assessment of various abatement measures applied in case ships (step-form cost-effectiveness curves)	Norway	2009	2 Case ships (74,000 DWT bulk carrier, 8000 TEU Container ship)	12 energy efficiency measures	[23]
	Organization for Economic Co-operation and Development and the International Transport Forum	2009	8500 TEU Container ship	Slow steaming	[24]
	Germany	2012	Container ship fleet	12 energy efficiency measures	[25]
	Netherlands	2015	10 Case ships	18 energy efficiency measures + 2 alternative fuels	[26]
	Singapore	2016	3 Case ships (bulk carrier, container ship, tanker)	14 energy efficiency measures (5 optional + 9 technical)	[27]
	Norway	2017	6 Case ships	8 alternative fuels	[28]
	China	2018	2 Case ships (feeder container ship, ferry)	LNG	[29]
	Czech	2020	9 Case ships	12 energy efficiency measures	[30]
	Norway	2020	4 Case ships	8 alternative fuels	[31]
	Cyprus	2021	4 Case ships	18 energy efficiency measures (3 groups) + 2 alternative fuels (LNG, biofuel)	[14]
	Germany	2022	Not specified	4 alternative fuels (LNG, methanol, ammonia, and hydrogen)	[9]
	Turkey	2022	Bulk carrier	Ammonia	[32]

1.2. Research Gaps and Limitations

Summarizing the previous literature in this field, although extensive studies have developed MACC models to investigate the cost-effectiveness and abatement potentials of representative greenhouse gas emission reduction measures, there are still some gaps and limitations in terms of abatement measures, GHG emission assessment methods, and market-based measure implications. Firstly, there is limited research focusing on both energy efficiency measures and alternative fuels. Currently, applying energy efficiency technologies will not be enough to meet the increasingly stringent emission reduction targets, and thus, the development trend is toward the integrated application of alternative fuels and energy efficiency measures to achieve a greater emission reduction potential [33]. Secondly, the international regulatory system for assessing greenhouse gas emissions from marine fuels is shifting from a vessel-based approach to a lifecycle approach. The MACC of marine alternative fuels calculated following the vessel-based approach will be no longer applicable to the upcoming lifecycle assessment method [34]. Thirdly, previous MACC studies did not adequately account for the influence of market mechanisms. The EU has currently introduced a package of market-based mechanisms to the shipping industry, including the Emissions Trading System (ETS) and FuelEU Maritime. At the IMO level, the carbon pricing mechanism linked to marine fuel GHG intensity is also under discussion. Therefore, it is necessary to investigate the impact of carbon pricing mechanism on the cost-effective performance of emission reduction measures.

1.3. Contribution of This Research

In view of these gaps and limitations, this paper develops a cost-effective performance model for typical energy efficiency measures and alternative fuels based on the second MACC method. The MACC model distinguishes between energy efficiency technologies and alternative fuels in the emission assessment system by using the Tank-to-Wake and Well-to-Wake approach, respectively. Moreover, the model also incorporates the influencing factor of the carbon pricing mechanism. Based on the MACC model, greenhouse gas emission reduction measures, including technological energy efficiency measures, operational efficiency measures, renewable energy utilization measures, and alternative fuels, are calculated and ranked for five typical vessel types: bulk carriers, container ships, product tankers, crude oil tankers, and Ro-Ro passenger ferries. Sensitivity analyses are carried out to analyze the impact of fuel prices on cost-effectiveness and to estimate the carbon emission pricing needed to bridge the cost gap between alternative fuels and traditional fuels. The proposed MACC models and results can provide insights into greenhouse gas emission reduction measure selection, investment decisions, and policy formulation in the shipping industry.

2. Methodology

2.1. Research Procedures

The MACC research requires determining each project's financial details and GHG abatement potential over the project's lifecycle. The research procedure as shown in Figure 1 comprises the five steps below:

- Conduct a comprehensive survey of various representative GHG abatement measures, and screen applicable measures for case study vessels.
- Develop a MAC model based on the survey data, and calculate the MAC of individual abatement measures.
- Rank the measures according to their cost-effectiveness, and construct a MACC based on the relationship between the MAC and the abatement potential of each measure.
- Perform a sensitivity analysis to investigate several input parameters' influence on the cost-effectiveness.
- Propose recommendations on the abatement measure application and policy development.

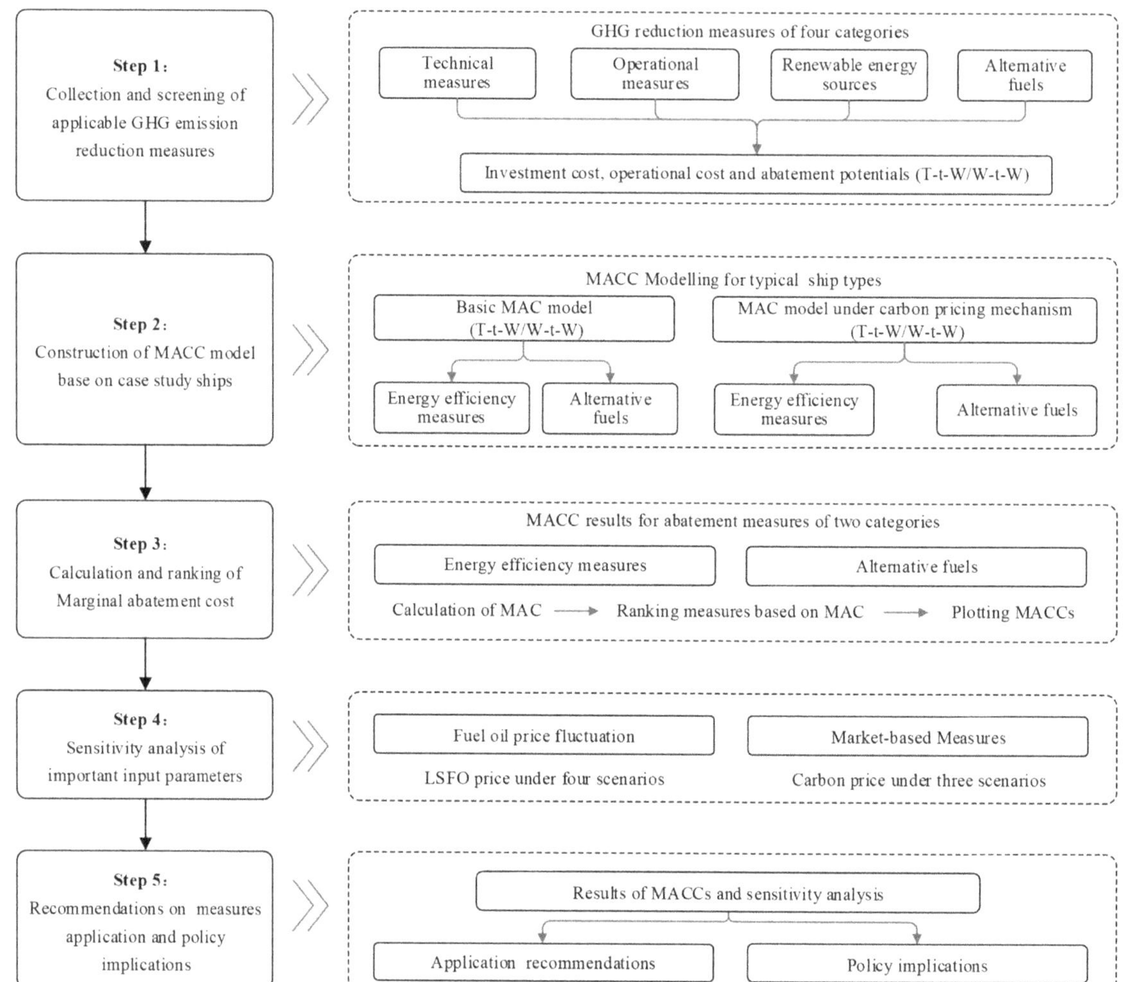

Figure 1. Research procedures of the MACC study.

2.2. Basic MAC Model

The method described in the IMO MEPC62 report is adopted in this paper to develop the MAC model for abatement measures [18]. The marginal abatement cost is defined as the quotient between net costs of implementing an abatement measure and its GHG emission abatement amount [7]. For the net costs, all costs of different categories need to be annualized for calculation [14,35]. Therefore, the MAC of an abatement measure can be calculated using the following equation:

$$MAC = \frac{\Delta NCOST}{\Delta CO_{2e}} \tag{1}$$

where

$\Delta NCOST$ represents the annual net cost of implementing an abatement measure compared with conventional ships (USD/year);
ΔCO_{2e} represents the annual GHG emission abatement amount (t CO_{2e}/MJ).

2.2.1. Net Cost

The net cost of implementing a mitigation measure can be defined as the following equation:

$$\Delta NCOST = IC + OC - CS \tag{2}$$

where

IC is the annualized investment cost of implementing the measure (USD/year);
OC is the operational cost related to using the measure (USD/year);
CS is the cost savings obtained by implementing the measure (USD/year).

The annualized investment cost can be calculated using the following equation:

$$IC = TC \times \frac{d}{1 - (1 + d)^{-L}} \tag{3}$$

where

TC is the total investment cost of implementing the measure;
d is the discount rate;
L is the lifetime of a vessel implementing the measure.

The operational cost needs to be calculated according to measure categories. For energy efficiency measures, the data for operational cost can be collected from the related literature. For alternative fuels, the operational costs are the primarily fuel costs. The cost function of using alternative fuels can be defined as the following equation:

$$OP = FP_{alt} \times FC_{LSFO} \times \frac{LHV_{LSFO}}{LHV_{alt}} \tag{4}$$

where

FP_{alt} is the price of the alternative fuel (USD/t);
FC_{LSFO} is the annual LSFO fuel consumption of the vessel (t/year);
LHV_{LSFO} and LHV_{alt} represent the lower heating value of LSFO and alternative fuel, respectively (MJ/kg).

The cost savings depend on the measure category:

$$CS = \begin{cases} FP_{LSFO} \times FC_{LSFO} \times AP & \text{for energy efficiency measures} \\ FP_{LSFO} \times FC_{LSFO} & \text{for alternative fuels} \end{cases} \tag{5}$$

where AP represents the abatement potential of energy efficiency measures (%).

2.2.2. Abatement Amount

The abatement amount of energy efficiency measures and alternative fuels are assessed based on the Tank-to-Wake and Well-to-Wake approaches, respectively. Therefore, the calculation of the abatement potentials of energy efficiency measures and alternative fuels should be distinguished. The GHG emission considered in this paper includes CO_2, CH_4, and N_2O, and they are accounted for based on the measurement unit of CO_2 equivalent [36].

The abatement amount for energy efficiency measures can be calculated using the following equation:

$$\Delta CO_{2e} = \begin{cases} GHG_{TtW} \times FC_{LSFO} \times LHV_{LSFO} \times AP_{eff} & \text{for energy efficiency measures} \\ GHG_{WtW} \times FC_{LSFO} \times LHV_{LSFO} \times AP_{alt} & \text{for alternative fuels} \end{cases} \tag{6}$$

where

GHG_{TtW} is the GHG emission factor of LSFO in the scope of Tank-to-Wake (g CO_{2e}/MJ);
GHG_{WtW} is the GHG emission factor of LSFO in the scope of Well-to-Wake (g CO_{2e}/MJ);
AP_{eff} is the abatement potential of the energy efficiency measure (%);

AP_{alt} is the abatement potential of the alternative fuel (%).

Based on the Well-to-Wake methodology in the IMO LCA guidelines [34], the Well-to-Wake GHG emission factor is calculated as follows:

$$GHG_{WtW} = GHG_{WtT} + GHG_{TtW} \qquad (7)$$

where GHG_{WtT} is the GHG emission factor of LSFO in the scope of Well-to-Tank (g CO_{2e}/MJ).

The Well-to-Tank GHG emissions factor is calculated according to following equation:

$$GHG_{WtT} = e_{ec} + e_l + e_p + e_{td} - e_c - e_{sca} - e_{ccs} - e_{ccu} \qquad (8)$$

where

e_{ec} is the emissions from the extraction or from the cultivation of raw materials, g CO_{2e}/MJ;
e_l is the annualized emissions from carbon stock changes caused by land-use change (over 20 years), g CO_{2e}/MJ;
e_p is the emissions from processing, including electricity generation, g CO_{2e}/MJ;
e_{td} is the emissions from transport and distribution, g CO_{2e}/MJ;
e_c is the emissions credits generated by biomass growth, g CO_{2e}/MJ;
e_{sca} is the emission savings from soil carbon accumulation via improved agricultural management, g CO_{2e}/MJ;
e_{ccs} is the emission savings from CO_2 capture and geological storage, g CO_{2e}/MJ;
e_{ccu} is the emission savings from CO_2 capture and utilization, g CO_{2e}/MJ.

The Tank-to-Wake GHG emission factors are calculated according to the following equation:

$$GHG_{TtW} = \frac{\left(1 - C_{slip}\right) \times \left(C_{fCO_2} + C_{fCH_4} \times GWP_{CH_4} + C_{fN_2O} \times GWP_{N_2O}\right) + \left(C_{slip} \times GWP_{CH_4}\right) - e_{occs}}{LHV_{alt}} \qquad (9)$$

where

C_{slip} is the coefficient accounting for fuel slip (% of fuel mass);
C_{fCO_2} is the CO_2 emission conversion factor (g CO_2/g fuel);
C_{fCH_4} is the CH_4 emission conversion factor (g CH_4/g fuel);
C_{fN_2O} is the N_2O emission conversion factor (g N_2O/g fuel);
GWP_{CH_4} is the Global Warming Potential of methane (g CO_{2e}/g CH_4);
GWP_{N_2O} is the Global Warming Potential of N_2O (g CO_{2e}/g N_2O);
e_{occs} is the emission savings from on-board CO_2 capture and geological storage (g CO_{2e}/MJ);
LHV_{alt} is the lower heating value of alternative fuel (MJ/g).

2.3. MAC Model under Carbon Pricing

To incentivize the adoption of GHG emission abatement measures, several carbon pricing mechanism proposals are under discussion [37]. The carbon pricing is a market-based instrument that sets a price on carbon dioxide (CO_2) or equivalent GHG emissions. When considering the potential impact of carbon prices, the MAC calculation formula can be modified to the following form:

$$MAC' = \frac{\Delta NCOST'}{\Delta CO_{2e}} \qquad (10)$$

where $\Delta NCOST'$ represents the new annual net cost taking into account the cost changes caused by the carbon price and ΔCO_{2e} is still given by Equation (6).

The new annual net cost will be reduced due to the lower cost of the market-based mechanism achieved by implementing emission abatement measures, and therefore,

$$\Delta NCOST' = IC + OC - CS - MS \tag{11}$$

$$MS = \Delta CO_{2e} \times CP \tag{12}$$

where

MS is the annual carbon price savings achieved by implementing the abatement measure (USD/year);
CP is the carbon price determined in the market-based mechanism (USD/t CO_{2e}).

By combining Equations (1), (10), (11) and (12), it can be derived that

$$MAC' = \frac{\Delta NCOST}{\Delta CO_{2e}} = \frac{\Delta NCOST - MS}{\Delta CO_{2e}} = \frac{\Delta NCOST - \Delta CO_{2e} \times CP}{\Delta CO_{2e}} = MAC - CP \tag{13}$$

2.4. MACC Construction

After the MAC value and abatement amount of each measure are calculated, a MACC can be constructed. The MACC is typically represented in a two-dimensional coordinate system in the form of a histogram, with the MAC value is displayed on the Y-axis and the annual GHG abatement amount on the X-axis. The abatement measures are arranged in ascending order of MAC value from left to right on the Y-axis [11]. Based on the results of the MAC value, abatement measures are typically classified into negative MAC measures and positive MAC measures. The negative MAC measures are below the X-axis, indicating that these measures can achieve an emission reduction while saving costs; whereas the positive MAC measures are above the X-axis, signifying that these emission reduction measures may require a cost input.

2.5. Sensitivity Analysis

The calculation results of the MACC model are affected by factors such as input parameters and assumptions. A sensitivity analysis provides a way to describe how the model result responds to changes in the input data and assumptions [20]. This study focuses on two important parameters: fuel price and carbon price. Fuel prices are significantly affected by global political and economic factors, and their fluctuations are difficult to predict. Carbon price may be subject to uncertainty due to market-based mechanism implementations, abatement technology development, and other factors.

3. Case Study

3.1. Ship Types

The primary focus of this paper is to examine the cost-effectiveness of applying greenhouse gas emission reduction measures to specific vessels. Therefore, it is necessary to select representative ship types as the subjects of study. To this end, five ship types featuring high GHG emission contributions and strong demand in the new-building market are considered: bulk carriers (61–63K DWT), container ships (8000 TEU), product tankers (115K DWT), crude oil tankers (315–320K DWT), and Ro-Ro passenger ferries (3500 DWT). The basic technical parameters and information of the five ship types are shown in Table 2 [19].

Table 2. Basic technical parameters and information of the five ship types.

Ship Type	Classification	Main Engine Power (kW)	Design Speed (kn)	Fuel Consumption (t/year)	Newbuilding Cost (Million USD)
Bulk carrier	Handysize (61–63K DWT)	10,000	14.6	4900	34.5
Container ship	8000 TEU	68,000	25.0	264,000	98
Product tanker	LR2 (115K DWT)	13,000	14.8	5400	63
Crude oil tanker	VLCC (315–320K DWT)	26,000	15.5	145,000	117.5
Ro-Ro passenger ferry	3500 DWT	16,000	20.3	7900	110

3.2. Abatement Measures

In this study, 19 typical emission reduction measures are screened for the MACC analysis by reviewing scientific studies and consulting industrial experts. The investigated 19 abatement measures can be classified into four categories: (1) 7 technical measures, (2) 3 operational measures, (3) 3 renewable energy utilization measures, and (4) 6 alternative fuels.

The majority of emission reduction measures are applicable to different ship types. However, several measures are influenced by factors such as ship type, tonnage, and main engine power, resulting in certain limitations on their applicability. To ensure a comprehensive assessment of the cost-effectiveness of different measures when applied to specific vessel types, the applicability of various measures is taken into consideration as extensively as possible. At the same time, some of these measures are considered to be mutually exclusive and are not suitable for simultaneous application, for instance, measures highly correlated in emission reduction mechanisms (such as Flettner rotor and rigid sail) or measures with incompatible applications (such as slow steaming and waste heat recovery).

Moreover, when considering applicable ship types for alternative fuels such as methanol and ammonia, the potential hazards and impacts of their toxicity on humans and the environment should be closely considered. Methanol has slight toxicity, and skin contact can cause irritation, inflammation, or burns. Methanol is not persistent in the environment and biodegrades quickly [38]. Methanol fuel has been practically applied on chemical tankers, Ro-Ro passenger ferries, and container ships, and the IMO guidelines for the safety of ships using methanol as fuel has been established. Therefore, the safety of methanol application on various ship types can be ensured. Ammonia is highly toxic, and contact can result in irritation, blindness, and even death [39,40]. Ammonia is also toxic to aquatic life and, because of its high solubility in water, can damage the marine ecology if large quantities are spilled. Current regulations do not permit the use of ammonia as a marine fuel due to its toxicity. The IMO is currently evaluating how the IGF Code needs to change to allow ammonia as fuel [41]. Accordingly, we consider that the applicable ship types for ammonia are mainly limited to cargo ships at the preliminary development stage.

Based on our literature review and expert consultation, basic information on ship type applicability, abatement potentials, investment costs, and operational costs of various energy efficiency measures and alternative fuels are collected and calculated, as detailed in Tables 3 and 4, respectively.

Table 3. Basic information for energy efficiency measures.

Category	Sub-Category	Abatement Measures	Applicability	Abatement Potential	Investment Cost (USD)	Operational Cost (USD)	Ref.
Energy efficiency measures	Operational measures	Slow steaming (SS, with 10% reduction)	All ship types except for cruise vessels and ferries	19%	N/A	N/A	[42]
		Optimization of Trim and Ballast (OTB)	All ship types	1.5–4%	26,700	N/A	[14]
		Propeller maintenance	All ship types	1%	3000–4500 (Maintenance at intervals of 5 years)	N/A	[17]
	Technical measures	Optimized water flow of hull openings (OWF)	All ship types	3%	42,000–240,000	N/A	[14,42]
		Air lubrication (AL)	• Bulk carriers and crude oil tanker > 60,000 dwt. • Container ships > 2000 TEU • LPG/LNG carriers	5–7%	Approx. 3% of shipbuilding cost	11,000	[14,42]
		Hull coating (HC)	All ship types	1.5%	Approx. 30 × DWT^(2/3) (Generally recoated at intervals of 5 years)	N/A	[42]

Table 3. *Cont.*

Category	Sub-Category	Abatement Measures	Applicability	Abatement Potential	Investment Cost (USD)	Operational Cost (USD)	Ref.
Energy efficiency measures	Technical measures	Propeller boss cap with fins (PBCF)	All ship types	2%	79,000–520,000	N/A	[14]
		Main engine tuning (MET)	All ship types	0.45%	27,000–48,000	N/A	[14,42]
		Waste heat recovery (WHR)	Main engine power ≥ 10,000 kW (slow steaming vessel would not be able to use WHR)	3–8%	327 USD/kW (Proportional to main engine power)	10,000–30,000/year	[43]
		Speed control of pumps and fans (SCPF)	All ship types	0.5%	100–200 USD/kW (Auxiliary engine power)	N/A	[44]
	Renewable energy sources	Flettner rotors (FR)	Bulk carriers, crude oil tankers, chemical tankers, and product tankers (above 10,000 DWT)	8.5%	2,000,000–4,000,000	N/A	[17,45]
		Rigid sails (RS)	Bulk carriers, crude oil tankers, chemical tankers, and product tankers (above 10,000 DWT)	3–5%	300,000–600,000	N/A	[43,45]
		Solar panels (SP)	Ships have sufficient deck space available (tankers, vehicle carriers, and Ro-Ro vessels)	0.2%	3400 USD/kW (Power of SP generally calculated as 1% of the auxiliary engine power)	N/A	[14,17]

Table 4. Basic information for alternative fuels.

Category	Sub-Category	Abatement Measures	Applicability	Abatement Potential	Investment Cost (USD)	Operational Cost (USD)	Ref.
Alternative fuels	Fossil fuel	LNG	All ship types	13.9%	Approx. 15–20% higher than conventional vessels	Mainly fuel cost	[34]
	Biofuels	Bio-LNG	All ship types	77.7%	Approx. 15–20% higher than conventional vessels	Mainly fuel cost	[34]
		Bio-methanol	All ship types	85.0%	Approx. 14.4% higher than conventional vessels	Mainly fuel cost	[34]
		Hydrotreated vegetable oil (HVO)	All ship types	82.1%	Equivalent to conventional vessels	Mainly fuel cost	[34]
	Electrofuels	E-methanol	All ship types	95%	Approx. 14.4% higher than conventional vessels	Mainly fuel cost	[34,46]
		E-ammonia	Cargo ships	100%	Approx. 21.2% higher than conventional vessels	Mainly fuel cost	[47]

3.3. Emission Factors

For alternative fuels with mature production technologies and emission factors well proven in the industry, such as LSFO, LNG and bio-LNG, the emission factors are calculated using the Well-to-Wake methodology specified in the IMO LCA guidelines. For alternative fuels whose production technology has not yet matured and default emission factors have not been specified in relevant regulations, such as bio-methanol, e-methanol, and e-ammonia, the emission factors in the relevant research literature are referenced. According to the sustainability criteria applied to alternative fuels in the IMO LCA guidelines, the Well-to-Wake GHG emission reduction potentials of various alternative fuels are evaluated with LSFO as the reference; this principle is also adopted in this study. The GHG emission factors of various marine fuels from the perspective of Well-to-Tank, Tank-to-Wake, and Well-to-Wake are summarized in Table 5.

Based on the results of Well-to-Wake emission factors, methanol or ammonia that is produced using fossil energy could lead to increased GHG emissions in a lifecycle perspective. Therefore, the fossil-based methanol and ammonia do not comply with the sustainability criteria and will not considered in the MACC analysis.

Table 5. GHG emission factors for alternative fuels.

Fuel Category	Fuel Type	Engine Type	GHG Emission Factors (g CO_{2e}/MJ)			Abatement Potential
			GHG_{WtT}	GHG_{TtW}	GHG_{WtW}	
Fossil fuels	LSFO	Diesel	13.2	76.8	90	Baseline
	LNG	DF Diesel	18.5	59.0	77.5	13.9%
	Methanol	DF Diesel	31.3	71.6	100.4	−14.3%
	Ammonia	DF Diesel	121	0	121	−34.4%
Biofuels	Bio-LNG	DF Diesel	−38.9	59.0	20.1	77.7%
	Bio-methanol	DF Diesel	−58.1	71.6	11.0	85.0%
	HVO	Diesel	−20.7	71.9	51.2	41.3%
Electrofuels	E-Methanol	DF Diesel	−67.1	71.6	4.5	95.0%
	E-Ammonia	DF Diesel	0	0	0	100%

3.4. Fuel Prices

As described in Table 3, the operational costs for using alternative fuels in the MAC calculation mainly consider the annual fuel cost. Considering that the operational lifecycle of a vessel typically spans 25 years, the calculation of annual fuel costs should be based on the average fuel prices for the next 25 years. Accordingly, the average price of each alternative fuel is derived from the fuel price prediction spanning from 2025 to 2050, as presented in Table 6 [48,49].

Table 6. Average price of alternative fuels (2025–2050).

Fuel Type	LSFO	LNG	Bio-LNG	Bio-Methanol	HVO	E-Methanol	E-Ammonia
Average price (USD/t)	455	425	1416	612	1750	1174	740

4. Results and Discussion

4.1. Results of MACC Basic Model

For bulk carriers (61–63K DWT), container ships (8000 TEU), product tankers (115K DWT), VLCCs (315–320K DWT), and Ro-Ro passenger ferries (3500 DWT), MACCs are developed for each vessel category over the period of 2025 to 2050, as shown in Figures 2–6. The MACC results for energy efficiency measures and alternative fuels are presented, respectively, in the left and right coordinate systems, as the emission reduction potential of the two category measures are evaluated based on the Tank-to-Wake and Well-to-Wake approaches, respectively.

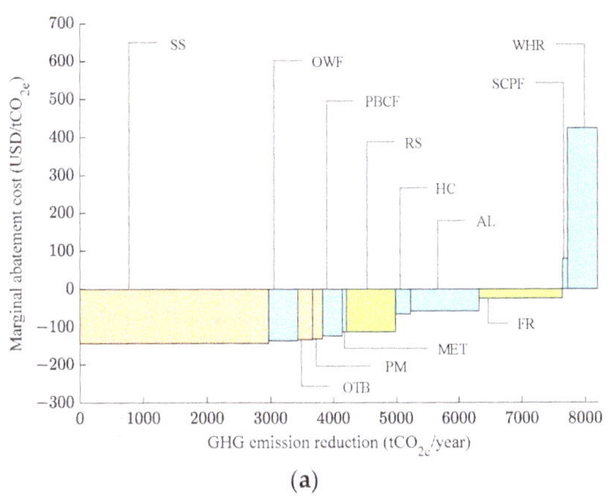

(a)

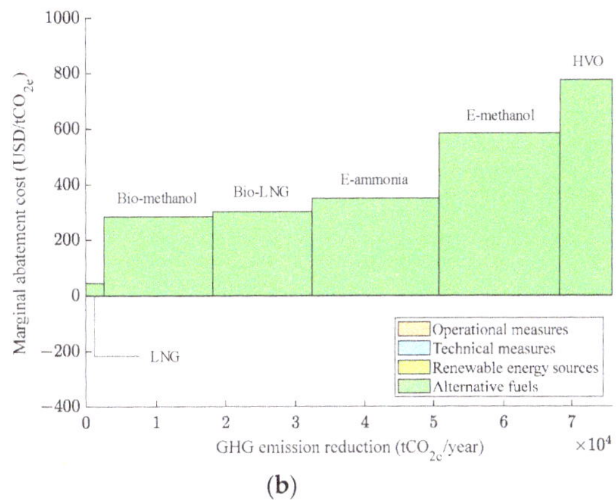

(b)

Figure 2. MACC results for bulk carriers (61–63K DWT). (**a**) Energy efficiency measures. (**b**) Alternative fuels.

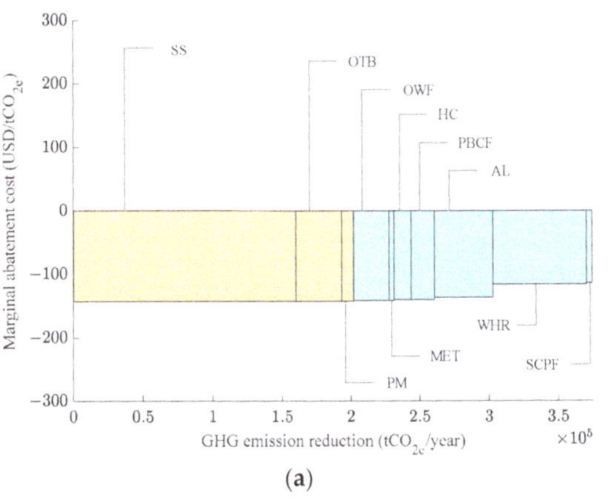

(a)

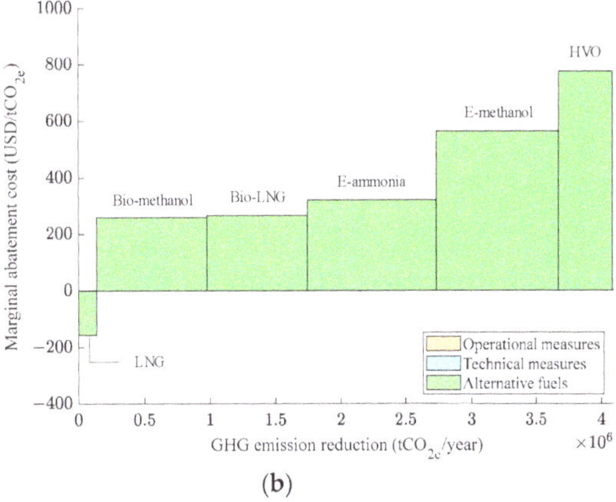

(b)

Figure 3. MACC results for container ships (8000 TEU). (**a**) Energy efficiency measures. (**b**) Alternative fuels.

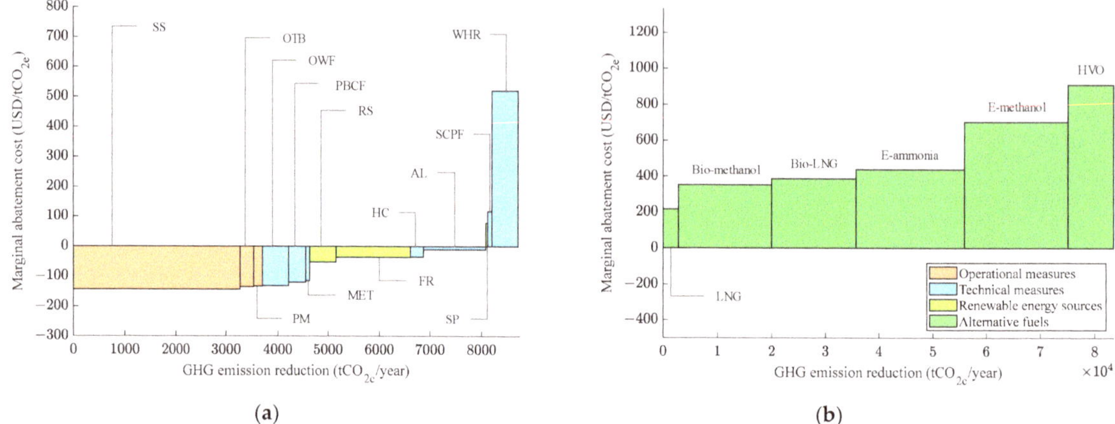

Figure 4. MACC results for product tank (115K DWT). (**a**) Energy efficiency measures. (**b**) Alternative fuels.

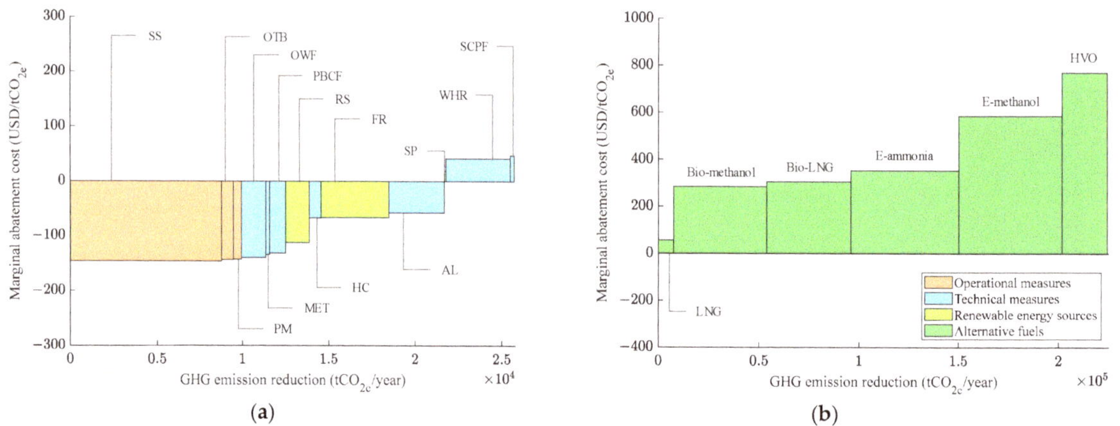

Figure 5. MACC results for VLCCs (315–320K DWT). (**a**) Energy efficiency measures. (**b**) Alternative fuels.

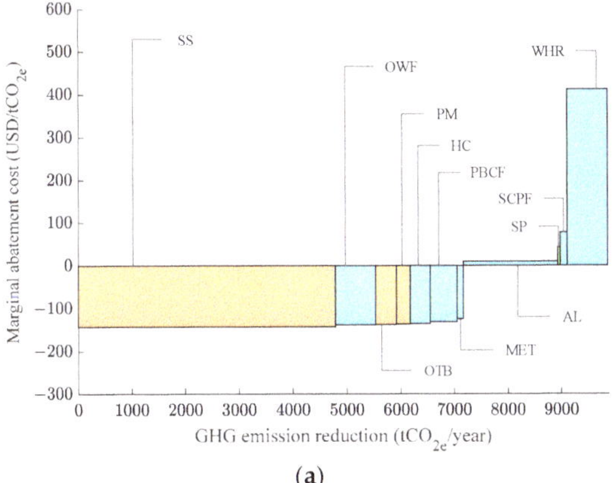
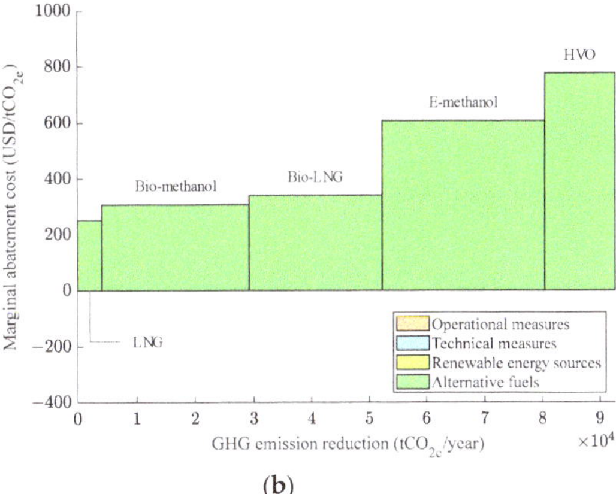

Figure 6. MACC results for Ro-Ro passenger ferries (3500 DWT). (**a**) Energy efficiency measures. (**b**) Alternative fuels.

Figures 2a, 3a, 4a, 5a and 6a show that, in terms of energy efficiency measures, the cost-effective performance of various energy efficiency measures generally follows the same order from the perspective of sub-category classification: operational measures, technical measures, and renewable energy utilization measures. Among the operational measures, slow steaming offers the most optimal cost-effectiveness and can achieve substantial emission reductions, while trim and ballast optimization, and propeller maintenance have relatively high cost-effectiveness but achieve lower emission reductions. Regarding the technical measures, the optimized water flow of hull openings, main engine tuning, hull coating, and propeller boss cap with fins are all cost-effective technologies when applied to the five ship types. Air lubrication is a cost-effective technology for the four cargo ships but becomes a cost-positive technology for the Ro-Ro passenger ferry. A waste heat recovery system and speed control of pumps and fans are cost-positive technologies for bulk carriers, product tankers, VLCCs, and Ro-Ro passenger ships and are a cost-effective technology when applied to container ships. Container ships have the highest fuel consumption, and thus, the investment involved in applying of a waste heat recovery system and speed control of pumps and fans can be recovered by saving fuel costs. As for renewable energy utilization measures, a rigid sail and Flettner rotors are both cost-effective technologies for applicable ship types, i.e., bulk carriers, product tankers, and VLCCs. A rigid sail presents better a cost-effective performance than Flettner rotors mainly due to its lower investment cost than the latter. A solar panel proved to be a cost-positive technology for applicable ship types, i.e., product tankers, VLCCs, and Ro-Ro passenger ferries, owing to its high investment cost and limited abatement potentials.

As can be seen from Figure 6a, the change pattern in the cost-effective performance of applying most energy efficiency measures for the Ro-Ro passenger ship is basically in line with the other four cargo ships, but there are still some differences in a few measures. Due to the high investment cost and relatively limited fuel consumption of the Ro-Ro passenger ship, it is more difficult to recover the investment by saving on fuel costs, so air lubrication is a cost-positive measure for Ro-Ro passenger ships. Furthermore, a rigid sail and Flettner rotors are not technically feasible measures for Ro-Ro passenger ships due to a lack of sufficient deck space for installation.

The results of the application of cost-effective energy efficiency measures are summarized in Table 7. Bulk carriers, container ships, product tankers, VLCCs, and Ro-Ro passenger ferries can achieve cumulative emission reductions of 40%, 36.5%, 38.5%, 38.5%, and 28.5%, respectively, cost-effectively. Accordingly, the resulting annual cost savings are USD 0.75 million, USD 43.34 million, USD 0.69 million, USD 2.18 million, and USD 1.0 million, respectively.

Table 7. Results of application of cost-effective energy efficiency measures.

Ship Type	Energy Efficiency Measures with Negative MAC Values		
	Ranking Orders	Abatement Potentials	Annual Cost Savings (million USD)
Bulk carrier (61–63K DWT)	SS, OWF, OTB, PM, PBCF, MET, RS, HC, AL	40%	0.75
Container ship (8000 TEU)	SS, OTB, PM, OWF, MET, HC, PBCF, AL, SCPF	36.5%	43.34
Product tanker (115K DWT)	SS, OTB, PM, OWF, PBCF, MET, RS, HC, AL	38.5%	0.69
VLCC (315–320K DWT)	SS, OTB, PM, OWF, MET, PBCF, RS, HC, AL	38.5%	21.8
Ro-Ro passenger ferry (3500 DWT)	SS, OWF, OTB, PM, HC, PBCF, MET	28.5%	1.0

Figures 2b, 3b, 4b, 5b and 6b show that, in terms of alternative fuels, the cost-effectiveness ranking of various alternative fuels applied to the four cargo ship types remains consistent, namely LNG, bio-methanol, bio-LNG, e-ammonia, e-methanol, and HVO. Compared with the cargo ships, the passenger ships are more sensitive to the application of fuels with high toxicity hazards. Considering that ammonia is highly toxic and can cause serious injuries and fatalities to humans depending on the level of ammonia concentration exposed [50], the e-ammonia option is not considered in the Ro-Ro passenger ferry. Therefore, the ranking of alternative fuels for the Ro-Ro passenger ferry is LNG, bio-methanol, bio-LNG, e-methanol, and HVO. The LNG fuel shows the best cost-effective performance for the investigated five typical ship types, but its lifecycle emission reduction potential is relatively lower as it is still from fossil resources. The utilization of LNG fuel on container ships represents a cost-effective technology, leading to greenhouse gas emission reductions while realizing cost savings of approximately USD 150 per ton of CO_{2e}. However, the application of LNG fuel to bulk carriers, product tankers, VLCCs, and Ro-Ro passenger ferries is a cost-positive technology. This is owing to the fact that applying LNG fuel involves a high investment cost, and it is difficult for ship types with limited fuel consumptions to counterbalance the investment by saving on fuel costs. Among the biofuels, bio-methanol performs slightly better than bio-LNG in cost-effective performance and lifecycle emission reduction potential. HVO presents the worst cost-effective performance among biofuels, owing to its lifecycle emission reduction potential being lower than that of bio-methanol and bio-LNG as well as its high fuel cost. For electrofuels, e-ammonia presents a better cost-effective performance than e-methanol. Although the lifecycle emission reductions of e-ammonia and e-methanol fuel are generally comparable, the cost of e-ammonia is significantly lower than that of e-methanol, thus making e-ammonia perform better in cost-effectiveness than e-methanol.

4.2. Sensitivity Analysis

The sensitivity analysis of bulk carriers (61–63K DWT) is presented in this section. The aim is to investigate the robustness of the cost-effective measures under the fluctuation scenarios of important input parameters such as fuel oil price and carbon price. For other vessel categories, the sensitivity analysis can be conducted following the same approach.

4.2.1. Fuel Oil Price

The MACC results of bulk carriers (61–63K DWT) for the changing average price of conventional fuel (LSFO) in the scenarios of a 50% decrease, a 50% increase, a 100% increase, and a 150% increase are illustrated in Figures 7–10. The changes in MAC value for various abatement measures under the sensitivity analysis on LSFO price carbon price are summarized in Table 8.

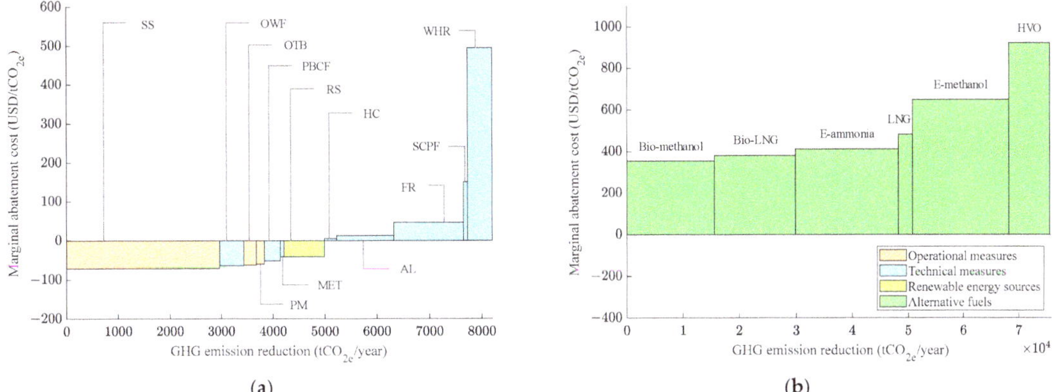

Figure 7. MACC results for bulk carriers (LSFO price decrease 50%). (**a**) Energy efficiency measures. (**b**) Alternative fuels.

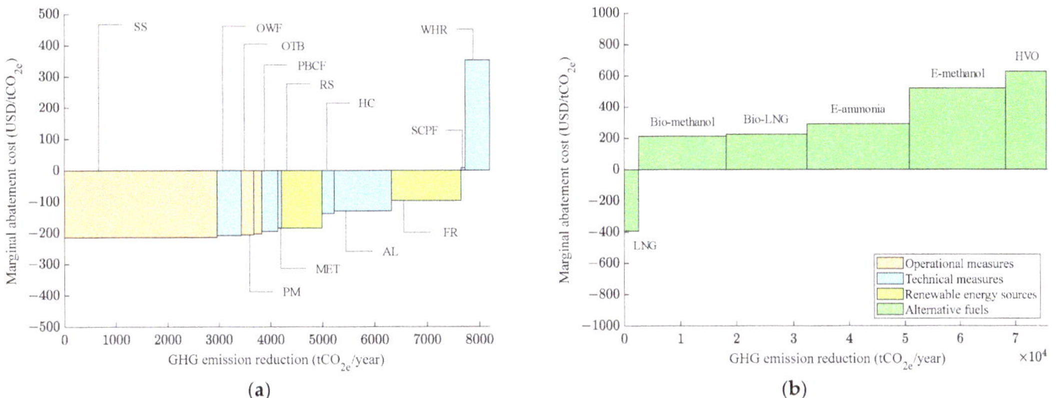

Figure 8. MACC results for bulk carriers (LSFO price increase 50%). (**a**) Energy efficiency measures. (**b**) Alternative fuels.

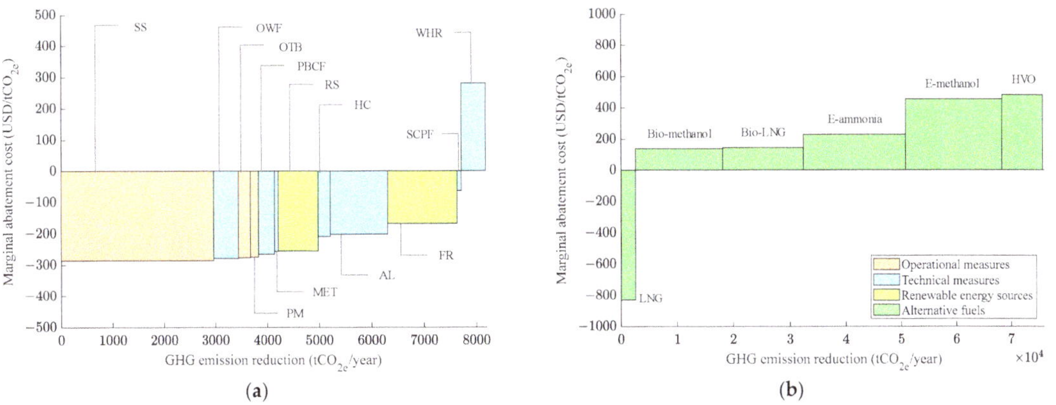

Figure 9. MACC results for bulk carriers (LSFO price increase 100%). (**a**) Energy efficiency measures. (**b**) Alternative fuels.

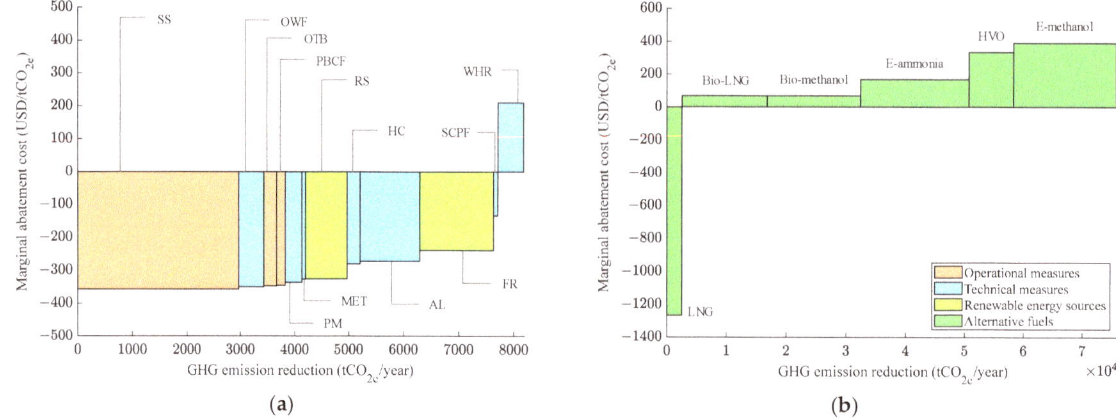

Figure 10. MACC results for bulk carriers (LSFO price increase 150%). (**a**) Energy efficiency measures. (**b**) Alternative fuels.

Table 8. Changes in MAC value for various abatement measures under sensitivity analysis on LSFO price.

Category	Ranking (Baseline)	Abatement Measures	LSFO Price (% Change from Baseline)				
			−50%	Baseline (455 USD/t)	+50%	+100%	+150%
			MAC (USD/tCO$_{2e}$)				
Energy efficiency measures	1	Slow steaming	−71	−143	−214	−285	−357
	2	Optimized water flow of hull openings	−64	−136	−207	−278	−350
	3	Optimization of Trim and Ballast	−62	−134	−205	−277	−348
	4	Propeller maintenance	−60	−131	−203	−274	−346
	5	Propeller boss cap with fins	−52	−123	−194	−266	−337
	6	Main engine tuning	−42	−113	−184	−256	−327
	7	Rigid sails	−41	−112	−183	−255	−326
	8	Hull coating	5	−66	−137	−209	−280
	9	Air lubrication	13	−59	−130	−201	−273
	10	Flettner rotors	46	−25	−96	−167	−239
	11	Speed control of pumps and fans	151	80	8	−63	−134
	12	Waste heat recovery	495	424	353	281	210
Alternative fuels	1	LNG	482	45	−392	−829	−1266
	2	Bio-methanol	355	284	212	141	68
	3	Bio-LNG	382	303	225	147	69
	4	E-ammonia	412	352	291	230	169
	5	E-methanol	648	584	520	456	334
	6	HVO	922	776	629	482	392

In the LSFO price decrease 50% scenario, the emission reduction potential from the application of cost-effective measures decreases from 40% to 32% compared with the baseline MACC. All of the energy efficiency measures experience a decrease in cost-effectiveness, while their ranking remains the same as the baseline scenario. Furthermore, hull coating, air lubrication, and Flettner rotors transform from cost-effective measures to cost-positive measures. For alternative fuels, the cost-effectiveness of all alternative fuels present a significant decrease, and the order of cost-effectiveness is as follows: bio-methanol, bio-LNG, e-ammonia, LNG, e-methanol, and HVO. The cost-effective performance of LNG fuel shows the most significant decrease, with its ranking dropping from the first to the fourth. This can be attributed to the fact that a 50% decrease in LSFO price leads to the LNG price becoming higher than the LSFO price. Consequently, using LNG fuel does not lead to fuel cost savings; instead, it results in a fuel cost increase.

In the LSFO price increase 50% scenario, the emission reduction potential from the application of cost-effective measures can reach 40%, remaining the same as the baseline MACC. The reason is that although improvements in cost-effectiveness for all the energy efficiency measures can be achieved, the speed control of pumps and fans as well as the waste heat recovery system still remain measures with positive costs, indicating that the cost-effective measures are the same as those in the baseline scenario. In terms of alternative fuels, the LNG fuel transforms from a cost-positive measure to a cost-effective measure compared with the baseline MACC results. Although LNG-powered vessels have high initial investment costs, a 50% increase in fuel prices would notably increase the price gap between LSFO and LNG, thereby effectively compensating for the higher investment costs through fuel cost savings and thus obtaining a negative MAC value. Bio-methanol, bio-LNG, e-ammonia, e-methanol, and HVO persist as cost-positive measures, but they all obtain improvements in cost-effectiveness.

In the LSFO price increase 100% scenario, the emission reduction potential from the application of cost-effective measures increases from 40% to 41% compared with the baseline MACC. This can be attributed to the fact that the speed control of pumps and fans transforms from a cost-positive measure to a cost-effective measure, thus making a contribution to the 1% increase in the emission reduction potential. The ranking order of all the energy efficiency measures remains the same as that for the baseline MACC scenario. In terms of alternative fuels, LNG, bio-methanol, bio-LNG, e-ammonia, e-methanol, and HVO persist as cost-positive measures, with the same ranking order as the baseline, but they all obtain improvements in cost-effectiveness.

In the LSFO price increase 150% scenario, the emission reduction potential from the application of cost-effective measures remains at 41% compared with the 100% price increase scenario. The reason is that the waste heat recovery system still remains a measure with a positive cost, and thus, the cost-effective measures and the associated emission reduction amount is the same as the 100% price increase scenario. In terms of alternative fuels, the ranking of cost-effective performance changed compared with that of the baseline scenario. Specifically, the cost-effective performance of bio-LNG surpasses that of bio-methanol, and the cost-effectiveness of HVO becomes superior to that of e-methanol. The 150% increase of LSFO price further narrow the price gaps between alternative fuels and LSFO, thereby decreasing the fuel costs for different fuel options. Consequently, the net costs and the corresponding cost-effectiveness ranking of different fuel schemes changed accordingly.

4.2.2. Carbon Price

In Figure 11, reference lines for the carbon prices of three scenarios (USD 100, 200, and 300 per ton of CO_{2e}) are illustrated based on the MACC baseline for bulk carriers (61–63K DWT). The changes in MAC value for various abatement measures under the sensitivity analysis on carbon price are summarized in Table 9. The comparison between the MACC value and the reference lines allows for an analysis of the impact of different carbon prices on the MACC.

For energy efficiency measures, in the scenario without a carbon pricing mechanism, the speed control of pumps and fans as well as the waste heat recovery are abatement measures with positive MACs. In the scenario of implementing a carbon price of USD 100 per ton of CO_{2e}, the MAC value of the speed control of pumps and fans becomes negative, while waste heat recovery still remains positive. In the scenario of a carbon price of USD 200 per ton of CO_{2e}, all of the measures can gain further improvements in cost-effectiveness. Due to the high investment costs of waste heat recovery systems, even implementing a carbon price of USD 300 per ton of CO_{2e} still cannot render them cost-effective measures. In general, the majority of energy efficiency measures are inherently cost-effective, and thus, shipping companies have a motivation to voluntarily apply these measures for economic considerations. Therefore, the application and promotion of cost-effective energy efficiency measures are not decisively influenced by the implementation of a carbon price.

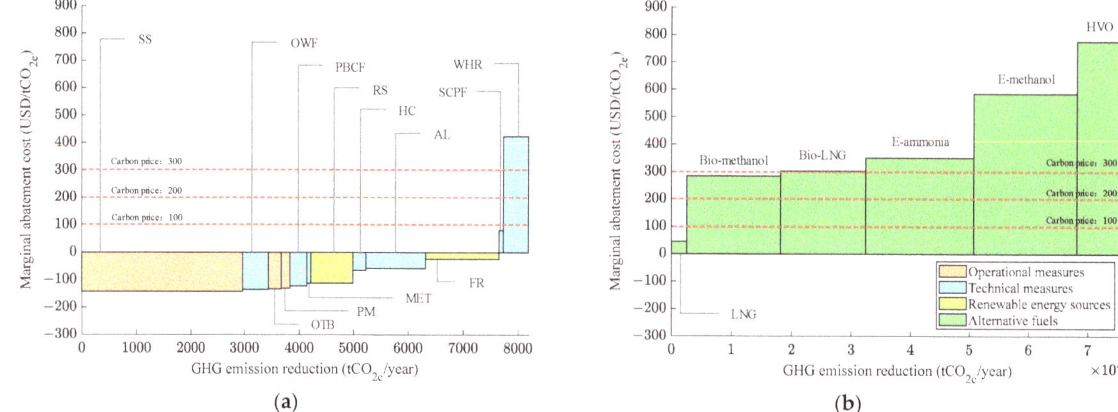

Figure 11. Impact of carbon price on MACC results (61–63K DWT bulk carrier). (**a**) Energy efficiency measures. (**b**) Alternative fuels.

Table 9. Changes in MAC value for various abatement measures under sensitivity analysis on carbon price.

Category	Ranking (Baseline)	Abatement Measures	Carbon Price (USD/t CO₂e)			
			Baseline	100	200	300
			MAC (USD/tCO₂e)			
Energy efficiency measures	1	Slow steaming	−143	−243	−343	−443
	2	Optimized water flow of hull openings	−136	−236	−336	−436
	3	Optimization of Trim and Ballast	−134	−234	−334	−434
	4	Propeller maintenance	−131	−231	−331	−431
	5	Propeller boss cap with fins	−123	−223	−323	−423
	6	Main engine tuning	−113	−213	−313	−413
	7	Rigid sails	−112	−212	−312	−412
	8	Hull coating	−66	−166	−266	−366
	9	Air lubrication	−59	−159	−259	−359
	10	Flettner rotors	−25	−125	−225	−325
	11	Speed control of pumps and fans	80	−20	−120	−220
	12	Waste heat recovery	424	324	224	124
Alternative fuels	1	LNG	45	−55	−155	−255
	2	Bio-methanol	284	174	74	−26
	3	Bio-LNG	303	203	103	3
	4	E-ammonia	352	252	152	52
	5	E-methanol	584	484	384	284
	6	HVO	776	676	576	476

In terms of alternative fuels, without the application of a carbon pricing mechanism, all fuel options are abatement measures with positive MACs. In the scenario of a carbon price of USD 100 per ton of CO_{2e}, the MAC value of LNG fuel becomes negative, while the MAC of other fuel options still remains positive. When implementing a carbon price of USD 300 per ton of CO_{2e}, bio-methanol turns into a cost-effective measure, although the margin is relatively small. Biomass LNG is positioned at the threshold between a cost-positive measure and a cost-effective one. For e-methanol, e-ammonia, and HVO, it can be observed that a carbon price of USD 400 to 800 per ton of CO_{2e} is required to turn it into a cost-effective measure. Generally, as alternative fuels commonly have high costs, a carbon price of USD 300–800 per ton of CO_{2e} is required to offset the cost gap with conventional fuel-powered vessels, making the use of alternative fuels economically feasible. However, taking the EU carbon market as an example, the average carbon price in 2023 is approximately USD 100 per ton of CO_2. This carbon price level may have some contribution in promoting the use of LNG fuel, yet it is still far from sufficient to stimulate the industry's scaled investment and application of alternative fuels such as biofuels and electrofuels.

5. Conclusions

In this study, the investigation of the economic performance of energy efficiency measures and alternative fuels for shipping GHG emission reduction was carried out using a cost-effectiveness model based on the MACC methodology. By innovatively introducing the Tank-to-Wake and Well-to-Wake emission assessment approaches into the MACC methodology, the model was capable of reflecting distinguished GHG emission abatement potentials for energy efficiency measures and alternative fuels from the down-stream and lifecycle basis, respectively. Representative ship types with significant GHG emission contributions, including bulk carriers (61–63K DWT), container ships (8000 TEU), product tankers (115K DWT), VLCCs (315–320K DWT), and Ro-Ro passenger ships (3500 DWT), were taken as research cases. MACCs were developed for each vessel type for various GHG abatement measures including operational measures, technical measures, renewable energy sources, and alternative fuels. The main conclusions derived from this work are summarized as follows:

- The energy efficiency measures that are cost-effective when applied to the five investigated ship types mainly include slow steaming, trim and ballast optimization, propeller maintenance, optimized water flow of hull openings, main engine tuning, hull coating, and propeller boss cap with fins. Ship owners can prioritize the adoption of these energy efficiency measures in their decarbonization strategies.
- The cost-effectiveness ranking of various alternative fuels applied to the typical ship types generally remains consistent, namely LNG, bio-methanol, bio-LNG, e-ammonia, e-methanol, and HVO.
- The cost-effective performance of LNG fuel is closely related to the application ship types and its fuel consumption. LNG fuel is a cost-effective option when applied to 8000 TEU container ships with an annual fuel consumption of 264,000 t. However, it becomes a measure with a positive MAC value for the other four investigated ship types with relatively lower fuel consumption.
- The adoption of alternative fuels including bio-methanol, bio-LNG, e-ammonia, e-methanol, and HVO on the investigated five typical ship types are proven to be measures with positive MAC values due to their high fuel costs.
- The cost-effective performance of energy efficiency measures will be influenced to varying degrees in different LSFO price scenarios, but the cost-effectiveness ranking of the various energy efficiency measures remains consistent.
- Fluctuations in fuel oil price significantly affect the cost-effective performance of different alternative fuels. Moreover, when fuel prices increase or decrease to a certain extent, the ranking of the cost-effective performance of different alternative fuels will change accordingly.
- A carbon pricing mechanism does not have a significant effect on most energy efficiency measures, but it has a certain stimulating effect on several cost-positive energy efficiency measures, such as waste heat recovery system, speed control of pumps and fans, and solar panels.
- A carbon pricing mechanism can effectively improve the cost-effective performance of alternative fuels with a high fuel cost. To bridge the fuel cost gap between the conventional fuels and alternative fuels such as bio-methanol, bio-LNG, e-ammonia, e-methanol, and HVO, a carbon price ranging from USD 300 to 800 per ton of CO_{2e} needs to be imposed.

This study developed a framework for evaluating the cost-effective performance of marine alternative fuels and energy efficiency measures. It can be used as a supporting tool for shipping companies to develop and optimize decarbonization strategies and for maritime authorities to plan and formulate market-based mechanisms on carbon pricing. A potential future research direction is to introduce more emerging and gradually matured abatement measures, such as an on-board carbon capture system and carbon-neutral fuels with novel production processes, into the evaluation framework so as to give a more comprehensive evaluation. In addition, it would be interesting to investigate the contribution of various cost-effective measures to the Energy Efficiency Existing Ship Index (EEXI) and Carbon Intensity Indicator (CII).

Author Contributions: Conceptualization, H.T. and Y.Z.; methodology, H.T.; software, H.T. and Z.L.; validation, Z.L. and Y.Z.; formal analysis, H.T. and Z.L.; investigation, H.T. and Y.Z.; resources, H.T., Z.L. and Y.Z.; data curation, H.T. and Y.Z.; writing—original draft preparation, H.T.; writing—review and editing, H.T., Z.L. and Y.Z.; visualization, H.T. and Z.L.; supervision, Y.Z.; project administration, H.T. All authors have read and agreed to the published version of the manuscript.

Funding: This research received no external funding.

Institutional Review Board Statement: Not applicable.

Informed Consent Statement: Not applicable.

Data Availability Statement: The data that support the findings of this study are available within the article.

Acknowledgments: We thank the anonymous reviewers whose comments helped improve and clarify this manuscript.

Conflicts of Interest: The authors declare no conflicts of interest.

References

1. IMO. *2023 IMO Strategy on Reduction of GHG Emissions from Ships*; MEPC: London, UK, 2023.
2. Paul, B.; James, B.; Chester, L.; Line, S.; Jamie, S.; Adam, H.; Iain, S. How to decarbonise international shipping: Options for fuels, technologies and policies. *Energy Convers. Manag.* **2019**, *182*, 72–88.
3. Zheng, W.; Abdel, E.M.; Yang, C.; Jiayuan, T. Decarbonizing the international shipping industry: Solutions and policy recommendations. *Mar. Pollut. Bull.* **2018**, *126*, 428–435.
4. Yuan, J.; Nian, V.; He, J.; Yan, W. Cost-effectiveness analysis of energy efficiency measures for maritime shipping using a metamodel based approach with different data sources. *Energy* **2019**, *189*, 116205. [CrossRef]
5. Cullinane, K.; Yang, J. Evaluating the Costs of Decarbonizing the Ship Industry: A Review of the Literature. *J. Mar. Sci. Eng.* **2022**, *10*, 946. [CrossRef]
6. Korberg, A.D.; Brynolf, S.; Grahn, M.; Skov, I.R. Techno-economic assessment of advanced fuels and propulsion systems in future fossil-free ships. *Renew. Sustain. Energy Rev.* **2021**, *142*, 110861. [CrossRef]
7. Maurício Aguilar Nepomuceno, D.O.; Alexandre, S.; David Alves Castelo, B. Implementation of Maritime Transport Mitigation Measures according to their marginal abatement costs and their mitigation potentials. *Energy Policy* **2022**, *160*, 112699. [CrossRef]
8. Hu, H.; Yuan, J.; Nian, V. Development of a multi-objective decision-making method to evaluate correlated decarbonization measures under uncertainty—The example of international shipping. *Transp. Policy* **2019**, *82*, 148–157. [CrossRef]
9. Wahl, J.; Kallo, J. Carbon abatement cost of hydrogen based synthetic fuels—A general framework exemplarily applied to the maritime sector. *Int. J. Hydrogen Energy* **2022**, *47*, 3515–3531. [CrossRef]
10. Jinchi, D.; Hui, W.; Lingyun, P.; Bofeng, C.; Hui, L.; Jinnan, W.; Lu, Y.; Chuyu, X.; Yang, C. Marginal abatement cost curves and mitigation technologies for petrochemical and chemical industries in China. *Environ. Eng.* **2021**, *39*, 32–40.
11. Jun, Y.; Szu Hui, N.; Weng Sut, S. Uncertainty quantification of CO_2 emission reduction for maritime shipping. *Energy Policy* **2016**, *88*, 113–130.
12. Shihping Kevin, H.; Lopin, K.; Kuei-Lan, C. The applicability of marginal abatement cost approach: A comprehensive review. *J. Clean. Prod.* **2016**, *127*, 59–71.
13. Elkafas, A.G.; Rivarolo, M.; Barberis, S.; Massardo, A.F. Feasibility Assessment of Alternative Clean Power Systems onboard Passenger Short-Distance Ferry. *J. Mar. Sci. Eng.* **2023**, *11*, 1735. [CrossRef]
14. Kyprianidou, I.; Worrell, E.; Charalambides, G.A. The cost-effectiveness of CO_2 mitigation measures for the decarbonisation of shipping. The case study of a globally operating ship-management company. *J. Clean. Prod.* **2021**, *316*, 128094.
15. Fabian, K.; Paul, E. Marginal abatement cost curves: A call for caution. *Clim. Policy* **2012**, *12*, 219–236.
16. IMO. *Second IMO GHG Study 2009*; International Maritime Organization: London, UK, 2009.
17. Jasper, F. Technical support for European action to reducing Greenhouse Gas Emissions from international maritime transport. *Transp. Syst. Logist.* **2009**, *37*, 1198–1209.
18. IMO. *Third IMO GHG Study 2014*; International Maritime Organization: London, UK, 2014.
19. IMO. *Fourth IMO GHG Study 2020*; International Maritime Organization: London, UK, 2020.

20. Magnus, S.E.; Tore, L.; Peter, H.; Øyvind, E.; Stig, B.D. Future cost scenarios for reduction of ship CO_2 emissions. *Marit. Policy Manag.* **2011**, *38*, 11–37.
21. Maddox Consulting. *Analysis of Market Barriers to Cost Effective GHG Emission Reductions in the Maritime Transport Sector*; Maddox Consulting, LLC: Atlanta, GA, USA, 2012.
22. Nadine, H.; Sonja, P. The potential contribution of the shipping sector to an efficient reduction of global carbon dioxide emissions. *Environ. Sci. Policy* **2014**, *42*, 56–66.
23. Magnus, S.E.; Øyvind, E.; Rolf, S.; Tore, L.; Sverre, A. Cost-effectiveness assessment of CO_2 reducing measures in shipping. *Marit. Policy Manag.* **2009**, *36*, 367–384.
24. Philippe, C. *Greenhouse Gas Emissions Reduction Potential from International Shipping, OECD/ITF Joint Transport Research Centre Discussion Papers*; OECD Publishing: Paris, France, 2009.
25. Pierre, C.S.; Martin, K. CO_2 emissions of the container world fleet. *Procedia-Soc. Behav. Sci.* **2012**, *48*, 1–11.
26. Lindstad, H.; Verbeek, R.; Blok, M.; Zyl, S.; Hübscher, A.; Kramer, H.; Purwanto, J.; Ivanova, O.; Boonman, H. *GHG Emission Reduction Potential of EU-Related Maritime Transport and on its Impacts*; Publications Office of the European Union: Luxembourg, 2015.
27. Yuan, J.; Ng, S. Emission reduction measures ranking under uncertainty. *Appl. Energy* **2017**, *188*, 270–279. [CrossRef]
28. DNV GL. *Low Carbon Shipping Towards 2050*; DNV GL: Høvik, Norway, 2017.
29. Wu, Y.-H.; Hua, J.; Chen, H.-L. Economic feasibility of an alternative fuel for sustainable short sea shipping: Case of cross-taiwan strait transport. In Proceedings of the 4th World Congress on New Technologies, Madrid, Spain, 19–21 August 2018.
30. Yegnidemir, G. Analysis of the Reduction of Emissions from Ships in Europe. Master Thesis, Czech Technical University, Prague, Prague, 2020.
31. DNV GL. *Ammonia as a Marine Fuel*; DNV GL: Høvik, Norway, 2020.
32. Ejder, E.; Arslanoglu, Y. Evaluation of ammonia fueled engine for a bulk carrier in marine decarbonization pathways. *J. Clean. Prod.* **2022**, *379*, 134688. [CrossRef]
33. Armstrong, V.N.; Banks, C. Integrated approach to vessel energy efficiency. *Ocean Eng.* **2015**, *110*, 39–48. [CrossRef]
34. IMO. *Guidelines on Life Cycle GHG Intensity of Marine Fuels*; International Maritime Organization: London, UK, 2024.
35. Liu, N.; Fan, L.; Chen, X. Marginal Abatement Cost Curve of Technology Oriented Under Carbon-Trading Mechanism—Taking Cement, Thermal Power, Coal and Iron and Steel Sectors as an Example. *Forum Sci. Technol. China* **2017**, *7*, 57–63. [CrossRef]
36. Comer, B.; Osipova, L. *Accounting for Well-to-Wake Carbon Dioxide Equivalent Emissions in Maritime Transportation Climate Policies*; International Council on Clean Transportation: Washington, DC, USA, 2011.
37. ITF. *Carbon Pricing in Shipping. Organisation for Economic Co-operation and Development*; OECD Publishing: Paris, France, 2022.
38. Brynolf, S.; Fridell, E.; Andersson, K. Environmental assessment of marine fuels: Liquefied natural gas, liquefied biogas, methanol and bio-methanol. *J. Clean. Prod.* **2014**, *74*, 86–95. [CrossRef]
39. McKinlay, C.; Turnock, S.; Hudson, D. Route to zero emission shipping: Hydrogen, ammonia or methanol? *Int. J. Hydrogen Energy* **2021**, *46*, 28282–28297. [CrossRef]
40. ABS. *Sustainability Whitepaper-Ammonia as a Marine Fuel*; American Bureau of Shipping: Houston, TX, USA, 2020.
41. SGMF. *Ammonia as a Marine Fuel-an Introduction*; Society for Gas as a marine fuel: London, UK, 2023.
42. IMarEST. *Marginal Abatement Costs and Cost-Effectiveness of Energy-Efficiency Measures*; MEPC: London, UK, 2011.
43. DNV GL. *Energy Efficiency Appraisal Tool for IMO.*; DNV GL: Høvik, Norway, 2016.
44. Greenvoyage2050. Available online: https://greenvoyage2050.imo.org/technology/frequency-controlled-electric-motors (accessed on 3 January 2024).
45. Clarkson Research. *Green Environmental Protection Development and Outlook*; Clarkson Research: London, UK, 2022.
46. MAN Energy Solutions. *Shipping en route to Paris Agreement overshoot*; MAN Energy Solutions: Copenhagen, Denmark, 2022.
47. MMM Center for Zero Carbon Shipping. *Maritime Decarbonization Strategy 2022*; Mærsk Mc-Kinney Møller Center for Zero Carbon Shipping: Copenhagen, Denmark, 2022.
48. MMM Center for Zero Carbon Shipping. *Industry Transition Strategy 2021*; Mærsk Mc-Kinney Møller Center for Zero Carbon Shipping: Copenhagen, Denmark, 2021.
49. Solakivi, T.; Paimander, A.; Ojala, L. Cost competitiveness of alternative maritime fuels in the new regulatory framework. *Transp. Res. Part D Transp. Environ.* **2022**, *113*, 103500. [CrossRef]
50. Jang, H.; Mujeeb-Ahmed, M.P.; Wang, H.; Park, C.; Hwang, I.; Jeong, B.; Zhou, P.; Mickeviciene, R. Regulatory gap analysis for risk assessment of ammonia-fuelled ships. *Ocean Eng.* **2023**, *287*, 115751. [CrossRef]

Journal of
*Marine Science
and Engineering*

MDPI

Article

Complex Use of the Main Marine Diesel Engine High- and Low-Temperature Waste Heat in the Organic Rankine Cycle

Sergejus Lebedevas and Tomas Čepaitis *

Marine Engineering Department, Faculty of Marine Technology and Natural Sciences, Klaipeda University, 91225 Klaipeda, Lithuania; sergejus.lebedevas@ku.lt
* Correspondence: tomas.cepaitis@wsy.lt

Citation: Lebedevas, S.; Čepaitis, T. Complex Use of the Main Marine Diesel Engine High- and Low-Temperature Waste Heat in the Organic Rankine Cycle. *J. Mar. Sci. Eng.* **2024**, *12*, 521. https://doi.org/10.3390/jmse12030521

Academic Editor: Leszek Chybowski

Received: 23 February 2024
Revised: 17 March 2024
Accepted: 18 March 2024
Published: 21 March 2024

Correction Statement: This article has been republished with a minor change. The change does not affect the scientific content of the article and further details are available within the backmatter of the website version of this article.

Abstract: The decarbonization problem of maritime transport and new restrictions on CO_2 emissions (MARPOL Annex VI Chapter 4, COM (2021)562) have prompted the development and practical implementation of new decarbonization solutions. One of them, along with the use of renewable fuels, is the waste heat recovery of secondary heat sources from a ship's main engine, whose energy potential reaches 45–55%. The organic Rankine cycle (ORC), which uses low-boiling organic working fluids, is considered one of the most promising and energy-efficient solutions for ship conditions. However, there remains uncertainty when choosing a rational cycle configuration, taking into account the energy consumption efficiency indicators of various low-temperature (cylinder cooling jacket and scavenging air cooling) and high-temperature (exhaust gas) secondary heat source combinations while the engine operates within the operational load range. It is also rational, especially at the initial stage, to evaluate possible constraints of ship technological systems for ORC implementation on the ship. The numerical investigation of these practical aspects of ORC applicability was conducted with widely used marine medium-speed diesel engines, such as the Wartsila 12V46F. Comprehensive waste heat recovery of all secondary heat sources in ORC provides a potential increase in the energy efficiency of the main engine by 13.5% to 21% in the engine load range of 100% to 25% of nominal power, while individual heat sources only achieve 3% to 8%. The average increase in energy efficiency over the operating cycle according to test cycles for the type approval engines ranges from 8% to 15% compared to 3% to 6.5%. From a practical implementation perspective, the most attractive potential for energy recovery is from the scavenging air cooling system, which, both separately (5% compared to 6.5% during the engine's operating cycle) and in conjunction with other WHR sources, approaches the highest level of exhaust gas potential. The choice of a rational ORC structure for WHR composition allowed for achieving a waste heat recovery system energy efficiency coefficient of 15%. Based on the studied experimental and analytical relationships between the ORC (generated mechanical energy) energy performance (P_{turb}) and the technological constraints of shipboard systems (G_w), ranges for the use of secondary heat sources in diesel operational characteristic modes have been identified according to technological limits.

Keywords: load mode; operational characteristic; waste heat recovery; ORC; energy efficiency; technological limitations

1. Introduction

Reduction in carbon dioxide emissions in the maritime sector is essential in addressing the global climate change problem and achieving sustainability goals. The maritime industry significantly contributes to the emission of greenhouse gases, and ships are responsible for a large portion of the world's CO_2 emissions. The reduction in greenhouse gas (GHG) emissions and air pollution from ship power plants is particularly relevant, as the maritime transport sector has become the first globally to establish regulatory decarbonization limits [1]. According to data from the International Maritime Organization

(IMO) [2], in 2020, maritime transport emitted approximately 2–3% of the world's total CO_2, and it is expected that this figure will increase if action is not taken.

In July 2011, the Convention for the Prevention of Air Pollution from Ships was further amended to include a new Chapter 4, supplementing the convention with regulations aimed at preventing air pollution from ships. These regulations are outlined in MARPOL 73/78 Annex VI [3], with a primary focus on addressing the issue of decarbonization by enhancing the energy efficiency of ships. The chapter defines the requirement for reducing greenhouse gas emissions for newly built ships—the Energy Efficiency Existing Ship Index (EEXI) [4]. Starting from 1 January 2023, the requirements for the indices came into effect: EEXI—for existing ships, and CII—Carbon Intensity Indicator. These norms align with the latest initiatives of the European Parliament COM (2021)562, 2021/0211/COD, which connect maritime transport decarbonization with the use of renewable and low CO_2-emitting fuels instead of fossil fuels [5,6].

Without the application of secondary emission prevention technologies, ensuring compliance with the environmental requirements of MARPOL 73/78 Annex VI and the written plans of the EU under COM (2021)562, 2021/0211/COD for maritime transport to become climate-neutral by 2050 becomes challenging. Key solutions for ensuring the decarbonization of maritime transport, as established by the International Maritime Organization (IMO), include the regulations governing the control and reduction in CO_2 emissions, such as the Energy Efficiency Existing Ship Index (EEXI) (MEPC.351 (78)), Carbon Intensity Indicator (CII) (MEPC.352 (78)), and the Energy Efficiency Design Index (EEDI) (MEPC.203 (62)) [7–9].

Decarbonization in marine transport faces complexities due to the power structure of operating ships. In the maritime transport sector, over 95% of ships rely on internal combustion engines for propulsion and electricity generation, predominantly marine diesel engines. Presently, the energy efficiency of these engines is approximately 50%, with the remaining 50% of energy lost to secondary heat sources or regenerated in outdated heat utilization systems that employ water as a working fluid (WF) with limited potential [10]. Based on scientific studies, the utilization of secondary heat can provide an additional 5 to 8% fuel consumption efficiency with corresponding technological solutions [11–14]. According to the latest forecasts by Det Norske Veritas, appropriate solutions for the ship propulsion system should improve the efficiency of ships from 5% to 20%, and waste heat recovery (WHR) systems are included in these solutions [15].

When evaluating the attractiveness of a WHR system on ships, it is necessary to take into account the quality of secondary heat sources; the higher the temperature of the secondary heat source, the more effectively it can be utilized [16]. Ships have access to secondary heat sources with a wide temperature range, with the majority consisting of high-temperature exhaust gases, whose temperature fluctuates between 280 °C and 360 °C. Medium- and low-temperature heat sources are available in auxiliary ship systems, such as scavenge air systems, engine internal cooling systems, and lubrication cooling systems. The achievable significant amount of heat enhances the attractiveness of WHR systems on ships [17]. The application of WHR systems is further increased by the relatively simple modernization of the ship. A series of scientific studies indicate that considering the best short-term or long-term investments, WHR systems are regarded as one of the best investments in new ships or in the modernization of existing ones [18–21].

WHR and cogeneration systems are well-known and proven practices in onshore power plants. From the late 1880s to the early 1900s, oil and gas-fired heat utilization technologies were increasingly used throughout Europe and the United States [22]. In maritime transport, WHR and cogeneration systems began to be employed in the 19th century, and the technological progress between 1970 and 1980 encouraged the adaptation of more sophisticated WHR systems on ships. Steam-based Rankine cycle systems were commonly used to recover secondary exhaust gas heat from main propulsion engines, converting it into electrical power and improving overall energy consumption efficiency. However, the significance of the steam Rankine cycle diminishes with modern marine diesel engines, whose exhaust gas temperatures decrease in regard to improved efficiency over time [23,24].

To maintain the attractiveness of a WHR system, instead of the steam Rankine cycle where water is the working fluid, it is reasonable to use an organic Rankine cycle with organic working fluids. The thermodynamic processes of the cycle remain the same as in a conventional Rankine cycle; the difference lies in the properties of the cycle. The wide range of organic fluids allows for the creation of customized thermodynamic designs that can match any heat source. Moreover, due to stringent environmental, economic, and safety regulations, new organic fluids are constantly being developed [25]. In addition to the organic Rankine cycle (ORC), Kalina and Brayton cycles are also used in practice. The Kalina cycle is a variant of the Rankine cycle (RC) and is a registered trademark of Global Geothermal. The originality of this thermodynamic cycle lies in the use of a mixture of two fluids as the working fluid. Initially, this mixture consisted of water and ammonia, and later, other mixtures emerged. Compared to the traditional Rankine cycle, this innovation introduces a temperature change in the previously isothermal phases of fluid vaporization and condensation. This increases the thermal efficiency of the cycle because the average temperature during heat addition is higher compared to a similar Rankine cycle, while the average temperature during heat rejection is lower. However, when comparing them to the application of ORC in maritime transport, the latter is superior in several aspects: high flexibility, safety, low maintenance requirements, and good thermal efficiency. ORC enables more efficient implementation of energy cogeneration from low-temperature heat sources with a simple, low-maintenance design [26–29].

Interest in the application of modern ORC systems in maritime transport has uplifted more intensive research. Díaz-Secades' bibliometric analysis and systematic review of waste heat recovery revealed that the ORC system was the most widely utilized. Following this review, the authors concluded that not one unique system would be optimal for maximizing waste heat recovery, but rather a blend of various devices would be necessary. Depending on the heat quality, a thermodynamic cycle could be paired with absorption refrigeration, thermoelectricity, and, when feasible, cold energy recovery. However, installing multiple systems to recover exhaust gas waste heat is not strongly recommended, as it could raise backpressure in the exhaust line and subsequently increase engine fuel consumption [30].

Song et al. [31] investigated the use of secondary heat with a 996 kW marine diesel engine in an ORC, and the results showed that a rational system configuration could increase the power plant efficiency by 10.2%. The application of ORC WHR systems using secondary heat in a ship demonstrated a fuel cost reduction ranging from 4 to 15%. The detailed application of ORC systems in maritime transport is reviewed in Ng's study [32]. Park et al. also provided a review focused on experimental ORC performance, analyzing and presenting key data on prototypes, developed systems, and trends [33]. Radica et al. [34] proposed an integrated heat and power system utilizing a supercritical organic Rankine cycle (ORC) with R123 and R245fa as the working fluids to fulfill both heat and electricity requirements for a Suezmax oil tanker. The findings indicated that the system adequately satisfies all heat and electricity needs at maximum capacity, leading to an overall enhancement in the thermal efficiency of the ship's power plant by over 5%. Baldi et al. [35] studied waste heat recovery (WHR) performance in relation to the operational profiles of marine vessels. The findings highlight the significant influence of ship types on WHR performance. Ozdemir studied the impact of the cogeneration cycle structure using a recuperative heat exchanger (RHE) to preheat the working fluid [36]. Konur et al.'s research on the ORC system was modeled thermodynamically for tanker ship diesel generators. The fuel-saving potential and resulting environmental benefits were assessed and discussed according to operation modes; organic Rankine cycle system integration provided a total fuel-saving of 15% from diesel generators and the total fuel consumption of the vessel was reduced by 5.16% [37,38]. In contrast to the ORC systems commonly used in geothermal applications on land, recovering waste heat from diesel engines aboard ships faces variability. Ships experience changing environmental conditions, such as ambient temperature changes, and operate under variable profiles, resulting in inconsistent loads. Ng's proposed method for characterizing the waste heat profile using a generic operational profile and a specifically designed diesel engine waste heat model represents a scientific novelty approach not typically found in mainstream ORC literature. This represents a significant advancement in the initial stages of ORC design [39]. Moreover, one study on Ng compared two cycle configurations, simple ORC and recuperative ORC, and the results showed that recuperative ORC offers an additional 16% extra net work output over simple ORC [40]. An operational profile-based thermal-economic evaluation model was established to provide an evaluation of the organic Rankine cycle used for marine engine waste heat in research. The results showed that operational condition has a great effect on system thermodynamic performance—the maximum thermal efficiency and net power output both decline with the decrease in engine load. The system can satisfy a 5-year payback with evaluated working fluids, except RC318 [41]. In Qu et al.'s study on recovering the waste heat of different energy levels in diesel engines, a slightly complicated waste heat recovery system was proposed, which included a power turbine unit, an SRC unit, and an ORC unit. Under the condition of a load of 100%, the total power generation reached 1079.1 kW. Among them, the maximum thermal efficiency and exergy efficiency of the SRC-ORC unit occurred at 90% load and the maximum thermal efficiency and exergy efficiency of the SRC-ORC unit occurred at 90% load, which reached 28.5% and 65.7%, respectively [42]. Baldasso et al. explored the design of an ORC system for recovering exhaust gas waste heat, concluding that design units with a minimal pinch point temperature approach could result in unfeasible WHR boiler designs [43]. A WHR system based on the steam SRC and ORC utilizing the heat of the exhaust gas and the jacket cooling water of a MAN B&We14K98 marine engine was evaluated in X. Liu's research. The results show that the proposed system could improve the thermal efficiency of the engine by 4.42% and reduce fuel consumption by 9322 tons per year at an engine load of 100% [44]. As scientific research expands, several ORC systems have been implemented on ships in practice, although their application is still rare. One of the first ships with an ORC WHR system from the manufacturer "Opcon" was the "M/V Figaro" in 2012, with a declared system power of 500 kW and achieving a fuel economy of 4–6%. From 2015 to 2018, only five ORC WHR systems were installed on ships with similar results; the efficiency of the systems improved fuel efficiency from 3% to 15% [45]. The most practical information is available on the implementation and operation of the ORC system in the "Arnold Maersk" ship project, where the heat from the engine's internal circuit is utilized. Several non-optimized solutions in the WHR system are noted, such as mismatches in working fluid flow regulation when the seawater temperature decreases [46]. ORC also offers tangible benefits, replacing three diesel generators with one SORC generator would decrease the weight by 12 tons, and would also decrease fuel consumption by 2.1 ton/day [47]. The integration of alternative WHR systems with the ORC has been examined in numerous instances, primarily employing supercritical cycles to capture higher-grade waste heat while reserving the ORC for lower-grade heat recovery [48–50]

Based on WHR ORC research and practical applications, the effective implementation of an organic Rankine cycle (ORC) on ships is more challenging than in onshore power plants because the secondary heat sources on a ship vary depending on the load conditions. Additionally, as the ship moves in different regions, the condensation parameters of the system change due to the fluctuating seawater temperature [51,52]. Optimal system applications require rational decisions, leading to a need for a broader adaptation of ORC system structures in maritime transport. This involves selecting a rational cycle structure and optimizing the system's operation to work more energy-efficiently and cost-effectively, considering the wide operating power plant load modes and environmental conditions of marine diesel engines.

The application of waste heat recovery (WHR) systems in ships, utilizing the energy potential of multiple heat sources characterized by high energy efficiency indicators, appears promising. However, according to available information sources, it remains narrowly studied. There are only a few ships in the global fleet that are equipped with such systems, but their functioning is based only on individual secondary heat sources, such as the mechanical energy regenerated by the hot or cold cooling circuit of the cylinders used in the shaft generator or electrical consumers of the ship. One of the reasons for the limited application of this technology in maritime transport, alongside technological constraints, is the absence of sufficient scientific research regarding the enhancement of energy efficiency in power plants utilizing low- and high-temperature WHR ORC systems. This is particularly pertinent when the engine operates within its operational load range, as specified in ISO 8178 [4]. No less relevant is the efficient utilization of secondary heat energy sources while considering the constraints of a ship's technological parameters. For the purpose of this decision and a number of other aspects of low-temperature heat sources of the ship power plant in a WHR cycle, comprehensive analytical and numerical studies of the organic Rankine cycle (ORC) were conducted at Klaipeda University.

The research was conducted under limiting conditions characteristic of ship operation and technological constraints, covering the analysis of the rational construction of the cycle and the mutual control principles of energy components. It also involved the separate and complex adaptability of secondary heat sources specific to maritime transport, and the impact of the choice of working fluids on the efficiency and performance indicators of the WHR cycle within the characteristic operational range of the ship's power plant. Furthermore, this study examined the influence of limiting factors and other related aspects. The chosen research object, applying ORC in a ship, was the widely used four-stroke, medium-speed, "Wärtsila" 12V46F marine diesel engine, operating within the 25–100% load range according to the ISO 8178 E3 cycle.

To conduct this study, the research consisted of two main stages:

- A study was conducted on the energy efficiency and performance indicators of the WHR cycle, considering variable operational conditions typical for maritime transport while the ship's power plant operated over a wide range of load conditions. This included numerical studies and comparative analysis evaluations of the impact of three characteristic ORC categories of working fluids. Additionally, it involved studying the principles of heat regeneration and power turbine regulation in the WHR cycle, with a rational adaptation to ship propulsion plant, as well as experimental numerical variation studies. These findings are presented in the authors' publication [53].
- In the second stage of research, the main goal was related to the formation of an information base to substantiate the rational choice of ORC structure based on energy consumption efficiency indicators within the operational load range, considering the limitations of ship technological systems.
- The results of this research stage are presented in the author's publication. The authors primarily attribute the scientific novelty and practical significance of the research to the ORC's applicability in utilizing secondary heat sources of the ship's power plant under various complex combinations in operating conditions, including 25–100% of the load range of the ship power plant. Moreover, the research determined the relationship between cycle energy performance with cycle structure and outboard water flow rate, which is considered one of the limitations of ORC applicability in ship technological systems.

2. Methodological Aspects of the Research

Research on the applicability of WHR systems and their cycles to the main engine of a ship was conducted based on the practical, approved, and certified internal combustion engine software "Impuls", as well as numerical methods and mathematical models in the thermodynamic software "Thermoflow version 31". The energy interaction between the characteristic energy parameters of the WHR cycle's energy balance and the system's limiting conditions was assessed graphically for practical estimation by analyzing the changes in pressure–enthalpy (p-h) of the working substances diagram. The p-h diagram graphically represents the thermodynamic properties of working substances. To ensure the reliability of the research, the obtained results were compared with the manufacturer's technical specifications, and the calculation ranges, such as the energy balance of the diesel engine, were performed using classical internal combustion engine theory methods.

2.1. Formation, Justification, and Identification of the ORC Research Cycle Structure (Complex form of WHR Cycle with Different Heat Sources)

In the course of research, the main focus was on energy generation from secondary heat sources characteristic of ships in a single-stage organic Rankine cycle. To analyze and assess the operation of this WHR cycle system, a simulation model was developed using the thermodynamics software "Thermoflow version 31". The first stage of the research was dedicated to evaluating the key aspects of ORC formation for the ship's propulsion complex, utilizing the primary heat source—exhaust gases, which contain the maximum thermal energy of 28.9% of total fuel energy (scavenge air 14.2%, cylinder cooling jacket 9.7%). Comparative studies were conducted based on the use of this secondary heat source to assess different types of working fluids, identifying a more rational choice based on ship conditions. The research also aimed to evaluate energy efficiency and performance, considering different power turbine designs and control aspects [53]. Building on the resolved aspects of rational ORC utilization, the second stage of the research, presented in the author's publication, is dedicated to comprehensive studies on different combinations of secondary heat sources (exhaust gases, internal cylinder cooling circuit, and scavenge air cooler) for ORC implementation. The thermal energy of secondary heat sources is converted into mechanical energy in the cycle's power turbine, and this mechanical energy is transformed into electrical energy in the generator. This generated electrical energy can be utilized to improve the efficiency and sustainability of the ship, contributing to the EU goals of achieving emissions neutrality in maritime transport.

The numerical study simulation model scheme of the "Thermoflow version 31" software is presented in Figure 1.

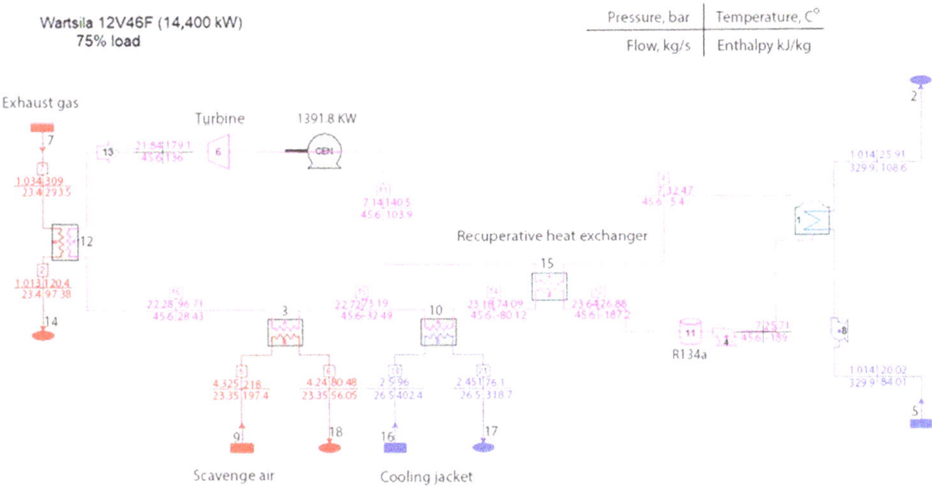

Figure 1. Scheme of single–phase organic Rankine cycle structure with complex secondary heat source inputs.

The following components forming the schematic diagram of the WHR cycle are shown in Figure 1: 1—condenser; 2—outboard seawater discharge; 3—heat exchanger of the scavenge air circuit; 4—feed pump; 5—seawater intake; 6—turbine; 7—exhaust gas heat source; 8—seawater pump; 9—secondary heat source of the scavenge air cooler; 10—heat exchanger of the cylinder internal cooling circuit; 11—working fluid tank; 12—exhaust gas heat exchanger; 14—atmosphere; 15—recuperative heat exchanger; 16—flow from engine cylinder internal cooling circuit; 17—flow to the cylinder internal cooling circuit to the engine; and 18—supply of cooled scavenge air to the engine. The presented data of ORC parameters in Figure 1 are provided as an example of one of the many variations. All relevant data are presented in Appendix B, Tables A3 and A4.

Figure 2 represents a p–h diagram illustrating the characteristic energy parameters and thermodynamic states of an organic Rankine cycle with a secondary heat source and Freon R134a working fluid. In the 1–2 segment, the working fluid (in liquid phase) pressure is raised to the required level corresponding to the turbine expansion parameters, directly linked to cycle efficiency. Upon reaching the set pressure, the primary heat supply into the cycle occurs from the RHE in the 2–3 segment, where the heat retained from the turbine outlet is transferred. At point 3, heat from the secondary heat source is supplied into the cycle. The working fluid is heated in the 3–4 segment, transitioning from unsaturated vapor to superheated vapor, thereby increasing cycle power and efficiency. Superheated vapors at point 4 reach the turbine, where expansion occurs until point 5, transforming heat into mechanical work. During this transformation, the turbine drives the generator and performs useful work, achieving cycle efficiency. After expansion in the turbine, the superheated vapor working fluid travels to the recuperative heat exchanger, where in the 5–6 segment, unused heat after the turbine is returned to heat the liquid phase of the working fluid, transitioning from superheated vapor to saturated vapor. In the condensation process in the 6–1 segment, the working fluid changes state from saturated vapor to saturated liquid, and the cycle repeats.

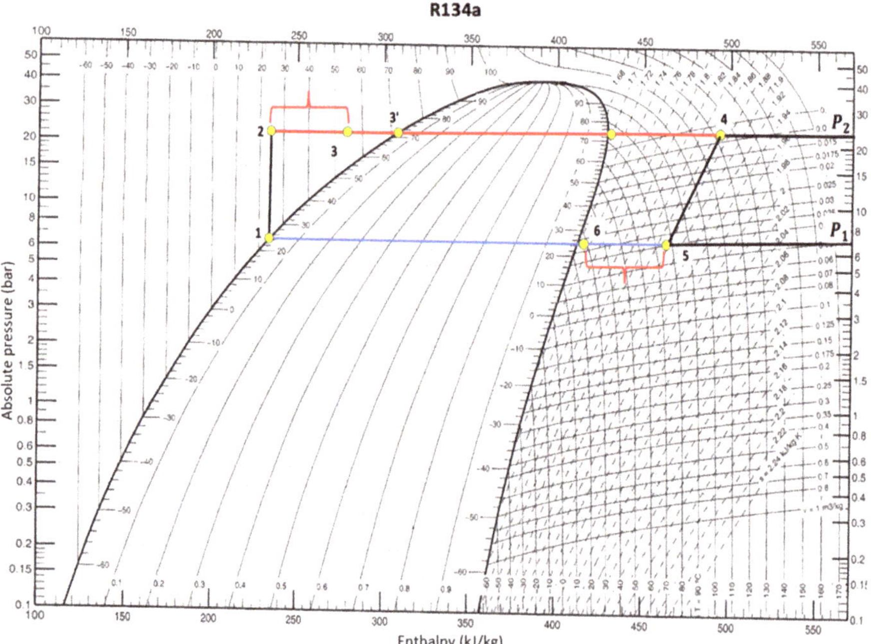

Figure 2. Working fluid Freon R134a p–h (Mollier) chart with characteristic points of energy parameters and thermodynamic states in the ORC cycle.

To increase the useful efficiency and energy efficiency, and optimize the ORC operation of the WHR cycle, an RHE is included in its structure (Figure 1, position 15). By using an RHE, it is possible to recover and utilize the waste heat in the cycle that would otherwise be discharged with the onboard water in the WF condenser.

The regenerative heat exchanger also has a secondary positive effect related to the auxiliary system of the WHR cycle, specifically the condenser (position 1). Due to the fixed degree of expansion of the turbine, the expansion ends at the superheated vapor region; therefore, a pre-cooler upstream of the condenser becomes necessary to cool the working substance in addition to the saturation temperature. This requirement would involve the need for an additional onboard water pump's efficiency and additional energy costs for the WHR cycle to function. As the working substance passes through the RHE, it enters the condenser at a lower temperature, resulting in a reduced load on the onboard pump.

The ORC operates on the following principle (see Figure 1): The WF pressure is raised by the pump (position 4) to the working pressure, and the substance is fed to the recuperative heat exchanger (position 15). In the RHE, the WF is heated by the still-superheated vapor downstream of the turbine. The WF undergoes a phase change from saturated liquid to superheated vapor. From the RHE, the superheated vapor of the WF is directed to the secondary heat exchangers of the ship's power plant, respectively, in increasing order of the heat source temperature: position 10—for the cylinder cooling circuit, position 3—for the scavenge air cooling circuit, and position 12—for the exhaust gas circuit. In these heat exchangers, the WF undergoes a phase change from superheated vapor to overheated vapor. The heat energy of the overheated WF vapor in the turbine (position 6) is converted into mechanical work P_{turb}; the turbine rotates the generator (position GEN), where electrical energy is generated, which is then used in the ship's power plant propulsion system or according to the needs of electrical consumers (assessment of the energy cycle efficiency based on P_{turb}, without considering further mechanical energy conversion into electrical energy by the turbine generator). In the form of vapor, the WF returns from the turbine to the RHE, where it releases some energy in heat form to the WF at the beginning of the cycle. The still-superheated vapor is directed to the condenser, where the WF is cooled by onboard water to the saturated liquid state, and the cycle repeats. The indicator of the WHR cycle's energy efficiency is determined by the ratio of the useful effect P_{turb} to the secondary heat source of the energy device Q_{SS}.

The secondary heat sources can be included or disconnected from the cycle by appropriately adjusting the valves in the pipelines. In this structure, various possibilities for connecting and disconnecting heat sources during the cycle can be achieved, allowing for a rational adaptation based on the load regime and environmental conditions. The research in this study examines a sequentially connected heat exchanger scheme based on its simplicity of implementation.

2.2. Selection of ORC Working Fluid and Formation of Physical and Energetic Indicators

Based on the previously conducted research by the authors [53], Freon R134a was chosen as the working fluid, which demonstrated the highest energy efficiency among the tested wet, dry, and isentropic working fluids (Freon products R134a, R141b, R142b, R245fa, and Isopentane). The selection of Freon R134a refrigerant among all assessed working materials is based on the operational reliability of the cycle's technological system concerning the fluid's saturation pressure and temperature [53]. Also, the provisions of EU and IMO regulatory regulations on the use of Freon products in refrigeration technology and conditioning systems were taken into account.

The International Maritime Organization restricts the use of hydrochlorofluorocarbons in ship systems, which cause depletion of the ozone layer. On new ships from 1 January 2020, the MARPOL banned the use of hydrochlorofluorocarbons (HCFCs) in refrigeration installations. The ban is documented in Regulation 12 "Ozone-depleting substances" Annex VI to MARPOL 73/78 [54]. Moreover, HCFCs have additional drawbacks including significant global warming potential (GWP), which is another reason why their use is regulated.

According to the revised EU Regulation 517/2014 [55], from 1 January 2025 (with various exceptions until 1 January 2026), the use of fluorinated greenhouse gases with a GWP of 2500 or more for servicing and repairing all refrigeration equipment is prohibited. However, this ban will not come into effect until 1 January 2032, and will apply to regenerated fluorinated greenhouse gases with a GWP of 2500 or more used for technical servicing or repair of existing air conditioning and heating equipment. From 1 January 2032, the use of fluorinated greenhouse gases with a GWP of 750 or more (up to 2500) is prohibited, except for regenerated fluorinated greenhouse gases used in the repair and maintenance of refrigeration equipment. Thus, in EU ports and after 1 January 2032, for existing installations with a refrigerant having a GWP of no more than 2500, replenishment is possible solely through regenerated or recycled products. Among HCFCs with a GWP below 2500, for the most part, the single-component refrigerant Freon R134A and the blend R407F are used on ships with class registration [56,57]. Based on this, taking into account the EU regulatory restrictions, as well as the results of the authors' first phase of research, Freon R134a with an ozone depletion potential of 1430 [55] was used for further research.

Based on the results of the initial research stage [53], for the operation of the KC, a "wet" type working fluid, Freon R134a, was chosen, and its properties are provided in Table 1.

Table 1. Main thermophysical properties of Freon R134a.

	Mass Percentage	Boiling Point ($^\circ$C)	Critical Pressure (kPa)	Critical Temperature ($^\circ$C)	Chemical Composition
R134a	100	-26.07	4060	101.06	CH_2FCF_3

2.3. Selection of the Research Object for ORC

To ensure a broad range of research, marine diesel engines chosen from the most popular engine manufacturers such as MAN B&W, Wärtsila, Sulzer, Cummins, and Hyundai were evaluated. Wärtsila is one of the leading companies in maritime technology, including environmental technology adaptation and development. According to the manufacturer's statistical data for the year 2022 [58], medium-speed Wärtsila marine four-stroke diesel engines account for 44% of the total market. The widespread use of the manufacturer's products in maritime transport is related to their known high reliability and efficiency, high fuel efficiency, flexible adaptation with a wide range of offerings, good environmental performance, service, and a solid reputation. Additionally, the increasing market share of four-stroke diesel engines is attributed to their attractive specific mass and size parameters, along with their existing close energy efficiency compared to two-stroke engines.

To conduct the research experiments of this study, the Wärtsila 12V46F four-stroke marine diesel engine was chosen due to its wide engine series and correspondingly broad nominal power (P_e) range. The construction of this engine is analogous to models offered by other popular marine diesel engine manufacturers, which expands the applicability of the research results. The cross-section of the engine is shown in Figure 2 and the main engine parameter is provided in Appendix B, Table A1. Existing research on engines from this manufacturer has demonstrated realistic development prospects.

The manufacturer provides the specifications and a guide for this engine in accessible sources, where design data and the adaptation of marine diesel systems in installations can be found. In this guide, the engine's technical specifications and key energy data are presented at 50%, 75%, 85%, and 100% load, while energy balance data are provided at 100% load only. When the engine operated at 25% load, energy parameters were modeled using the Impuls mathematical model.

Depending on the type of ship where the engine is installed, a specific operational cycle is determined for the engine. For example, ferry-type ships operate in the E3 operational cycle mode, and the main engine data are provided according to ISO 8178—Table 2.

Table 2. "Wärtsila" 12V46F general parameters operating at ISO 8178 operational cycle E3.

Load Modes	P_e, kW	n, rmp	b_e, g/kWh	G_{air}, kg/s	G_f, kg/s	$T_{exh.g.}$, °C
100%	1200	600	178.7	26.1	0.72	366
75%	900	545	188.7	23.35	0.54	309
50%	600	478	190.6	18.8	0.384	273
25%	300	378	197.0	14.5	0.2	255

The decision to use diesel fuel during engine operation was alternatively considered in studies alongside investigations into the energy efficiency of the engine operating with renewable and low-carbon-dioxide-generating fuel (LCA) types. With the onset of fleet decarbonization, its plans are linked to LCA expansion. According to experts, during the current to mid-2030s period, mainly next-generation biodiesels will be used, gradually transitioning to bio-LNG as LNG is replaced [59,60]. The expansion of ammonia and methanol in the fleet, related to infrastructure and necessary development aspects, is foreseen in later stages. In the absence of these fuel types developed in shipping (except for separate pilot study models), there is a lack of engine energy data essential for ORC studies [15,61]. Therefore, it is rational to limit comparative assessment to diesel, biodiesel, and LNG (Bio-LNG).

According to numerous scientific studies, due to relatively minor differences in chemical elemental composition compared to diesel (an increase of 1–11% in oxygen content at the expense of carbon), the components of the engine's heat balance structure change insignificantly, accounting for approximately 3–4% in ORC studies.

The evaluations of LNG utilization were conducted based on the specifications of a Wartsila manufacturer's engine, specifically the 12V46DF model, operating with two fuel types (diesel and LNG). When the engine switches from diesel to LNG operation, structural changes occur in the heat balance: heat generated by fuel combustion is regenerated into effective work, resulting in a 20% decrease in WHR cooling systems and a 10% decrease in charge air cooling. Additionally, as WHR from exhaust gases increases (due to the conversion from a heterogeneous to a homogeneous combustion model when switching to LNG), it increases proportionally. Therefore, it is reasonably anticipated that the energy cycle efficiency and corresponding energy efficiency indicators of the ORC will reach similar values with separate WHR regeneration, and their comprehensive adaptability to ORC will not cause changes affecting operation.

2.4. Mathematical Model of Numerical Studies of Engine Parameters

The energy balance calculation of the selected "Wärtsila" 12V46F engine was performed using classical combustion engine methods [53], relying on the results of energy parameter modeling.

For engine energy parameter modeling studies, a single-phase mathematical model was applied using the "Impuls" software. The choice of the "Impuls" program for research is based on its effective application in creating and modifying high-speed transport engines [53,58,62]. This software has been continuously improved by adding secondary models to assess fuel and air mixture formation and combustion, fuel injection dynamics, vaporization, flame propagation, and different chemical compositions of fuel. Many phenomenological sub-models implemented in this program share similarities with another widely used software, "AVL BOOST 2023R2" [63]. The main version of the program includes 18 secondary models allowing the simulation of closed-cycle diesel engine models with a turbocharger. This modeling is based on quasi-static thermodynamics and gas equations, taking into account various factors such as design parameters of the exhaust system, variable efficiency coefficients of the gas turbine and compressor, heat losses to the engine cooling system, and environmental air parameters. These software tools provide methodological foundations for modeling and analyzing engine performance based on the principle of energy balance sustainability (differently from multi-zone models). The continuous improvement in the program and the integration of advanced sub-models contribute to the progress of engine technologies and a better understanding of the complex engine processes.

The algorithm of the model simulates the closed energy cycle model of a diesel engine with a turbocharger. It is based on quasi-static thermodynamics and gas equations, considering the design parameters of the exhaust system, variable efficiency coefficients of the gas turbine and compressor, heat losses to the engine cooling system, and environmental air parameters. The processes occurring in the engine cylinder are described by a system of differential equations, consisting of the first law of thermodynamics equation (Equation (1)) (energy conservation law), the mass conservation equation of the working substance (Equation (2)), and the state equation of the working substance (Equation (3)):

$$\frac{dU}{d\tau} = \frac{dQ_{re}}{d\tau} - \frac{dQ_e}{d\tau} - p \cdot \frac{dV}{d\tau} + h_s \cdot \frac{dm_s}{d\tau} - h_{ex} \cdot \frac{dm_{ex}}{d\tau}, [\text{kJ/s}] \tag{1}$$

$$\frac{dm}{d\tau} = \frac{dm_s}{d\tau} + \frac{dm_{inj}}{d\tau} - \frac{dm_{ex}}{d\tau}, [\text{kg/s}] \tag{2}$$

$$\frac{dp}{d\tau} = \frac{m \cdot R}{V} \cdot \frac{dT}{d\tau} + \frac{m \cdot T}{V} \cdot \frac{dR}{d\tau} + \frac{R \cdot T}{V} \cdot \frac{dm}{d\tau} - \frac{p}{V} \cdot \frac{dV}{d\tau}, [\text{Pa/s}] \tag{3}$$

where U—internal energy of the working substance within the engine cylinder, kJ; τ—time, s; Q_{re}—the rate of heat transfer into the system (energy added to the system through heat), kJ/s; Q_e—the rate of heat transfer out of the system (energy removed from the system through heat), kJ/s; p—pressure within the engine cylinder, Pa; V—volume of the engine cylinder, m^3; h_s—specific enthalpy of the working substance entering the system, kJ/kg; m_s—mass flow rate of the working substance entering the system, kg/s; h_{ex}—specific enthalpy of the working substance exiting the system, kJ/kg; m_{ex}—mass flow rate of the working substance exiting the system, kg/s; m—total mass of the working substance within the system, kg; R—ideal gas constant, measured in J/(kg·K); T—temperature within the engine cylinder, K; and dm_{inj}—mass flow rate of injected substance (fuel), kg/s.

The heat release was determined using the Wiebe model [64] with G. Woschni's additions [65,66], which are widely applied in studies modeling internal combustion engine working processes [67,68]. The "TEPLM" software was used for experimental indicator analysis. It employs a closed thermodynamic cycle energy balance model to evaluate heat transfer through the cylinder walls.

2.5. Calculation of the Energy Balance during the Operation of the Diesel Engine in the Operational Characteristic Modes

The energy balance of the engine is one of the most important factors in evaluating the operation of a WHR system because the energy results of the cycle depend on it. The freely available specification of the "Wärtsila" 12V46F diesel engine has been used to form energy indicators. The provided specification data for the P_e range of load modes are limited (data are provided for 100% load) and do not include all secondary heat sources. Therefore, to assess the use of the WHR cycle over a wide range of load modes, energy balance parameter calculations for the missing data of the WHR cycle were performed, simultaneously aligning them with the engine specification. Calculations were performed with the engine operating in propulsion mode and presented in a logical sequence. The relevant balance of secondary heat sources for the WHR system consists of the heat values of exhaust gases, scavenge air cooling, internal cylinder cooling, and lubrication system cooling and are presented in Equation (4):

$$Q_{SS} = Q_{exh.} + Q_{sc.air} + Q_{cil} + Q_{oil} \tag{4}$$

Based on the classic structure of the ship's cooling and lubrication system, the heat components Q_{oil} (kJ/s) and Q_W (kJ/s) in the ORC modeling cycle are the integral parts of the cylinder block cooling balance and are combined into a single secondary heat source. Finally, the variations examined include the heat source of exhaust gases, scavenge air, and the cylinder cooling (along with Q_{oil} (kJ/s)). The calculations were performed using classical calculation methods and the results are presented in Appendix B, Table A2.

2.6. ORC Energy Efficiency and Its Structure Unit Parameters

During numerical simulations, useful efficiency coefficients, η, are evaluated, reflecting the efficiency of the WHR cycle. The energy efficiency indicators of ORC utilization on a ship include the following:

η_e—ship powerplant efficiency,

η_{eRC}—ship powerplant efficiency with ORC,

η_{RC}—ORC efficiency,

$\eta_{eRC_{cikl}}$—ship power plant efficiency with ORC with ISO 8178 operational cycle.

The assessment of different organic working fluids in the Rankine cycle is carried out based on the change in the energy efficiency indicator of the engine. The effective $\delta\eta_{eRC}$ efficiency coefficient (EE) stands for the relative increase/change in engine power with and without ORC. The useful efficiency coefficient η_e of the ship's power plant indicates the part of the energy effectively utilized compared to the amount of fuel energy used in the technological process. Two analytical expressions of η_{eRC} are used to assess the efficient operation of the WHR cycle:

1. Evaluating how much the effective useful coefficient η_{eRC} of the main engine increased with the WHR system cycle, using secondary heat in it to generate electrical energy.
2. Evaluating the efficiency of energy use in the WHR cycle itself η_{RC}.

The useful efficiency coefficient of the ship's power plant indicates how efficiently the thermal energy of the fuel is used to perform useful work. The EE of the engine without the WHR system is calculated using the ICE theory classical Equation (5):

$$\eta_e = \frac{P_e}{H_u \cdot G_f} \tag{5}$$

The cycle with heat input from three secondary heat sources is examined as follows:

- Exhaust gas secondary heat source;
- Internal cylinder cooling circuit secondary heat source;
- Scavenge air cooling circuit secondary heat source.

Determination of η_{eRC} for a power plant with an organic Rankine cycle with all three heat sources is presented in Equation (6):

$$\eta_{eRC} = \frac{(P_{turb} + P_e)}{H_u \cdot G_f} \tag{6}$$

P_{turb} is formed from the supplied heat from the three secondary heat sources according to Equation (7):

$$Q_\in = Q_{exh.} + Q_{cil.} + Q_{sc.air} = Q_f \cdot q_{exh.} \cdot \eta_{t.exh.} \cdot \Psi_{t.exh} + Q_f \cdot q_{cil.} \cdot \eta_{t.cil.} \cdot \Psi_{t.cil.} + Q_f \cdot q_{sc.air} \cdot \eta_{sc.air} \cdot \Psi_{t.sc.air} \tag{7}$$

Here, $\eta_{t.cil}$ and $\eta_{sc.air}$ are the thermal efficiency coefficients of the heat exchangers, $\Psi_{t.cil}$ and $\Psi_{t.sc.air}$ are energy utilization factors (similar to the case of exhaust gases and expansion turbines) for decreasing temperatures in the heat exchanger to the level specified in the engine specifications: inflatable air—up to 50 °C; cylinder cooling circuit—up to 75 °C.

In general, secondary heat sources form P_{turb} as described in Equation (8):

$$P_{turb} = Q_\in (\eta_{t.ad} \cdot \eta_m \cdot \Psi_{turb}) \tag{8}$$

where $Q_\in$ is the heat supplied to the turbine. Heat transformations also occur in the turbine, $\Psi_{t.cil} = \frac{h_{w1}-h_{w2}}{h_{w1}-h_w'}$, when $h_{w1} - h_{w2}$ results in a real decrease, and the $h_{w1} - h_w'$ decrease is necessary according to the specification.

The energy utilization factor for inflatable air is evaluated similarly. The specific heat of secondary heat sources is described by Equations (9) and (10):

$$q_{cil.} = \frac{Q_{cil.}}{H_u \cdot G_f}, \tag{9}$$

$$q_{sc.air} = \frac{Q_{sc.air}}{H_u \cdot G_f} \tag{10}$$

Fuel heat release during combustion in the engine is described by Equation (11):

$$Q_f = H_u \cdot G_f \tag{11}$$

It is expedient to describe P_{turb} in two forms.

The power equation of the turbogenerator, in terms of the efficiency of the WHR cycle, is described by Equation (12):

$$P_{turb} = ((Q_{exh} \cdot \eta_{t.exh} \cdot \Psi_{exh}) + (Q_{cil} \cdot \eta_{t.cil} \cdot \Psi_{cil}) + (Q_{sc.air} \cdot \eta_{t.sc.air} \cdot \Psi_{sc.air})) \cdot \eta_{t.ad} \cdot \eta_m \cdot \Psi_{turb}. \tag{12}$$

The utilization coefficient of secondary heat sources Ψ_i is one of the essential parameters determining the utilization of enthalpy and temperature of the sources in the cooling circuits of cylinders and the cooling circuit of the compressed air up to the values regulated by the engine manufacturer. This includes the temperature of the exhaust gases relative to the dew point temperature of the exhaust gases, aiming to avoid sulfuric acid condensation in the exhaust tract when the engine operates with sulfur-containing fuel. The parameter $\Psi = 1,0$, for the exhaust gas heat source could be considered when the temperature of the exhaust gases decreases to the air temperature at the engine intake (i.e., 50 °C). However, in such a case, the system does not ensure conditions related to the concentration of H_2SO_4. Therefore, alternatively, it is assumed that $\Psi = 1,0$, when the temperature decreases to the dew point.

The relative efficiency coefficient parameter $\Psi_{turb.}$ of the turbogenerator illustrates the ratio of the actual decrease in the working substance's enthalpy in the turbogenerator to the potential decrease in enthalpy down to the boiling temperature. The efficiency of the turbine is characterized by the adiabatic efficiency parameter $\eta_{T_{ad}}$ and the mechanical efficiency parameter η_m. Expressing $Q_\in = G_f \times H_u \times q_\in$ and $P_e = G_f \times H_u \times \frac{\eta_e}{3600}$, we obtain Function (2.15) in terms of the dimensionless quantities.

As a result, the overall efficiency of the power plant with an organic Rankine WHR cycle, introducing three secondary heat sources, is determined by the Formulas (13) and (14):

$$\eta_{eRC} = \frac{P_e + Q_\in(\eta_{t.ad} \cdot \eta_m \cdot \Psi_{turb})}{H_u \cdot G_f} \tag{13}$$

$$\eta_{eRC} = [\eta_e + (q_{exh} \cdot \eta_{t.exh} \cdot \eta_{t.exh} + q_{cil} \cdot \eta_{t.cil} \cdot \eta_{t.cil} + q_{sc.air} \cdot \eta_{sc.air} \cdot \eta_{sc.air}) \cdot (\eta_{t.ad} \cdot \eta_m \cdot \Psi_{turb})] \tag{14}$$

ORC cycle efficiency is determined by Equation (15), respectively:

$$\eta_{RC} = \frac{H_u \cdot G_f (q_{exh} \cdot \eta_{t.exh} \cdot \Psi_{t.exh} + q_{cil} \cdot \eta_{t.cil} \cdot \Psi_{t.cil} + q_{sc.air} \cdot \eta_{sc.air} \cdot \Psi_{sc.air}) \cdot (\eta_{t.ad} \cdot \eta_m \cdot \Psi_{turb})}{H_u \cdot G_f (q_{exh} + q_{cil} + q_{sc.air})} \tag{15}$$

Then, using the P_{turb} evaluation form to identify and improve the factors influencing the efficiency of the WHR cycle, the operational parameters of the cycle power turbine can be optimized and their relationship with a rational choice can be determined.

The power generated in a propulsion turbine is defined by Equation (16):

$$P_{turb} = G_{wm}\left(h_{tg_1} - h_{tg_2}\right) \tag{16}$$

The turbine nozzle apparatus design is indicated by the parameter π_T, which shows the degree of pressure reduction (the pressure ratio before and after the turbine) of the WF before and after the turbine. The power generated in the turbogenerator is also expressed by Equation (17):

$$P_{turb} = G_{WF} \cdot t_{WF_1} \cdot c_{pWF}\left[1 - \pi_T^{\frac{K-1}{K}}\right] \cdot \eta_{t.ad} \cdot \eta_m \cdot \beta \tag{17}$$

β—energy input impulse coefficient, which is 1,0 under WHR cycle conditions;
k—specific heat ratio.

Then, the change in the total generated mechanical energy of the engine's cycle, taking into consideration the generated P_{turb}, is expressed by Equation (18):

$$\delta\eta_{eRC} = \left(\frac{\eta_{eRC}}{\eta_e} - 1\right) = \left(\frac{\frac{P_e + P_{turb}}{H_u \cdot G_f}}{\frac{P_e}{H_u \cdot G_f}} - 1\right) = \frac{P_{turb}}{P_e}; \tag{18}$$

In order to identify and improve the factors influencing the efficiency of the WHR cycle, the operational parameters of the turbogenerator can be optimized and their relationship with a rational selection can be determined.

We evaluated the WHR cycle efficiency, η_{RC}, and determined how efficiently the secondary heat sources are transformed into the turbine mechanical work and further converted into electricity in the generator.

3. Results

In the ship's power plant, various secondary heat sources with different temperatures and heat quantities prevail. Heat sources with an average quality are considered when their temperature is $\geq$230 °C. At such temperatures, the utilization of thermal energy is not complicated and is typically employed in steam boilers. However, the utilization of low-quality secondary heat sources (temperature—30–200 °C) in steam boilers is challenging, and they are often ignored. On the other hand, WHR systems with an organic Rankine cycle open up possibilities for the utilization of low-temperature secondary heat sources. In ship power plants, a wide range of secondary heat sources is present, with the majority of energy being lost through exhaust gases, which can reach temperatures of 220–400 °C. In modern marine diesel engines, the waste heat dissipated by the compressed air cooler is slightly lower than the temperature of the exhaust gases, which can range from 170–250 °C. The third in size is usually the internal cylinder cooling circuit of the engine, with a temperature ranging from 75–95 °C in modern diesel engines. The heat quantity from other sources (oil cooling, heat radiation, etc.) is not significant, making their rational use impractical.

Complex Form WHR Cycle with Different Heat Sources

Comparative numerical studies of the cycle energy efficiency were conducted using different combinations of individual and complex secondary heat sources. The secondary heat sources were supplied in a sequential order in the WHR cycle structure, from the lowest temperature to the highest. Finally, the heat supply to the cycle was carried out in the following order: supply of heat from the cylinder cooling circuit (96 °C) $\rightarrow$ supply of heat from the compressed air cooler (220 °C) $\rightarrow$ supply of heat from the exhaust gases (364 °C). The cycle combinations were implemented using the working fluid Freon R134a, with which the best energy efficiency parameters were achieved in a wide power plant range in the first stage of dissertation research. This substance is also widely used in land-based cogeneration plants. The expansion of the working fluid in the cycle and the mechanical work were ensured by a turbo-generator with a variable geometry turbine, which achieved 25–30% better results than the results of the first stage of the research with a turbine with a fixed geometry. The results of individual secondary heat sources in ORC are presented in Appendix B, Table A3.

The obtained results of the engine's energy balance calculation indicate the distribution of heat source quantities, where $Q_{cil.} < Q_{air} \rightarrow Q_{cil.} + Q_{air} \approx Q_{exh.g.}$, demonstrating the importance of low-temperature sources. Typically, the heat potential is evaluated based on the ratio of energy supplied with fuel. However, when evaluating a variable and wide power plant load range, it is more rational to assess it based on the power of the plant under the corresponding load conditions. A graph (Figure 3) depicting the heat quantity ratio between secondary heat sources and the engine's effective power allows for the assessment of each source according to the P_e characteristic.

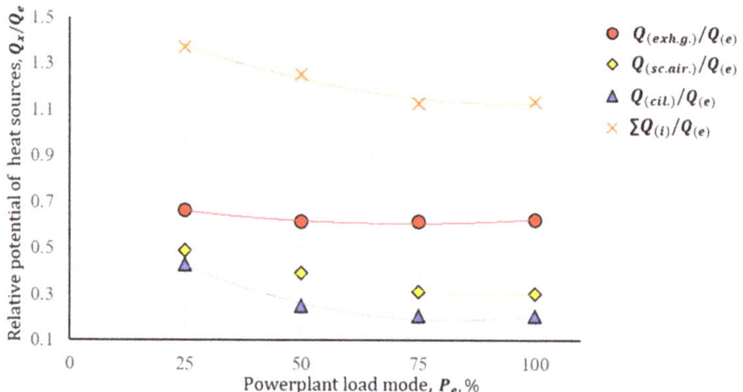

Figure 3. The relative potential of secondary heat sources with the effective power of the engine.

The highest heat potential, especially under low and moderate load conditions, is associated with the use of a variable geometry turbine in the WHR cycle [53]. To assess the attractiveness of each secondary heat source, experimental studies were conducted on the variation in individual heat sources in the WHR cycle, and the results obtained are presented graphically as follows:

- The power generated by the secondary internal cylinder cooling circuit heat source P_{turb} in the WHR cycle ranges from 160 kW to 310 kW. The seawater flow rate for condensation ranges from 55 kg/s to 108 kg/s, corresponding to load conditions of 25–100% of the engine, at a seawater temperature of 20 °C. Graphically, it can be observed that the most significant change in the WHR system's useful efficiency coefficient is achieved under low load conditions (Figure 4).

- The power generated by the secondary scavenge air cooling circuit heat source (P_{turb}) in the WHR cycle is higher than that of the cylinder cooling circuit, ranging from 234 kW to 477 kW. The seawater flow rate for condensation ranges from 74 kg/s to 152 kg/s, corresponding to load conditions of 25–100% of the engine, at a seawater temperature of 20 °C. Graphically, it can be observed that the most significant change in the WHR system's useful efficiency coefficient is also achieved under low load conditions (Figure 5).

- The power generated by the secondary exhaust gas circuit heat source P_{turb} in the WHR cycle, under low load conditions, is slightly lower than the compressed air source, producing 249 kW. However, when there is a high load, the generated power is almost twice as much as the compressed air source, reaching 898 kW. The seawater flow rate for condensation in the cycle varies from 74 kg/s to 152 kg/s, corresponding to load conditions of 25–100% of the engine, at a seawater temperature of 20 °C. Graphically, it can be observed that the greatest positive change in the useful efficiency coefficient of the WHR system is achieved under low load conditions (Figure 6).

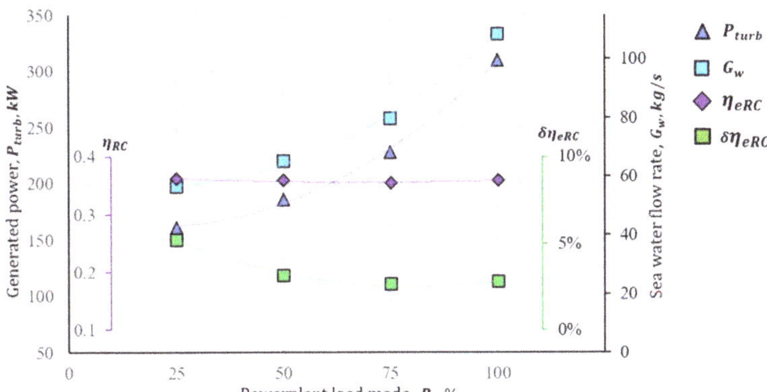

Figure 4. Results of cylinder block cooling water secondary heat source energy parameters in the WHR cycle.

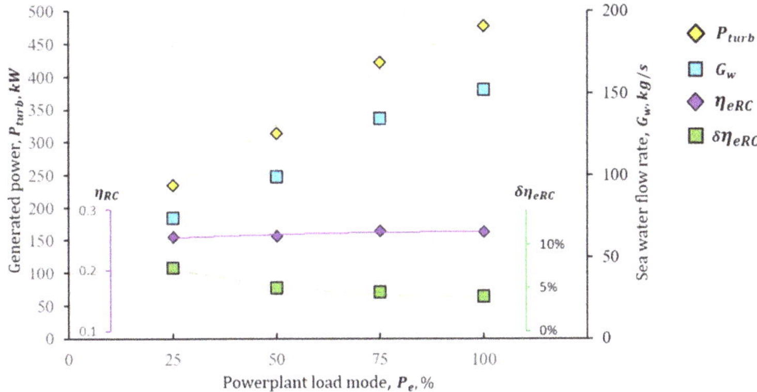

Figure 5. Results of scavenging air cooling secondary heat source energy parameters in the WHR cycle.

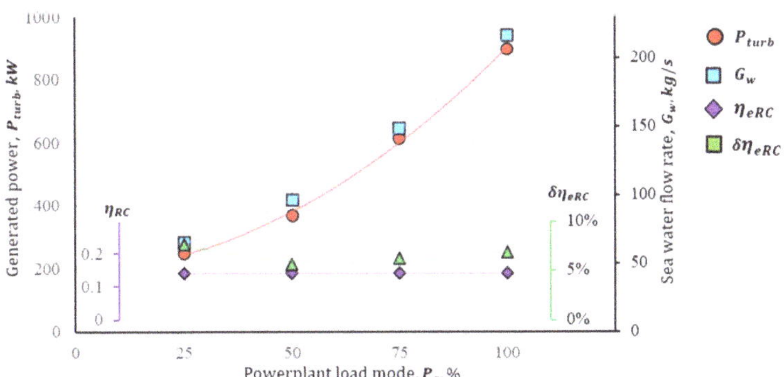

Figure 6. Results of exhaust gas secondary heat source energy parameters in the WHR cycle.

Based on the obtained results, it is noted that supplying a higher amount of heat (exhaust gas > compressed air heat > cylinder cooling circuit heat) results in higher system efficiency and a greater positive change in the WHR system's useful efficiency coefficient. In all cases, there is an observed relationship between efficiency and the seawater flow rate, which can be a crucial limitation when choosing the heat source in the WHR cycle. The results of the complex secondary heat source in ORC are presented in Appendix B, Table A4.

The results of the numerical variational studies allow for the formulation of methodological foundations for the WHR cycle structure. The results of studying various heat source options in the WHR cycle demonstrate a direct connection between the power generated by the turbine P_{turb} and the seawater flow rate G_w, which is used to condense the WF. The evaluation of G_w is important in the context of assembling the WHR cycle structure because it is related to the selection of seawater pumps; thus, it is rational to assess the inclusion of ballast purpose pumps already used on the ship in the auxiliary condensation system of the WHR cycle, considering their efficiency and thus saving costs and space for separate pumps.

In the case of technological feasibility, the ballast (or other) pumps of the ship could ensure the inflow of board water into the condenser, thus saving space on the ship and project costs. According to statistical data, the average efficiency of ballast water pumps for a research ship's propulsion engine "Wärtsila" 12V46F, of a similar size, in the operating fleet and manufacturer specifications, ranges from 300 to 600 m^3 per hour. To ensure the condensation of the working fluid in the WHR cycle without the efficiency of ballast pumps, it is necessary to use separate board water pumps. Looking at the statistical data of the "DESMI" pumps widely used in practice, the maximum proposed pump flow for ORC operation is 6000 m^3/h (see Appendix B, Figure A1), or when converted to mass units—1667 kg/s. However, the use of pumps with such efficiency on board is complicated due to mass and size limitations.

Therefore, the maximum flow rate becomes another limiting factor for the energy generated by the ORC. To optimize the generated mechanical energy, P_{turb}, the energy efficiency indicators η_{RC}, $\delta\eta_{eRC}$, and G_w, a mathematical modeling experiment, and theoretical solutions are used to establish a connection between P_{turb} and G_w.

The establishment of analytical expressions between P_{gen} and G_w is based on the heat balance equation of the WHR cycle. The energy balance of the WHR cycle, expressed in a generalized form, is given by energy balance in the form of heat as follows:

$$Q_h = Q_T + Q_w, \text{ or, per unit mass of WF,} \tag{19}$$

$$G_{WF}q_h = G_{WF}\cdot q_T + q_w G_{WF}, \tag{20}$$

where:

Q_h—total heat transferred per unit mass of working fluid.

q_h—specific WHR heat transferred per unit mass of the working fluid in the heat exchanger (heat exchange in the regenerative heat exchanger is not considered in the balance due to the assumed equality between the heat transferred and received by the working fluid in the RHE).

q_t—energy, in the form of specific heat, transformed into mechanical work in a power turbine.

q_w—transferred specific heat from the working fluid to the overboard water.

On the other hand, the energy balance in the condenser is $G_{WF}\cdot q_w = q^{\Delta}{}_w G_w$, or $G_{WF}\cdot q_w = q^{\Delta}{}_w G_w$, where $q^{\Delta}{}_w$ is gained heat from seawater cooling, and G_w is the seawater flow rate (heat losses in the condenser are not estimated). As a result, the formula $G_{WF}\cdot q_{SS} = G_{WF}.q_T + q_w Gw$ is derived.

Substituting $G_{WF}q_T$ to $G_{WF}q_{SS}\cdot K_1$ (where K_1 is the cumulative efficiency of the power turbine), the energy balance equation transforms into the expression $G_{WF}\cdot q_{SS} = G_{WF}\cdot q_{SS}\cdot K_1 + q_w G_w$, and after simplification into the following form:

$$\frac{G_{WF}\cdot q_{SS}}{G_{WF}\cdot q_{SS}\cdot K_1} = 1 + \frac{q_w G_w}{G_{WF}\cdot q_{ss}\cdot K_1} = 1 + \frac{q_w G_w}{Q_T}, \tag{21}$$

$$\frac{G_{WF}}{G_{WF} \cdot K_1} = 1 + \frac{q_w G_w}{Q_T} = \frac{1}{K_1} \tag{22}$$

Since for the evaluated technological embodiment of the WHR cycle power turbine indicators K_1 and q_w (when $T_w = const.$) are evaluated as constants, the constant is also the ratio $\frac{G_w}{Q_T}$.

Equation (22) is characterized by universality, making it applicable to any combination of secondary heat sources if Q_{RHE} is sufficient to heat the working substance to the beginning of vaporization in the WHR cycle, which occurs in the recuperative heat exchanger. Otherwise, the condition $\frac{G_w}{Q_T} = const$ is specific to a particular regeneration variant. This characteristic is revealed in Figure 7, which shows individual graphs of $\frac{G_w}{P_{turb}}$ for internal water circuit and air cooling heat regeneration, and the heat regeneration of exhaust gases both separately and in combination with other heat sources, are characterized by the same $\frac{G_w}{P_{turb}}$ dependence, which is graphically represented in Figure 7.

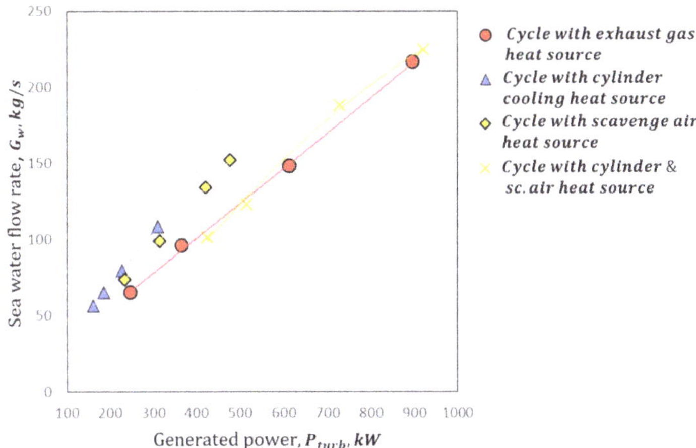

Figure 7. The relationship between the WHR cycle performance and the outboard water flow rate.

The obtained result provides a theoretical basis to standardize the principles of forming a rational structure for the WHR cycle. This involves evaluating the potential of secondary heat sources in the context of the research engine module and its connection with the technological constraints of shipboard water systems. Methodological aspects and the logical sequence of the research are presented schematically in Appendix A, Figure A1.

4. Conclusions

In order to systematize the rational selection of low-temperature (internal cylinder and scavenge air cooling circuits) and high-temperature (exhaust gas) secondary heat sources in a WHR cycle for energy efficiency priorities, complex, analytical, and numerical variations were carried out in a study of the medium-speed diesel engine "Wärtsila" 12V46F's organic Rankine cycle (ORC) with Freon R134a working fluid within the operational range of 25–100% engine load according to the ISO 8178 E3 cycle.

Compared with the practical applicability of a single heat source ORC in a marine environment, the comprehensive potential for utilizing secondary energy to increase ship power plant engine efficiency η_{RC} was evaluated up to 13.5–21% in the range of 100–25% load modes. The obtained results for the external heat balance structure typical for a main medium-speed diesel engine can be summarized as follows:

- The rational distribution of ORC heat exchangers based on the increasing characteristic temperature of secondary heat sources (operating in the range of 25–100%

- engine load, engine cooling jacket, scavenge air, and exhaust gas WHR secondary heat cooling circuits) ensures close proximity to the energy potential of the heat sources: internal cylinder cooling circuit—95%; scavenge air cooling circuit—84%; and exhaust gases—99% (with $\Psi \cong 1,0$);

- The results indicate that it is rational to use ORC throughout the typical operational range of the engine, as reducing the nominal power from 100% to 25% leads to an improvement in the effective efficiency increase using ORC $\delta\eta_{eRC}$ as follows: WHR for exhaust gases from 6.9% to 7.7%; charge air cooling circuit from 4% to 7.3%; and cylinder block cooling circuit from 2.8% to 5.2%.

- Specifically, the high efficiency of increasing η_{eRC} at low engine loads determines the most crucial operational $\delta\eta_{eRC}$ average values for the entire load cycle, respectively, 6.6%; 4.8%; and 3.1%.

- The comprehensive composition of a WHR system with various combinations ensures $\delta\eta_{eRC_{cikl}}$ increase over the operational load cycle ranging from 14.8% (all three secondary heat source WHR cycle) to 3% (only low-temperature WHR cycle).

- Attention is drawn to the relatively high energy efficiency of the implementation of the scavenge air cooling WHR system—the difference in $\delta\eta_{eRC}$ compared to exhaust gas WHR is only about 1.5%: approximately ~5% versus 6.5%, respectively. In combination with a relatively straightforward technical implementation, this allows considering this WHR as one of the effective components of the ship engine's ORC, both in its standalone and combined applications with other WHR systems.

- ORC variation study data indicate that the application of secondary heat sources in the marine power plant in the operational characteristic range alternatively ensures the total power plant efficiency and improves energy performance. Variations in the complex use of exhaust gases, internal cylinder cooling circuit, and scavenge air cooler heat guarantee $P_{turb} \approx 500$–1842 kW, $\delta\eta_{eRC} \approx 21.4$–$7.0\%$, indicator values.

- In pursuit of complex heat source application in the cycle, an experimentally identified and analytically supported linear relationship between the cycle's energy efficiency P_{turb} and the efficiency of the condensation system's pump (G_w) is established. Based on this, the selection of ORC heat source complex utilization strategies is limited by data from one of the sources and is evaluated in terms of the technologically achievable efficiency of G_w pumps in relation to P_{turb}.

Based on the conducted comprehensive analytical studies and expanded numerical experiments, the main task of the next stage of ongoing research is related to formulating the methodological principles for the rational application of ORC in marine power plants. The practical implementation of these principles constitutes the main objective of the current phase of our research.

Author Contributions: Conceptualization, S.L.; methodology, S.L.; software, T.Č.; validation, T.Č.; analysis, S.L. and T.Č.; investigation, S.L.; data curation, S.L. and T.Č.; writing—original draft preparation, S.L. and T.Č.; writing—review and editing, S.L. and T.Č.; visualization, T.Č.; funding acquisition, T.Č. All authors have read and agreed to the published version of the manuscript.

Funding: The project was financed by Research Council of Lithuania and the Ministry of Education, Science and Sport Lithuania (Contract No. S-A-UEI-23-9).

Institutional Review Board Statement: Not applicable.

Informed Consent Statement: Not applicable.

Data Availability Statement: Data is contained within the article.

Acknowledgments: The authors are grateful to the company "Thermoflow Inc" for the thermal engineering software "Thermoflow" and the possibility to run cycle simulations for the experimental research. The authors are grateful to the developers of the thermal engineering software "Thermoflow" for the opportunity to run cycle simulations for the experimental research.

Conflicts of Interest: The authors declare no conflicts of interest.

Nomenclature

b_e	Specific fuel consumption, g/kWh.
$c_{p_{WF}}$	Specific isobaric heat of the working fluid, kJ/(kgK).
G_{air}	Charge air flow before entering the engine cylinder, kg/s.
G_{WF}	Flow rate of working fluid, kg/s.
G_f	Hourly engine fuel consumption, kg/s.
G_w	Seawater flow rate, kg/h.
H_u	Lower fuel calorific value, kJ/kg.
h_{tg_i}	Enthalpy of the working material before and after the turbogenerator, kJ/kg.
$h_{w1}; h_{w2}$	Enthalpy of the working before and after cylinder cooling jacket heat exchanger, kJ/kg.
h_w'	Enthalpy value which is necessary according to engine manufacturer specification, kJ/kg.
k	Specific heat ratio.
K_1	The cumulative efficiency of the power turbine.
n	revolutions, min^1.
p	Pressure, Pa
P_e	Main engine power, kW.
P_{turb}	The power generated by the turbogenerator of the WHR system, kW.
t_{WF_1}	The temperature of the working fluid, °C.
$T_{exh.g.}$	Exhaust gas temperature, °C.
$Q_{exh.}$	Power plant exhaust gas energy part of heat balance, kJ/s.
Q_f	Total fuel energy, kW.
$Q_{sc.air}$	Power plant scavenges air cooling energy part of heat balance, kJ/s.
Q_{oil}	Power plant lubricating oil cooling energy part of heat balance, kJ/s.
$Q_{cil.}$	Power plant cylinder cooling jacket energy part of heat balance, kJ/s.
Q_w	WHR cycle heat dissipation through overboard water, kJ/s.
Q_h	Total heat transferred per unit mass of working fluid, kJ/s.
Q_{SS}	Secondary heat source transferred heat, kJ/s.
Q_T	Transformed heat in the turbine into mechanical work, kJ/s.
$q_{exh.}; q_{cil.}; q_{sc.air}$	Specific heat of secondary heat sources, kJ/kg.
q_h	Heat transferred from the working substance to the condenser, kJ/kg.
q_{SS}	Transferred specific heat from secondary heat sources to WF, kJ/kg.
q_w	Transferred specific heat from the working material to the overboard water kJ/kg.
$q^{\wedge}_w$	Gained heat from seawater cooling.
π_T	The degree of pressure drop in the turbine.
$T_{exh.g.}$	Exhaust gas temperature, °C.
η_e	Coefficient of performance of the main power plant.
η_{eRC}	The total coefficient of performance of the ship's main power plant with a WHR system.
η_{RC}	Coefficient of performance of the WHR cycle.
$\delta\eta_{eRC}$	Relative change in ship power plant efficiency with and without ORC.
$\eta_{eRC_{cikl}}$	Ship power plant efficiency with ORC with ISO 8178 operational cycle.
$\eta_{t.sc.air}; \eta_{t.cil.}; \eta_{t.exh.}$	Thermal efficiency coefficient of the secondary heat source exchangers.
$\eta_{t.ad}$	Internal (adiabatic) efficiency of the turbogenerator.
η_m	Mechanical efficiency of the turbogenerator.
$\Psi_{t.cil}; \Psi_{t.sc.air}; \Psi_{t.exh}$	Energy utilization factors of secondary heat sources.
β	pulse energy input factor.
t_{WF_1}	Temperature of the WF before the turbine, °C.

Abbreviations

CII	Carbon intensity indicator
CO_2	Carbon dioxide
EE	Efficiency coefficient
EEDI	Energy efficiency design index
EEXI	Existing energy efficiency index
EU	European Union

GHG	Greenhouse gases
GWP	Global warming potential
HCFC	Hydrochlorofluorocarbons
IMO	International Maritime Organization
LCA	Low-carbon-dioxide-generating fuel
MARPOL	International Convention for the Prevention of Pollution from Ships
MEPC	Marine Environment Protection Committee
ORC	Organic Rankine cycle
SRC	Steam Rankine cycle
RHE	Recuperative heat exchanger
WF	Working fluid
WHR	Waste heat recovery

Appendix A

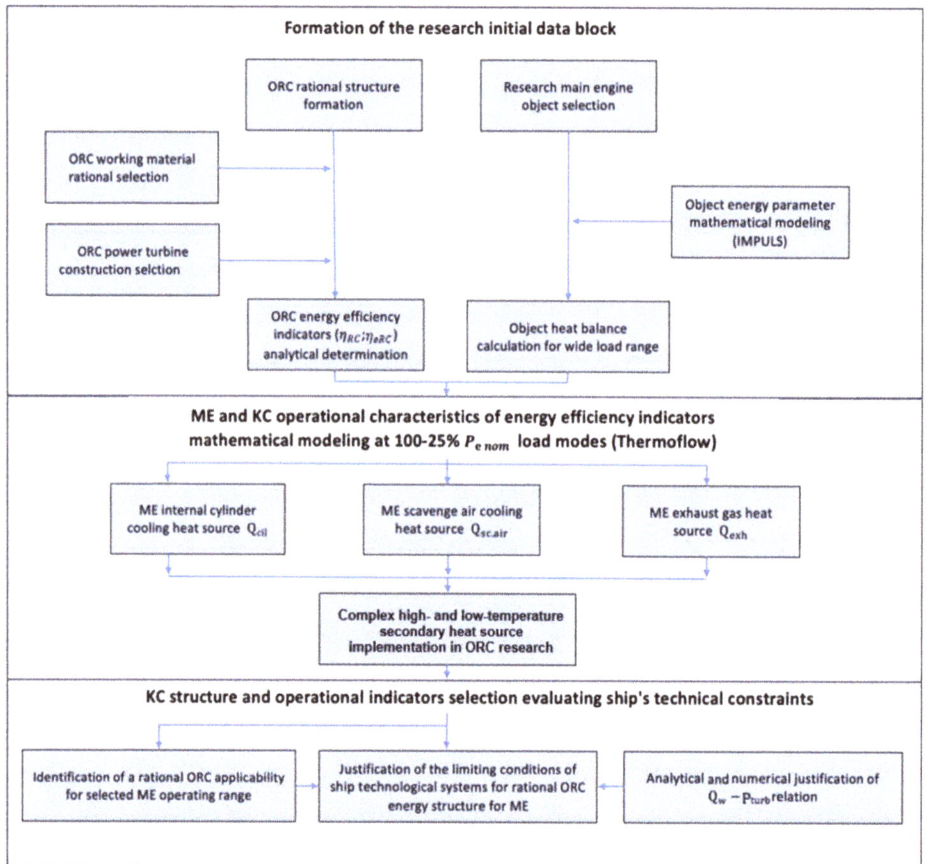

Figure A1. Block algorithm for the formation of the cogeneration cycle structure and identification of parameters [53].

Appendix B

Table A1. Research object "Wärtsila" 12V46F main engine general parameters.

Parameter	Data	Dimension
Manufacturer, type	WÄRTSILA 12V46, trunk type	-
Year of manufacture	2008	Year
Piston stroke	580	Mm
Average piston speed	9.7	m/s
Cylinder diameter	460	mm
Number of cylinders	12	vnt.
Nominal power	12,000	kW
Possibility of reversal	Non-reversal	-
Type	4 stroke	-
Number of valves	48	pcs.
Crankshaft revolutions	350–600	rpm
Type of fuel used	IFO 380 heavy fuel oil, diesel	-
Compression pressure	56	bar
Maximum combustion pressure	135	bar
Specific fuel consumption	174	g/kWh

Table A2. Energy balance indicator calculation results.

	Load Mode %				
	100%	**85%**	**75%**	**50%**	**25%**
P_e, cil. kW	1200	1020	900	600	300
G_f, kg/s	0.72	0.59	0.55	0.38	0.20 *
η_e	0.469	0.483	0.459	0.44	0.425 *
α_e	2.5	2.68	298	3.38	3.8 *
G_{air}, kg/s	26.1	23.35	23.35	18.8	14.5 *
ρ_{air}, kg/m^3	4.51	4.26	4.44	4.1	3.98 *
P_K, bar	4.24	4.01	4.17	3.86	3.75
t_g, °C	366	316	309	273	255 *
P_K', bar	4.45	4.2	4.38	4.06	3.93
t_k' / c_p', °C	220 / 29.344	211 / 29.324	218 / 29.34	205 / 29.311	200 / 29.3
M_1, mol	1.25	1.34	1.49	1.69	1.9
M_{CO_2}, CO_2 kg fuel			0.0725		
M_{H_2O}, H_2O kg fuel			0.063		
M_{O_2}, O_2 kg fuel	0.156	0.174	0.206	0.248	0.291
M_{N_2}, N_2 kg fuel	0.99	1.06	1.18	1.338	1.51
M_2, mol	1.28	1.37	1.52	1.72	1.996
mC_V', kj/kmolK			20.795		
mC_p', kj/kmolK			29.11		
mC_V'', kj/kmolK	22.31	22.11	21.93	21.67	21.46
mC_p'', kj/kmolK	30.63	30.43	30.25	29.96	29.78
$Q_{exh.g.}$, kW	8990	6622	6622	4411	2387
Q_f, kW	30,744	25,193	23,485	16,226	8540
Q_e, kW	14,400	12,240	10,800	7200	3600
$Q_{sc.air}$, kW	4369	3629	3851	2814	1010
$Q_{cil} + Q_{oil}$, kW	2985	2702	2212	1801	1543
Q_{rad}, kW	420		Not applicable		

* extrapolation.

Table A3. Individual secondary heat source in ORC results.

Working Material	Load, %	Working Fluid Enthalpy (pos. 12), kJ/kg		Exhaust Gas Temperature (pos. 12), C		Working Material Temperature (pos. 6)		Working Fluid Enthalpy (pos. 6), kJ/kg		Working Material Flow, kg/s	Pressure, Bar (pos. 6)		Pressure decrease Ratio (in Turbine, pos. 6)	Power, kW	Scavenge Air Temperature (pos.3)		Scavenge Air Flow, kg/s	Cylinder Cooling Temp. (poz 10)		Cylinder Cooling Flow, kg/s	η_e	$\delta\eta_{eRC}$	$\eta_{eRC_{cikl}}$	η_{RC}	$\Psi_{t.exh}$	$\Psi_{t.sc.air}$	$\Psi_{t.cil}$
EXHAUST GAS																										
R134a	100	−86.16	132.3	364	123.2	179.5	137.3	132.3	100.5	29.5	21.84	7.14	3.059	897.7	N/A		N/A	N/A		N/A	0.469	6.87%	0.48	0.142	0.994	N/A	N/A
R134a	75	−86.17	132.3	309	121.5	175.9	137.3	132.3	100.5	20.3	21.84	7.14	3.059	613.6	N/A		N/A	N/A		N/A	0.459	6.25%		0.142	0.994	N/A	N/A
R134a	50	−86.18	132.3	273	121.3	175.9	137.3	132.3	100.5	13.15	21.84	7.14	3.059	367.7	N/A		N/A	N/A		N/A	0.44	5.66%		0.142	0.992	N/A	N/A
R134a	25	−86.16	132.3	255	121.7	175.9	137.3	132.3	100.5	8.9	21.84	7.14	3.059	248.7	N/A		N/A	N/A		N/A	0.425	7.65%		0.141	0.988	N/A	N/A
SCAVENGE AIR																										
R134a	100	N/A		N/A		103.5	61.77	47.49	23.57	21	21.84	7.14	3.059	477.3	220	55.42	26.1	N/A		N/A	0.469	3.93%	0.47	0.263	N/A	0.9975	N/A
R134a	75	N/A		N/A		103.2	61.38	47.06	23.19	18.6	21.84	7.14	3.059	422	218	55.08	23.35	N/A		N/A	0.459	4.46%		0.265	N/A	0.9995	N/A
R134a	50	N/A		N/A		105	63.37	49.27	25.17	13.7	21.84	7.14	3.059	313.8	205	55.64	18.8	N/A		N/A	0.44	4.91%		0.257	N/A	0.9958	N/A
R134a	25	N/A		N/A		105.5	63.92	49.87	25.71	10.2	21.84	7.14	3.059	234.3	200	55.72	14.5	N/A		N/A	0.425	7.25%		0.255	N/A	0.9951	N/A
CYLINDER COOLING																										
R134a	100	N/A		N/A		87.62	43.9	27.55	5.815	15	21.84	7.14	3.059	309.9	N/A		N/A	96	75.26	35.7	0.469	2.76%	0.47	0.35889	N/A	N/A	0.988
R134a	75	N/A		N/A		87.62	43.9	27.55	5.813	11	21.84	7.14	3.059	227.2	N/A		N/A	96	75.51	26.5	0.459	2.65%		0.35455	N/A	N/A	0.976
R134a	50	N/A		N/A		87.62	43.89	27.54	5.81	9	21.84	7.14	3.059	185.9	N/A		N/A	96	75.33	21.5	0.44	3.13%		0.35784	N/A	N/A	0.984
R134a	25	N/A		N/A		87.61	43.89	27.54	5.81	7.8	21.84	7.14	3.059	161.1	N/A		N/A	96	75.18	18.5	0.425	5.20%		0.36139	N/A	N/A	0.994

Table A4. Complex secondary heat source in ORC results.

Working Material	Load, %	Working Fluid Enthalpy (pos. 12), kJ/kg Before	After	Exhaust Gas Temperature (pos. 12), C Before	After	Working Material Temperature (pos. 6) Before	After	Working Fluid Enthalpy (pos. 6), kJ/kg Before	After	Working Material Flow, kg/s	Pressure, Bar (pos. 6) Before	After	Pressure decrease Ratio (in Turbine, pos. 6)	Power, kW	Scavenge Air Temperature (pos.3) Before	After	Scavenge Air Flow, kg/s	Cylinder Cooling Temp. (poz 10) Before	After	Cylinder Cooling Flow, kg/s	η_e	$\delta\eta_{eRC}$	$\eta_{eRC_{cikl}}$	η_{RC}	$\Psi_{t.exh}$	$\Psi_{t.sc.air}$	$\Psi_{t.cil}$
EXHAUST GAS + SCAVENGE AIR + CYLINDER COOLING																											
R134a	100	28.76	135.7	364	120.5	178.8	140.3	135.7	103.6	60.4	21.84	7.14	3.059	1842.2	220	80.59	26.1	96	76.09	35.7	0.469	0.520	13.46%	0.132	0.996	0.846	0.948
	75	38.43	136	309	120.4	179.1	140.5	136	103.9	45.6	21.84	7.14		1391.8	218	80.48	23.35	96	76.1	26.5	0.459		13.49%	0.131	0.994	0.845	0.948
	50	47.19	136.3	273	120.1	179.3	140.7	136.3	104.2	32.5	21.84	7.14		992.7	205	79.82	18.8	96	76.1	21.5	0.44		14.39%	0.130	0.993	0.835	0.948
	25	55.55	136.5	255	120.6	179.5	140.9	136.5	104.4	24.2	21.84	7.14		739.6	200	79.57	14.5	96	76.09	18.5	0.425		21.39%	0.132	0.991	0.831	0.948
EXHAUST GAS + SCAVENGE AIR																											
R134a	100	72.29	178	364	120.3	178	139.4	134.8	102.8	45.9	21.84	7.14	3.059	1426.9	220	80.59	26.1	N/A			0.469	0.504	10.56%	0.133	0.996	0.846	N/A
	75	75.79	178.3	309	120.1	178.3	139.8	135.2	103.1	35.6	21.84	7.14	3.059	1084.2	218	80.48	23.35				0.459		10.63%	0.131	0.994	0.845	
	50	80.37	178.5	273	120.8	178.5	139.9	135.4	103.3	24.3	21.84	7.14	3.059	740.5	205	79.82	18.8				0.44		10.87%	0.130	0.993	0.835	
	25	83.79	178.6	255	120.5	178.6	140.1	135.5	103.4	17.2	21.84	7.14	3.059	524.3	200	79.57	14.5				0.425		15.36%	0.131	0.991	0.831	
EXHAUST GAS + CYLINDER COOLING																											
R134a	100	-12.1	134.5	364	120.3	177.8	139.2	134.5	102.5	44.1	21.84	7.14	3.059	1340.8				96	75.19	35.7	0.469	0.500	9.96%	0.141084	0.999	N/A	0.991
	75	-8.334	134.6	309	120.6	177.9	139.3	134.6	102.6	31.1	21.84	7.14	3.059	945.8				96	75.19	26.5	0.459		9.34%	0.140796	0.997		0.991
	50	3.048	135	273	120.5	178.2	139.6	135	102.9	21.9	21.84	7.14	3.059	666.6				96	75.19	21.5	0.44		9.84%	0.140583	0.997		0.991
	25	15.9	135.3	255	120.7	178.5	139.9	135.3	103.3	16.4	21.84	7.14	3.059	499.7				96	75.2	18.5	0.425		14.68%	0.1402	0.995		0.991
SCAVENGE AIR + CYLINDER COOLING																											
R134a	100					170.7	132	126.2	94.9	31	21.84	7.14	3.059	921.9	220	83.14	26.1	96	75.19	35.7	0.469	0.490	7.04%	0.132168	N/A	0.831	0.991
	75					152.4	113.3	104.9	75.5	26	21.84	7.14	3.059	727.6	218	79.63	23.35	96	75.19	26.5	0.459		7.31%	0.14799		0.850	0.991
	50					176.8	138.2	133.4	101.5	17	21.84	7.14	3.059	515.3	205	79.89	18.8	96	75.19	21.5	0.44		7.73%	0.112181		0.645	0.991
	25					177	138.4	133.6	101.7	14	21.84	7.14	3.059	424.7	200	79.23	14.5	96	75.19	18.5	0.425		12.58%	0.124063		0.680	0.991

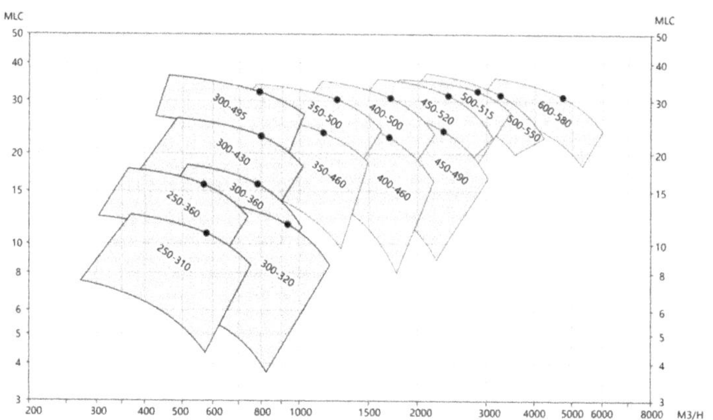

Figure A2. DESMI DSL centrifugal water pump specification statistics.

References

1. United Nations Framework Convention on Climate Change (UNFCCC). World Trade and the Reduction of CO2 Emissions, International Chamber of Shipping. 2014. Available online: https://www.ics-shipping.org/wp-content/uploads/2014/08/shipping-world-trade-and-the-reduction-of-co2-emissions-min.pdf (accessed on 23 November 2023).
2. International Maritime Organization (IMO). Third IMO Greenhouse Gas Study 2014. Available online: https://www.imo.org/en/ourwork/environment/pages/greenhouse-gas-studies-2014.aspx (accessed on 9 November 2023).
3. Maritime Industry Authority. MARPOL 73/78. 1973. Available online: https://marina.gov.ph/wp-content/uploads/2020/10/MARPOL-73_78.pdf (accessed on 9 November 2023).
4. MARPOL Annex VI, Resolution MEPC. 308(73). 2018. Available online: https://wwwcdn.imo.org/localresources/en/Knowledge-Centre/IndexofIMOResolutions/MEPCDocuments/MEPC.308(73).pdf (accessed on 23 November 2023).
5. Guidelines on Survey and Certification of the Attained Energy Efficiency Existing Ship Index (EEXI), Resolution MEPC. 351(78). 2022. Available online: https://wwwcdn.imo.org/localresources/en/KnowledgeCentre/IndexofIMOResolutions/MEPCDocuments/MEPC.351(78).pdf (accessed on 23 November 2023).
6. Interim Guidelines on Correction Factors and Voyage Adjustments for CII Calculations, Resolution MEPC.355(78). 2022. Available online: https://wwwcdn.imo.org/localresources/en/KnowledgeCentre/IndexofIMOResolutions/MEPCDocuments/MEPC.355(78).pdf (accessed on 23 November 2023).
7. EU Monitor. Use of Renewable and Low-Carbon Fuels in Maritime Transport, COM(2021)562. Available online: https://www.eumonitor.eu/9353000/1/j4nvhdfdk3hydzq_j9vvik7m1c3gyxp/vlkimqggznzr (accessed on 23 November 2023).
8. European Commision, Monitor. Directive of the European Parliament and of the Council, 2021, COM(2021)551. Available online: https://eur-lex.europa.eu/resource.html?uri=cellar:618e6837-eec6-11eb-a71c-01aa75ed71a1.0001.02/DOC_1&format=PDF (accessed on 1 November 2023).
9. International Renewable Energy Agency. *A Pathway to Decarbonise the Shipping Sector by 2050*; International Renewable Energy Agency: Abu Dhabi, United Arab Emirates, 2021; ISBN 978-92-9260-330-4.
10. Maersk Mc-Kinney Moller Center for Zero Carbon Shipping. The Shipping Industry's Fuel Choices on the Path to Net Zero. 2023. Available online: https://www.globalmaritimeforum.org/content/2023/04/the-shipping-industrys-fuel-choices-on-the-path-to-net-zero_final.pdf (accessed on 1 December 2023).
11. Theotokatos, G.; Livanos, G.A. Techno-economic analysis of single pressure exhaust gas waste heat recovery systems in marine propulsion plants. *Proc. Inst. Mech. Eng. Part M J. Eng. Marit. Environ.* **2013**, *227*, 83–97.
12. Livanos, G.; Theotokatos, G.; Pagonis, D. Techno-economic investigation of alternative propulsion plants for ferries and roro ships. *Energy Convers. Manag.* **2014**, *79*, 640–651. [CrossRef]
13. Dimopoulos, G.; Georgopoulou, C.; Kakalis, N. Modelling and optimization of an integrated marine combined cycle system. In Proceedings of the 24th International Conference on Energy, Cost, Optimization, Simulation and Environmental Impact of Energy Systems (ECOS), Novi Sad, Serbia, 4–7 July 2011; pp. 1283–1298.
14. Hountalas, D.T.; Katsanos, C.; Mavropoulos, G.C. Efficiency improvement of large scale 2-stroke diesel engines through the recovery of exhaust gas using a rankine cycle. *Procedia-Soc. Behav. Sci.* **2012**, *48*, 1444–1453. [CrossRef]
15. Maritime Forecast 2050, Energy Transition Outlook, DNV. 2022. Available online: https://www.dnv.com/maritime/publications/maritime-forecast-2023/ (accessed on 10 December 2023).
16. Dharmalingam, P.; Kumar, J.N.; Suryanarayanan, R.; Kumar, S.S. Waste heat recovery. In *The National Certification Examination for Energy Managers and Energy Auditors*, 2nd ed.; Bureau of Energy Efficiency: New Delhi, India, 2004; pp. 1–18.
17. Woodyard, D. *Pounder's Marine Diesel Engines and Gas Turbines*, 9th ed.; Elsevier: Oxford, UK, 2009.

18. Olaniyi, E.O.; Prause, G. Investment Analysis of Waste Heat Recovery System Installations on Ships' Engines. *J. Mar. Sci. Eng.* **2020**, *8*, 811. [CrossRef]
19. Baldi, F.; Gabrielii, C. A feasibility analysis of waste heat recovery systems for marine applications. *Energy* **2015**, *80*, 654–665. [CrossRef]
20. Altosole, M. Waste heat recovery systems from marine diesel engines: Comparison between new design and retrofitting solutions. In *Maritime Technology and Engineering*; CRC Press: Boca Raton, FL, USA, 2014; pp. 735–742.
21. Díaz-Secades, L.A.; González, R.; Rivera, N.; Montañés, E.; Quevedo, J.R. Waste heat recovery system for marine engines optimized through a preference learning rank function embedded into a Bayesian optimizer. *Ocean Eng.* **2023**, *281*, 114747. [CrossRef]
22. Annual Report to the Congress for 1982 (March 1983) Industrial and Commercial Cogeneration (February 1983). Available online: https://www.princeton.edu/~ota/disk3/1983/9565/9565.PDF (accessed on 3 December 2023).
23. Hatchman, J.C. Steam cycles for waste heat recovery: A case study. *R&D J.* **1991**, *7*, 32–38.
24. Korlak, P.K. Comparative analysis of the heat balance results of the selected Tier III-compliant gas-fuelled two-stroke main engines. *Combust. Engines* **2023**, *193*, 24–28. [CrossRef]
25. Tchanche, B.F.; Bertrand, M.; Papadakis, G. Heat resources and organic Rankine cycle machines. *Renew. Sustain. Energy Rev.* **2014**, *39*, 1185–1199. [CrossRef]
26. Tchanche, B.F.; Lambrinos, G.; Frangoudakis, A.; Papadakis, G. Low-grade heat conversion into power using organic Rankine cycles—A review of various applications. *Renew. Sustain. Energy Rev.* **2011**, *15*, 3963–3979. [CrossRef]
27. Tian, H.; Shu, G.; Wei, H.; Liang, X.; Liu, L. Fluids and parameters optimization for the organic Rankine cycles (ORCs) used in exhaust heat recovery of Internal Combustion Engine (ICE). *Energy* **2012**, *47*, 125–136. [CrossRef]
28. Wei, D.; Lu, X.; Lu, Z.; Gu, J. Performance analysis and optimization of organic Rankine cycle (ORC) for waste heat recovery. *Energy Convers. Manag.* **2007**, *48*, 1113–1119. [CrossRef]
29. Konur, O.; Yuksel, O.; Korkmaz, S.A.; Colpan, C.O.; Saatcioglu, O.Y.; Muslu, I. Thermal design and analysis of an organic rankine cycle system utilizing the main engine and cargo oil pump turbine based waste heats in a large tanker ship. *J. Clean. Prod.* **2022**, *368*, 133230. [CrossRef]
30. Díaz-Secades, L.A.; González, R.; Rivera, N. Waste heat recovery from marine engines and their limiting factors: Bibliometric analysis and further systematic review. *Clean. Energy Syst.* **2023**, *6*, 100083. [CrossRef]
31. Song, J.; Song, Y.; Gu, C. Thermodynamic and lysis and performance optimization of an Organic Rankine Cycle (ORC) waste heat recovery system for marine diesel engines. *Energy* **2015**, *82*, 976–985. [CrossRef]
32. Ng, C.W.; Tam, I.C.K.; Wu, D. *Study of a Waste Energy Driven Organic Rankine Cycle Using Free Piston Linear Expander for Marine Application*; The National Technical University of Athens: Athens, Greece, 2019.
33. Park, B.-S.; Usman, M.; Imran, M.; Pesyridis, A. Review of Organic Rankine Cycle experimental data trends. *Energy Convers. Manag.* **2018**, *173*, 679–691. [CrossRef]
34. Grljušić, M.; Medica, V.; Radica, G. Calculation of efficiencies of a ship power plant operating with waste heat and power production. *Energies* **2015**, *8*, 4273–4299. [CrossRef]
35. Baldi, F.; Larsen, U.; Gabrielii, C. Comparison of different procedures for the optimisation of a combined diesel engine and organic Rankine cycle system based on ship operational profile. *Ocean Eng.* **2015**, *110*, 85–93. [CrossRef]
36. Necmi, O. Thermodynamic Analysis of Gas Turbine Cogeneration Power Plants. *Int. J. Eng. Sci. Res. Technol.* **2016**, *5*, 736–742. [CrossRef]
37. Konur, O.; Yuksel, O.; Korkmaz, S.A.; Colpan, C.O.; Saatcioglu, O.Y.; Koseoglu, B. Operation-dependent exergetic sustainability assessment and environmental analysis on a large tanker ship utilizing Organic Rankine cycle system. *Energy* **2023**, *262*, 125477. [CrossRef]
38. Konur, O.; Colpan, C.O.; Saatcioglu, O.Y. A comprehensive review on organic Rankine cycle systems used as waste heat recovery technologies for marine applications. *Energy Sources Part A Recover. Util. Environ. Eff.* **2022**, *44*, 4083–4122. [CrossRef]
39. Ng, C. *Modelling and Simulation of Organic Rankine Cycle Waste Heat Recovery System with the Operational Profile of a Ship*; Newcastle University: Newcastle upon Tyne, UK, 2021.
40. Ng, C.; Tam, I.C.K.; Wetenhall, B. Waste Heat Source Profiles for Marine Application of Organic Rankine Cycle. *J. Mar. Sci. Eng.* **2022**, *10*, 1122. [CrossRef]
41. Shu, G.; Liu, P.; Tian, H.; Wang, X.; Jing, D. Operational profile based thermal-economic analysis on an Organic Rankine cycle using for harvesting marine engine's exhaust waste heat. *Energy Convers. Manag.* **2017**, *146*, 107–123. [CrossRef]
42. Qu, J.; Feng, Y.; Zhu, Y.; Zhou, S.; Zhang, W. Design and thermodynamic analysis of a combined system including steam Rankine cycle, organic Rankine cycle, and power turbine for marine low-speed diesel engine waste heat recovery. *Energy Convers. Manag.* **2021**, *245*, 114580. [CrossRef]
43. Baldasso, E.; Mondejar, M.E.; Andreasen, J.G.; Rønnenfelt, K.A.T.; Nielsen, B.; Haglind, F. Design of organic Rankine cycle power systems for maritime applications accounting for engine backpressure effects. *Appl. Therm. Eng.* **2020**, *178*, 115527. [CrossRef]
44. Liu, X.; Nguyen, M.Q.; Chu, J.; Lan, T.; He, M. A novel waste heat recovery system combing steam Rankine cycle and organic Rankine cycle for marine engine. *J. Clean. Prod.* **2020**, *265*, 121502. [CrossRef]
45. Ng, C.; Tam, I.C.K.; Wu, D. Thermo-Economic Performance of an Organic Rankine Cycle System Recovering Waste Heat Onboard an Offshore Service Vessel. *J. Mar. Sci. Eng.* **2020**, *8*, 351. [CrossRef]

46. Sellers, C. Field operation of a 125kW ORC with ship engine jacket water. *Energy Procedia* **2017**, *129*, 495–502. [CrossRef]
47. Mohammed, A.G.; Mosleh, M.; El-Maghlany, W.M.; Ammar, N.R. Performance analysis of supercritical ORC utilizing marine diesel engine waste heat recovery. *Alex. Eng. J.* **2020**, *59*, 893–904. [CrossRef]
48. Ouyang, T.; Su, Z.; Zhao, Z.; Wang, Z.; Huang, H. Advanced exergo-economic schemes and optimization for medium–low grade waste heat recovery of marine dual-fuel engine integrated with accumulator. *Energy Convers. Manag.* **2020**, *226*, 113577. [CrossRef]
49. Ouyang, T.; Su, Z.; Yang, R.; Li, C.; Huang, H.; Wei, Q. A framework for evaluating and optimizing the cascade utilization of medium-low grade waste heat in marine dual-fuel engines. *J. Clean. Prod.* **2020**, *276*, 123289. [CrossRef]
50. Su, Z.; Ouyang, T.; Chen, J.; Xu, P.; Tan, J.; Chen, N.; Huang, H. Green and efficient configuration of integrated waste heat and cold energy recovery for marine natural gas/diesel dual-fuel engine. *Energy Convers. Manag.* **2020**, *209*, 112650. [CrossRef]
51. Wang, X.; Shu, G.-Q.; Tian, H.; Liang, Y.; Wang, X. *Simulation and Analysis of an ORC-Desalination Combined System Driven by the Waste Heat of Charge Air at Variable Operation Conditions*; SAE Technical Paper 2014-01-1949; SAE International: Warrendale, PA, USA, 2014.
52. Chen, W.; Fu, B.; Zeng, J.; Luo, W. Research on the Operational Performance of Organic Rankine Cycle System for Waste Heat Recovery from Large Ship Main Engine. *Appl. Sci.* **2023**, *13*, 8543. [CrossRef]
53. Lebedevas, S.; Čepaitis, T. Research of Organic Rankine Cycle Energy Characteristics at Operating Modes of Marine Diesel Engine. *J. Mar. Sci. Eng.* **2021**, *9*, 1049. [CrossRef]
54. *Report of the Marine Environment Protection Committee on its Seventy-Sixth Session*; MEPC 76/15/ADD.1; MEPC: London, UK, 24 August 2021.
55. ES 2022/0099(COD). Proposal for a Regulation of the European Parliament and of the Council on Fluorinated Greenhouse Gases, Amending Directive (EU) 2019/1937 and Repealing Regulation (EU) No 517/2014. 19 October 2023. Available online: https://eur-lex.europa.eu/legal-content/EN/TXT/?uri=celex:52022PC0150 (accessed on 9 November 2023).
56. Hafner, A.; Gabrielii, C.H.; Widell, K. *Refrigeration Units in Marine Vessels—Alternatives to HCFCs and high GWP HFCs*; Nordic Council of Ministers: Copenhagen, Denmark, 2019.
57. Møller-Hansen, T. *Alternatives to HCFC as Refrigerant in Shipping Vessels*; TemaNord 2000:573; Nordic Council of Ministers: Copenhagen, Denmark, 2000.
58. Wärtsila "Marine market shares 2022". 2022. Available online: https://www.wartsila.com/docs/default-source/investors/market-shares/ship-power-market-share/marine-market-share-2022.pdf?sfvrsn=37f13b43_1 (accessed on 28 December 2023).
59. Lam JS, L.; Piga, B.; Zengqi, X. *Role of Bio-LNG in Shipping Industry Decarbonization*; Nanyang Technological University: Singapore, 2022.
60. BioLNG in Transport: Making Climate Neutrality a Reality A joint White Paper about bioLNG Production and Infrastructure as Enabler for Climate Neutral Road and Maritime Transport. EBA, GIE, NGVA & SEA-LNG. 2020. Available online: https://www.europeanbiogas.eu/wp-content/uploads/2020/11/BioLNG-in-Transport_Making-Climate-Neutrality-a-Reality.pdf (accessed on 28 December 2023).
61. Alternative Fuels for Containerships. DNV. 2023. Available online: https://www.dnv.com/maritime/publications/alternative-fuels-for-containerships-methanol-and-ammonia-download/ (accessed on 28 December 2023).
62. Lebedev, S.; Nechayev, L. *Sovershenstvovaniye Pokazateley Vysokooborotnykh Dizeley Unifitsirovannogo Tiporazmera*; Akademiya Transporta RF: Barnaul, Russia, 1999; p. 48.
63. Lebedev, S.; Lebedeva, G.; Matievskij, D.; Reshetov, V. *Formirovanie Konstruktivnogo Rjada Porshnej Dlja Tipaža Vysokooborotnyh Forsirovannyh Dizelej*; Akademiya Transporta RF: Barnaul, Russia, 2003; p. 49.
64. Kreienbaum, J. *The Energy Crises of the 1970s as Challenges to the Industrialized World*; Historisches Institut, Universität Rostock: Rostock, Germany, 2013.
65. Grabowski, L.; Wendeker, M.; Pietrykowski, K. *AVL Simulation Tools Practical Applications*; Lublin University of Technology: Lublin, Poland, 2012.
66. Wiebe, I. *Brennverlauf und Kreisprozesse Vonverbrennungsmotoren*; VEB Verlag Technik: Berlin, Germany, 1970; 280p.
67. Woschni, G. *Eine Methode zur Vorausberechnung der Änderung des Brenverlaufs Mittelschnellaufender Dieselmotoren bei Geanderten Betriebsbedigungen*; Springer Fachmedien: Wiesbaden, Germany, 1973; pp. 106–110.
68. Woschni, G. *Verbrennungsmotoren*; TU Munchen: Munich, Germany, 1988.

Journal of
Marine Science and Engineering

Article

A Fault-Tolerant Control Method Based on Reconfiguration SPWM Signal for Cascaded Multilevel IGBT-Based Propulsion in Electric Ships

Fan Zhang [1], Zhiwei Zhang [2], Zhonglin Zhang [1], Tianzhen Wang [1,*], Jingang Han [1] and Yassine Amirat [3,*]

[1] Logistics Engineering College, Shanghai Maritime University, Shanghai 201306, China; zhangfan@shmtu.edu.cn (F.Z.); 202030210046@stu.shmtu.edu.cn (Z.Z.); jghan@shmtu.edu.cn (J.H.)
[2] Shanghai Power Industrial & Commerical Co., Ltd., Shanghai 200001, China; dianjing@21cn.com
[3] ISEN Yncréa Ouest, L@bISEN, 29200 Brest, France
* Correspondence: tzwang@shmtu.edu.cn (T.W.); yassine.amirat@isen-ouest.yncrea.fr (Y.A.)

Abstract: Electric ships have been developed in recent years to reduce greenhouse gas emissions. In this system, inverters are the key equipment for the permanent-magnet synchronous motor (PMSM) drive system. The cascaded insulated-gated bipolar transistor (IGBT)-based H-bridge inverter is one of the most attractive multilevel topologies for modern electric ship applications. Usually, the fault-tolerant control strategy is designed to keep the ship in operation for a certain period. However, the fault-tolerant control strategy with hardware redundancy is expensive and slow in response. In addition, after fault-tolerant control, the ship's PMSM may experience shock and overheating, and IGBT life is reduced due to uneven switching frequency distribution. Therefore, a stratified reconfiguration carrier disposition Sinusoidal Pulse Width Modulation (SPWM) fault-tolerant control strategy is proposed. The proposed strategy can achieve fault tolerance without any extra hardware. A reconfiguration carrier is applied to improve the fundamental amplitude of inverter output voltage to maintain the operation of the ship's PMSM. In addition, the available states of faulty H-bridge are fully used to contribute to the output. These can improve the life of IGBTs by reducing and balancing the power loss of each H-bridge. The principles of the proposed strategy are described in detail in this study. Taking a cascaded H-bridge seven-level inverter as an example, simulation and experimental results verify that the proposed strategy, in general, has a potential future application on electric ships.

Keywords: electric ships; cascaded multilevel inverter; stratified reconfiguration; fault-tolerant control; ecological sustainable development

Citation: Zhang, F.; Zhang, Z.; Zhang, Z.; Wang, T.; Han, J.; Amirat, Y. A Fault-Tolerant Control Method Based on Reconfiguration SPWM Signal for Cascaded Multilevel IGBT-Based Propulsion in Electric Ships. *J. Mar. Sci. Eng.* **2024**, *12*, 500. https://doi.org/10.3390/jmse12030500

Academic Editors: Jian Wu, Xianhua Wu and Lang Xu

Received: 29 January 2024
Revised: 13 March 2024
Accepted: 15 March 2024
Published: 18 March 2024

1. Introduction

In recent years, maritime departments around the world have gradually begun to pay attention to the issue of ecologically sustainable development and initiated a series of measures to reduce the emissions of pollutants, carbon, and sulfide. Among them, the development and utilization of new energy have attracted the most attention. These mainly focus on the in-depth development of sustainable natural energy sources such as solar, wind, and ocean energy. Accordingly, the electric propulsion system has attracted researchers' interest [1,2].

Today, inverters are widely used in ships, carbon-free energy generation, transportation, motor drives, and other fields [3–8]. A high-boost Z-source inverter is proposed in an inland river cruise ship supplied by a fuel cell (FC) as the main power source and a supercapacitor (SC) as the auxiliary power source [3]. A new model using state-averaged models of the inverter and a hybrid model of the rectifier is developed to give an effective solution combining accuracy with the speed of the simulation and an appropriate interface to the electrical network model [4]. A control power module for hybrid inverter systems is implemented to drive electric propulsion ships [5]. Inverters can also be applied in marine photovoltaic on/off-grid systems [7,8].

Multilevel inverters have more advantages because of the lower voltage of their components, their lower switching frequency, and lower switching power loss, among others. The topologies of multilevel inverters mainly include diode-clamped inverters [9], flying-capacitor inverters [10], modular multilevel converters [11], and cascaded inverters [12]. The multilevel inverters are also applied on ships [13–16]. The novel symmetric and asymmetric multilevel inverter topologies with a minimum number of switches are proposed for the high voltage of an electric ship's propulsion system [13]. An innovative single-phase and three-phase H-bridge-derived multilevel inverter topology is being proposed in marine ships [14]. A cascaded H-bridge multilevel inverter is used to implement a proportional–integral speed current controller algorithm in the driving circuit of the Brushless Direct Current Motor (BLDC) motor for electric propulsion ships using a power analysis program [15]. To minimize of total harmonic distortion in multilevel inverters, teaching–learning-based optimization (TLBO) is used for marine propulsion systems [16].

Among them, cascaded H-bridge inverters are the most popular due to their characteristics of easy modularization and convenient expansion of level numbers. However, many semiconductor devices are required for cascaded H-bridge multilevel inverters. And it is a fact that the power semiconductor device is one of the most fragile components in electric ships' propulsion systems [17,18]. Therefore, the reliability of cascaded H-bridge inverters is relatively low, which underscores the importance of fault-tolerant control.

Most of the research on inverter fault-tolerant control focuses on open-circuit and short-circuit faults in semiconductors [19]. Short circuits are catastrophic failures that immediately trip or damage the system [20]. Therefore, a protection circuit is usually designed. In this way, the short-circuit fault can be regarded as the open-circuit fault. In cases of open-circuit faults, the inverter will skip the corresponding voltage level, which leads to the distortion of the voltage waveform. It may affect other devices, such as the motor and the grid. In other words, the fault may spread to other systems and cause subsequent failures through the power system. Thus, it is attractive to quickly eliminate the influence of the open-circuit fault of the cascaded H-bridge. The whole process can be divided into fault diagnosis and fault-tolerant control. Fault diagnosis is an essential step that is responsible for determining fault information and activating the corresponding fault-tolerant control method. Some effective diagnosis methods have been proposed in the field of ships [21–23]. This paper mainly focuses on the research on fault-tolerant control of open-circuit faults.

There are several studies that discuss fault-tolerant control of inverters on ships [24–26]. A new PMSM without the neutral point was modified to realize fault-tolerant control [24]. A modified lookup table was designed to improve the functioning of the fault-tolerant direct torque control (DTC) for off-shore ship propulsion [25]. Two TRIACs were added to pass faulty devices, and two switches were added to the fault-tolerant control of inverters on ship [26]. These methods either require a motor redesign, which may not work for a PMSM, or add redundancy in electric ships. Because there are few studies on IGBT conduction frequency in fault-tolerant control of inverters on ships, it is necessary to learn from the studies on power electronics on land.

To ensure the reliable operation of the system, the fault-tolerant method in [27,28] bypasses the faulty cell. In addition, the health cell in the other phase should be bypassed to achieve voltage balance. This means that three redundant cells need to be reserved even if only one IGBT fails. In contrast, the topology structure in [29] adds one backup H-bridge cell, three fast-blowing fuses, and three electromechanical relays. The relay R1 is turned on by the control circuitry, blowing the fuse and adding the auxiliary cell to the faulty phase. However, the complexity and cost of the circuit have improved due to the redundant backup bridge. To solve this problem, a fault-tolerant method without a redundant backup bridge is proposed in [30]. The topology will be reconfigured by the pilot switch when a fault occurs. Then, the triangle carrier needs to be reconfigured to apply to the reconfigured topology. It not only isolates the fault IGBT but also retains the healthy power supply, which keeps the output voltage amplitude of the inverter. However, this method is only designed for the cascaded five-level inverter. In [31], a new fault-tolerant control topology is proposed. The proposed topology adds four relays in each module, which can perform the isolation and elimination of the fault module from the whole circuit. It can be applied to higher-level inverters. In contrast, the method in [32] uses fewer relays to bypass the faulty cell. And this method connects the batteries to the healthy cell to achieve fault-tolerant control, which can be applied to an asymmetric mode. In [33], a serial fault-tolerant topology based on sustainable reconfiguration is proposed to achieve serial fault-tolerant control. However, the switches used in those articles are electromechanical switches, which are slow to respond compared to semiconductor devices. In addition, the method in [34,35] proposes a novel inverter topology. They can also quickly respond when a fault occurs. Partial voltage levels can be output through two disjoint loops. This means that when the switch in one loop fails, the voltage can be output through the other loop. However, the proposed topologies need to add many switching devices. The method in [36] proposes a fault-tolerant control strategy based on the divided voltage modulation algorithms. It can also achieve fault-tolerant control of the induction motor drive system without hardware redundancy or algorithm redundancy. However, the fundamental amplitude of the output voltage decreases. The method in [37] proposes a strategy to achieve a higher utilization ratio of healthy IGBTs and sinusoidal output voltage. However, a conduction state is not used. In addition to the points above, closed-loop control of a permanent magnet synchronous motor leads to IGBT module control signal characteristics that are not obvious. After the fault occurs, the system will work in an abnormal state, which brings huge security risks to the whole system.

To solve these problems, a stratified reconfiguration carrier disposition the SPWM fault-tolerant control strategy is proposed. The main contribution of this article is to improve the performance of inverter output voltage in post-fault operation, reduce and balance the power loss of the H-bridge, and improve the reliability of the system. The problem is analyzed in Section 2. The operating principle is illustrated in Section 3. Simulation and experimental results verify the theoretical analysis in Section 4. Finally, this paper is concluded in Section 5.

2. Problem Description

The electric propulsion system topology of a ship's DC network is shown in Figure 1

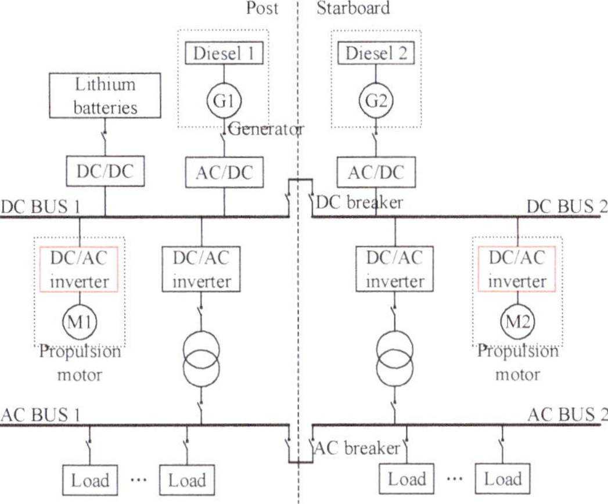

Figure 1. Ship DC electric propulsion system topology.

The network is divided into two subsystems: the Port DC network and the Starboard DC network. Each DC network subsystem includes four types of converters: AC/DC rectifiers of generators (G1 and G2), bidirectional DC/DC converters of batteries or solar panel energy storage, DC/AC inverters of loads, and DC/AC invertors of propulsion motors (M1 and M2).

The DC side of the above four types of converters takes the DC bus (post and starboard DC bus are, respectively, recorded as DC bus 1 and DC bus 2) as the common connection point. The post and starboard DC bus can be divided or closed through the DC breaker to achieve networking or independent operation. On the load side, DC/AC inverters change the DC bus voltage into an AC voltage with adjustable amplitude and frequency. They then connect the transformer to the three-phase 380V AC bus to power the ship's loads. The left and starboard AC buses (respectively recorded as AC bus 1 and AC bus 2) are divided or closed through the AC breaker, which can also realize network operation or independent operation.

A three-phase cascaded multilevel inverter is adopted in the DC/AC inverter of the propulsion motor. The propulsion motor drives the propeller to overcome the load resistance of the ship. The load characteristics of propeller are complicated. The relationship between the parameters is as follows:

$$\begin{cases} J' = \dfrac{V_P}{\sqrt{V_P^2 + D^2 n^2}} \\ K'_P(J') = \frac{1}{2}a_{0P}T_0(J') + a_{1P}T_1(J') + \cdots\cdots + a_{nP}T_n(J') \\ K'_T(J') = \frac{1}{2}a_{0T}T_0(J') + a_{1T}T_1(J') + \cdots\cdots + a_{nT}T_n(J') \\ P_e = K'_P\rho n^2(1 - t_{p0})(V_P^2 + D^2 n^2) \\ T = K'_T\rho D^3(V_P^2 + D^2 n^2) \\ V_P = V_S(1 - \omega) \end{cases} \tag{1}$$

where J' is the bounded form of the advance ratio of the propeller; K'_P is the torque coefficient; K'_T is the thrust coefficient; P_e is the thrust; T is the torque; V_p is the speed of the propeller relative to the water; V_s is the ship's speed; and n is the propeller's speed.

The simulation results of direct startup are shown in Figure 2.

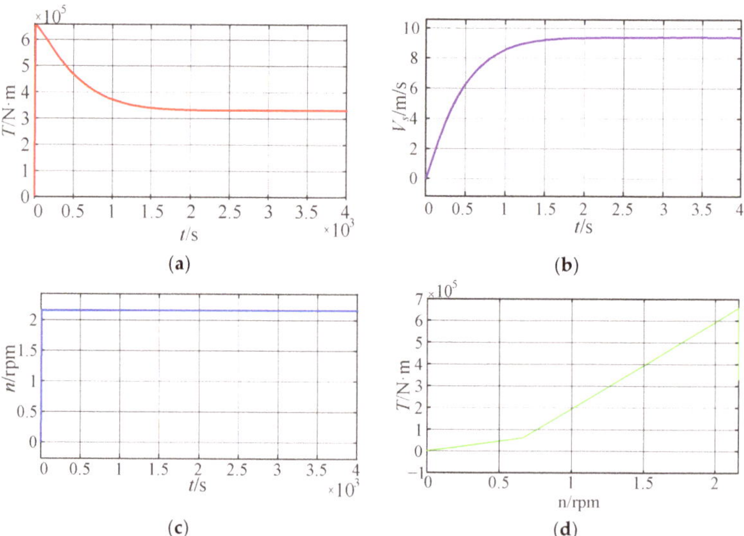

Figure 2. The simulation results of direct startup: (**a**) torque waveform over time; (**b**) relative speed over time; (**c**) propeller speed over time; (**d**) torque waveform with speed change.

When direct starting is adopted, the propeller's speed and torque increase rapidly. When the propeller's speed reaches the maximum value, the torque reaches the peak value. As the speed increases and the advance ratio increases, the torque of the paddle will continue to decrease until it is stable. The torque of the propeller is proportional to the speed. k times the product of torque and speed is equal to power. Therefore, in this paper's simulation, the propeller propelled by the motor is regarded as a fan in load characteristic.

The cascaded multilevel inverter consists of several H-bridges. Every H-bridge includes four IGBTs and an independent DC power supply. Due to the identical structure of the three phases in the multilevel inverter, only one phase is analyzed.

2.1. Operation State

The single phase of a cascaded seven-level inverter is shown in Figure 3.

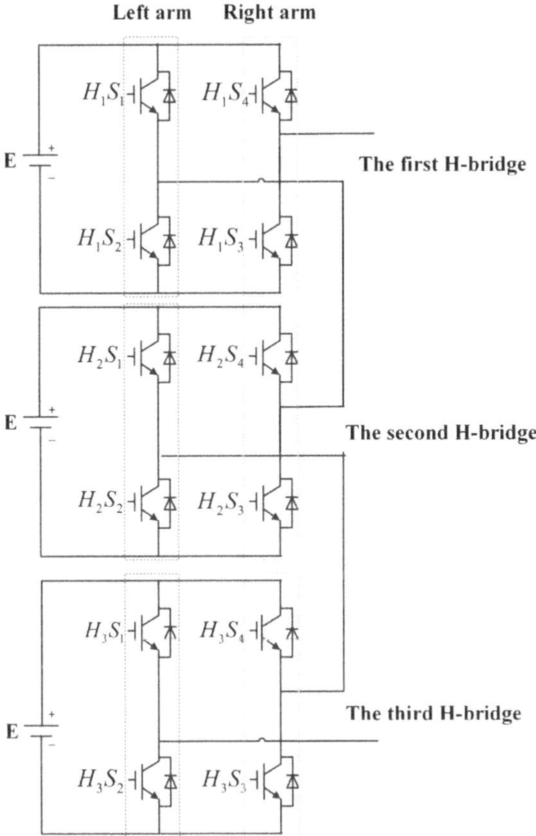

Figure 3. Cascaded H-bridge seven-level inverter.

As shown in Figure 4, carrier disposition modulation is used to output the Pulse Width Modulation (PWM) signal for the inverter. The bridge arm is driven to switch, in turn, in a determined period. Each H-bridge provides a different output voltage at different times. The total output voltage is the sum of the output voltages of each H-bridge. Therefore, the inverter composed of N H-bridges can generate 2N + 1 levels.

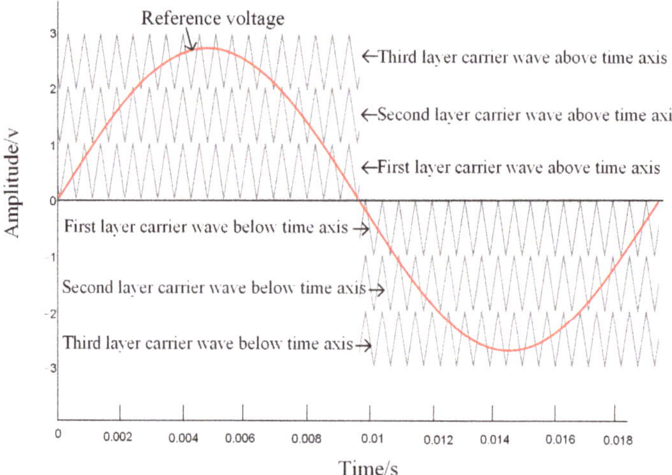

Figure 4. Carrier disposition modulation.

The switching pattern and output voltage waveforms are shown in Figure 4.

As shown in Figure 5, each H-bridge independently manages distinct segments of the sinusoidal reference. Consequently, even though the modulation technique of each H-bridge is fundamentally similar, their respective output voltage waveforms are not identical. When an open-circuit fault occurs in an IGBT, it will lead to failure to drive. The conduction state of the faulty H-bridge will be affected, which means one voltage level is skipped. The total output voltage also skips the parts that the H-bridge is responsible for. The voltage wave will be asymmetric and distorted due to the loss of voltage level. In addition, the phase voltage will become unbalanced in the three-phase system, such as the motor drive system.

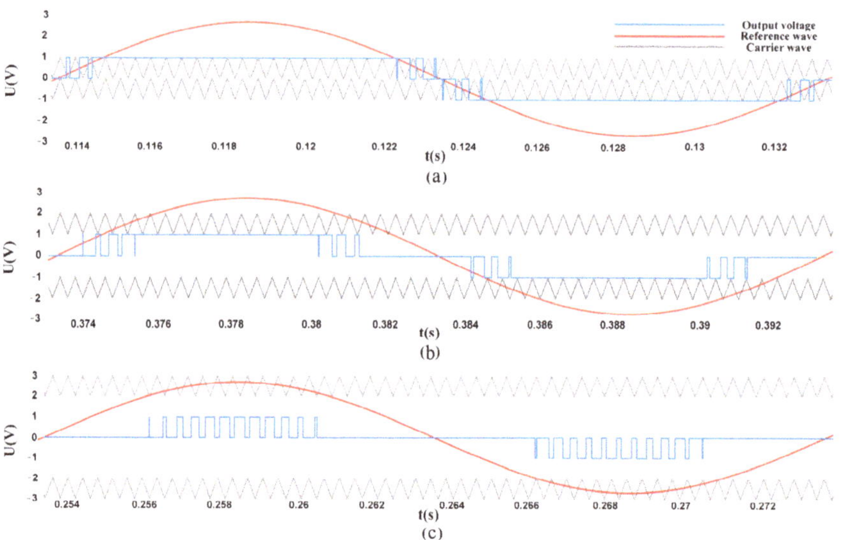

Figure 5. Switching pattern and output voltage waveform: (**a**) the first H-bridge; (**b**) the second H-bridge; (**c**) the third H-bridge.

2.2. Fault-Tolerant State

To maintain the normal operation of the inverter system, the faulty H-bridge must be removed from the system. In [38], the faulty H-bridge is directly isolated by an isolation switch in parallel. Once one IGBT faults, the whole H-bridge cell will be cut off, and other healthy H-bridge cells will be reconfigured to work in the five-level operation. The other method in [34] involves the PWM control forcing the corresponding IGBTs on or off. The faulty H-bridge is in a forward or reverse bypass state. In contrast, the response time is accelerated because there is no need to use isolation switches to short circuit the faulty H-bridge. However, the adjustment range of the motor's torque and speed is reduced when the inverter uses those methods in the motor drive system because the state of the H-bridge is lost, which causes a reduction in output voltage capability of inverter. The load capacity is reduced due to the reduction in the fundamental amplitude. The increase in duty cycle can increase the amplitude. However, the power loss will increase with the increase in duty cycle, and the junction temperature will increase due to the increase in IGBT power loss. This will cause the IGBT failure rate to increase, thus reducing the reliability of the system.

For fault diagnosis in a ship's DC electrical system, there has been a lot of research developments, such as a convolutional-neural-network-based method [21], Res-BiLSTM [22], and a layering linear discriminant analysis [23].

For fault-tolerant control in the ship's system, a five-phase fifteen-slot four-pole interior PMSM without the neutral point was modified [24]. This method requires a redesign of the motor and is costly. A modified lookup table, flux, and torque hysteresis bands are designed to improve the functioning of the fault-tolerant DTC of five-phase induction motor (FPIM) drive [25]. This method may not be suitable for a PMSM in electric ships. The method in [26] solves this problem by adding two TRIACs to pass faulty devices. And two additional switches are added to the circuit. This can be interpreted as faulty healthy devices being replaced by redundant devices. This method adds additional devices.

For these unresolved doubts, a stratified reconfiguration carrier disposition SPWM fault-tolerant strategy is proposed. The faulty device is bypassed without an isolating switch. The fundamental voltage amplitude is improved as much as possible after fault tolerance. And the reliability of the system is also increased.

3. A Stratified Reconfiguration Carrier Disposition SPWM Fault-Tolerant Control Strategy

Figure 6 shows the schematic diagram of the stratified reconfiguration disposition SPWM fault-tolerant control strategy. There are no additional devices to achieve fault-tolerant control.

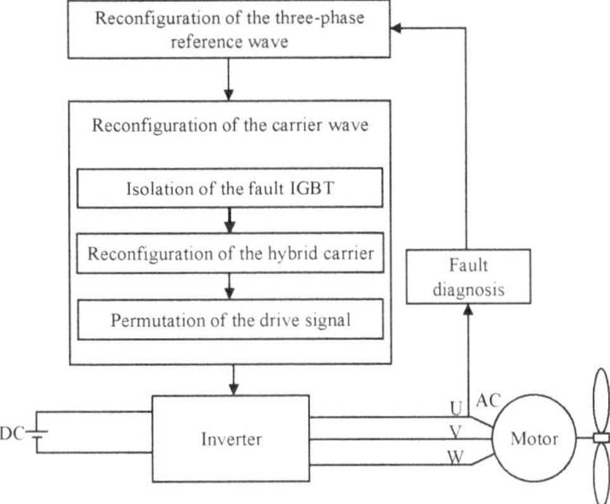

Figure 6. The entire fault-tolerant control system.

In Figure 6, the output voltage information of the inverter is used for real-time fault diagnosis. When the fault information is detected, the fault-tolerant control method based on reconstructed SPWM signal is activated. This fault-tolerant control method can be divided into modulating wave reconfiguration and carrier reconfiguration. The carrier reconfiguration includes the isolation of the fault IGBT, the reconfiguration of the faulty H-bridge carrier, the reconfiguration of the healthy H-bridge carrier, and the permutation of the drive signal. This method can alleviate the influence of the fault.

3.1. Reconfiguration of the Three-Phase Reference Wave

In the cascaded multilevel inverter, voltage references for phase A, B, and C can be expressed as follows:

$$\begin{cases} u_{Aref} = a \times m_a \sin(100\pi t) \\ u_{Bref} = b \times m_a \sin(100\pi t + \theta_{AB}) \\ u_{Cref} = c \times m_a \sin(100\pi t - \theta_{AC}) \end{cases} \tag{2}$$

where a, b, c are the amplitude of the reference voltage of phases A, B and C, respectively, and θ_{AB}, θ_{AC}, θ_{BC} are the phase angles between AB, AC, and BC, respectively. During the normal operation of the inverter, $a = b = c = n$, n is number of cascaded H-bridges, $\theta_{AB} = \theta_{AC} = \theta_{BC} = 120°$.

The vector diagram of the three-phase reference voltage is presented in Figure 7a; it shows that the output voltage of the inverter is balanced. However, equivalent voltage cannot be output according to the reference wave due to fault, as shown in Figure 7b–d.

In Figure 7b, when the fault occurs in phase A, it will be $a < b = c$. The inverter will not be able to output a three-phase balanced voltage waveform that matches the reference modulation waveform, as shown in Figure 7a. In Figure 7c, when the fault occurs in phases A and B, it will be $a = b < c$. The inverter will not be able to output a three-phase balanced voltage waveform either. Therefore, the phase of the three-phase reference voltage must be reconfigured. The reconfiguration algorithm is as follows:

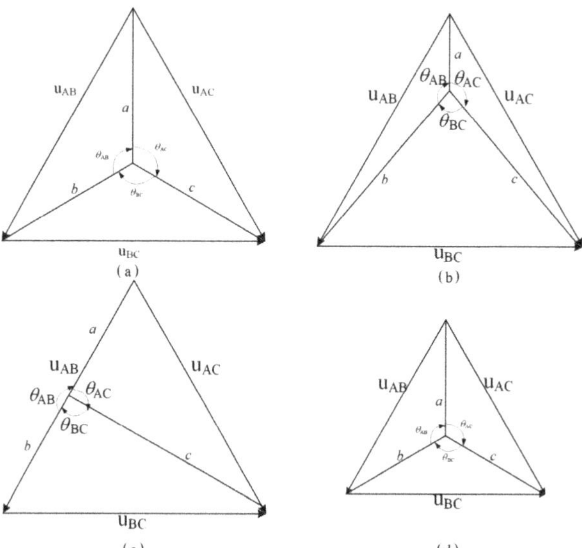

Figure 7. Reconfiguration of the three-phase reference wave: (**a**) normal state, (**b**) the fault occurs in a single phase, (**c**) the fault occurs in two phases, (**d**) the fault occurs in three phases.

When the fault of the H-bridge occurs in a single phase, to reconfigure the phase of the H-bridge after isolating the faulty H-bridge, the algorithm is as follows:

$$\begin{cases} a^2 + b^2 - 2ab\cos(\theta_{AB}) = a^2 + c^2 - 2ac\cos(\theta_{AC}) = b^2 + c^2 - 2bc\cos(\theta_{BC}) \\ \theta_{AB} + \theta_{AC} + \theta_{BC} = 360° \end{cases} \tag{3}$$

If the fault of the H-bridge occurs in two phases, to reconfigure the phase of the H-bridge after isolating the faulty H-bridge, the algorithm is as follows:

$$\begin{cases} c = \sqrt{a^2 + b^2 + ab}, a = b = \frac{n}{2} \\ \theta_{AB} + \theta_{AC} + \theta_{BC} = 360° \\ \theta_{AC} = \theta_{BC} = \cos^{-1}(a - b) \end{cases} \tag{4}$$

If the fault of the H-bridge occurs in three phases, to reconfigure the phase of the H-bridge after isolating the faulty H-bridge, the algorithm is as follows:

$$\begin{cases} c = a = b = \frac{n}{2} \\ \theta_{AB} + \theta_{AC} + \theta_{BC} = 360° \\ \theta_{AB} = \theta_{AC} = \theta_{BC} \end{cases} \tag{5}$$

When the three-phase voltage is unbalanced, the line voltages will also be unbalanced. The imbalance line voltage of the three-phase system may lead to an unstable operation of the power equipment, reduced power factor, energy loss, and other problems. Therefore, when a fault occurs, the three-phase voltage should be balanced first.

3.2. Reconfiguration of the Carrier Signal

After three-phase voltage balance reconfiguration, the equivalent output voltage of each phase is determined. To achieve the desired output, three steps are presented: (i) isolation of the fault IGBT; (ii) reconfiguration of the hybrid carrier; and (iii) redistribution of the drive signal.

3.2.1. Isolation of the Fault IGBT

When a signal IGBT fails, the fault diagnosis method is used to detect the fault location. If the fault occurs in the IGBT of the reverse conduction circuit, $k = 1$. If the fault occurs in the IGBT of the forward conduction circuit, $k = 0$. The carrier signal of the faulty H-bridge is modified to the following:

$$\begin{cases} C_+^* = \frac{n}{n-j}C_+ + (-1)^k(n + 1 - i) \\ C_-^* = \frac{n}{n-j}C_- - (-1)^k(n + 1 - i) \end{cases} \tag{6}$$

where C_+, C_- are the carrier signals above the time axis and below the time axis of the faulty H-bridge, respectively. C_+^*, C_-^* are the corresponding carrier signal after fault-tolerant control. The two indexes j and i are the number and location of faulty H-bridges.

If a fault occurs in a cascaded H-bridge seven-level inverter, the reference and carrier signals will be reconfigured to make the drive signal of the faulty IGBT set to zero and the remaining topology can be regarded as a cascaded five-level inverter. If the second fault occurs in the same conduction loop in different H-bridge, the remaining topology can be seen as a cascaded three-level inverter.

If the second fault occurs in the different conduction loop in a different H-bridge, it means that there is a healthy positive bridge arm in one faulty H-bridge and a healthy reverse bridge arm in another faulty H-bridge. These remaining healthy bridge arms in the faulty H-bridges will be fully utilized. In other words, for one fault, the faulty H-bridge with a healthy positive conduction loop can output two voltage levels of +E and 0. For another fault, another faulty H-bridge with a healthy negative conduction loop can output two voltage levels of −E and 0. Then, when the two faulty H-bridges are combined as far as possible, the output voltage wave can be reduced by only two voltage levels, that is, the seven levels will be five levels. Especially when more H-bridges are cascaded, the on-state of the H-bridges is utilized as much as possible to increase the amplitude of the output voltage after fault tolerance.

3.2.2. Reconfiguration of the Hybrid Carrier

The reconfiguration of the remaining carriers can be carried out using traditional level-shifted pulse width modulation. However, the degradation of the voltage will cause a reduction in the fundamental amplitude. Therefore, the hybrid carrier will be used to improve the duty cycle.

In order to increase the duty cycle, the top-most and bottom-most carriers are replaced by a triangular–trapezoidal signal, and the triangular–trapezoidal signal has the same frequency as the replaced triangular signal. The H-bridge is a conduction state when the modulated signal is bigger than that of the trapezoidal carrier signal. Therefore, the conduction time is longer because carriers are triangular–trapezoidal signals. That is, the H-bridge is a conduction state for a longer time in one cycle. The duty cycle can be increased by this hybrid carrier.

3.2.3. Redistribution of Drive Signal

A larger duty cycle can provide a higher average power of the inverter output but it also increases the loss and temperature of the IGBT. The loss may cause a reduction in IGBT reliability. Therefore, in order to decrease the duty cycle of the healthy bridge, the healthy devices in faulty bridges are used to contribute to the output voltage through the redistribution of the drive signal.

When a single IGBT fault occurs in the H-bridge, the forward or reverse conduction will be blocked. However, the fault cell is in the bypass state after the isolation of the fault IGBT, and the remaining conduction state is not used. To reduce the power loss, the remaining conduction state in the faulty cell will be used to contribute to the output.

As shown in Figure 8, the drive signal is alternated at different periods. It can make the remaining conduction state of the faulty H-bridge output the highest and lowest levels. The power loss will be evenly distributed. Meanwhile, the switching times of the switch tube are equal in a half period, which also balances the power among the output modules.

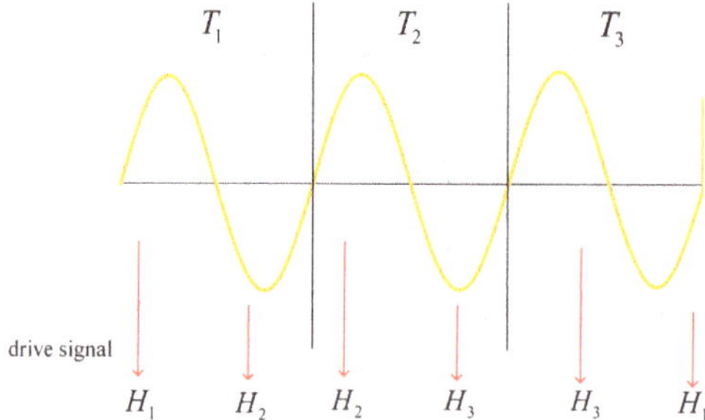

Figure 8. Redistribution of drive signal.

4. Simulation Results and Analysis

4.1. Simulation Results

4.1.1. Seven-Level Inverter

To verify the efficiency of the proposed method, this section will show its simulation results. Because this paper only focuses on fault-tolerant control, the simulation model is built based on the assumption that the fault has been detected and diagnosed correctly.

A cascaded H-bridge seven-level inverter system is built using MATLAB/Simulink R2022b software. Every input DC voltage value is set to 65 V. A series-connected R-L impedance of 21 (Ω) and 8 (mH) is taken into consideration, which is connected to the terminals of the inverter as a load. As shown in Figure 9, an open fault is generated in H_3S_2 in the first H-bridges at $t = 0.02$ s. The voltage in that phase loses a level and becomes asymmetrical.

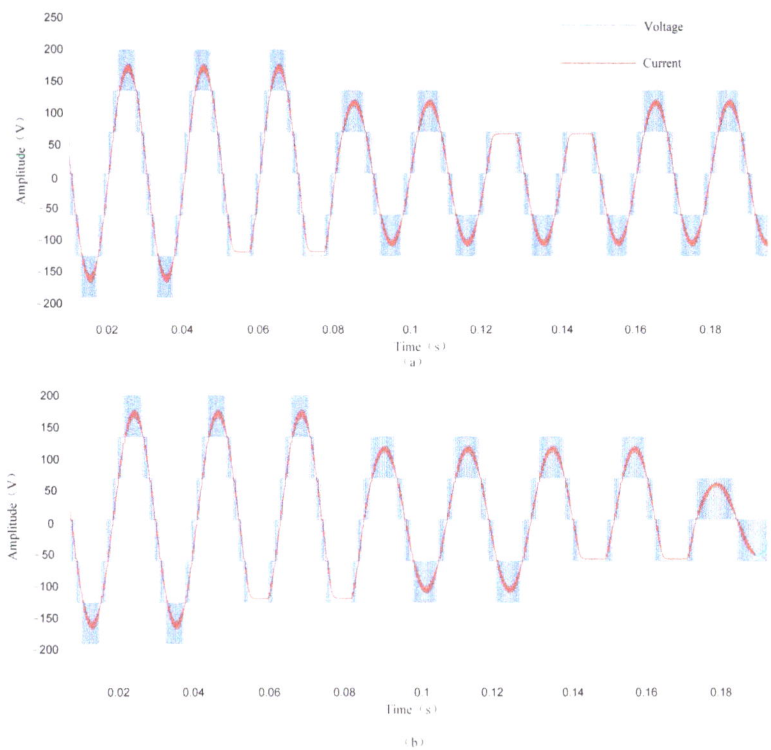

Figure 9. Simulation results of output voltage: (**a**) fault in H_3S_2 and H_2S_1; (**b**) fault in H_3S_2 and H_2S_4.

At $t = 0.8$ s, the reference and carrier signals are reconfigured. The symmetry of the phase voltage is restored. However, the phase voltage is degraded from seven levels to five levels. And the switching times are increased due to the reconfiguration of the modulation signal. Therefore, the drive signal is redistributed. As shown in Figure 10a, the faulty H-bridge is used to output voltage. The number of switches in a period is reduced, which causes the reduction in the power loss of the healthy H-bridge, as shown in Figure 10b,c.

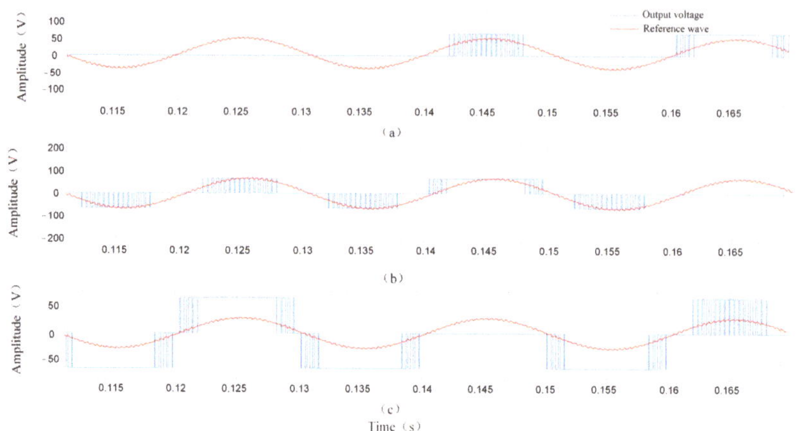

Figure 10. Simulation results of output voltage of each H-bridge: (**a**) the first H-bridge; (**b**) the second H-bridge (**c**) the third H-bridge.

At $t = 0.12$ s, the second fault occurs in H_2S_1 (Figure 9a) and H_2S_4 (Figure 9b), respectively. As shown in Figure 9a, because H_2S_1 is on the forward conduction loop, the output voltage keeps the five-level in post-fault operation. As shown in Figure 9b, the output is reduced to three levels due to the fact that H_2S_4 is in the reverse conduction state. In conclusion, the performance of the output voltage has improved as much as possible.

4.1.2. Motor Drive System

To verify the effectiveness of the proposed strategy, a simulation model of the PMSM drive system is built, and the proposed method is also applied. The parameters of IGBT/Diode are shown in Table 1. The DC voltage is 35 V. The PMSM parameters are shown in Table 2.

Table 1. Parameters of IGBT/Diode.

Components	Value
Internal resistance	$R_{on} = 1 \times 10^{-3}\ \Omega$
Snubber resistance	$R_s = 1 \times 10^5\ \Omega$
Snubber capacitance	$Cs = \text{inf}$

Table 2. PMSM parameters of MATLAB simulation.

Parameters	Number of Pole Pairs	Stator Inductance		Stator Resistance	Flux Linkage	Moment of Inertia	Damping Coefficient
	P_n	L_q	L_d	R	ψ_f	J	B
Value	4	8.5 (mH)	8.5 (mH)	2.875 (Ω)	0.175 (Wb)	0.003 (kg·m^2)	0.008 (N·m·s)

(a) The parameters of Proportional Integral (PI) controller in the rotational speed loop

In order to facilitate the parameter setting of the speed loop PI controller, the motor motion equation of PMSM is written as follows:

$$J\frac{d\omega_m}{dt} = T_e - T_L - B\omega_m \tag{7}$$

$$T_e = \frac{3}{2} p_n i_q [i_d(L_d - L_q) + \varphi_f] \tag{8}$$

where ω_m is the mechanical angular speed of the motor; J is the moment of inertia; B is the damping coefficient; and T_L is the load torque. Active power damping is used to adjust the parameters of the speed loop PI controller, and the active power damping is defined as follows:

$$i_q = i_q' - B_a \omega_m \tag{9}$$

When the control strategy ($i_d = 0$) is adopted and the motor is assumed to start under no-load ($T_L = 0$), the following expression can be derived:

$$\frac{d\omega_m}{dt} = \frac{1.5 p_n \varphi_f}{J}(i_q' - B_a \omega_m) - \frac{B}{J}\omega_m \tag{10}$$

By assigning the poles of (10) to the desired closed-loop bandwidth β, the transfer function of the speed relative to the Q-axis current can be obtained as follows:

$$\omega_m(s) = \frac{1.5 p_n \varphi_f / J}{s + \beta} i_q'(s) \tag{11}$$

The coefficient of active power damping can be obtained by (10) and (11):

$$B_a = \frac{\beta J}{1.5 p_n \varphi_f} \tag{12}$$

Then, the expression of the speed loop controller is as follows:

$$i_q^* = \left(k_{pw} + \frac{k_{iw}}{s}\right)(\omega_m^* - \omega_m) - B_a \omega_m \tag{13}$$

Therefore, the proportional gain and integral gain of the PI controller can be adjusted by the following formula:

$$\begin{cases} K_{pw} = \frac{\beta J}{1.5 p_n \varphi_f} \\ K_{iw} = \beta K_{pw} \end{cases} \tag{14}$$

where β is the expected frequency band bandwidth of the speed loop. The bandwidth of the speed ring is selected as 50 rad/s. The parameters of the speed loop PI controller are calculated by (14) and the parameters of the motor.

(b) The parameters of PI controller in the current loop

The conventional PI controller is combined with the feedforward decoupling control strategy. The voltage of the d–q-axes can be obtained as follows:

$$\begin{cases} u_d = \left(k_{pd} + \frac{k_{id}}{s}\right)(i_d^* - i_d) - \omega_e L_q i_q \\ u_q = \left(k_{pq} + \frac{k_{iq}}{s}\right)(i_q^* - i_q) + \omega_e(\varphi_f + L_d i_d) \end{cases} \tag{15}$$

where K_{pd} and K_{pq} are the proportional gains of the PI controller, and K_{id} and K_{iq} are the integral gains of the PI controller. Internal model control has the advantages of a simple structure and a single parameter. Therefore, the internal model control strategy is used to design and adjust the parameters of the PI controller.

Figure 11a shows a typical internal model control block diagram, where $\hat{G}(s)$ is the internal model; $G(s)$ is the controlled object; and $C(s)$ is the internal model controller. According to the classic automatic control principle, the block diagram shown in Figure 11b can be obtained through an appropriate equivalent transformation shown in Figure 11a, and its equivalent controller is as follows:

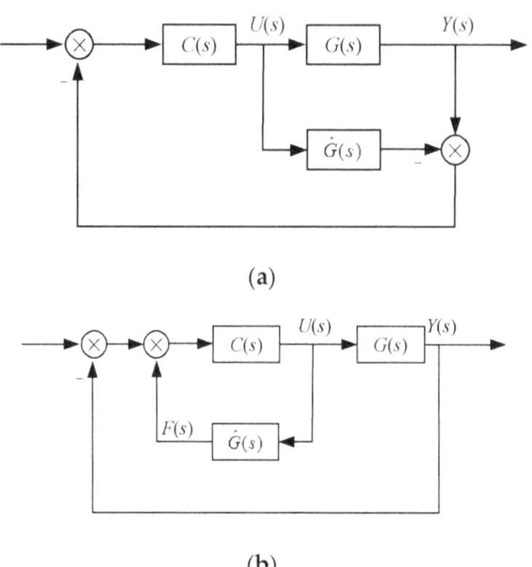

Figure 11. Internal model control strategy structure: (**a**) internal model control block diagram; (**b**) internal model control equivalent block diagram.

$$F(s) = [I - C(s)\hat{G}(s)]^{-1}C(s) \tag{16}$$

where I is the identity matrix. If the internal model modeling is accurate, $\hat{G}(s) = G(s)$. Then, there is no feedback in the system, and the transfer function of the system is as follows:

$$G_c(s) = G(s)C(s) \tag{17}$$

To ensure the stability of the system, $G(s)$ and $C(s)$ need to be stable. The current loop of a control system can be approximated as a first-order system. $C(s)$ is defined by $\hat{G}(s) = G(s)$:

$$C(s) = \hat{G}^{-1}(s)L(s) = G^{-1}(s)L(s) \tag{18}$$

where $L(s) = \alpha I/(s + \alpha)$, and α is the design parameter. By substituting (18) into (16), the designed internal model controller can be obtained, which is as follows:

$$F(s) = \alpha \begin{bmatrix} L_d + \frac{R}{s} & 0 \\ 0 & L_q + \frac{R}{s} \end{bmatrix} \tag{19}$$

By substituting (19) into (16), $G_c(s)$ can be calculated as follows:

$$G_c(s) = \frac{\alpha}{\alpha + s}I \tag{20}$$

By comparing (20) and (15), it can be seen that the adjustment parameters of the PI controller meet the following:

$$\begin{cases} K_{pd} = \alpha L_d \\ K_{id} = \alpha R \\ K_{pq} = \alpha L_q \\ K_{iq} = \alpha R \end{cases} \tag{21}$$

According to the parameters of the motor, $\alpha = 1000$ rad/s. According to (21), the parameters of the PI controller in the current loop can be calculated. The calculated parameters of the PI controller may not be optimal. In the process of simulation, it is necessary to further debug the parameters to achieve the best control effect. The controller parameters obtained through debugging are shown in Table 3.

Table 3. The parameters of PI controller.

Parameter Type	Value
Speed loop parameters	$K_{p\omega} = 0.35$, $K_{i\omega} = 0.85$
D-axis current loop parameters	$K_{pd} = 0.0085$, $K_{id} = 2.875$
Q-axis current loop parameters	$K_{pq} = 8.5$, $K_{iq} = 2.875$

The PI controller parameters in Table 2 determined based on a stability analysis are entered into the MATLAB model. When the load torque T_L is replaced by the fan load torque, the speed is as shown in Figure 12. The load characteristics of the fan are as follows:

$$T_L = T_f + kn^2 \tag{22}$$

where T_f is friction torque on the bearing; and k is proportional coefficient.

As shown in Figure 12, the initial speed of the propulsion motor PMSM is 800 rpm. The load torque of the propeller is suddenly loaded to the PMSM at $t = 0.2$ s. The motor speed oscillates accordingly and then becomes stable. IGBT faults happen in the first and third H-bridges at $t = 0.6$ s. The effect of the speed will become worse, and the speed will fluctuate between 790.9 rpm and 804.5 rpm and show the characteristics of periodic changes. After a delay for fault detection and diagnosis, the proposed method is put into use at $t = 0.8$ s. It will be noted that the neutral point cannot be offset because the motor drive system requires phase voltage balance. Therefore, the three-phase reference signals are reconfigured by (4). Isolating the faulty IGBT stabilizes the speed of the motor. However, it decreases from 800 rpm to 530 rpm due to the reduction in voltage level. At $t = 1.5$ s, the SPWM strategy of the hybrid carrier is used. The motor speed increases to 590 rpm. Therefore, the proposed fault-tolerant control is realized to achieve a three-phase voltage balance and constant frequency in the seven-level or a five-level voltage of the motor drive system. In addition, the IGBT power loss of the H-bridge is reduced.

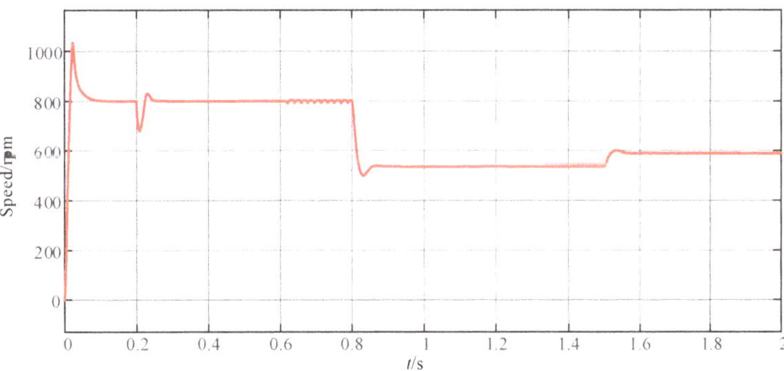

Figure 12. Motor speed simulation results.

To solve the voltage drop problem, there are two ways to restore speed. One is to increase the voltage value of the DC bus. However, the ship's DC bus cannot be added arbitrarily. Another one is to reduce the excitation current of the motor and thereby reduce the excitation flux. The motor speed will be increased under the condition of ensuring the voltage balance. The excitation flux of PMSM is provided by a permanent magnet, and this flux is constant. If the magnetic flux strength is expected to be reduced, the air gap magnetic flux can only be weakened by increasing the demagnetization component of the stator current. In this way, flux-weakening control can be achieved, just like for the separately excited DC motor. Under these circumstances, the d-axis current i_d must be maintained at a negative value to shift the operating point laterally into the operable region. The negative d-axis current is the so-called flux-weakening current, and flux-weakening control is responsible for driving the flux-weakening current such that the motor always operates inside the operable region even when the operating conditions vary [39].

Flux-weakening control is performed at 1.5 s. The speed and torque waveforms are shown in Figure 13.

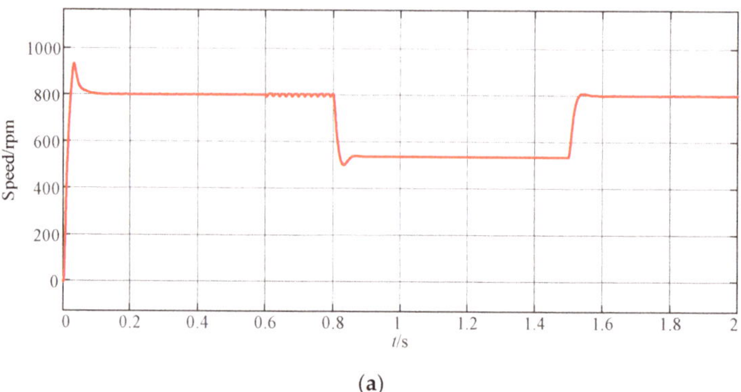

(**a**)

Figure 13. *Cont.*

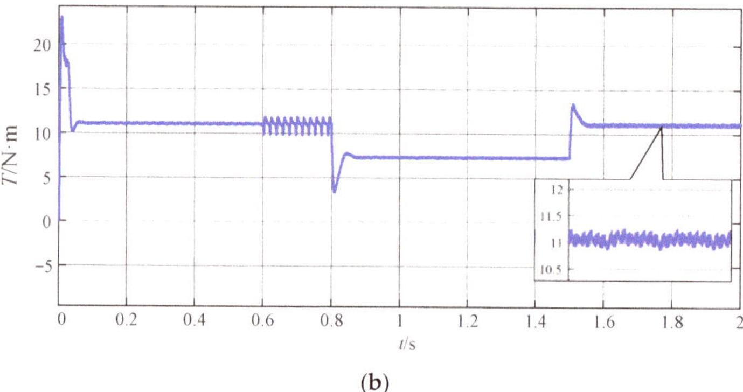

(**b**)

Figure 13. Simulation results with flux-weakening control: (**a**) speed waveform; (**b**) torque waveform.

As shown in Figure 13, speed and torque can be restored. Therefore, the possibility of uninterrupted operation of the motor drive system is improved as much as possible.

4.2. Experimental Results

Figure 14 depicts the experimental platform. It consists of a cascaded H-bridge seven-level inverter as a power stage, and the control strategy is implemented in the MicroLabBox dSPACE system that generates the signals of the switch gates. Due to the limitations of the experimental conditions, the DC source is used to simulate the ship's DC bus, and PMSM is replaced by RL. Table 4 shows the main parameters of the system. We have selected the IGBT IKW50N65F5 as the power switch transistor, which includes a built-in reverse diode.

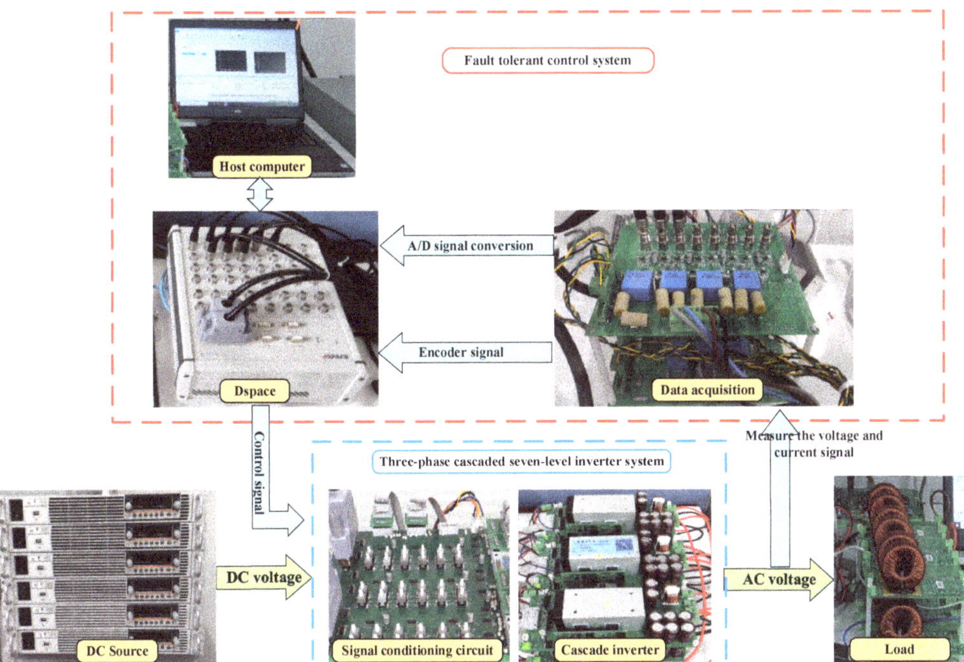

Figure 14. Experiment's platform.

Table 4. Experiment's parameters.

Components	Value
DC source voltage	V_{dc} = 56 V
Resistive load	R = 21 Ω
Inductive load	L = 12 mH
Reference frequency	50 Hz
Switching frequency	3 kHz

As shown in Figure 15, the three-phase cascaded seven-level inverter is supplied by power sources. And the control signal given by the controller dSPACE needs to go through the signal conditioning circuit to the three-phase cascaded five-level inverter. Meanwhile, the signal conditioning circuit has the function of setting faults. The output voltage signal of the three-phase cascaded seven-level inverter is directly sent to the load. The voltage and the current sensor are used to monitor the output signal of the inverter in real time. The collected voltage and current signals are sent to the controller dSPACE as feedback signals. The collected phase voltage signal is used as the monitoring signal of fault diagnosis, and it is used to realize fault-tolerant control.

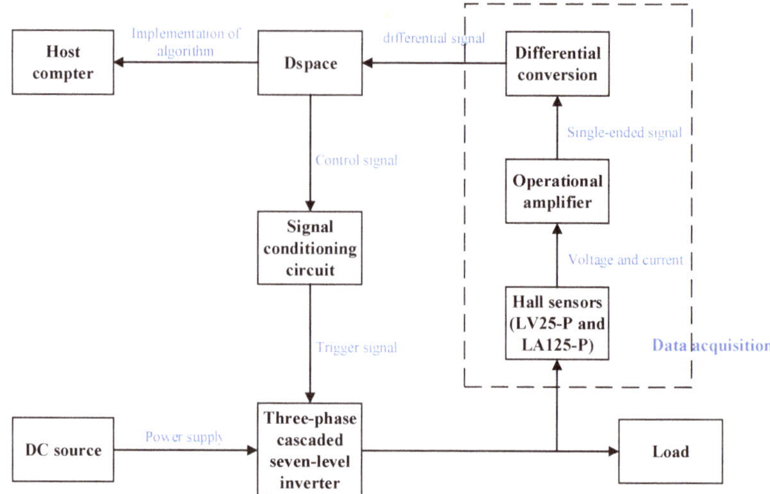

Figure 15. Connection diagram of experiment.

Figure 16 shows the changes in the output voltage and output current under fault-tolerant control of the proposed strategy. This process can be divided into five states: normal operation, IGBT failure, fault-tolerant control, second IGBT failure, and second fault-tolerant control.

As shown in Figure 16a:

(1) In the first stage, the inverter is in a normal working state. The output voltage of the inverter is a symmetrical seven-level voltage waveform, with the output current being a sine wave.

(2) In the second stage, due to the IGBT (H_3S_4) on the H_3 reverse conduction circuit having an open-circuit fault, there is a level reduction in the output voltage on the negative half axis. As a result, the total harmonic distortion of the output voltage increases and the output current is distorted.

(3) In the third stage, a fault-tolerant control method is adopted based on the reconstructed SPWM signal. Although the amplitude of the output voltage is reduced and the voltage level is reduced compared to the normal state, the inverter can output a symmetrical five-level voltage waveform. This means the total harmonic distortion of the output voltage is reduced during faults, and the output current is restored to a sinusoidal waveform.

(4) In the fourth stage, the IGBT (H_2S_4) on the H_2 reverse conduction circuit has an open-circuit fault. As a consequence, the output voltage of the five levels loses one voltage level on the negative half axis, and the total harmonic distortion increases further.

(5) In the fifth stage, the inverter can output a symmetrical three-level voltage by the proposed fault-tolerant control strategy, and the output current is restored to a sine wave.

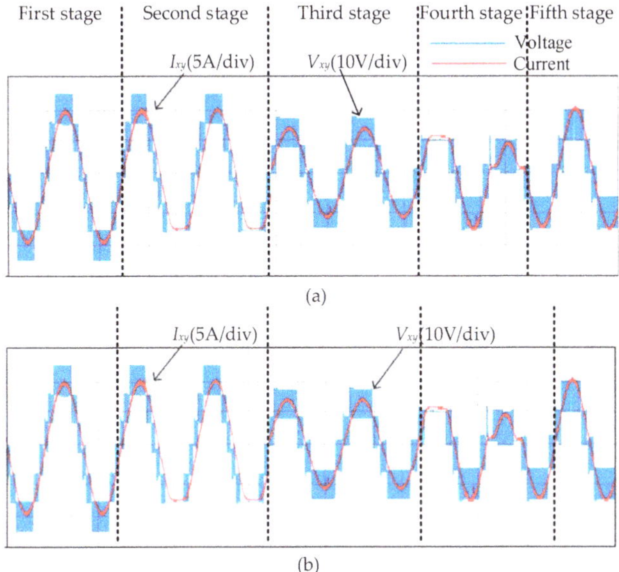

Figure 16. Experimental results of output voltage: (**a**) fault in H_3S_2 and H_2S_4; (**b**) fault in H_3S_2 and H_2S_1.

It can be seen that both single IGBT faults and double IGBT faults can be achieved using the proposed method for fault-tolerant control.

As shown in Figure 16b, identical to the first three stages of the above analysis, the proposed fault-tolerant control method was used to achieve merely one-level reduction. Nevertheless, the fourth stage is unlike the above situation. The IGBT (H_2S_1) on the H_2 forward conduction circuit leads to an open-circuit fault, causing a level reduction in the first half cycle of the output voltage. In the fifth stage, by using the proposed fault-tolerant control method again, it can be found that fault-tolerant control can be achieved without reducing the level, and the total harmonic distortion of the output voltage can be improved.

As a consequence, the proposed fault-tolerant control method not only effectively achieves fault-tolerant control, but also improves the performance of the output voltage in terms of the fault types that cause different conduction circuit blockages for different H-bridges.

As shown in Figure 17, the faulty H-bridge is used to output the forward voltage in T_1. Meanwhile, the second and third H-bridges are in the conduction state, which can reduce power loss. In addition, for the whole period, the number of switches on each H-bridge is the same as in the half period. This can balance the power loss of each H-bridge as much as possible.

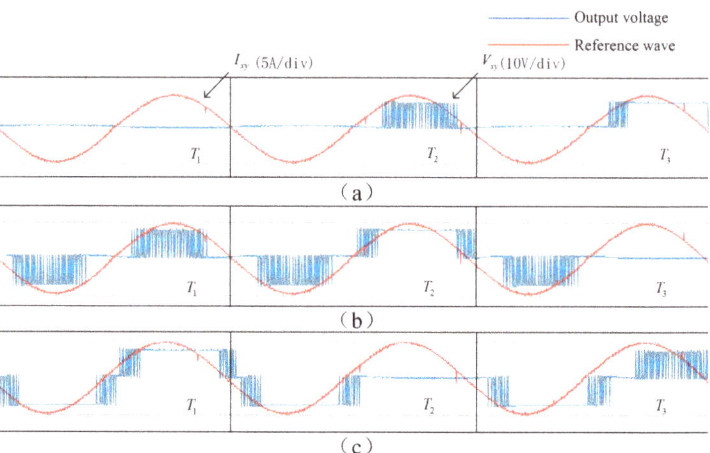

Figure 17. Experimental results of output voltage of each H-bridge: (**a**) the first H-bridge; (**b**) the second H-bridge (**c**) the third H-bridge.

In the experimental results, the performance of the output voltage is improved as much as possible in post-fault operation, and the power loss of the healthy H-bridge is reduced, which can improve the reliability of the ship's power system.

4.3. Comparison with Other Methods

The fault-tolerant control effect of [31] in the three-phase grid-connected experimental platform is shown in Figure 18. The whole process is divided into three states: healthy operation state, H-bridge fault state, and fault-tolerant control state.

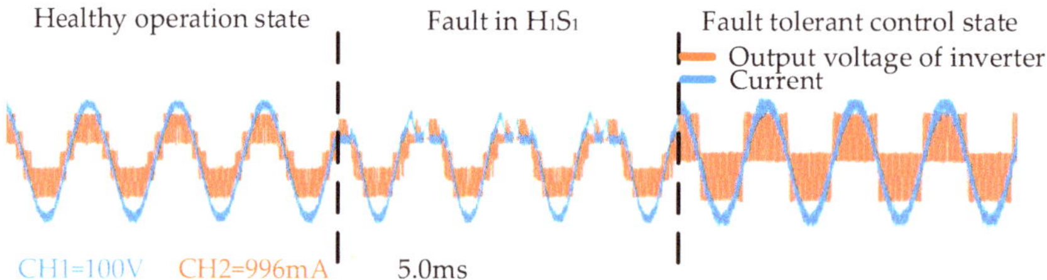

Figure 18. Experimental results of fault-tolerant control method proposed in [24].

During H-bridges' healthy operation, the inverter can output a symmetrical five-level voltage waveform. Nevertheless, when an open-circuit fault occurs in H_1, the output level cannot be maintained at five. Then, the waveform of the inverter voltage is distorted, and the current of the grid also shows asymmetric distortion. After fault-tolerant control in [31], the inverter only outputs three-level voltage although the peak value of the output voltage is essentially unchanged, and the waveform of the output voltage is symmetrical. This fault-tolerant method reduces the output voltage by three levels after only one fault in the H-bridge. Although limited to the experimental conditions, the motor experiment not being able to be carried out. The results of the grid-connected experiment with RL show the current's harmonic increase. This is a disaster for motor operation. Compared with [31], the proposed fault-tolerant control method can both be applied to higher level inverters and maintain the level number of output voltage as much as possible. Thus, the total harmonic distortion of the current may be reduced.

The fault-tolerant control effect in [40] is similar to that of the proposed method in this paper. However, when a single IGBT fault occurs in the third state, the healthy bridge arm of the faulty H-bridge is not utilized. The switching frequency of some IGBT devices in the third state is reduced by the proposed method based on reconfiguration SPWM, and thus the IGBT power loss is reduced.

The torque ripple is estimated to be 1.6 Nm in [25]. The torque ripple of the proposed method is around 0.42 Nm as shown in Figure 13b, which is about 73% lower than in the scheme proposed by Chikondra et al. [25].

More investigations that can adapt to the existing systems in ships are needed, including research on factors which have not been considered by the proposed method. The following are some examples:

(a) The DC power supply is replaced by the actual ship's DC bus;
(b) Wind disturbance factors should be considered, such as mean wind pressure, variable wind pressure, the ship's absolute heading angle, absolute wind angle, drift angle, etc.;
(c) Wave interference factors should be considered, such as irregular wave drift force and moment, wave and ship encounter angle, drift force coefficient, wave force interference coefficient, etc.;
(d) The electromagnetic interference generated by the inverters needs to be dealt with.

5. Conclusions

A stratified reconfiguration carrier disposition SPWM fault-tolerant control strategy for a ship's PMSM drive system is proposed. Compared to the inverter topology with additional devices [26–29,31,34,35], the proposed method is only based on software. This makes it useful for ships where space is limited. Thus, the proposed hybrid carrier could improve the duty cycle of SPWM. The experimental results show that the decrease problem [37] of the fundamental amplitude of the output voltage is solved in faulty conditions. Because the conduction state of the healthy bridge arm in the faulty H-bridge is fully utilized, the performance of the output voltage is improved. Furthermore, the switching times of the IGBT are identical over a period. That is to say, the power losses of IGBTs are even. Therefore, the proposed method is suitable for motor drive applications that do not require maintaining voltage amplitude.

Author Contributions: Conceptualization, F.Z. and Z.Z. (Zhonglin Zhang); methodology, F.Z., Z.Z. (Zhiwei Zhang) and Z.Z. (Zhonglin Zhang); software, Z.Z. (Zhonglin Zhang); formal analysis, F.Z., Z.Z. (Zhiwei Zhang), Z.Z. (Zhonglin Zhang), J.H., T.W. and Y.A.; writing original draft preparation, F.Z., Z.Z. (Zhiwei Zhang), Z.Z. (Zhonglin Zhang), J.H., T.W. and Y.A.; writing—review and editing, T.W. and Y.A. All authors have read and agreed to the published version of the manuscript.

Funding: This research received no external funding.

Institutional Review Board Statement: Not applicable.

Informed Consent Statement: Not applicable.

Data Availability Statement: Data are contained within this article.

Conflicts of Interest: Author Zhiwei Zhang was employed by Shanghai Power Industrial & Commerical Co., Ltd. The remaining authors declare that the research was conducted in the absence of any commercial or financial relationships that could be construed as a potential conflict of interest.

References

1.   Gao, D.; Wang, X.; Wang, T.; Wang, Y.; Xu, X. Optimal Thrust Allocation Strategy of Electric Propulsion Ship Based on Improved Non-Dominated Sorting Genetic Algorithm II. *IEEE Access* **2019**, *7*, 135247–135255. [CrossRef]
2.   He, Y.; Fan, A.; Wang, Z.; Liu, Y.; Mao, W. Two-phase energy efficiency optimisation for ships using parallel hybrid electric propulsion system. *Ocean. Eng.* **2019**, *238*, 1–30. [CrossRef]
3.   Gan, S.H.; Shi, W.F.; Xu, X.Y. Proposed Z-Source Electrical Propulsion System of a Fuel Cell/Supercapacitor-Powered Ship. *J. Mar. Sci. Eng.* **2023**, *11*, 1543. [CrossRef]
4.   Apsley, J.M.; Gonzalez-Villasenor, A.; Barnes, M.; Smith, A.C.; Williamson, S.; Schuddebeurs, J.D.; Norman, P.J.; Booth, C.D.; Burt, G.M.; McDonald, J.R. Propulsion Drive Models for Full Electric Marine Propulsion Systems. *IEEE Trans. Ind. Appl.* **2009**, *45*, 676–684. [CrossRef]

5. Kim, M.-G.; Choi, J.-L.; Song, T.-H.; Kim, Y.-S.; Lee, S.-G. A control power module for a hybrid inverter systems to drive electric propulsion ships. *J. Adv. Mar. Eng. Technol.* **2018**, *42*, 75–81. [CrossRef]

6. Kanellos, F.D.; Tsekouras, G.J.; Prousalidis, J. Onboard DC grid employing smart grid technology: Challenges, state of the art and future prospects. *IET Electr. Syst. Transp.* **2015**, *5*, 1–11. [CrossRef]

7. Che, Y.B.; Xue, S.Y.; Wang, D.M. An AC Voltage Sensor-less Control Strategy for Ocean Ship Photovoltaic Grid-connected Inverter with LCL Filter. *J. Coast. Res.* **2018**, *83*, 109–115. [CrossRef]

8. Yao, N.; Zhang, Z.; Zhang, Z.Y. Application of Photovoltaic Off-Grid Inverter in Marine Engineering. *J. Coast. Res.* **2020**, *110*, 193–196. [CrossRef]

9. Rodr´ıguez, J.; Bernet, S.; Wu, B.; Pontt, J.O.; Kouro, S. Multilevel voltage-source-converter topologies for industrial medium-voltage drives. *IEEE Trans. Ind. Electron.* **2007**, *54*, 2930–2944. [CrossRef]

10. Rodríguez, J.; Bernet, S.; Steimer, P.K.; Lizama, I.E. A survey on neutral-point-clamped inverters. *IEEE Trans. Ind. Electron.* **2010**, *57*, 2219–2230. [CrossRef]

11. Mathew, J.; Rajeevan, P.P.; Mathew, K.; Azeez, N.A.; Gopakumar, K. A multilevel inverter scheme with dodecagonal voltage space vectors based on flying capacitor topology for induction motor drives. *IEEE Trans. Power Electron.* **2013**, *28*, 516–525. [CrossRef]

12. Peng, H.; Hagiwara, M.; Akagi, H. Modeling and analysis of swithing-ripple voltage on the DC link between a diode rectifier and a modular multilevel cascaded inverter. *IEEE Trans. Power Electron.* **2013**, *28*, 75–84. [CrossRef]

13. Menaka, S.; Muralidharan, S. Novel Symmetric and Asymmetric Multilevel Inverter Topologies with Minimum Number of Switches for High Voltage of Electric Ship Propulsion System. *Indian. J. Geo-Mar. Sci.* **2017**, *46*, 1920–1930.

14. Kumar, A.; Bhatia, R.S.; Nijhawan, P. Performance investigation of an innovative H-bridge derived multilevel inverter topology for marine applications. *Indian J. Geo-Mar. Sci.* **2020**, *49*, 441–449.

15. Jeon, H.-M.; Hur, J.-J.; Yoon, K.-K. Control Method for Performance Improvement of BLDC Motor used for Propulsion of Electric Propulsion Ship. *J. Korean Soc. Mar. Environ. Saf.* **2019**, *25*, 802–808. [CrossRef]

16. Muralidharan, S.; Gnanavadivel, J.; Muhaidheen, M.Y.S.; Menaka, S. Harmonic Elimination in Multilevel Inverter Using TLBO Algorithm for Marine Propulsion System. *Mar. Technol. Soc. J.* **2021**, *55*, 117–126. [CrossRef]

17. Townsend, C.D.; Summers, T.J.; Watson, A.J.; Betz, R.E.; Clare, J.C. A Simplified Space Vector Pulse Width Modulation Implementation in Modular Multilevel Converters for Electric Ship Propulsion Systems. *IEEE Trans. Transp. Electr.* **2019**, *5*, 335–342.

18. Chen, Y.; Li, Z.Y.; Zhao, S.S.; Wei, X.G.; Kang, Y. Design and Implementation of a Modular Multilevel Converter with Hierarchical Redundancy Ability for Electric Ship MVDC System. *IEEE J. Emerg. Sel. Top. Power Electron.* **2017**, *5*, 189–202. [CrossRef]

19. Wang, B.; Cai, J.; Du, X.; Zhou, L. Review of power semiconductor device reliability for power converters. *CPSS. Trans. Power. Electron. Appl.* **2017**, *2*, 101–117. [CrossRef]

20. Yaghoubi, M.; Moghani, J.S.; Noroozi, N.; Zolghadri, M.R. IGBT open-circuit fault diagnosis in a quasi-Z-source inverter. *IEEE Trans. Ind. Electron.* **2019**, *66*, 2847–2856. [CrossRef]

21. Yan, G.; Hu, Y.; Shi, Q. A Convolutional Neural Network-Based Method of Inverter Fault Diagnosis in a Ship's DC Electrical System. *Pol. Marit. Res.* **2022**, *29*, 105–114. [CrossRef]

22. Xie, J.; Shi, W.; Shi, Y. Research on Fault Diagnosis of Six-Phase Propulsion Motor Drive Inverter for Marine Electric Propulsion System Based on Res-BiLSTM. *Machines* **2022**, *10*, 736. [CrossRef]

23. Jin, G.; Wang, T.; Amirat, Y.; Zhou, Z.; Xie, T. A Layering Linear Discriminant Analysis-Based Fault Diagnosis Method for Grid-Connected Inverter. *J. Mar. Sci. Eng.* **2022**, *10*, 939. [CrossRef]

24. Park, J. *Multi-Phase Fault Tolerant PMSM Drive System*; Texas A&M University: College Station, TX, USA, 2015.

25. Chikondra, B.; Muduli, U.; Behera, R. An Improved Open-Phase Fault-Tolerant DTC Technique for Five-Phase Induction Motor Drive Based on Virtual Vectors Assessment. *IEEE Trans. Ind. Electron.* **2021**, *68*, 4598–4609. [CrossRef]

26. Zhang, Z.; Wang, T.; Chen, G.; Amirat, Y. A Serial Fault-Tolerant Topology Based on Sustainable Reconfiguration for Grid-Connected Inverter. *J. Mar. Sci. Eng.* **2023**, *11*, 751. [CrossRef]

27. Xia, Y.; Xu, Y.; Gou, B. A Data-Driven Method for IGBT Open-Circuit Fault Diagnosis Based on Hybrid Ensemble Learning and Sliding-Window Classification. *IEEE Trans. Industr. Inform.* **2019**, *16*, 5223–5233. [CrossRef]

28. Song, W.; Huang, A.Q. Fault-tolerant design and control strategy for cascaded H-bridge multilevel converter-based STATCOM. *IEEE Trans. Ind. Electron.* **2010**, *57*, 2700–2708. [CrossRef]

29. Yazdani, A.; Sepahvand, H.; MCrow, L.; Ferdowsi, M. Fault detection and mitigation in multilevel converter STATCOMs. *IEEE Trans. Ind. Electron.* **2011**, *58*, 1307–1315. [CrossRef]

30. Aleenejad, M.; Iman-Eini, H.; Farhangi, S. Modified space vector modulation for fault-tolerant operation of multilevel cascaded H-bridge inverters. *IET Power Electron.* **2013**, *6*, 742–751. [CrossRef]

31. Zhao, Z.; Wang, T.; Benbouzid, M. A fault-tolerant reconfiguration system based on pilot switch for grid-connected inverters. *Microelectron. Reliab.* **2022**, *131*, 114511. [CrossRef]

32. Haji-Esmaeili, M.; Naseri, M.; Khoun-Jahan, H.; Abapour, M. Fault-tolerant structure for cascaded H-bridge multilevel inverter 538 and reliability evaluation. *IET Power Electron.* **2017**, *10*, 59–70. [CrossRef]

33. Jahan, H.K.; Panahandeh, F.; Abapour, M.; Tohidi, S. Reconfigurable Multilevel Inverter with Fault-Tolerant Ability. *IEEE Trans. Power Electron.* **2018**, *33*, 7880–7893. [CrossRef]

34. Mhiesan, H.; McCann, R.; Farnell, C.; Mantooth, A. Novel Circuit and Method for Fault Reconfiguration in Cascaded H-Bridge Multilevel Inverters. In Proceedings of the 2019 IEEE Applied Power Electronics Conference and Exposition (APEC), Anaheim, CA, USA, 17–22 March 2019; pp. 1800–1804.
35. Maddugari, S.K.; Borghate, V.B.; Sabyasachi, S.; Karasani, R.R. A Linear-Generator-Based Wave Power Plant Model Using Reliable Multilevel Inverter. *IEEE Trans. Ind. Appl.* **2019**, *55*, 2964–2972. [CrossRef]
36. Jalhotra, M.; Sahu, L.K.; Gautam, S.P.; Gupta, S. Reliability and energy sharing analysis of a fault-tolerant multilevel inverter topology. *IET Power Electron.* **2019**, *12*, 759–768. [CrossRef]
37. Geng, C.; Wang, T.; Tang, T.; Benbouzid, M.; Qin, H. Fault-tolerant control strategy of the cascaded five-level inverter based on divided voltage modulation algorithm. In Proceedings of the 2016 IEEE IPEMC ECCE Asia, Hefei, China, 22–26 May 2016.
38. Zhang, J.H.; Liu, Z.; Wang, T.; Benbouzid, M.E.H.; Wang, Y. An Arm Isolation and Reconfiguration Fault Tolerant Control Method Based on Data-driven Methodology for Cascaded Seven-level Inverter. In Proceedings of the 2018 IEEE 7th Data Driven Control and Learning Systems Conference (DDCLS), Enshi, China, 25–27 May 2018; pp. 939–943.
39. Chen, J.J.; Chin, K.P. Minimum copper loss flux-weakening control of surface mounted permanent magnet synchronous motors. *IEEE Trans. Power Electron.* **2003**, *18*, 929–936. [CrossRef]
40. Wang, T.Z.; Zhang, J.H.; Wang, H.; Wang, Y. A Multi-Mode Fault-Tolerant Control Strategy for Cascaded H-Bridge Multilevel Inverters. *IET Power Electron.* **2020**, *13*, 3119–3126. [CrossRef]

Journal of
Marine Science and Engineering

MDPI

Article

Methodological Aspects of Assessing the Thermal Load on Diesel Engine Parts for Operation on Alternative Fuel

Sergejus Lebedevas and Edmonas Milašius *

Faculty of Marine Technology and Natural Sciences, Klaipeda University, 91225 Klaipeda, Lithuania;
sergejus.lebedevas@ku.lt
* Correspondence: edmonas.m@protonmail.com; Tel.: +370-60117362

Abstract: The decarbonization of maritime transport has become a crucial strategy for the adoption of renewable low-carbon fuels (LCFs) (MARPOL 73/78 (Annex VI) and COM (2021) 562-final 2021/0210 (COD)). In 2018, 98% of operated marine diesel engines ran on fossil fuels. The application of LCFs, according to expert assessments (DNV GL), is considered the most effective solution to the decarbonization challenge in the maritime sector. This publication presents methodological proposals related to assessing the reliability of operational diesel engines when transitioning to low- carbon fuels. The proposed methodology implements an interconnected assessment of the combustion cycle parameters and the limiting reliability factors of the thermal load on the most critical components of the cylinder–piston group. The optimization of the combustion cycle parameters for the indicators of energy and the environmental efficiency of low-carbon fuel applications was combined with the evaluation and assurance of permissible values of the thermal load factors on the components to determine the overall reliability of the engine. Thus, the possibility of overload and engine failures was already eliminated at the retrofitting design stage. The algorithm for the parametric analysis was grounded in the practical application of established α-formulae for the heat exchange intensity, such as those of the Central Diesel Engine Research Institute and G. Woschni. This approach was combined with modeling the combustion cycle parameters by employing statistical or single-zone mathematical models such as IMPULS and AVL BOOST. The α-formulae for low carbon fuels were verified based on the thermal balance data. The structure of the solutions for the effectiveness of the practical implementation of this methodology was comprehensively oriented towards diesel "families", as exemplified by the models 15/15 (p_{mi} = 1.2, 1.4, and 1.6 MPa). The long-term goal of the obtained results in the structure of comprehensive decarbonization research was to assess the factors of the reliable operation of characteristic groups of medium-speed (350–1000 rpm) and high-speed (1000–2100 rpm) marine engines for reliable operation in the medium term on ammonia.

Keywords: decarbonization of operational diesel fleet; combustion cycle; mechanical and thermal load factors; parametric analysis

Citation: Lebedevas, S.; Milašius, E. Methodological Aspects of Assessing the Thermal Load on Diesel Engine Parts for Operation on Alternative Fuel. *J. Mar. Sci. Eng.* **2024**, *12*, 325. https://doi.org/10.3390/jmse12020325

Academic Editors: Jian Wu, Xianhua Wu and Lang Xu

Received: 21 December 2023
Revised: 8 February 2024
Accepted: 11 February 2024
Published: 13 February 2024

Correction Statement: This article has been republished with a minor change. The change does not affect the scientific content of the article and further details are available within the backmatter of the website version of this article.

1. Introduction

The most important direction for the development of maritime shipping is decarbonization and increasing the energy efficiency of marine diesel engines. Strategic plans for decarbonization are defined in the regulations of the International Maritime Organization, MARPOL 73/78 (Annex VI) [1]; the European Parliament and the Council, COM (2021) 562-final 2021/0210 (COD) [2] COM (2021) 559-final 2021/0223 (COD) [3]; and many others.

The maritime transport sector was the first sector globally to implement greenhouse gas (GHG) emission monitoring; additionally, it boasts reduction factors regulated in the form of environmental normative indices, both for newly built ships and, as of 2023, for entire operating fleets [1]. The energy efficiency design index, mandatory for all newly built ships, was introduced on 1 January 2013. The CO_2 reduction level of the first stage (grams of CO_2 per ton of cargo per nautical mile) was set at 10% below the baseline, and it will be progressively increased to achieve a 40% reduction by 2030 compared with the reduction levels of 2008 [1,4]. The suggestions for achieving the desired energy efficiency and decarbonization level include reducing the resistance of the ship hull, using waste heat recovery systems, increasing the energy efficiency of auxiliary power units, and using low-carbon fuels (LCFs) [1,5]. Recent legislation for reducing GHG emissions has considered reducing the CH_4 and N_2O levels, in addition to the CO_2 levels, as they are up to 50 and 250 times more harmful, respectively, to the ozone layer compared with CO_2 [2]. The energy efficiency existing ship index and the annual operation carbon intensity indicator were introduced on 1 January 2023, with the aim of giving ratings as soon as 2024. For ships in service, the energy efficiency can be improved by optimizing the propellers, reducing the speed by limiting the engine power, and implementing technological systems for the regeneration of alternative energy sources: utilizing wind energy (parachutes and Magnus-effect devices) while gradually transitioning to LCFs [5]. Experts consider using LCFs to be the most effective strategy for achieving decarbonization goals, with its benefits estimated at 20–22% [5] according to the classification society DNV GL, whereas other measures offer, on average, only a 4–7% improvement [5].

The attractiveness of using LCFs is determined by the technological features of the internal combustion engine (ICE) of the ship, unlike other types of transport. According to the IMO's fourth GHG study [6], 98.4% of all engines used in the fleet in 2018 were conventional fuel-oil engines; therefore, to achieve the decarbonization targets set for the transition period of 2030–2035, the use of LCFs, such as biofuels and liquefied natural gas (LNG), is one of the options, which can be later gradually switched to bio-LNG, ammonia, e-diesel, and other LCFs [2,5].

According to LR/UMAS (2020) [7], maritime infrastructure is ready for second- and third-generation biodiesel, e-diesel, bio-LNG, and e-LNG. However, the price of such fuel is significantly higher compared with that of fossil fuels [8]; according to 2020 data, the prices of e-diesel, biodiesel, and bio-LNG are 1280%, 260–400%, and 500% higher compared with that of fossil fuels, respectively. The use of methanol on ships is in the initial stage of practical solutions [2,5], and ammonia and hydrogen are in the stage of pilot technological solutions [5,7]. Since LNG is a fossil fuel with a lower CO_2 emission factor per kg of fuel, LNG is considered a transition fuel in the short term (until 2030). Major marine engine manufacturers such as MAN and Wartsila have developed and produced a range of dual-fuel (DF) engines that run on both diesel and LNG, with power outputs of up to 20,700 kW [9]. These engines are optimized to run on LNG and diesel, and their engine resources are guaranteed by the manufacturer. Simultaneously, the use of LNG in ships already in service has been widely studied, primarily in terms of engine energy efficiency [10,11], environmental aspects [12], and economic aspects [13,14].

However, research on engine reliability when transitioning to LCFs is lacking. The major research question in this aspect is whether the changes in mechanical and thermal stresses are at acceptable levels.

Research on decarbonization and engine reliability when transitioning from diesel-based to LCF-based engines has begun at Klaipeda University, and it includes developing main and auxiliary engines in the Lithuanian fleet; in the initial stage, basic methodological solutions and tool formats for their practical implementation are being considered.

This study provides a methodological basis for implementing interrelated solutions for optimizing the combustion cycle parameters and the mechanical and element thermal stress factors when operating engines on LCFs. This method is intended for implementation in low-, medium-, and high-speed marine engine groups operating mainly on biodiesel and LNG (bio-LNG); it is expected to cover the use of ammonia.

Factors Influencing Thermal Loading of Engine Parts and Their Impact on Reliability

The reliability of an ICE is determined by several interrelated factors: the design and material properties of the components, or sub-assemblies, and the mechanical and thermal loading factors that form the temperature and stress field [15,16]. To convert engines to LCF operation, due to limited retrofitting capabilities, changes in the design and materials are assumed to not be foreseen. In the cylinder–piston group (CPG) parts, which experience the most stress and limit the engine reliability factor, the thermal stresses amount to more than 50% [17], and up to 90% in the old-generation engine models. Therefore, to ensure that the mechanical stress level designed by the CPG is not exceeded during retrofitting (engine operation with LCFs), the maximum cycle pressure (P_{max}), the main influencing factor, was assumed as a constant. Thus, engine reliability for LCF operation can be ensured by assessing the thermal stresses; further, the thermal stresses can be kept under permissible levels by optimizing the characteristic parameters of the combustion cycle. When retrofitting an engine for LCF, significant attention is given to monitoring and ensuring the effective operation of the engine cooling system. The alteration of heat dissipation in the cylinder during LCF combustion should not lead to the engine overheating due to an insufficient thermal performance of the cooling system. This aspect, along with the optimization of the LCF combustion cycle, is increasingly attracting the attention of researchers. To avoid catastrophic failures, it is assumed that a well-designed cooling system, capable of handling the engine's nominal power on diesel fuel over extended periods, is in place. For example, Cabuk A.S.'s [18] study focuses on using IoT technology to minimize maintenance costs and prevent failures in ship cooling systems. The system integrates smart sensors and Node-Red, allowing remote monitoring of ship engines and cooling pumps. It detects and analyzes real-time data, providing valuable insights into current, temperature, and vibration to ensure the effective operation of the system. Since most marine cooling pumps are currently centrifugal, and ongoing studies are exploring alternative, more efficient solutions. To ensure adequate performance reserves in the cooling system and thereby avoid overload, Fatigati et al. [19] explore and optimize a Low-Speed Sliding Vane Rotary Pump (LS SVRP) used in internal combustion engine cooling. Employing a model-based design approach, the research reduces revolution speed to enhance volumetric capability without increasing pump dimensions significantly. The LS SVRP prototype, tested across various conditions, achieves overall efficiencies nearing 60%, surpassing centrifugal pumps, and the SVRP demonstrates an efficiency much less dependent on operating conditions. Another study by Di Giovine et. al. [20] introduces a lumped parameter model for a triple-screw pump, experimentally validated to explore its potential as an efficient alternative to centrifugal pumps in internal combustion engine-cooling circuits. The model achieves a mean error of 0.6% in flow rate calculations and demonstrates satisfactory volumetric efficiency compared to experimental data. Mechanical efficiency calculations, influenced by variable friction coefficients, yield a global efficiency of up to 70%. The model is applied to estimate the efficiency of a triple-screw pump as a cooling pump for an F1C IVECO 3l engine, showing an 8% improvement over a standard centrifugal pump. This aspect of retrofitting the cooling system should also be carefully considered as a means of ensuring acceptable thermal loads on the components of a diesel engine during its operation on LCF, thereby enhancing overall efficiency.

On the other hand, the thermal load on a part is characterized by the maximum temperature according to the material properties, the limiting temperature in the characteristic zones of the part (e.g., in the area of the upper compression ring of the piston), and the thermal stresses. For the CPG parts of ICEs, the thermal load (without changing the structure or boundary conditions of heat dissipation from the part) is determined by the coefficient of heat release from the working body (α_{gas}) and the temperature (T_{gas}). The existing analytical descriptions of α_{gas}, which determines the heat transfer intensity, can be divided into two groups of simplified mathematical models: those based on similarity theory criteria, and those based on the classical theoretical foundations of extended gas dynamics. Among these, the simplified mathematical models (MMs) of the first group are the most widely used models. The analytical solutions by Woschni 1967 [21], Hohenberg 1979 [22], Annand 1963 [23], Sitkei 1972 [24], Chang 2004 [25], Wu 2009 [26], and others operate with relatively easy-to-determine combustion cycle parameters. Typically, the classical criteria of the similarity theory of heat transfer (Nusselt number (Nu), Reynolds number

(Re), and Prandtl (Pr)) are used as the analytical basis for formulating αgas formulae; the constants in the formulae are determined from statistical summaries of the experimental data. These formulae are widely applicable owing to the statistical generalizations of the experimental data for different types of engines and the accuracy of the obtained results. Recently, classical α_{gas} solutions have been modified based on the results of a limited experiment or MM of a single-engine model [21–23], including an engine running on LCFs.

For example, Rabeti et al. [27] investigated the heat transfer coefficient of a homogeneous charge compression ignition engine running on natural gas. The study determined the errors in the heat transfer coefficient calculations and compared them to the 3D model results using calibrated coefficients from classical MMs. A single-cylinder Waukesha engine (82.55/114.3) at 800/1100/1400 rpm and with different inlet pressures was used for the tests. The heat transfer coefficient calculations were performed using a zero-dimensional, single-zone model to compare the accuracy of the MM, as no experimental studies were available at that time to verify the baseline. A 3D computational fluid dynamics (CFD) model with detailed chemical kinetics simulation results was used for verification. In turn, the reliability of the 3D model simulation was verified by calculating the heat transfer coefficient when the engine was running on petrol and comparing it with the experimental data. Several operating modes were simulated with varying engine speeds and intake pressures. The results showed that single-zone MMs overestimated the heat release coefficients compared with the 3D model using natural gas and standard coefficients, and calibrated α_{gas} scaling coefficients were recommended. With calibrated scaling coefficients, the closest results were obtained by Assanis [25] and Hohenberg [22], with average errors of 14.3% and 16.3%, respectively. Moreover, an evaluation of the solutions of [27] revealed that single-zone MMs are sensitive to the fuel type and combustion characteristics.

Depcik et al. [28] evaluated the possibility of applying classical heat transfer coefficient calculation models to small-displacement ICEs. Small ICEs of <100 cm^3 have a combustion chamber with a relatively large external surface area; this directly affects the heat loss to the cylinder wall and reduces the overall energy efficiency. Therefore, using conventional heat transfer correlations, their study attempted to adapt the α_{gas} calculations developed for larger engines to the study subject, which was a 3W-55i (44/35) engine with a power output of 4.4 kW and a maximum speed of 8500 min^{-1}. The results confirmed the applicability of the calculations of large-ICE heat transfer coefficients to small engines with parameter optimization.

Hassan et al. [29] investigated the applicability of heat transfer coefficients using the Woschni [21] equations for an engine running on gasoline fuel; 1D and 3D MMs were used. With a four-stroke, single-cylinder, 114.8 cm^3, high-speed, 1500–9500 min^{-1} motorcycle engine as the study subject, they investigated the temperature conditions of the intake and exhaust valves and the heat balance of the engine. The valve temperatures were measured using thermocouples and compared with the simulation data. Reportedly, the temperature errors at the intake and exhaust valves were 3.73% and 0.17% at 2500 min^{-1} and 4.12% and 0.70% at 5500 min^{-1}, respectively. The highest temperature regions were concentrated around the combustion surface, the highest heat flow was transmitted through the exhaust valve, the highest temperature was recorded at the exhaust valve neck, and the highest temperature of the intake valve was recorded at the combustion surface. Through these studies, the MM was determined, and the engine heat balance and thermal stresses were experimentally verified.

In summary, the applicability of α_{gas} MMs developed based on the classical similarity theory has been confirmed in new-generation ICEs of different sizes and purposes. In the absence of new proposals for analytical α_{gas} solutions, classical α_{gas} MMs are widely used to solve practical problems. However, their practical testing is limited in the case of LCFs, which makes research in this area relevant and in demand when solving problems involving the decarbonization of transport.

The second group of α_{gas} MMs is based on the classical theoretical foundations of extended gas dynamics after applying them to the physical processes in the ICE cylinder during the combustion cycle. Petrichenko [30] proposed such an MM, as follows:

$$\frac{\partial p}{\partial t} + \mathrm{div}\ \rho_v = 0; \quad \begin{cases} \rho\left(\frac{\partial v_i}{\partial x} + v_{i,j}v_i\right) = -\frac{1}{H_i}\frac{\partial p}{\partial q_i} + \tau_{ij,j}; \\ \rho c_p\left(\frac{\partial T}{\partial t} + \frac{v_j}{H_j}\frac{\partial T}{\partial q_j}\right) + div\ q = q_{v.} \end{cases} \tag{1}$$

where ρ is the density; t is the time; v is the fluid motion vector with component v_i (i = 1, 2, 3); H_i is the corresponding metric coefficient (Lamé coefficients); $v_{i,j}$ is the absolute derivative of component v_i in coordinate q_j; p is the pressure; T is the temperature; $-\tau_{ij,j}$ is the absolute derivative of the tangent friction tensor in coordinate q_j; c_p is the specific heat capacity of fluids; q is the heat flow density vector; and q_v is the volumetric density of internal heat.

However, the practical application of these models, particularly for diesel engines in operation, is challenging because they are based on numerous differential equation variables, so they require extensive research under laboratory conditions and detailed graphical material of the design geometry. This type of analytical solution is partially used exclusively in multizone combustion cycle study models, such as AVL Fire and KIVA-3V [31,32].

Taking a closer look at the application of LCFs to an already operational fleet of ship engines, two main directions of scientific research emerge. The primary focus lies in researching and optimizing energy efficiency and environmental indicators. Another ongoing area of research pertains to the economic aspects of the operational fleet's transition to low-carbon fuels. Typically, addressing the failures of engine components identified during operation involves changing the materials, optimizing the construction, and employing detailed temperature field modeling for heat transfer coefficients while operating with traditional petroleum-derived fuels. The authors propose a methodological solution that combines a variation optimization stage for engine operation with low-carbon fuels and simultaneously evaluates the factors most affecting the reliability of the most thermally loaded components, such as pistons. Such solution principles were not found in the available literature.

For the purposes of this study, a parametric analysis of the interrelationship between the thermal state of the components and the parameters of the combustion cycle involved, using the analytical descriptions of the first group's α_{gas}, was developed based on the similarity theory; these include the main design parameters of the engine under study, the chemical properties of the working body, and the thermodynamic parameters characterizing the operating loads of the engines.

2. Methodological Aspects: Selection and Justification of the Analytical Form of α_{gas} Calculation

According to the set goal in the publication, the methodological aspects became the main issues to be addressed. Therefore, the second methodological section is dedicated to identifying and justifying one of the most rational parameter structures determining the heat exchange intensity for practical use. Based on the solutions presented in the second section, the third section develops solutions for the created methodology, describes the algorithms and their implementation using IT tools, and provides an application example.

Most of the solutions for using simplified α_{gas} models have been extensively validated on different types of ICEs operating over a wide load range and on different fuels. In particular, the model proposed by Woschni et al. [21] is commonly used to solve practical problems. Further, the analytical solution in [21] has been widely used in single-zone combustion cycle models such as AVL BOOST [28,30]. Woschni et al. and Merker et al. [21,33] made the theoretical assumption of a stationary, fully turbulent tube flow. Their formulae were based on the similarity theory criteria, which describe the intensity of heat release (Nu), the intensity of fluid flow around the parts (Re), and the physical and chemical properties of fluids (Pr). Furthermore, they distinguished the peculiarities of using this formula for different strokes and validated it for various engine models. For the dimensionless heat transfer coefficient Nu, a semi-empirical equation was obtained from a dimension analysis, as follows, essentially filling the structure with the characteristic parameters of ICEs:

$$Nu = C * Re^{0.8} * Pr^{0.4}, \tag{2}$$

were $Nu = \frac{\alpha D}{\lambda}$, with λ denoting the thermal conductivity coefficient, α denoting the heat transfer coefficient, and D denoting the cylinder diameter; $Re = \frac{\rho w D}{\eta}$, with ρ denoting the density, η denoting the dynamic viscosity, and w denoting the speed, assumed to be the average piston speed; and $Pr = \frac{\eta}{\rho} \alpha$.

For practical use, (2) was simplified to parameters that can be easily determined in a single-zone combustion model, thus allowing the heat transfer coefficient of the combustion cycle to be determined as follows:

$$\alpha_{gas} = 127.93 * D^{-0.2} * \rho^{0.8} * w^{0.8} * T^{-0.53}, W/(m^2K) \tag{3}$$

Similarly, widely validated analytical solutions based on Equation (4) have been performed at the Central Diesel Engine Research Institute (CDERI), St Petersburg, by Molodtsov et al. [34]. As experimentally determined in [34,35], the heat transfer coefficient α_{gas} and CPG temperature depend primarily on the cylinder diameter, the average piston speed, the composition of the working body, the pressure, and the temperature. The structural equation proposed in [34] describes the variation character of the heat transfer coefficient from the listed parameters.

$$\alpha_{gas} = A * P_r^{0.4} * \frac{\lambda_{p\,gas}}{\mu_{gas}^m} * \frac{C_m^m}{D^{1-m}} * \left(\frac{P_{gas}10^4}{R_{gas}\,g\,T_{gas}}\right)^m, W/\left(m^2K\right) \tag{4}$$

The first part of the above equation, $A\,P_r^{0.4}\,\frac{\lambda_{gas}}{\mu_{gas}^m}$, reveals the gas composition and the influence of the parameters, where $A = 2.75 + 58.6\,(D/c_m)$, according to the static generalization of the parameters, and c_m is the average piston velocity. The second part, $\frac{C_m^m}{D^{1-m}}$, describes the dependence of α_{gas} on the main engine design parameters and operating mode. The third part, $\left(\frac{P_{gas}10^4}{R_{gas}\,g\,T_{gas}}\right)^m$, represents the thermodynamic parameters of the combustion cycle of the working process. Thus:

$$\alpha_{gas} = \left(2.75 + 58.6\frac{D}{C_m}\right) * \frac{\lambda_{p\,gas}}{\mu_{gas}^{0.5}} * \left(\frac{C_m}{D}\right)^{0.5} * \left(\frac{P_{gas}10^4}{R_{gas}\,g\,T_{gas}}\right)^{0.5}, W/\left(m^2K\right) \tag{5}$$

where α_{gas} is the current heat transfer coefficient during the cycle; $\lambda_{p\,gas}$ is the gas heat conductivity coefficient (W/(m K)); μ_{gas} is the gas viscosity coefficient (kg/(m·s)); P_{gas} and T_{gas} are the current gas pressure (N/m^2) and temperature (K) during the cycle, respectively); c_m is the average piston velocity (m/s); D is the cylinder diameter (m); R_{gas} is the gas constant (J/(kg·K)); and g is the free-fall acceleration.

In determining the numerical values of the constants, A and m, generalized test data from a wide range of high-speed and medium-speed diesel engines with volumetric mixture formation, both turbocharged and naturally aspirated, were utilized: 12/14, H18/22, 25/34, and 26/26. The experience of the successful application of this equation in investigating the thermomechanical state of components in the cylinder–piston group of engines such as 15/15, 15/18, 16.5/18.5, and 16.5/15.5 was also considered [36].

Thus, the α_{gas} calculations by Woschni et al. and CDERI are typically used to calculate each crankshaft rotation angle. However, the CDERI model has also been adapted for parallel generalized use in an integral form when $\alpha_{gas\,av}$ is calculated for the entire heat exchange cycle $\alpha_{gas} = (\varphi)$. This feature is significantly advantageous, particularly in the Klaipeda University studies, during which the combustion cycle parameters, which determine the energy efficiency, are optimized. These parameters are also evaluated in the integral form of the entire combustion cycle. The proposed methodological solution using $\alpha_{gas\ av}$ has been successfully validated in the research and design of ICE models tuned for various average pressures [33,34]. Based on these fundamentals, $\alpha_{gas\ av}$ [34,35] was used to form an interrelated combustion cycle algorithm.

3. Results of Research: Parametric Analysis of Combustion Cycle Indicators and Thermal Load

The relationship among the average heat transfer coefficient $\alpha_{gas\ av}$, the average temperature of the entire engine working cycle $T_{gas\ av}$, and the combustion cycle parameters is based on a mathematical engine combustion cycle model. Here, note that $\alpha_{gas\ av\ T}^{com}$ and $T_{gas\ av\ T}^{com}$ denote the average heat transfer coefficient and the average temperature for the part of the cycle section when heat is transferred to the component, respectively. The use of a single-zone MM enables the calculation of the current values of the parameters of the working body and the characteristics of heat release during the cycle, along with the characteristic parameters of the combustion cycle T_{gas}, P_{gas}, $X = f(\lambda,\ \varepsilon,\ \lambda_p,\ P_k,\ T_k)$, where ε is the compression ratio and λ_p is the degree of pressure increase. Thus, the dependence of the heat transfer coefficient [34,35] can be simplified when applying it to a specific engine model for transferring it to LCFs, as follows:

$$\alpha_{gasav} = \left(2.75 + 58.6\frac{D}{C_m}\right) * \frac{\lambda_{p\ gas}}{\mu_{gas}^{0.5}} * \left(\frac{C_m}{D}\right)^{0.5} * \left(\frac{P_{gas}}{R_{gas}T_{gas}}\right)^{0.5},\ W/\left(m^2 K\right) \quad (6)$$

For structurally similar engines at a constant working speed (c_m), the multipliers $\left(2.75 + 58.6\frac{D}{C_m}\right)\left(\frac{C_m}{D}\right)^{0.5}$ can be converted to a constant (N):

$$\alpha_{gas\ av} = N * \frac{\lambda_{p\ gas}}{\mu_{gas}^{0.5}} * \left(\frac{P_{gas}}{R_{gas}T_{gas}}\right)^{0.5},\ W/\left(m^2 K\right) \quad (7)$$

The universality of the formula, when applied to both petroleum-derived fuel and the use of low-carbon fuels (LCFs), is related to the existing structural parameters $\lambda_{p\ gas}$ and μ_{gas} (corresponding to the conductivity and the viscosity of the combustion products in the engine cylinder, respectively), which differ for various fuel types. The peculiarities of the LCF combustion cycle also result in different values of the parameter X, which influences both $\lambda_{p\ gas}$ and the engine's modeled energy indicators. This structural connection is revealed by a system of Equation (8).

The transition to an interrelated parametric analysis with combustion cycle parameters is represented by the following system of equations:

$$\begin{cases} T_{gas},\ P_{gas},\ X = f\left(\lambda, \varepsilon, \lambda_p,\ P_k, T_k\right) \\ \lambda_{p\ gas},\ \mu_{gas} = f\left(\lambda, X, T_{gas}\right) \\ \alpha_{gas\ av} = N * \frac{\lambda_{p\ gas}}{\mu_{gas}^{0.5}} * \left(\frac{P_{gas}}{R_{gas}T_{gas}}\right) \end{cases} \quad (8)$$

The first Equation in (8) shows the influence of the characteristic indicators of the combustion cycle on the thermodynamic parameters of the cycle and the heat release characteristics; the second equation shows the formation of the physical properties of the working body; and the third equation is based on the first two equations.

The combined solution of these equations allows us to express the dependence of the similarity conditions of the heat transfer parameters from the working body to the component, as follows (P_{max}-fixed):

$$\alpha_{gas\ av},\ T_{gas\ av}\left(\alpha_{gas\ av\ T}^{com},\ T_{gas\ av\ T}^{com}\right) = f\left(\lambda, \varepsilon, \lambda_{p\ gas}, \mu_{gas}, P_k, T_k\right) \quad (9)$$

The transition to the average of the heat exchange process is $\alpha_{\text{gas av T}}^{\text{com}}$, $T_{\text{gas av T}}^{\text{com}}$. The alternative solution is implemented in two ways: the first one is based on the analysis of statistical multivariate data [36,37], where significant factors are identified and separated using the influencing factor. This enables an assessment of the thermal load level on the components for various combustion cycle organization strategies and the selection of the most rational one when optimizing energy and environmental performance. For practical applications, the dependencies of $\alpha_{\text{gas av T}}^{\text{com}}$, $T_{\text{gas av T}}^{\text{com}}$ are presented as a function of the interrelated parameters of the diesel engine, i.e., $P_{\max}/P_k$, λ, ε, λ_p, and T_k, and the methodological approach can be applied for choosing the most rational combination. The influence of the charge air parameters on cylinder filling is reflected by additional factors. Thus, the conditions to perform a simultaneous parametric analysis between the characteristics of the combustion cycle process and the thermal and mechanical stress of the piston group from the function of the values $P_{\max}/P_k$, λ, and ε or ε, λ, λ_p, P_k, and T_k with respect to $P_{\max}/P_k = \varepsilon^n \lambda_p$ are met.

The second is based on modeling the combustion cycle of a specific engine model or an "engine family", preferably using a single-zone mathematical model. In this process, the heat transfer coefficients to the walls from the working fluid of the engine components are calculated, and based on these coefficients, the heat losses to the cooling system are determined. The choice of methods is determined by the research tasks and the availability of necessary parameters for calculating $\alpha_{\text{gas av}}$ and $T_{\text{gas av}}$ for the studied object. However, in both cases, an assessment of the adequacy of the obtained heat exchange parameters for practical use can be performed by comparing them with the engine's thermal balance data.

For greater clarity, a block diagram of the combined parametric analysis model is presented in Figure 1. The implemented algorithm serves as the foundation for addressing the planned research tasks aimed at expanding the use of LCFs in the main and auxiliary engines of operational vessels.

The features of the method allow for an interrelated parametric analysis. The method is adaptive for a wide range of engine loads based on the average indicator (effective) pressure p_{mi} (p_{me}). This is relevant to the practical application of the method in the context of different in-service engine models of the same engine type and addresses the challenge of using LCFs in the study of engines within a so-called "family", the models of which differ in their level of boost, configuration of the main systems, and structure of the operating loading cycle on the vessel.

The combinations of the sets λ, ε, λ_p, P_k, and T_k can be identified to calculate $\alpha_{\text{gas av T}}^{\text{com}}$ and $T_{\text{gas av T}}^{\text{com}}$ and plot the resulting values on a graph in parallel with the energy efficiency (η_i). As the ε, λ, P_k, and T_k values are available for optimization for the chosen simple retrofitting, further studies will be conducted by changing ε and recalculating $\alpha_{\text{gas av T}}^{\text{com}}$ and $T_{\text{gas av T}}^{\text{com}}$ for all of the selected operating parameters. The aim is to simulate the duty cycle process for an engine running on LCFs with similar effective power indicators p_{me} as an engine running on fossil fuel, with a maximum achievable η_i. The LCFs (line shape) combination of $\alpha_{\text{gas av T}}^{\text{com}}$ and $T_{\text{gas av T}}^{\text{com}}$ in the graph must not exceed the calculated thermal stresses of the engine running on fossil fuel. Studies based on a similar methodology and the set goals of boosting engines and justifying the design solutions for parts have been conducted [37,38]. The reliability of various engine structural components was investigated for promising model types by increasing the average indicator pressure p_{mi}.

In this context, there are practically no differences in the implementation of the successfully tested combined analysis method, both in terms of the tasks related to diesel engine boosting and in the planned retrofitting of operational models within a certain power range of a family of diesel engines. Below is a proposed example of implementing such an approach. For the studied models of the "family" of operational engines in the characteristic power range of $p_{mi} = 1.2$, 1.4, and 1.6 MPa, the possible retrofitting options for organizing the combustion cycle combinations of λ, P_k, λ_p, and ε (four combinations for the fixed levels of $\lambda = 1.75$, 2.0, 2.25, and 2.5) are determined within the technological accessibility constraints. This is achieved with limitations on $P_{\max}$ and a selected variant of the supercharged air-cooling system under the condition T_K = const. The maximum P_{max} should be set based on the real strength reserves of the components of the cylinder–piston group (CPG) and the crankshaft mechanism (CM) of the "family" models declared by the manufacturer for operation on diesel fuel.

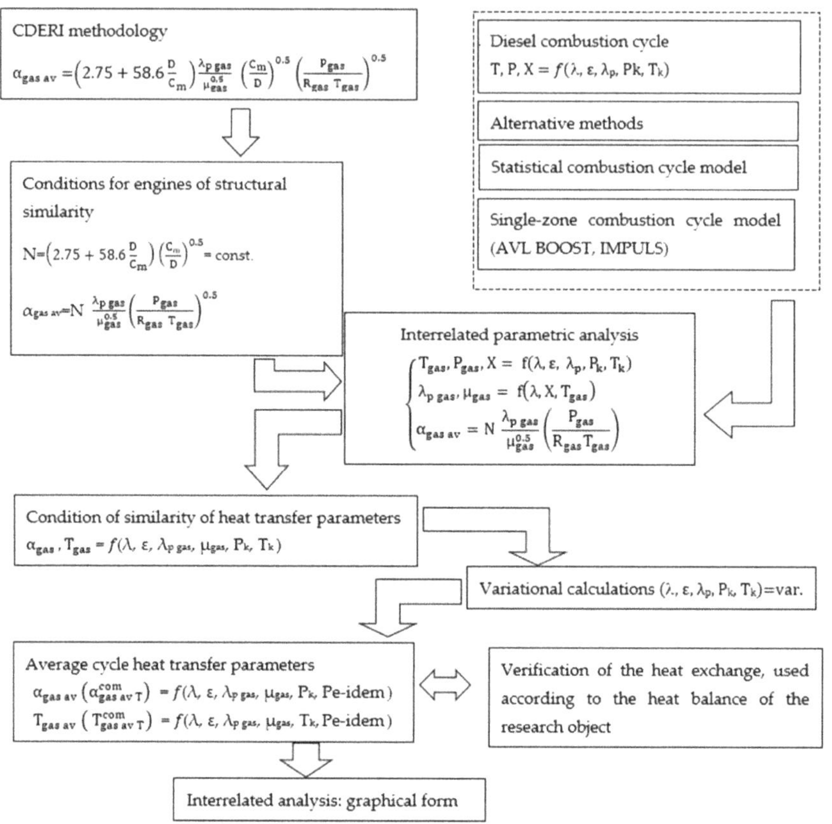

Figure 1. Block diagram of the combined parametric analysis model.

Due to technical allowances in the manufacturing of components and the regulation of diesel assemblies, the value of P_{max} for the same models of a "family" of diesel engines varies within a specific range. Typically, its magnitude does not exceed ±2–3% of the nominal value of P_{max} that is regulated by technical documentation. This circumstance is assumed to be taken into account when performing a parametric analysis of the indicator process using the developed algorithm. Its implementation includes the following sequence of operations:

1. For the analyzed power range of "family" models, p_{mi}, given the a priori values of P_{max} and T_k based on the specification or measurement on the vessel (in accordance with the mechanical strength reserves of the components of the cylinder–piston group and the chosen supercharged air-cooling system), ranges of rational changes in the defining parameters of the combustion cycle organization λ and ε are formed (specifically for the considered example, λ = 1.75, 2.0, 2.25, and 2.5).

2. Based on the theoretical relationship between diesel parameters for the identified values of λ, the initial data for determining the heat exchange indicators are calculated, relying on statistical data or combustion cycle modeling.

3. After determining the assessed levels of p_{mi}, combinations of λ, λ_p, ε, P_k, and T_k are calculated according to the methodology [38], providing the values of the average heat exchange parameters in the cylinder $\alpha^{com}_{gas\ av\ T}$ and $T^{com}_{gas\ av\ T}$.

4. The obtained combinations $\alpha^{com}_{gas\ av\ T}$ and $T^{com}_{gas\ av\ T}$ are plotted in the nomogram field $\alpha^{com}_{gas\ av\ T}$, $T^{com}_{gas\ av\ T}$ (see Figure 2). As a result, each of the analyzed levels of forcing by p_{mi} correspond to its own local area in the nomogram field, delimited by eight combinations of $\alpha^{com}_{gas\ av\ T}$ and $T^{com}_{gas\ av\ T}$. For the clarity and convenience of the subsequent analysis, lines of the fixed values of λ = 1.75, 2.0, 2.25, and 2.5 are displayed in the field of forcing areas (based on the possibility of increasing λ through a corresponding adjustment to the supercharging unit or its replacement).

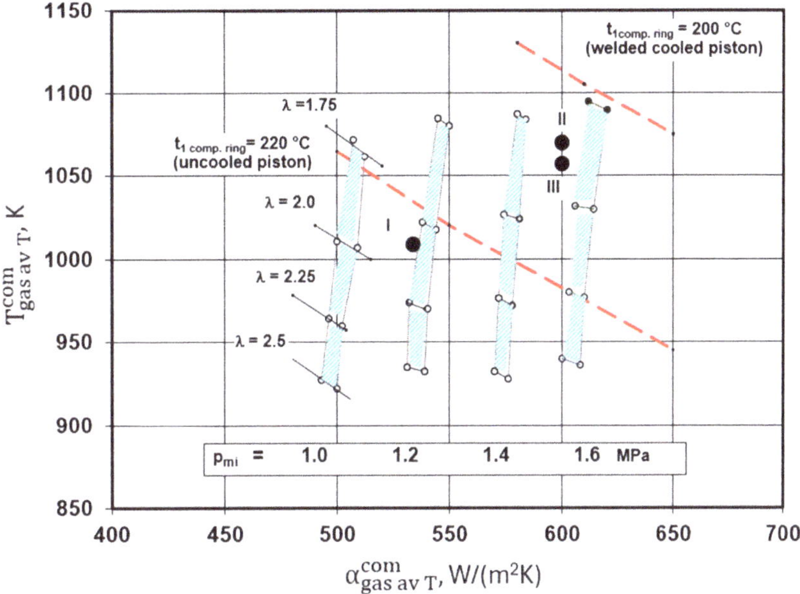

Figure 2. Fragment of an example illustrating combined parameters and thermal loads on the components of the cylinder–piston group of the 15/15 diesel "family" models during operation on the studied low-carbon fuel.

In terms of the fuel economy indicators, all of the considered combinations of λ, P_k, λ_p, and ε are invariant since, with a fixed p_{mi}, they fulfill the condition of a close fuel economy. To choose their rational combinations based on the thermal stress indicators, the indicators limiting the reliable operation of the examined model's piston when using diesel fuel (maximum surface temperature, temperature of characteristic zones of construction T_{com}, and stresses σ_{com}) were plotted on the nomogram in the form of isolines of constant values. The introduction of multiple criteria into the analysis allows the limiting factors to be determined and concentrates attention on them during the optimization of the indicator process parameters. In Figure 2, the reliability of the uncooled piston's operation is associated with the level of the characteristic temperature in the zone of the first compression ring: its maximum value when using oils with additives and considering operational conditions is accepted as $t_{1comp\cdot ring}$ = 220 °C.

It is important to note that the developed approach is not limited to the framework of a quasi-stationary piston-loading analysis. Evaluations of piston construction that consider the fatigue strength reserves for the total stresses arising from the action of the stationary temperature field and variable loads from the impact of P_{max} are possible. In this case, it is rational to use the results of the piston fatigue tests, generalized in the form of a

Goodman–Soderberg diagram, as exemplified by the company Wellworthy Piston Rings Limited and in the materials of several other studies.

The analysis of the mutual position of the isolines T_{com}, σ_{com} = const. with the fields of investigated forcing levels by p_{mi} allows for a multivariate assessment of the rational organization of the combustion cycle when transitioning the operation of the examined engine "family" models to a wide range of LCFs. Figure 2 provides an example of using the developed algorithm for the combined parametric analysis method for the "family" of 15/15 models.

The Roman numerals I and II (III) correspond to the level of forcing of two models of the "family", p_{mi} = 0.95 MPa and p_{mi} = 1.17 MPa, respectively. With an equal thermal piston load, the reserve of applying LCFs for the "family" models in the characteristic power range p_{mi}, compared to the diesel fuel baseline, is approximately 20%: 1.4 MPa versus 1.15 MPa. Evidence of this is also provided by the position of the point corresponding to p_{mi} = 0.95 MPa for the standard configuration of the diesel in the field of prospective forcing at p_{mi} = 1.2 MPa. The predominant influence on the temperature state of the piston assembly is exerted by λ. Theoretically, its increase from 1.75 to 2.25 (due to the readjustment of the turbocharger or its replacement during retrofitting) allows a nearly constant heat load to be maintained on the piston in the investigated p_{mi} range and limits the use to the uncooled piston modification. Under real conditions, ensuring reliable starting characteristics of diesel engines of this size restricts the boundaries of the possible reduction in ε to the level of 13–14 units. Accordingly, the range of possible changes in λ narrows, as otherwise, considering the low dynamics of the indicator process, the condition for limiting P_{max} would not be met. As a result, the theoretical transition to an oil-cooled piston design, in case of its technological implementation with the accepted position of the thermal stress limit line $t1_{comp\cdot ring}$ = 220 °C on the nomogram and the implementation of λ = 2.0, becomes necessary for "family" models with a p_{mi} > 1.2 MPa.

Thus, the provided example of using the applied method illustrates the principle of influencing the magnitude of the thermal load with the aim of fully utilizing the potential of different models within the operated diesel family for transitioning their operation to LCFs.

One of the key conditions for the effective use of the developed parametric analysis is the justified assignment of the restrictive levels $\alpha^{com}_{gas\ av\ T}$ and $T^{com}_{gas\ av\ T}$ in the nomogram field (Figure 2). In contrast to the possibilities of research under laboratory conditions and at engine manufacturers, direct measurements of the temperature and stress of parts are not possible for operating models. In this case, a comparison of the data obtained will be used when the engine is running on base-oil fuel, for which the manufacturer guarantees certain reliability indicators (lifetime before overhaul, failure rates, etc.), and LCFs. This means that, after verifying the mathematical model of the object under study on the basis of a specification or experimental data, modeling the indicators and parameters of the heat exchange cycle of the traditional diesel fuel combustion cycle will be carried out. The resulting boundary isolines of the combinations after they are plotted in the parametric analysis nomogram field are supplemented by the results of similar calculations of LCF combustion cycle indicators in the form of local combinations of $\alpha^{com}_{gas\ av\ T}$ and $T^{com}_{gas\ av\ T}$. Based on a comparative analysis of the data obtained, it will be possible to make decisions about the possible optimization of combustion cycle indicators. At this stage, a statistical analysis of literary sources and our own research findings will be used [39–41].

Possible uncertainty in the results of the parametric analysis is associated with the use of dependencies ($\alpha^{com}_{gas\ av\ T}$) for cases where LCFs are used, although the authors of a number of analytical dependencies declare their use for many types of fuels [42,43]. However, the research algorithm also provides for the possibility of adjusting the structure of the calculated λ dependencies, primarily the multiplier constants, based on the comparison of the experimental and calculated items of the engine heat balance [43], as tested by the authors. The predicted accuracy of the methodology hinges on the practical application of its results, evaluating the range limits of diesel engine boosting by increasing the p_{me} (p_{mi}) [36]. The use of the α formula, widely validated for diesel fuel, ensured that the methodology's error did not surpass 2–3%. In the case of LCF usage, addressing the uncertainty arising from the applicability of the α formula involves verifying the formula based on engine heat balance data [43], resulting in an estimated error of 2–4%.

Based on previous authors' research on the four-stroke, four-cylinder diesel engine 79.5/95.5 running on diesel and dual Diesel/NG fuel, a proposed methodological use case fragment can be provided. When the engine operates on diesel fuel with a specified injection angle of 1 crank angle degrees (CAD) before top dead center (TDC), the energy efficiency index η_e reaches 0.365 at a load of p_{me} = 8 bar (2000 rpm). However, when switching the engine to dual fuel D20/NG80 (with the maximum experimentally achieved NG ratio), the η_e drops by 9–45% across the entire tested load range from 2 bar to 8 bar. Consequently, fuel overdraft exacerbates the reduction in harmful emissions due to the NG effect (NG forced into the cylinder during low-pressure intake strokes along with the air). To enhance engine efficiency (η_e) and ecological indicators without significant modernization, the pilot diesel injection timing was advanced to 13 degrees CAD before TDC. This resulted in a significantly smaller decline in energy efficiency compared to diesel, down to 19–5%, while notably enhancing ecological indicators (in all instances, the mechanical engine components' load factor P_{max} did not surpass the specified limits). The results of the thermal load assessment of the engine components temperature of the first compression ring zone are presented in Figure 3. The thermal load restriction of the piston is conditionally accepted based on p_{me} = 8 bar when operating on diesel (considering that experimental and computational studies of the piston's temperature field have not been conducted). When the engine operates on D20/NG80 with a specific injection angle of φ_{inj} = 1 CAD before TDC (blue line), the piston thermal load $\alpha_{gas\ av\ T}^{com}$ and $T_{gas\ av\ T}^{com}$ form does not exceed the set limit (dashed line). It is worth noting that according to G. Woshni's methodology, α_{gas} results were refined based on engine heat balance data. However, increasing φ_{inj} to 13 CAD before TDC ($\alpha_{gas\ av\ T}^{com}$, $T_{gas\ av\ T}^{com}$) already exceeds the allowable limits at p_{me} ~7, bar. The main reason is the significant increase in characteristic $T_{gas\ av}$ temperature. Based on this, one rational measure could be to increase the air–fuel equivalence ratio λ. The results of evaluating the obtained effect can also be rationally represented in the form of $\alpha_{gas\ av\ T}^{com}$, $T_{gas\ av\ T}^{com}$ diagrams. Decarbonization indicators in the maritime sector are declared in accordance with a certain dynamic of fossil fuel substitution by LCF [1,2]. Therefore, among other decisions acceptable for real-world operation, is a change in the dual diesel/NG fuel composition, for example, reducing the NG portion to D40/NG60 (which was also experimentally investigated and shown in Figure 3, green curve). The adverse impact of the change in η_e and harmful emissions, when compared to diesel, was smaller and is similar to D20/NG80 at 13 CAD before TDC, and the piston heat load also does not exceed the specified limit.

However, in some cases, the use of a single-zone module alone may not effectively solve the optimization problems of engine parameters for operation on LCF. This may primarily be due to the lack of adapted analytical solutions for the investigated LCF.

As acknowledged, the accuracy of single-zone models relies heavily on precise heat release characteristic (HRC) data. Therefore, optimization solutions involve various technologies and combinations thereof, particularly changing the characteristics and delivery strategies into the LCF cylinder, among others. Significant changes also occur in the HRC simultaneously. Therefore, in cases where verified analytical solutions for a specific fuel type are lacking, it is reasonable to consider the possibility of additional application of a multi-zone model over a single-zone model. For the assessment of engine HRC while operating on LCF, multi-zone models are employed initially. Subsequently, the analytically derived and generalized HRC data obtained can be effectively incorporated into a single-zone model through extensive numerical investigations.

Moreover, ECU experimental indicator diagrams (controlled by ECU in modern medium- and low-speed ship engines) can also serve as a source of HRC determination. Furthermore, the selected α model (e.g., G. Woschni) is refined based on engine heat balance data.

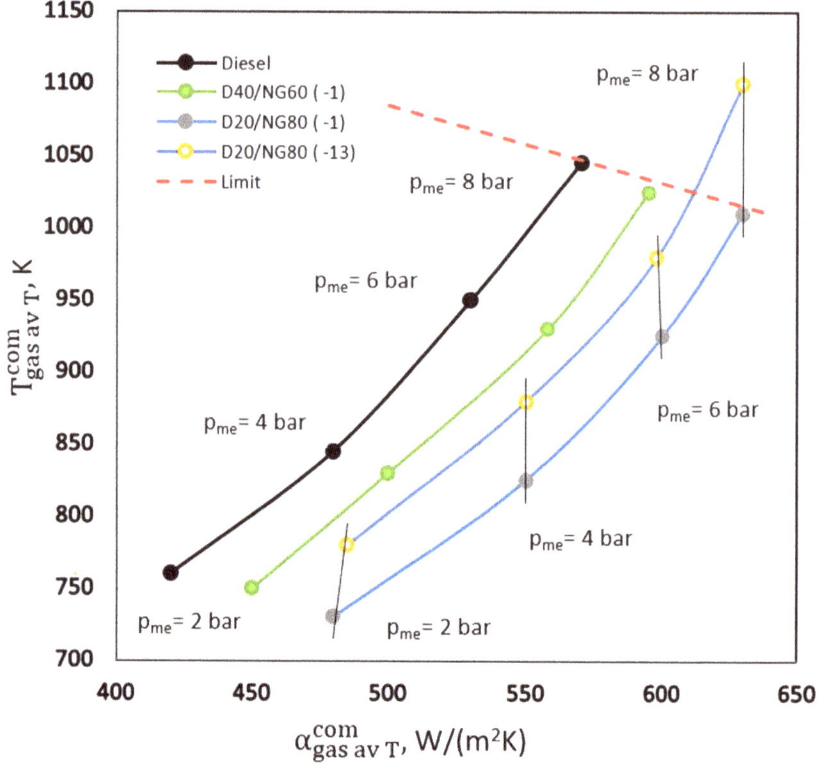

Figure 3. Engine (79.5/95.5) piston thermal load factor optimization fragment (n = 2000 rpm).

4. Conclusions

The presented methodological solutions addressing the decarbonization challenges in maritime transport aim to implement a comprehensive approach to enhance the performance of operational marine diesel engines when transitioning from fossil fuels to renewable and low-carbon alternatives. The proposed algorithm for a combined parametric analysis enables the assessment of an acceptable level of thermal load on the most stressed components of the cylinder–piston group. This assessment is conducted while optimizing the combustion cycle parameters for LCFs to improve the energy efficiency and reduce the emissions of toxic and greenhouse gases.

The implementation of a combined analysis of the heat exchange parameters in the cylinder and the combustion cycle is based on the use of well-established α-formulae for heat exchange intensity (CDERI, G. Woschni). Alternatively, the modeling of the engine's combustion cycle involves the application of a statistical model developed by the authors, oriented towards identifying the factors influencing the thermal load in different combustion cycle organizations. It also includes known single-zone models such as AVL BOOST and IMPULS. In the latter case, the tasks of identifying the permissible power ranges for the "family" for LCF operation are solved.

The graphical environment for visualizing the obtained results is considered sufficiently versatile for conducting comprehensive research, as it is not limited to a specific optimization strategy for combustion cycle parameters. The authors associate the long-term goal of the obtained results with the application of methodological solutions for a parametric analysis in ongoing studies on the use of ammonia in medium-speed marine diesel engines.

Author Contributions: Conceptualization, S.L. and E.M.; methodology, S.L.; writing—original draft preparation, E.M.; visualization, E.M.; supervision, S.L. All authors have read and agreed to the published version of the manuscript.

Funding: The project was financed by Research Council of Lithuania and the Ministry of Education, Science and Sport Lithuania (Contract No. S-A-UEI-23-9).

Institutional Review Board Statement: Not applicable.

Informed Consent Statement: Not applicable.

Data Availability Statement: No new data were created in this study. Data sharing is not applicable to this article.

Acknowledgments: The authors would like to thank the Klaipeda University Faculty of Marine Technology and Natural Sciences for their support.

Conflicts of Interest: The authors declare no conflicts of interest.

Abbreviations

The following abbreviations are used in this manuscript:

LCF	low-carbon fuel
MM	mathematical model
GHG	greenhouse gas
IMO	International Maritime Organization
DNV	Det Norske Veritas
ICE	internal combustion engine
CPG	cylinder–piston group
CM	crankshaft mechanism
DF	dual fuel
CAD	crank angle degrees
TDC	top dead center
CDERI	Central Diesel Engine Research Institute, St. Petersburg

References

1. European Commission. COM(2021) 562 Final 2021/0210 (COD). 2021. Available online: https://opac.oireachtas.ie/Data/Library3/Documents%20Laid/2021/pdf/TRAdocslaid110821_110821_152830.pdf (accessed on 10 December 2023).
2. International Maritime Organization. Resolution MEPC.328(76). 2021. Available online: https://www.imo.org/en/OurWork/Environment/Pages/Index-of-MEPC-Resolutions-and-Guidelines-related-to-MARPOL-Annex-VI.aspx (accessed on 10 December 2023).
3. European Commission. COM(2021) 559 Final 2021/0223 (COD). 2021. Available online: https://eur-lex.europa.eu/resource.html?uri=cellar:dbb134db-e575-11eb-a1a5-01aa75ed71a1.0001.02/DOC_1&format=PDF (accessed on 10 December 2023).
4. The Marine Environmental Protection Committee. Resolution MERC.377(80). 2023. Available online: https://wwwcdn.imo.org/localresources/en/MediaCentre/PressBriefings/Documents/Resolution%20MEPC.377(80).pdf (accessed on 10 December 2023).
5. DNV. Achieving the IMO decarbonization Goals. Containership Insights. MARITIME IMPACT. 23 July 2020. Available online: https://www.dnv.com/expert-story/maritime-impact/How-newbuilds-can-comply-with-IMOs-2030-CO2-reduction-targets.html (accessed on 10 December 2023).
6. International Maritime Organization. Fourt IMO Greenhouse Gas Study. 2020. Available online: https://wwwcdn.imo.org/localresources/en/OurWork/Environment/Documents/Fourth%20IMO%20GHG%20Study%202020%20-%20Full%20report%20and%20annexes.pdf (accessed on 10 December 2023).
7. LR/UMAS. Techno Economic Assessment of Zero Carbon Fuels. 2020. Available online: https://www.lr.org/en/about-us/pressroom/press-release/lr-and-umas-publish-techno-economic-assessment-of-zero-carbon-fuels/ (accessed on 10 December 2023).
8. Solakivi, T.; Paimander, A.; Ojala, L. Cost competitiveness of alternative maritime fuels in the new regulatory framework. *Transp. Res. Part D Transp. Environ.* **2022**, *113*, 103500. [CrossRef]
9. MAN. Official Website. Available online: https://www.man-es.com/docs/default-source/document-sync/man-v51-60g-eng.pdf?sfvrsn=3776bc19_1 (accessed on 10 December 2023).
10. Yao, S.; Li, C.; Wei, Y. Design and optimization of a zero-carbon emission system integrated with the utilization of marine engine waste heat and LNG cold energy for LNG-powered ships. *Appl. Therm. Eng.* **2023**, *231*, 120976. [CrossRef]
11. Díaz-Secades, L.A.; González, R.; Rivera, N. Waste heat recovery from marine engines and their limiting factors: Bibliometric analysis and further systematic review. *Clean. Energy Syst.* **2023**, *6*, 100083. [CrossRef]

12. Chountalas, T.D.; Founti, M.; Tsalavoutas, I. Evaluation of biofuel effect on performance & emissions of a 2-stroke marine diesel engine using on-board measurements. *Energy* **2023**, *278*, 127845. [CrossRef]

13. Schroer, M.; Panagakos, G.; Barfod, M.B. An evidence-based assessment of IMO's short-term measures for decarbonizing container shipping. *J. Clean. Prod.* **2022**, *363*, 132441. [CrossRef]

14. Balcombe, P.; Staffell, I.; Kerdan, I.G.; Speirs, J.F.; Brandon, N.P.; Hawkes, A.D. How can LNG-fueled ships meet decarbonization targets? An environmental and economic analysis. *Energy* **2021**, *227*, 120462. [CrossRef]

15. Langari, J.; Aliakbari, K.; Nejad, R.M. Failure analysis of CK45 steel crankshaft in light truck diesel engine. *Eng. Fail. Anal.* **2023**, *145*, 107045. [CrossRef]

16. Venkatachalam, G.; Kumaravel, A. Experimental Investigations on the Failure of Diesel Engine Piston. *Mater. Today Proc.* **2019**, *16*, 1196–1203. [CrossRef]

17. Lu, Y.; Zhang, X.; Xiang, P.; Dong, D. Analysis of thermal temperature fields and thermal stress under steady temperature field of diesel engine piston. *Appl. Therm. Eng.* **2017**, *113*, 796–812. [CrossRef]

18. Cabuk, A.S. Experimental IoT study on fault detection and preventive apparatus using Node-RED ship's main engine cooling water pump motor. *Eng. Fail. Anal.* **2022**, *138*, 106310. [CrossRef]

19. Fatigati, F.; Di Bartolomeo, M.; Cipollone, R. Development and experimental assessment of a Low Speed Sliding Rotary Vane Pump for heavy duty engine cooling systems. *Appl. Energy* **2022**, *327*, 120126. [CrossRef]

20. Di Giovine, G.; Mariani, L.; Di Battista, D.; Cipollone, R.; Fremondi, F. Modeling and experimental validation of a triple-screw pump for internal combustion engine cooling. *Appl. Therm. Eng.* **2021**, *199*, 117550. [CrossRef]

21. Woschni, G. *A Universally Applicable Equation for the Instantaneous Heat Transfer Coefficient in the Internal Combustion Engine*; SAE Technical Paper; SAE International: Warrendale, PA, USA, 1967; p. 670931.

22. Hohenberg, G.F. *Advanced Approaches for Heat Transfer Calculations*; SAE Technical Paper; SAE International: Warrendale, PA, USA, 1979; p. 790825. [CrossRef]

23. Annand, W.J.D. Heat Transfer in the Cylinders of Reciprocating Internal Combustion Engines. *Proc. Inst. Mech. Eng.* **1963**, *177*, 973–996. [CrossRef]

24. Sitkei, G.; Ramanaiah, G. *A Rational Approach for Calculation of Heat Transfer in Diesel Engines*; SAE Technical Paper; SAE International: Warrendale, PA, USA, 1972; p. 720027. [CrossRef]

25. Chang, J.; Güralp, O.; Filipi, Z.; Assanis, D.; Kuo, T.W.; Najt, P.; Rask, R. New Heat Transfer Correlation for an HCCI Engine Derived from Measurements of Instantaneous Surface Heat Flux. *SAE Trans.* **2004**, *113*, 1576–1593. [CrossRef]

26. Wu, Y.Y.; Chen, B.C.; Hsieh, F.C.; Ke, C.T. Heat transfer model for small-scale spark-ignition engines. *Int. J. Heat Mass Transf.* **2009**, *52*, 1875–1886. [CrossRef]

27. Rabeti, M.; Ranjbar, A.A.; Jahanian, O.; Ardebili, S.M.S.; Solmaz, H. Investigation of important semi-empirical heat transfer models for a natural gas fueled HCCI engine. *Energy Rep.* **2021**, *7*, 8652–8666. [CrossRef]

28. Depcik, C.; Alam, S.S.; Madani, S.; Ahlgren, N.; McDaniel, E.; Burugupally, S.P.; Hobeck, J.D. Determination of a heat transfer correlation for small internal combustion engines. *Appl. Therm. Eng.* **2023**, *228*, 120524. [CrossRef]

29. Hassan, M.A.S.M.; Razlan, Z.M.; Bakar, S.A.; Rahman, A.A.; Rojan, M.A.; Wan, W.K.; Ibrahim, Z.; Ishak, A.A.; Ridzuan, M.J.M. Derivation and validation of heat transfer model for Spark-Ignition engine cylinder head. *Appl. Therm. Eng.* **2023**, *225*, 120240. [CrossRef]

30. Petrichenko, R.M. Convective heat transfer in piston machines. In *Mechanical Engineering*; Leningrad Shipbuilding Institute: Leningrad, Russia, 1979.

31. Asadi, A.; Zhang, Y.; Mohammadi, H.; Khorand, H.; Rui, Z.; Doranehgard, M.H.; Bozorg, M.V. Combustion and emission characteristics of biomass derived biofuel, premixed in a diesel engine: A CFD study. *Renew. Energy* **2019**, *138*, 79–89. [CrossRef]

32. Choi, M.; Song, J.; Park, S. Modeling of the fuel injection and combustion process in a CNG direct injection engine. *Fuel* **2016**, *179*, 168–178. [CrossRef]

33. Merker, G.P.; Schwarz, C.; Stiesch, G.; Otto, F. *Simulating Combustion—Simulation of Combustion and Pollutant Formation for Engine-Development*; Springer: Berlin/Heidelberg, Germany, 2006; pp. 148–155.

34. Molodtsov, N.I.; Sokolov, P.V.; Vlasov, Y.Y. The influence of structural regulation indicators and augmentation level on heat release coefficient of gases to the surface of CPG parts. *Leningr. CDERI Proc.* **1975**, *69*, 3–17.

35. Molodtsov, N.I.; Sokolov, P.V.; Chirin, M.L. Experimental installations, methods, and results of studies of thermal parameters of diesel engines. *Leningr. CDERI Proc.* **1975**, *69*, 33–54.

36. Lebedevas, S.; Matievskij, D.D.; Lebedeva, G. Formation of power ranges for the use of a design range of pistons on Ch15/15 diesel engines. *Ser. Mech. Eng.* **2004**, *2*, 121–127. Available online: http://vestnikmach.ru/catalog/powgen/hidden/352.html (accessed on 10 December 2023).

37. Lebedevas, S.; Lebedeva, G. Mathematical model of combined parametrical analysis of Indicator process and thermal loading on the diesel engine piston. *Transp. Vilnius Tech.* **2004**, *19*, 108–118. [CrossRef]

38. Lebedevas, S.; Lebedeva, G.; Pikūnas, A.; Spruogis, B. A simultaneous parametric analysis of the in-cylinder processes for diesel engines. *Int. J. Veh. Syst.* **2011**, *18*, 18–45. [CrossRef]

39. Lebedevas, S.; Raslavičius, L.; Rapalis, P. Simulation of microalgae oil spray characteristics for mechanical fuel injection and CRDI systems. *Biomass Conv. Bioref.* **2022**. [CrossRef]

40. Lebedevas, S.; Čepaitis, T. Parametric Analysis of the Combustion Cycle of a Diesel Engine for Operation on Natural Gas. *Sustainability* **2021**, *13*, 2773. [CrossRef]
41. Kavtavaradze, R. Mathematical model of the complex heat exchange—Convection and radiation in the combustion chamber of the diesel engine. *Tech. Mech.* **1989**, *10*, 175–177.
42. Woschni, G. *Internal Combustion Engines*, 2nd ed.; TU Miunhen: Munich, Germany, 1988; 303p.
43. Dauksys, V.; Lebedevas, S. Comparative research of the dual-fuel engine exergy balance. *Energetics Vilnius Lith. Acad. Sci.* **2019**, *65*, 113–121.

Journal of
Marine Science and Engineering

Article

Optimization of Combustion Cycle Energy Efficiency and Exhaust Gas Emissions of Marine Dual-Fuel Engine by Intensifying Ammonia Injection

Martynas Drazdauskas * and Sergejus Lebedevas

Faculty of Marine Technology and Natural Sciences, Klaipeda University, 91225 Klaipėda, Lithuania; sergejus.lebedevas@ku.lt
* Correspondence: medienabaldams@gmail.com

Citation: Drazdauskas, M.; Lebedevas, S. Optimization of Combustion Cycle Energy Efficiency and Exhaust Gas Emissions of Marine Dual-Fuel Engine by Intensifying Ammonia Injection. *J. Mar. Sci. Eng.* **2024**, *12*, 309. https://doi.org/10.3390/jmse12020309

Academic Editor: Leszek Chybowski

Received: 31 December 2023
Revised: 4 February 2024
Accepted: 7 February 2024
Published: 9 February 2024

Correction Statement: This article has been republished with a minor change. The change does not affect the scientific content of the article and further details are available within the backmatter of the website version of this article.

Abstract: The capability of operational marine diesel engines to adapt to renewable and low-carbon fuels is considered one of the most influential methods for decarbonizing maritime transport. In the medium and long term, ammonia is positively valued among renewable and low-carbon fuels in the marine transport sector because its chemical elemental composition does not contain carbon atoms which lead to the formation of CO_2 emissions during fuel combustion in the cylinder. However, there are number of problematic aspects to using ammonia in diesel engines (DE): in-tensive formation of GHG component N_2O; formation of toxic NO_x emissions; and unburnt toxic NH_3 slip to the exhaust system. The aim of this research was to evaluate the changes in combustion cycle parameters and exhaust gas emissions of a medium-speed Wartsila 6L46 marine diesel engine operating with ammonia, while optimizing ammonia injection intensity within the limits of P_{max}, T_{max}, and minimal engine structural changes. The high-pressure dual-fuel (HPDF) injection strategy for the D5/A95 dual-fuel ratio (5% diesel and 95% ammonia by energy value) was investigated within the liquid ammonia injection pressure range of 500 to 2000 bar at the identified optimal injection phases (A $-10°$ CAD and D $-3°$ CAD TDC). Increasing ammonia injection pressure from 500 bar (corresponding to diesel injection pressure) in the range of 800–2000 bar determines the single-phase heat release characteristic (HRC). Combustion duration decreases from 90° crank angle degrees (CAD) at D100 to 20–30° CAD, while indicative thermal efficiency (ITE) increases by ~4.6%. The physical cyclic deNO$_x$ process of NO_x reduction was identified, and its efficiency was evaluated in relation to ammonia injection pressure by relating the dynamics of NO_x formation to local combustion temperature field structure. The optimal ammonia injection pressure was found to be 1000 bar, based on combustion cycle parameters (ITE, P_{max}, and T_{max}) and exhaust gas emissions (NO_x, NH_3, and GHG). GHG emissions in a CO_2 equivalent were reduced by 24% when ammonia injection pressure was increased from 500 bar to 1000 bar. For comparison, GHG emissions were also reduced by 45%, compared to the diesel combustion cycle.

Keywords: marine engine; ammonia; dual fuel; combustion cycle; HRC; combustion optimization; thermal efficiency; GHG emissions

1. Introduction

In 2021, the maritime transport sector of the European Union accounted for 3–4% of all EU CO_2 emissions, which are one of the main components of greenhouse gas emissions [1]. To reduce CO_2, IMO purposefully introduced regulatory measures for newly built and operated ships [2,3]. In 2011, the Energy Efficiency Design Index (EEDI) was introduced for newly built ships, increasing energy efficiency through technological solutions while reducing emissions. Since 2023-01, ships in operation over 400 GT are indexed according to the Energy Efficiency Existing Ship Index (EEXI) and must meet the minimum require-ments of the energy efficiency standard. Moreover, ships in operation over 5000 GT are obliged to collect and declare data of efficient energy use, and in case of insufficient energy efficiency assessment (CII), owners have to take corrective actions [3]. Addressing the climate change issue, IMO updated their GHG reduction strategy in 2023, and aim to reduce GHG emissions by 70–80% by 2040 compared to 2008 levels, and to reach zero GHG by 2050 [4]. In parallel, the EU has also set long-term targets for maritime transport to reduce GHG emissions by 90% by 2050 compared to 1990 levels according to SEC (2021) 562 directive [1]. To achieve the set targets, a significant reduction of CO_2 emissions in maritime transport sector is required, mainly by increasing energy efficiency use and using renewable and low-carbon fuel (hydrogen, ammonia). Logistic, hydrodynamic, and technological measures can potentially reduce GHG emissions from ships by 5–20% using renewable and low-carbon-dioxide-generating fuel—up to 100%, according to DNV [5]. In 2023-07, the EU adopted a regulation on additional measures related to the use of renewable and low-carbon fuels (FuelEU Maritime) [1]. The envisaged measures will ensure that the greenhouse gas intensity of the fuel used in the maritime transport sector gradually decreases over time from 2% in 2025 to 80% in 2050.

Different evaluations have found that existing IMO, EEDI, and EEXI (CII) measures are insufficient to achieve ambitious EU and IMO targets. It is expected that the use of zero or almost zero GHG technologies and fuel in international fleets will reach at least 5% by 2030, according to the 2023 IMO strategy plan [4]. The average operational age of ships in operation is 21.1 years in the economies of developed countries, and 28.6 years in the economies of developing countries [6]. Recently, on average, 79% of newly built ships annually choose traditional fuel, while 98.4% of ships in operation use petroleum fuel [7]. After evaluating the facts, it becomes obvious that the achievement of EU and IMO GHG reduction targets is related to retrofitting of newly built ships and operating ship power plants with renewable and low-carbon fuels such as ammonia.

Ammonia in shipping among renewable and low-carbon fuel species is assessed prospectively for the short-term 2025–2030 and the long-term 2050 period. Ammonia stores 50% more energy by volume than hydrogen [8], which is important considering limited fuel tank volume on board. Ammonia is also valued positively from a decarbonization perspective, as its chemical elemental composition does not contain carbon atoms which lead to the formation of CO_2 emissions when fuel is burned. However, there are a number of problematic aspects to using ammonia in maritime transport: due to nitrogen atoms in the chemical composition of ammonia, NO_x emissions are more intensively formed during combustion than compared to diesel [9]. In addition, a fraction of unburnt NH_3 emissions slip to the exhaust system during exhaust stroke. Ammonia emissions contribute to air pollution and can have detrimental effects on ecosystems, including acidification of soil and water bodies, which can harm plant and aquatic life. Moreover, ammonia emissions can lead to various human health issues. When individuals are exposed to elevated levels of ammonia over time or in concentrated forms, it can cause irritation and inflammation of the respiratory tract. Therefore, a solution to reduce these emissions is a priority in order to use ammonia in DE. Using ammonia in DE also requires solutions due to its unfavorable physical properties. In particular, ammonia combustion characteristics differ significantly from those of petroleum-based fuels and most renewable fuels (see Table 1). Due to a low fuel cetane number (5–7 units) and high auto-ignition temperature (650 °C), ammonia ignition is possible at an engine compression ratio of 35:1 and more, which is unrealistic to achieve considering the geometrical parameters of marine engines [10]. Therefore, to ensure ammonia combustion in diesel engines whose compression ratio usually reaches up to 20:1, pilot fuel with good auto-ignition characteristics is required (diesel, biodiesel).

Table 1. Comparison of fuel physical properties [9–11].

Fuel Property	Ammonia	Hydrogen	Methanol	LNG	Diesel
Formula	NH_3	H_2	CH_3OH	CH_4	$C_{12}H_{23}$
Density when liquified, kg/m^3	602.8	70.8	786.3	430	832
Calorific value, MJ/kg	18.8	120	19.7	38.1	42.7
Octane number	110	130	113	107	30
Cetane number	5–7	-	5–8	-	40–55
Ignition temperature, °C	651	585	385	540	254–285
Laminar flame speed, m/s	0.07–0.14	2.70	0.50	0.38	0.87
Stoichiometric air fuel ratio	6.06	34.32	6.45	17.2	14.5
Heat of vaporization, kJ/kg	1370	461	1103	510	232

Ammonia has a rapid and even kinetic combustion phase due to an efficiently mixed and evenly distributed ammonia and air mixture. On the other hand, it has more intense heat release compared to diesel, due to the kinetic ammonia combustion phase which leads to an increased maximum cyclic pressure (P_{max}). For example, with dual-fuel ratio D39/A61 (diesel 39% and ammonia 61%) [12], maximum cycle pressure reaches 86.2 bar when D100—76.3 bar. This has a negative impact on mechanical and thermal piston-rod group part loading, which reduces engine reliability. Reviewed studies in the literature are divided into two categories of dual ammonia diesel fuel injection into the cylinder: low-pressure dual-fuel strategy (LPDF), when ammonia in gaseous phase is introduced into the cylinder through the gas valve at low 2–3 bar pressure together with compressed air to intake manifold; and high-pressure dual-fuel strategy (HPDF), when ammonia in liquid phase is directly injected into the cylinder at high pressure through a separate fuel injector. Studies show that in both cases, engine ITE is close to that of a diesel engine, and with a fuel ratio of D20/A80 (diesel 20%, ammonia 80%) with LPDF strategy, ITE increases by 3.5% [13]. Nadimi et al. [12] have also found that higher engine ITE is achieved using ammonia, and that the difference in ITE compared to diesel engine mode is 17% (D100 (100% diesel) ITE—32%; D16/A84 ITE—37.6%). Indicative thermal efficiency increases when the engine is running on ammonia using LPDF strategy due to several reasons. According to the authors of [12], first of all, ammonia tends to have an ignition delay due to its high octane number and high autoignition temperature. Therefore, when transitioning to ammonia, an advanced start of diesel injection is necessary to achieve heat release characteristics similar to a diesel engine. Due to the advanced start of diesel injection, diesel has enough time to be evenly distributed in the combustion chamber. This results in a short and intense homogeneous heat release of the ammonia-diesel mixture. As a result, a lower combustion cycle temperature is achieved, leading to reduced heat losses through the cylinder walls to the cooling system. Heat loss decreased from 320 J/cycle when the engine was running only on diesel to 240 J/cycle at the fuel ratio A84/D16, when engine work did not change [12]. Also, less heat was lost through exhaust system as the exhaust gas temperature at the fuel ratio A84/D16 decreased by 132 °C compared to D100 [12]. Therefore, the distribution of the engine's heat balance components towards an increase in thermal efficiency is taking place. Aaron Reiter et al. [9] determined the highest ammonia diesel dual-fuel ratio at D5/A95 by energy value using LPDF strategy. However, engine fuel efficiency and ITE results under these conditions were very low, with ITE reaching 18.9%. Therefore, according to Aaron Reiter et al., this mode of engine operation is not rational. Nadimi et al. [12] used a wide range of ammonia ratios (0–84%) in a dual-fuel balance using LPDF strategy. However, at a higher percentage of ammonia, according to the authors, the engine lost its starting properties and did not start. Numerical studies by Tie Li and Xinyi Zhou et al. [13] also showed that at 90% ammonia in the dual-fuel balance, the combustion process became unstable when LPDF strategy was applied, and the mass fraction of unburnt NH_3 in the exhaust system increased more than six times. As a result, using LPDF strategy, the share of ammonia in the dual-fuel balance is limited to 80–84%. On the contrary, with HPDF strategy, the optimal proportion of ammonia in the dual-fuel balance is 95–97% [13,14]. In Tie Li and Xinyi Zhou et al., numerical studies [13] using HPDF strategy at the fuel ratio A97/D3 ITE practically did not change, and reached 45.3% while at D100—45.4%. Using LPDF strategy and the ratio, A80/D20 ITE reached 47.0% [13]. The increase of ITE is associated with lower heat losses to the cooling system due to the reduced interaction between the flame and cylinder walls close to top dead center (TDC) [13].

In addition to positive increase of ITE, GHG harmful components and CO_2 emissions are also higher with LPDF strategy. Using LPDF, more CO_2 emissions are released during the combustion cycle than compared to HPDF, as LPDF has a relatively large share of pilot fuel (diesel) in the dual-fuel balance ~20%. The amount of released CO_2 emissions during combustion depends solely on injected diesel mass [9,13]. It was also observed that N_2O, one of the GHG components, decreases in parallel with the increase of ammonia ratio in the dual-fuel balance. In the literature [12,15], it is hypothesized that during ammonia combustion, N_2O from the elemental chemical composition of ammonia is formed mainly in low-temperature zones during expansion stroke when NH_3 stuck in the gap between piston crown and cylinder liner turns into NH_2 and reacts with NO_2. This means that unburnt NH_3 and N_2O emissions correlate [15]. Numerical studies by Tie Li and Xinyi Zhou et al. [16] evaluated the differences between low and high dual-fuel injection strategies. NO_x emissions were found to be on average three times higher with LPDF than with HPDF strategy. Meanwhile, unburnt NH_3 emissions using HPDF injection strategy reached ~0.02 mg/kWh, while LPDF resulted 10–480 mg/kWh, depending on the fuel injection start angle. In continued numerical studies, Tie Li and Xinyi Zhou et al. [13], at fuel ratio A97/D3 using HPDF strategy, recorded NO_x emission levels approximately four times lower than at the fuel ratio A80/D20 with LPDF, and NH_3 emissions were also up to seven times lower. The reduction of NO_x emissions is associated with thermal deNO_x process of nitrogen oxides, during which active NH_2 radicals react with NO to form $N_2 + OH$ at 1000–1400 K cylinder temperatures [17]. The reduction of NH_3 emissions is attributed to more efficient combustion characteristics due to liquid ammonia penetration to the pilot fuel spray flame zone [13]. As a result, HPDF compared to LPDF injection strategy does not make a significant difference in terms of ITE, but in terms of emissions, HPDF injection strategy emits less NH_3, NO_x, and CO_2 emissions in all cases. HPDF injection strategy also has the potential to reduce CO_2 and NO_x emissions compared to a diesel engine.

One of the important aspects of ongoing research is the arrangement of ammonia and diesel injector nozzle holes. The literature analysis shows that arrangement angle of diesel and ammonia nozzle holes has no significant effect on ITE, but ecological indicators differ. Tie Li and Xinyi Zhou et al. [16] found that, when ammonia and diesel fuel injector nozzle holes are overlapped ($0°$ angle), a more efficient combustion process takes place. As a result, ammonia penetrates more efficiently into the pilot fuel combustion zone from the start of injection, and the induction period (from the start of injection till combustion) is shorter, and due to which the emission level of NO_x and NH_3 is lower. Meanwhile, ITE practically did not change and reached 51.6% when nozzle holes were overlapping, and 51.8% when nozzle holes were separated [16]. Valentin Scharl and Thomas Sattelmayer et al. [14], in their experimental studies of ammonia and diesel nozzle holes' arrangement influence on the induction period using HPDF injection strategy (2000 bar), also determined the optimal ($0–7.5°$) hole overlap range at which the shortest induction period and the most efficient fuel combustion, in terms of unburnt NH_3, were observed.

In summary, direct diesel engine transition to ammonia is limited due to ammonia's unfavorable physical characteristics, specifically, high exhaust gas emissions. Therefore, solutions to improve energy efficiency and reduce exhaust gas emissions while the engine is operating with ammonia is necessary. The optimization of the combustion cycle primarily involves adjusting fuel injection pressure, injection phase, and duration. The evolution of trends in modern diesel engines is grounded in numerous theoretical and experimental studies and justifications. Therefore, the systematic solutions they offer, including the identification of the main optimized parameters and their determining factors, are also rational in the case of ammonia use. Since the period of strategic thermal efficiency parameters increase for diesel engines, studies [18–21] have shown that under the condition P_{max} = const, heat release forming in the diesel engine cylinder practically does not affect ITE. The main factor influencing ITE is the heat release process duration. Based on these principles, the optimization of DE combustion cycle towards reducing the combustion duration primarily involves increasing fuel injection pressure. The increase in maximum combustion pressure is constrained by adjusting fuel injection timing towards TDC while simultaneously raising compression ratio. This approach was executed in MTU 396 series engines. Increased injection pressure was matched by adjusting the start of fuel injection to 4° CAD before TDC, and an increase in compression ratio (CR) from 15 to 17.8. Consequently, a 20% reduction in heat release duration was achieved, which resulted in improved fuel efficiency and NO_x reduction by 35% [21]. Thus, one of the most effective ways to influence heat release intensity is to increase fuel injection pressure. At the same time, increasing heat release intensity reduces combustion duration and increases combustion cycle dynamics and ITE. In parallel, adjusting the start of the injection angle closer to TDC, together with fuel injector design and parameters optimization, allows to improve ITE without the exceeded P_{max} limitation. Since the 1990s, a trend for optimizing the combustion process of market-leading diesel engines has emerged. During this period, studies were conducted on the influence of fuel injection pressure on fuel ignition and combustion dynamics [22–25]. Additionally, the ACE Company and the Japan Automotive Research Institute compared the effects of the duration and intensity of the initial fuel injection stage on air swirl parameters. Furthermore, comprehensive improvements were made to exhaust gas toxicity indicators by companies such as MTU, YaMZ, and Fev Motorentechnik GmbH & Co., KG [22,24,26].

The application of combustion cycle optimization trends, such as advancing the start of pilot fuel injection, changing injection rate and pressure, and organizing multi-stage injection, have enabled market-leading companies (Wärtsilä, MAN B&W, Caterpillar, etc.) to develop dual-fuel engines with LNG. These advancements have allowed them to achieve thermal efficiency similar to diesel engines and reduce PM and NO_x emissions by up to 90% compared to diesel engines [27–29]. Therefore, to increase ITE and reduce emissions by optimizing combustion cycle parameters, this research is based on fuel injection intensification by increasing injection pressure.

Considering the wide variety of diesel engines and models of ship power plants in operation, it is rational to base marine transport sector decarbonization with engine retrofitting based on numerical studies to reduce time and financial costs. Engine retrofitting for operation with other types of fuel by numerical methods is basically related to research and optimization of combustion cycle characteristics. Numerical research tasks for the ship's main propulsion diesel engine's operation on ammonia fuel are based on multi-zone mathematical models (MM), for example, using simulation software "AVL FIRE M". The use of multi-zone MM allows us to study combustion cycle physical processes with sufficient accuracy for solving practical problems. Klaipeda University conducted comprehensive marine diesel engine decarbonization research, including solutions for the rational use of renewable and low-carbon fuels [30–32], the use of secondary heat sources in engine cogeneration cycle [33], etc.

In this article, the research of ammonia combustion was performed to identify physical combustion process conditions and to provide rational technological solutions for ammonia applicability in ship power plants. The novelty of this article lies in the organization of combustion process. Marine engines operate under different conditions compared to automotive engines. One significant difference is in how the fuel interacts with the engine components. Unlike automotive engines, where the fuel film along the cylinder walls is common, marine engines are designed to avoid direct contact between the fuel jet and the cylinder walls. This difference in design and operation significantly alters the combustion characteristics of marine engines. Therefore, the purpose of this article is to evaluate the changes in combustion cycle parameters and exhaust gas emissions for the ship's propulsion diesel engine operating on ammonia, and to determine the limits of combustion cycle regulation parameters (NH_3 injection intensity when increasing the injection pressure). At the time of writing, the authors are not aware of any similar published articles, making this approach unique in optimizing the ammonia combustion cycle. The presented research approach could provide valuable insights for combustion cycle optimization during the transition of marine diesel engines to ammonia, requiring minimal changes to the engine structure. In addition, the NH_3 combustion cycle optimization strategy by injection intensification is related to optimization of the pilot diesel and ammonia injection phases, and is planned for continuous studies.

2. Materials and Methods

2.1. Research Object

A marine medium-speed four-stroke Wartsila 6L46 diesel engine (Wartsila: Helsinki, Finland) was selected as the research object. Medium-speed (300–1000 rpm) four-stroke diesel engines are widespread in ship propulsion systems, especially in smaller cargo ships, as well as in larger specialized ships, such as cruise ships, ferries, and ro-ro cargo ships [34]. Medium-speed four-stroke diesel engines' popularity is due to their higher power-to-weight and power-to-space ratio, easier periodic maintenance, and acquisition costs compared to low-speed two-stroke diesel engines [35]. Market-leading engine manufacturers' (BERGEN, WARTSILA, MAN) medium-speed four-stroke diesel engines thermal efficiency reach 46–49% when emission level meets IMO Tier II regulation, while using SCR technology meets IMO Tier III [36–38]. Research object selection and simulation model verification were carried out according to real DE data obtained during operation to bring research results as close as possible to practical application for maritime transport decarbonization. Table 2 presents the main research object structural parameters.

Table 2. Wartsila 6L46 engine data.

Parameter	Data
Bore, mm	460
Stroke, mm	580
Connecting rod length, mm	650
Compression ratio	12.5
Engine speed, rpm	500
Compressed air pressure at inlet valve close, bar	3.43
Compressed air temperature at inlet valve close, K	372
Inlet valve closing angle, CAD	120° BTDC
Exhaust valve closing angle, CAD	128° ATDC
The total number of injection holes for ammonia and diesel injectors	10
The location of ammonia and diesel injectors	Centre
Piston surface type	Bowl-In Piston

2.2. Research Strategy

Guidelines for the Wartsila 6L46 engine model creation to run on ammonia for numerical studies using AVL FIRE M simulation software (version 2022 R2) is based on findings of conducted research in the literature. Due to lower concentrations of NO_x and NH_3 emissions during combustion cycle, ammonia fuel injection is organized in liquid phase. Due to greater CO_2 emission reduction effect, the selected dual-fuel ratio of diesel and ammonia is D5/A95 (5% diesel and 95% ammonia according to energy value). To ensure the shortest induction period and the most effective combustion, the diesel and ammonia injectors' nozzle holes are overlapped (0° angle). Simulations were performed with the same amount of heat input. Inlet air pressure and temperature were unchanged for D100 and D5/A95. Initial simulation data are presented in Table 3.

Table 3. Initial data used for simulation.

Parameter	Diesel (D100)	Ammonia + Diesel (D5/A95)
Start of injection, CAD	710°	710 NH$_3$; 717 Pilot
Diesel injection duration, CAD	26°	3
Ammonia injection duration, CAD	-	26°
Injected mass (Diesel), g	1.87	0.1037
Injected mass (Ammonia), g	-	4.016
Injection pressure (Diesel), bar	500	500
Injection pressure (Ammonia), bar	-	500–2000
Diesel calorific value, MJ/kg	42.5	
Ammonia calorific value, MJ/kg	18.8	

2.3. Mathematical Model and Verification

Numerical mathematical studies of combustion cycle characteristics were performed by transferring a ship's propulsion main diesel engine Wartsila 6L46 (B/S = 460/580 mm, 500 rpm, 4 stroke, 6 cylinders, 6300 kW) to ammonia operation. Engine combustion chamber mesh for simulation was created using "AVL FIRE ESE DIESEL" software (version 2021 R1). To reduce simulation time, it was decided to perform the simulation for 1/5 part of the total combustion chamber (360°/5 = 72). Despite the fact that the simulation was performed on a 1/5 combustion chamber, emission results are presented for a full engine, i.e., six cylinders. Total fuel injection nozzle quantity for full chamber is 10; therefore, for 1/5 part, the injection nozzle quantity was set to 2. Injected ammonia and diesel mass was reduced to 1/5 part. The diesel and ammonia injection nozzles spray cones are overlapped, meaning 0° angle distribution. An example of nozzle distribution is presented in Figure 1a. The total number of cells was 97,076, average cell size was 3.80 mm, number of boundary layers was 2, thickness of boundary layers was 0.60 mm, and number of subdivisions in angular direction was 50 (Figure 1b).

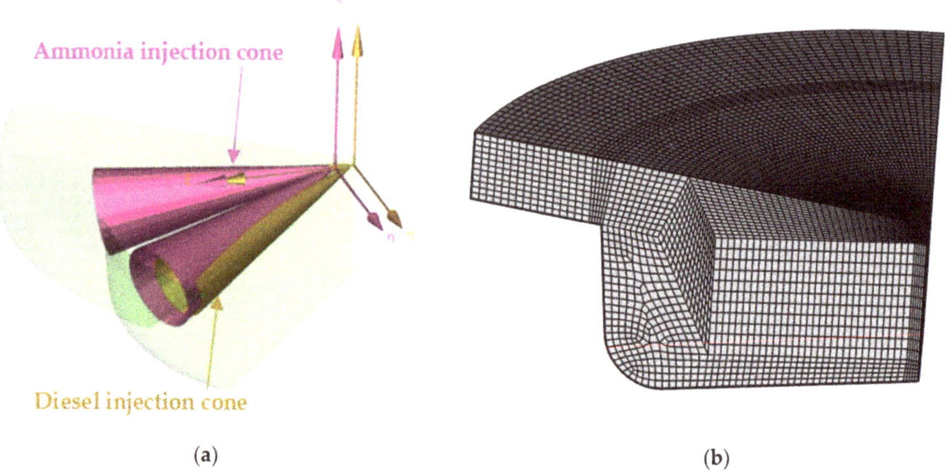

Figure 1. (a) Injector nozzle hole arrangement (0° angle) inside combustion chamber sector (pink—NH$_3$, gold—diesel); (b) 1/5 (72°) combustion chamber sector mesh.

Numerical studies were conducted using the multi-zone mathematical model developed by AVL company, specifically designed to investigate the physical processes involved in the ammonia combustion cycle. The gas phase reaction model, combining combustion and emission models, is capable of solving diesel and ammonia combustion reactions. General gas phase reaction model includes H, O, C, N, HE, and AR chemical elements and 54 numbers of species. The model is based on P. Glarborg's methodology for modeling nitrogen chemistry in combustion [39]. The experimental verification of the model was conducted by AVL company. Since combustion model is not yet commercially publicly available by AVL decision, the description of the general gas phase reaction model is not provided.

The spray module for calculating droplets in the simulation region uses the Lagrangian approach [40]. The droplets are tracked in a Lagrangian way through the computational grid used for solving the gas phase partial differential equations. Additional spray sub-models, such as the the Schiller–Naumann injection drag model [41] and Abramzon–Sirignano evaporation model [42], were selected for this simulation.

AVL FIRE M simulation software focuses on detailed analysis of physico-chemical processes taking place in the cylinder. However, it does not provide the combustion cycle (IMEP, BSFC, P_i, ITE) parameters. Therefore, these parameters are calculated from the array of AVL FIRE M simulation results according to given formulas.

Indicated mean effective pressure, IMEP, is calculated according to trapezoidal approximation Formula (1) [43]:

$$\text{IMEP} = \frac{1}{V_{cyl}} \sum_{k=\theta_0}^{\theta_f} \frac{P_{k+1} + P_k}{2} \cdot (V_{k+1} - V_k) \tag{1}$$

where V_{cyl} is cylinder volume [m^3]; P_k and P_{k+1} are consecutive cylinder pressure readings [bar]; V_k and V_{k+1} are cylinder volume measurements corresponding to P_k and P_{k+1} [m^3]; and the sum in increments of k from a crank angle degree value θ_0 to a value θ_f are calculated.

Indicated power P_i is calculated as per below Formula (2) [44]:

$$P_i = \frac{\text{IMEP} \cdot V_{cyl} \cdot n \cdot N}{z \cdot 60} \tag{2}$$

where n is cylinders number; N is engine speed [rpm]; and z is coefficient (z = 1 for 2-stroke engines, z = 2 for 4-stroke engines).

Specific fuel consumption, BSFC, is calculated according to the presented Formula (3) [44]:

$$\text{BSFC} = \frac{m_f}{P_i} \tag{3}$$

where m_f is fuel consumption [kg/h].

Engine indicative thermal efficiency, ITE, is calculated as per the below Formula (4) [44]:

$$\text{ITE} = \frac{3600 \cdot P_i}{m_f \cdot H_u} \tag{4}$$

where H_u is calorific value of kilogram fuel [kJ/kg].

An indicator diagram was measured for all six cylinders and averaged when the engine was running on D100 at 75% load. The combustion cycle parameters (IMEP, BSFC, P_i, P_{max}, ITE) were calculated from the indicator diagram using Formulas (1)–(4). The Wartsila 6L46 simulation model was created using AVL FIRE M simulation software. The simulation was performed when the engine was running on D100 at 75% load for model verification. A simulation indicator diagram was matched with the real operating engine indicator diagram for the combustion cycle from $-120°$ (intake valve closing, corresponding to $600°$ in the software) to $+128°$ (exhaust valve opening, corresponding to $848°$ in the software) crankshaft rotation angles when TDC was at $720°$ (Figure 2). The error of simulation meant indicative pressure compared to real engine value reached 2.4%. Indicator combustion cycle thermodynamic temperature and emission (CO_2, N_2O, CH_4, NO_x, NH_3) diagrams were calculated in parallel. A verified diesel engine combustion cycle simulation model was considered as the base engine operating mode for this research, and will be used for further comparison.

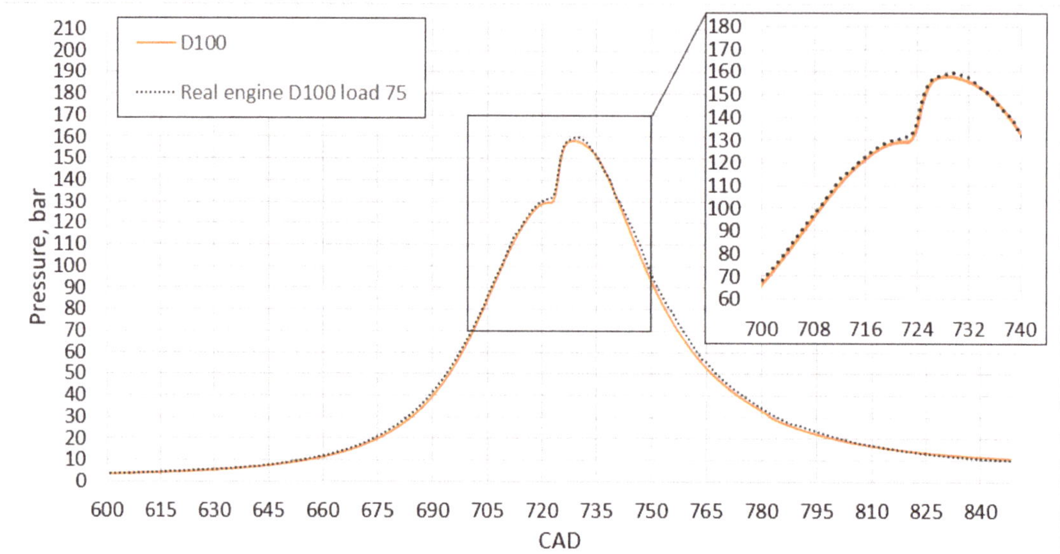

Figure 2. Verification of simulation indicator diagram with real operating engine indicator diagram.

Ammonia combustion and emission model validation through experiment was not feasible on operating the marine diesel engine. Therefore, the model validation relies on similar simulation results found in the literature [13,16]. In the literature, studies primarily focus on examining ammonia utilization technologies for relatively small cylinder diameter high-speed engines for automotive applications, with the exception of [16]. There is a notable absence of information regarding the application of ammonia to medium- and low-speed marine engines, despite ammonia being regarded as one of the primary alternative fuels in this sector. To justify the verification of the mathematical model, a comparative analysis of obtained results was performed with two articles which were discovered in the literature involving diesel engines running on ammonia using HPDF strategy and a similar dual-fuel ratio alongside ammonia injection pressure. A direct comparison of the combustion cycle parameters and exhaust gas emissions of this research to the literature would provide inaccurate results due to differences in engine types and combustion cycle organization. Thus, a comparison in the relative change in parameters compared to the diesel combustion cycle was selected. A detailed comparison of these parameters is provided in Table 4. In Case 1, ammonia injection pressure is set at 500 bar for the presented simulation results and 600 bar for simulation results in the literature [13], while Case 2 involves a 1000 bar ammonia injection pressure for both the present and the literature results [16]. Despite differences in engine speed and fuel start of injection angles, ammonia combustion in both cases resulted in a similar ignition delay. Moreover, the heat release rate matches with the literature as it progresses towards the expansion stroke in a similar manner. P_{max} also exhibits good agreement with the literature at 500 and 1000 bar ammonia injection pressures. For example, for Case 1 P_{max} decreases by 17% and 8%, while for Case 2, it increases by 9% and 10%. T_{max} mostly corresponds to the literature, while thermodynamic temperature (T) at 60° CAD after TDC shows a decreasing trend. Furthermore, exhaust gas emissions correspond to the literature in the same direction. In conclusion, ammonia combustion and emission model provide similar relative changes in combustion cycle parameters and exhaust gas emissions compared to the literature.

Table 4. Comparison of dual ammonia–diesel fuel combustion cycle research results with results from the literature review [13,16].

	Case 1		Case 2	
Parameter	**P_{inj} 500 bar**	**Li, T. et al. [13]**	**P_{inj} 1000 bar**	**Li, T. et al. [16]**
Engine type	4-stroke	4-stroke	4-stroke	2-stroke
Bore, mm	460	95	460	340
Stroke, mm	580	102	580	1600
Engine speed, RPM	500	1000	500	157
Dual-fuel ratio	D5/A95	D3/A97	D5/A95	D3/A97
Diesel injector nozzle hole number	10	8	10	4
Ammonia injector nozzle hole number	10	8	10	4
Diesel and ammonia injector nozzle hole angle	0°	0°	0°	0°
Diesel injection pressure, bar	500	600	500	200
Ammonia injection pressure, bar	500	600	1000	1000
Start of diesel injection, CAD	−3° TDC	−8° TDC	−3° TDC	−4° TDC
Start of ammonia injection, CAD	−10° TDC	−5° TDC	−10° TDC	−2° TDC
Ammonia ignition delay after diesel ignition, CAD	7°	4°	4°	1°
* P_{max}, %	>17	>8	<9	<10
* T_{max}, %	>2	>7	<10	0
* T at 60° CAD ATDC, %	>5	>7	>7	>20
* ITE, %	<2.2	0	<4.6	>1.1
* CO_2, %	>94	>96	>94	>96
* NO_x, %	>72	>50	>5	>46
N_2O, g/kWh	from 0.003 to 1.67	from 0 to 25 ppm	from 0.003 to 1.22	from 0.0007 to 0.0016
Unburned NH_3, g/kWh	from 0.011 to 8.94	from 0 to 130 ppm	from 0.011 to 1.51	N/A

* Relative change in parameters compared to diesel (D100) combustion cycle results. Symbol '<' refers to increase and '>' refers to decrease. N/A: value was too small to determine.

3. Results

3.1. Combustion Cycle Parameters

Several studies with ammonia used in DE are associated with poor combustion characteristics and therefore high emissions of unburnt NH_3 [9,10,12,13,15,16] Ammonia injection pressure ranges of 500–2000 bar are chosen for numerical studies to improve combustion characteristics and evaluate the dependence of harmful substances on fuel injection pressure. In this case, the start of injection for ammonia is constant at 710° CAD, while for pilot diesel it is 717° CAD (720° = TDC). Result analysis shows (see Figure 3) that diesel induction period is 4° CAD (observed from start of fuel injection until the first visible increase in heat release). Meanwhile, the ammonia induction period (represented by the second peak) lasts 13°, 14°, 15°, 16°, and 18° CAD at 2000, 1500, 1000, 800, and 500 bar, respectively. Increasing ammonia injection pressure shortens the induction period during which fuel is mixed with air and vaporized. When evaluating the induction period, it is useful to separate the diesel combustion and subsequent ammonia combustion, as the latter leads to a delay in the heat release characteristic. At injection pressures of 2000–800 bar, ammonia combustion delay after ignition of pilot fuel at 721° CAD is 2–5° CAD, while at 500 bar the delay reaches 7° CAD. In this case, a significant difference in combustion phases was determined from the differential heat release characteristic due to a delay in ammonia ignition. The first increase in heat release rate chart (Figure 3), especially at 500 bar ammonia injection pressure, is associated with the diesel combustion phase, and the second one with ammonia. As a result, the heat release characteristic became double phase. Delayed ignition of ammonia also leads to structural changes of the combustion cycle parameters and exhaust gas emissions. The primary increase in P_{max} resulting from the increase in ammonia injection pressure is fundamentally related to structural changes in the ammonia jet. Specifically, the fineness and uniformity of ammonia droplets increase [45,46]. Consequently, the evaporation time of the droplets is shorter, leading to a reduction in the induction period. Additionally, due to the uniformity of the droplets in the jet (i.e., a narrower range of differences in droplet diameters), when the flame covers a large volume of the jet, it triggers a much more intense heat release. On the other hand, a shorter induction period advances the ammonia combustion start phase to earlier angles, while maintaining the same diesel ignition phase (Figure 3). As a result, before the phase change of P_{max} at ~720° CAD, the amount of heat released after TDC, which determines P_{max} and corresponding T_{max} values, increases intensively. As a result (Figure 4), using ammonia in DE increases maximum cyclic temperature (T_{max}) in all cases. An increase in maximum cyclic pressure and temperature has a negative impact on the mechanical and thermal loading of the piston-rod group and, as a result, on engine reliability. Therefore, P_{max} and T_{max} are limited to research object design P_{max}—160 bar and T_{max}—1566 K, considering the D100 combustion cycle parameters. In the presented indicator diagram (Figure 5), P_{max} increases to 194 bar at 2000 bar injection pressure, but decreases to 132 bar at 500 bar pressure. Correspondingly, at 2000 bar injection pressure in the thermodynamic cycle temperature chart (Figure 4), T_{max} reaches 1779 K, while at 500 bar—1540 K. A summary evaluation of P_{max}, T_{max} and ITE parameters in relation to ammonia injection pressure and D100 cycle is shown in Figure 6.

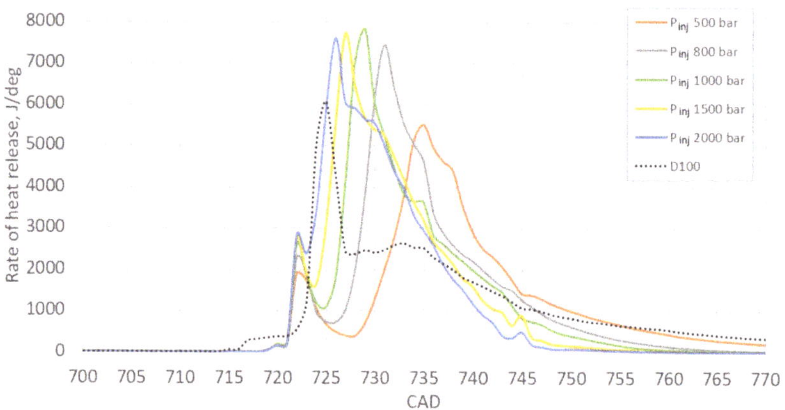

Figure 3. Heat release rate of ammonia injection pressure variations.

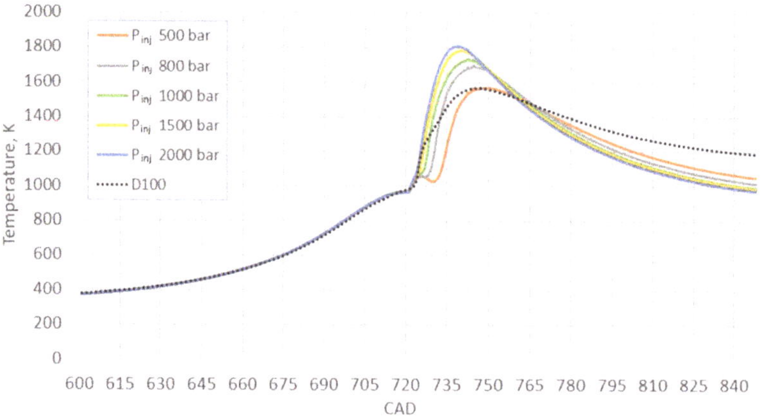

Figure 4. Temperature diagram of ammonia injection pressure variations.

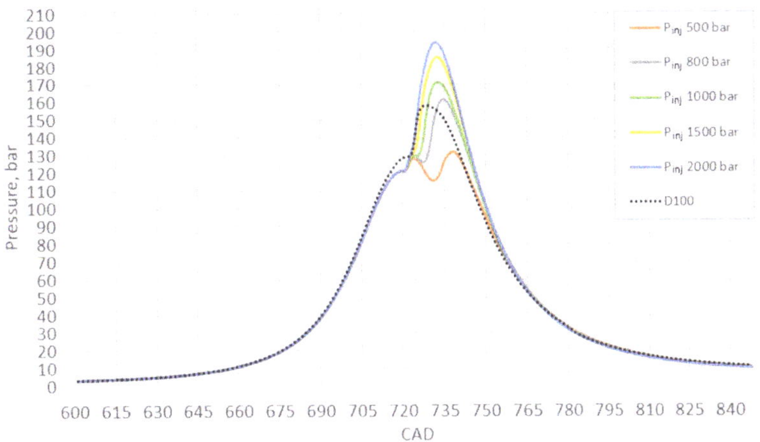

Figure 5. Indicator diagram of ammonia injection pressure variations.

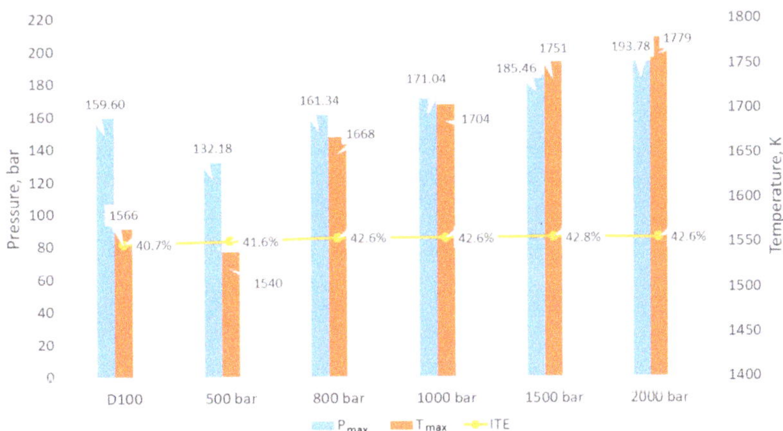

Figure 6. Comparison of P_{max}, T_{max}, and ITE variations of ammonia injection pressure with D100.

When comparing combustion cycle results at various ammonia injection pressures, it was found that ammonia injection pressure has no significant effect on indicative thermal efficiency. Only at 500 bar injection pressure is there a noticeable difference. Achieved ITE is 2.4% lower than at the remaining injection pressures. The reason was the delayed combustion in expansion stroke, at which the cylinder wall area increased and, accordingly, the heat loss to the cooling system increased. This can be validated by evaluating the integral of heat release (Figure 7). At 500 bar ammonia injection pressure, combustion duration reached ~60° CAD according to 95% of total heat release introduced with the fuel (corresponding to the limit of 75,000 J). The decrease in ITE was also influenced by a decrease in P_{max} and T_{max}, which determined the low combustion efficiency, which is determined by increase of unburnt NH_3 emissions. As a result, it was found that part of the heat introduced with the fuel was not released by evaluating the integral of heat release characteristic results at the end of the cycle. Meanwhile, gradually increasing the injection pressure above 800 bar results in a short combustion duration of 20–30° CAD, while at D100, combustion duration is ~90° CAD. Therefore, gradually increasing the ammonia injection pressure above 800 bar resulted in a high ITE of 42.6–42.8%, when D100 reached only 40.7%. In a general assessment, when the test engine was transferred to dual-fuel operation with ammonia, a 4.6% increase in ITE compared to the D100 combustion cycle was determined. The ITE increase coincides with other authors' results [9,12,13]. The increase in ITE is associated with a short and intense heat release close to TDC, due to which heat losses through the cylinder walls are lower.

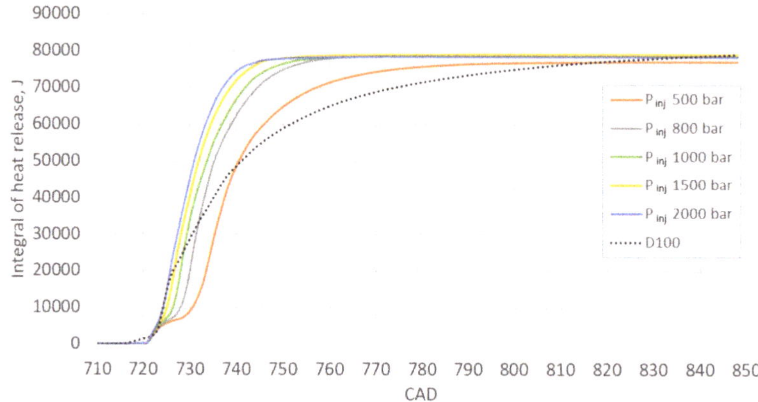

Figure 7. Integral of heat release of ammonia injection pressure variations.

To conclude, it can be stated that when optimizing ammonia and pilot fuel injection pressure, it is necessary to achieve a short combustion, and, respectively, a single-phase heat release, which is ensured by a short induction period. In this way, high ITE and combustion efficiency are achieved. However, increasing injection pressure also increases P_{max} and T_{max}, which are limited to research engine design data.

3.2. Ecological Indicators

Analyzing the release of harmful substances at the end of the combustion cycle, it was confirmed that CO_2 does not depend on injection pressure, since ammonia does not have carbon atoms in its chemical composition, so the release of CO_2 is mainly from injected pilot diesel mass. As a result, at 5% diesel mass by energy value, CO_2 emissions decreased up to 17 times and reached 33 g/kWh, when D100—564 g/kWh. CH_4 emissions, which also depend on pilot diesel mass, are estimated to have a 29.8 times greater contribution to the greenhouse effect than CO_2. During the ammonia combustion cycle, CH_4 decreased from 4.01 g/kWh to 0.00249 g/kWh at 500 bar and to 0.00023 g/kWh at 2000 bar. In this case, ammonia injection pressure has a significant influence on CH_4, since CH_4 as an incomplete combustion product depends on combustion efficiency.

In contrast to the decrease of CO_2 and CH_4 emissions, the N_2O component of GHG, which is estimated to have 273 times greater contribution to greenhouse effect than CO_2, is increasing. Published studies show that the N_2O formation mechanism during ammonia combustion takes place at temperature lower than 1400 K [12]. During combustion, NO and NO_2 react with NH and NH_2 radicals to form N_2O according to chemical reactions (5) and (6):

$$NH + NO \rightarrow N_2O + H \tag{5}$$

$$NH_2 + NO_2 \rightarrow N_2O + H \tag{6}$$

As a result, N_2O directly depends on combustion chamber temperature and ammonia fuel mass. This resulted in N_2O emissions of 1.67 g/kWh at lower combustion chamber temperatures at 500 bar injection pressure and 0.59 g/kWh at 2000 bar, with D100 at just 0.003 g/kWh. N_2O formation is analyzed by comparing visual results of N_2O mass fraction distribution in the combustion chamber with temperature fields at 770° CAD (Figure 8). Visual results analysis shows that N_2O is formed around the ammonia flame field only in 1000–1400 K temperature zone, while no N_2O formation was observed outside this temperature zone. Therefore, changing the ammonia injection pressure, which affects combustion temperature, can change the amount of N_2O emissions at the end of the combustion cycle. After comparing N_2O mass fraction distribution in the combustion chamber (Figure 8) at 500, 1000, and 2000 bar injection pressure, it was found that, in all cases with a similar temperature field around the flame field, the intensity of N_2O mass fraction distribution differs. At 1000 and 2000 bar, the intensity of N_2O mass fraction distribution is obviously lower, which can be attributed to unburnt NH_3 mass fraction distribution. NH_3 mass fraction in studied areas at 1000 and 2000 bar injection pressure is also lower due to more effective combustion. It can be reasonably stated that N_2O formation depends not only on temperature, but also is inseparable from ammonia fuel concentration in the combustion chamber.

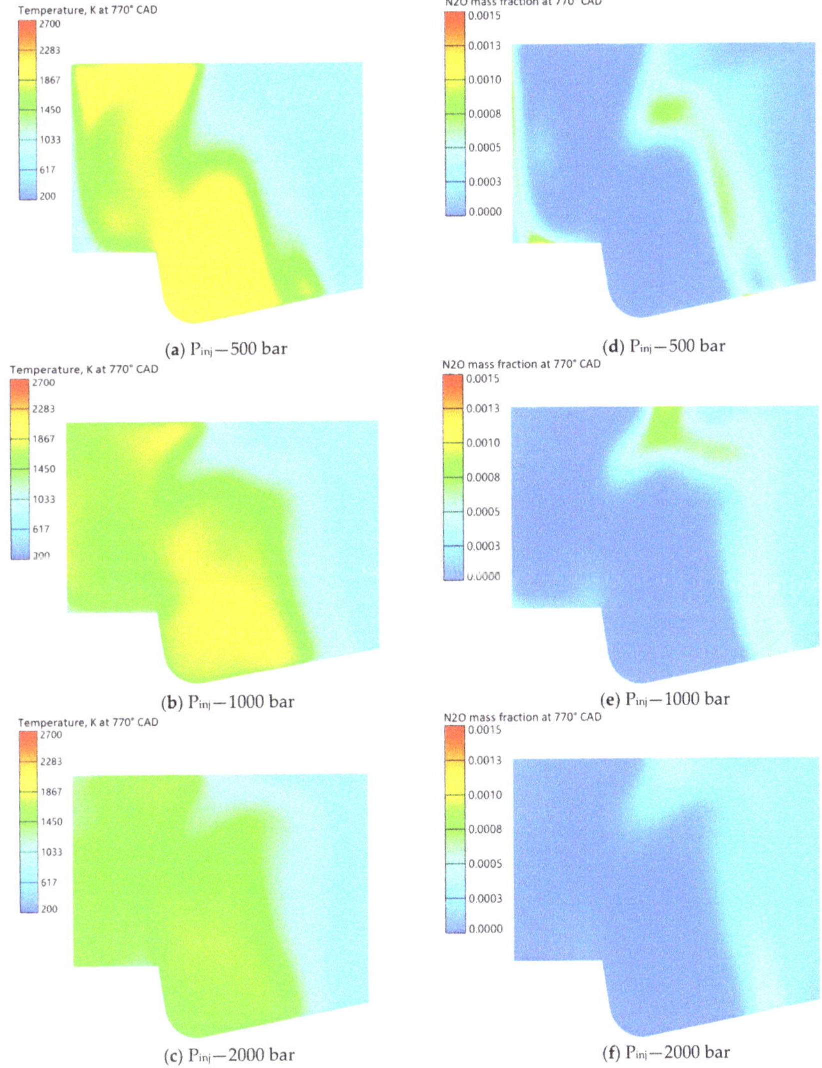

Figure 8. Comparison of N_2O cyclic mass fraction in combustion chamber with temperature fields at 770° CAD at 500, 1000, and 2000 bar ammonia injection pressure. (**a**–**c**) Temperature; (**d**–**f**) N_2O mass fraction.

On the other hand, NO_x emissions formation, which are also partially dependent on N atoms amount in fuel mixture, is also inseparable from injected ammonia mass. However, NO_x also depends on the combustion chamber temperature. Usually, thermal NO_x is formed at a combustion chamber temperature higher than 1600–1700 K, and with the increasing temperature, especially up to 2000 K and above [17,47], NO_x concentration increases exponentially. As a result, the highest amount of NO_x (9.42 g/kWh) were observed at the highest combustion temperature (T_{max}—1779 K) at 2000 bar injection pressure, while at 500 bar (T_{max}—1540 K) NO_x was lower (1.32 g/kWh). However, thermodynamic temperature reveals NO_x formation conditions quite conditionally; therefore, further evaluation of NO_x formation is carried out in relation to the structure of the combustion chamber local temperature field. Visual analysis of NO_x mass fraction distribution in the combustion chamber compared to temperature fields at 735° and 745° CAD (Figure 9) perfectly reflects thermal NO_x formation conditions. It was observed that the largest part of NO_x mass is emitted precisely in the highest temperature zones. However, observed insignificant NO_x mass fraction formation zones below 1700K combustion chamber temperature indicate fuel bound NO_x emissions. Therefore, at 500 bar ammonia injection pressure with smaller high temperature areas, NO_x formation was reasonably lower compared to 1000 or 2000 bar injection pressure. An interesting result was also observed, that despite high concentration of N atoms in dual-fuel balance, NO_x was formed 3.6 times less at 500 bar ammonia injection pressure than at D100, at practically the same maximum temperatures (D100 T_{max}—1566 K, D5/A95 500 bar T_{max}—1540 K). This phenomenon can be explained by the deNOx process, when at the cylinder temperature in range of 1000–1400 K, active NH_2 radicals react with NO to form N_2 + OH [17]. This process can be observed in the chart of NO_x cyclic mass fraction dependence on CAD (Figure 10). During heat release, NO_x emissions increase exponentially, peaking at the highest combustion chamber temperature. However, as temperature decreases, NO_x also begins to decrease due to the above chemical reaction. It was found that when ammonia injection pressure is 500 bar and T_{max} reaches 1540 K, the deNO$_x$ process is three times more intense compared to 2000 bar and T_{max} 1779 K. This is because, at 500 bar injection pressure, the open temperature window duration of 1000–1400 K is longer.

Unburnt NH_3 that slip into that exhaust system also raises concerns in the scientific community for large scale use of ammonia in DE due to ammonia toxicity to living organisms. Emissions of unburnt ammonia indicate combustion efficiency. NH_3 emissions were found to be dependent on ammonia injection pressure. At lower injection pressure, a strong increase in unburnt NH_3 was recorded, which is related to relatively poor ammonia vaporization in the combustion chamber volume. The visual results of NH_3 mass fraction distribution in the combustion chamber at 770° CAD (Figure 11) show that, at 500 bar injection pressure, burnt ammonia areas (blue color) are uneven and do not cover most of the combustion chamber area compared to NH_3 results at 2000 bar injection pressure. At 500 bar injection pressure, the level of NH_3 emission reaches 8.94 g/kWh, while at 2000 bar–0.67 g/kWh. It is worth mentioning that at 1000 and 1500 bar, NH_3 emissions were 1.51 and 1.05 g/kWh, respectively. Between 1000 and 2000 bar, the difference in NH_3 emissions is not significant, which can be attributed to a negligible difference in T_{max} between 1731 and 1779 K. In conclusion, fuel injection pressure has a significant effect on combustion efficiency, which in turn affects unburnt NH_3 emissions.

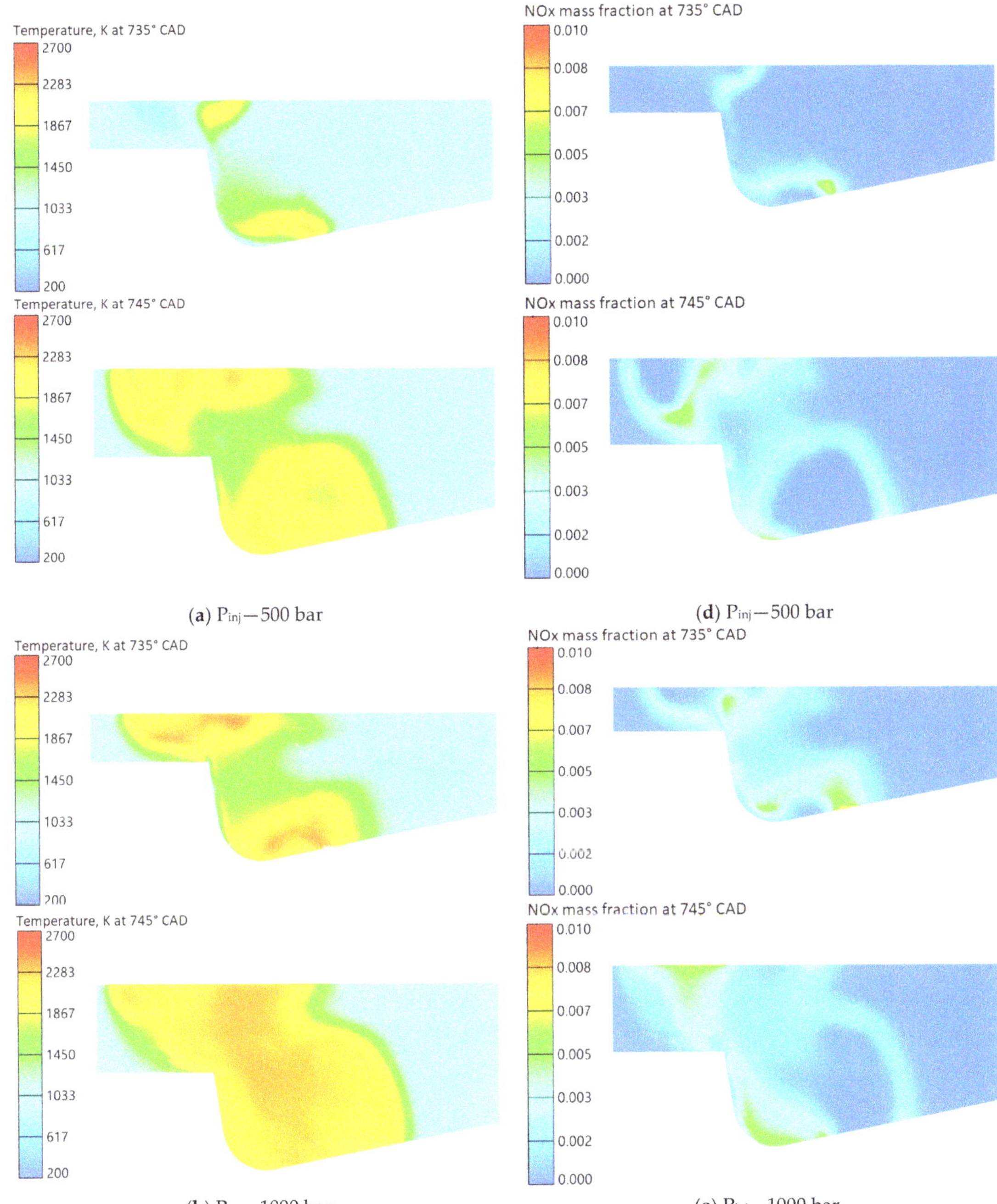

(**a**) P$_{inj}$—500 bar

(**d**) P$_{inj}$—500 bar

(**b**) P$_{inj}$—1000 bar

(**e**) P$_{inj}$—1000 bar

Figure 9. *Cont.*

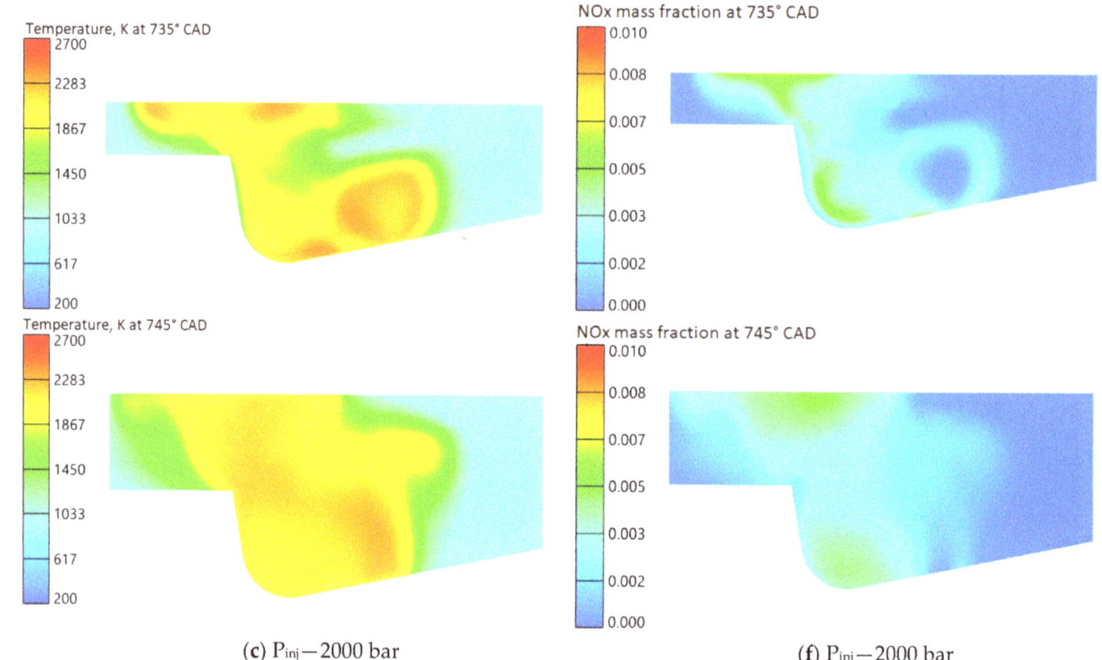

Figure 9. Comparison of NO_x cyclic mass fraction in combustion chamber with temperature fields at 735° and 745° CAD at 500, 1000, and 2000 ammonia injection pressure. (**a–c**) Temperature; (**d–f**) NO_x mass fraction.

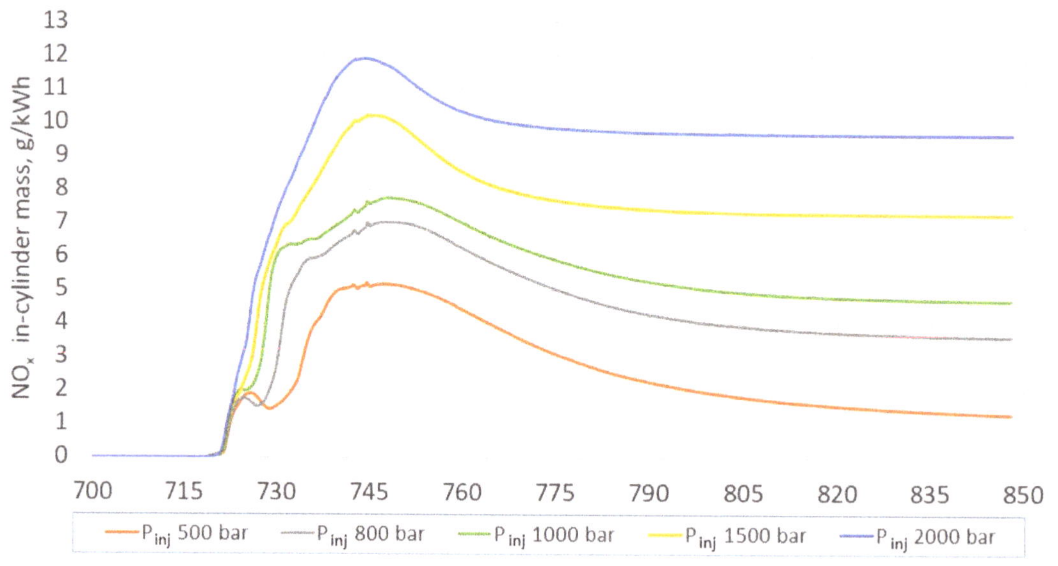

Figure 10. Comparison of NO_x formation during the combustion cycle at different ammonia injection pressures.

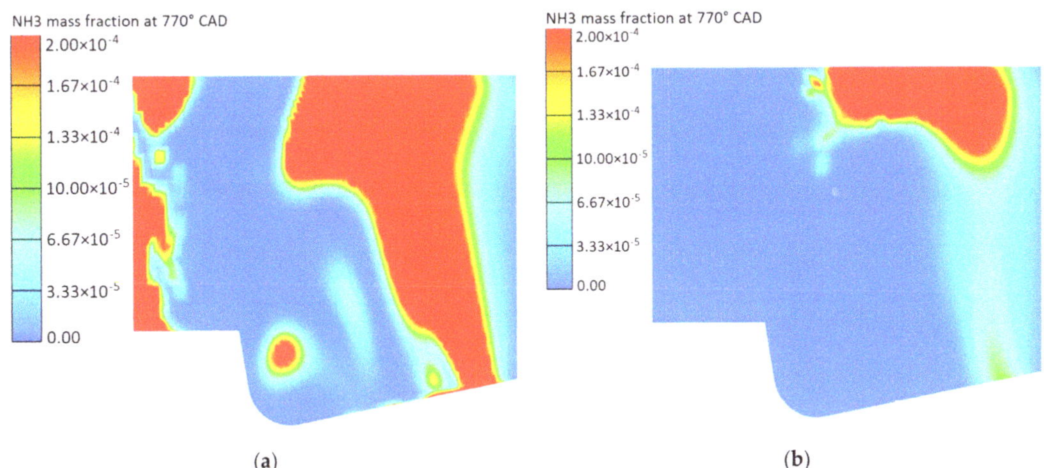

Figure 11. Comparison of the NH$_3$ cyclic mass fraction in the combustion chamber at 770° CAD. (**a**) Ammonia injection pressure 500 bar; (**b**) ammonia injection pressure 2000 bar.

In terms of GHG emissions when using ammonia in DE, N$_2$O has the biggest impact according to CO$_2$ potential (GHG = 273 × N$_2$O + 29.8 × CH$_4$ + CO$_2$) (Figure 12). At 2000 bar injection pressure at high combustion temperature outside the N$_2$O formation field, GHG emissions are lowest, 195 g/kWh, while at 500 bar–490 g/kWh. In all cases, using ammonia can reduce GHG emissions, as the D100 combustion cycle GHG is 684 g/kWh. However, GHG reduction in the reverse direction increases NO$_x$ emissions. On the other hand, NO$_x$ reduction leads to higher unburnt NH$_3$ emissions. Therefore, balance between main GHG (N$_2$O), NO$_x$, and NH$_3$ emissions should be selected.

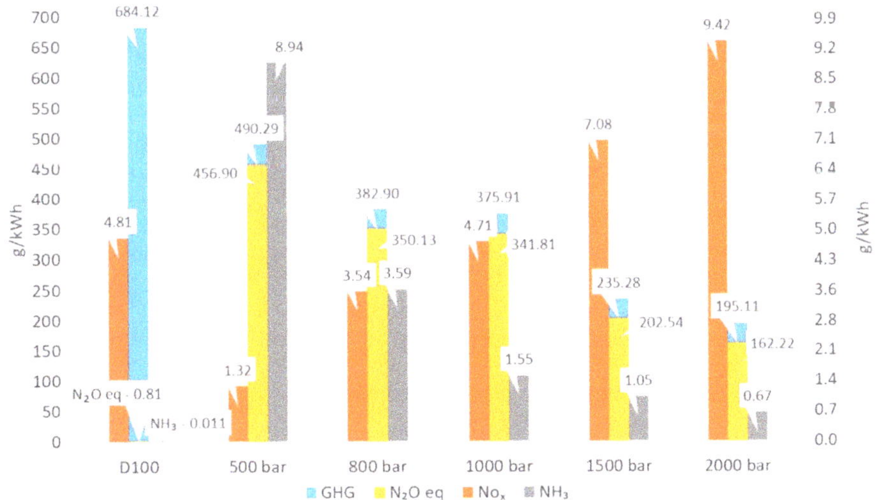

Figure 12. Comparison of emissions (GHG, N$_2$O eq, NO$_x$, and NH$_3$) formation during the combustion cycle at different ammonia injection pressures.

Summarizing the results of ammonia injection pressure dependence on combustion cycle parameters and exhaust gas emissions, the optimal injection pressure due to P$_{max}$ and T$_{max}$ design limitation is 800–1000 bar. In addition, ITE at these injection pressures compared to the 2000 bar was practically unchanged. In terms of toxic emissions, the balance between GHG, NO_x, and NH_3 is also at 800–1000 bar injection pressure. At 1000 bar injection pressure, GHG emission is 4.5% lower and NH_3 is 137% lower, although NO_x is 30% higher compared to 800 bar injection pressure. Therefore, the optimal ammonia injection pressure for the research object is 1000 bar, considering changes in combustion cycle parameters.

Summarized research main combustion cycle parameters and exhaust gas emission values are given in Table 5.

Table 5. Simulation results of ammonia injection pressure variations.

Parameter	D100 (P_{inj} 500 bar)	P_{inj} 500 bar	P_{inj} 800 bar	P_{inj} 1000 bar	P_{inj} 1500 bar	P_{inj} 2000 bar
T_{max}	1566.00	1540	1668	1731	1751	1779
P_{max}	159.60	132.18	161.34	175.38	185.46	193.78
IMEP	17.24	17.86	18.29	18.28	18.36	18.30
ITE	0.407	0.416	0.426	0.426	0.428	0.426
BSFC(D)	208.04	11.13	10.87	10.88	10.83	10.87
BSFC(NH_3)	-	434.84	424.68	425.01	422.97	424.53
P_i (complete engine)	4314.37	4191.17	4291.45	4288.09	4308.83	4293.00
CO_2	563.95	33.32	32.72	33.24	32.72	32.88
N_2O	0.0030	1.67	1.28	1.22	0.74	0.59
CH_4	4.01	0.00249	0.00132	0.000505	0.00037	0.00023
GHG	684.12	490.29	382.90	366.60	235.28	195.11
NO_x	4.81	1.32	3.54	4.59	7.08	9.42
NH_3	0.011	8.94	3.59	1.51	1.05	0.67

4. Conclusions

To optimize the dual ammonia and diesel fuel combustion cycle, combustion cycle parameters and exhaust gas emissions optimization strategy was chosen based on DE development trends and intensification of ammonia injection using high-pressure injection to reduce combustion duration. GHG emissions in CO_2 equivalent were reduced by 24% when ammonia injection pressure was increased from 500 bar to 1000 bar. For comparison, GHG emissions were also reduced by 45% compared to the diesel combustion cycle. Based on this research's results, detailed changes in the combustion cycle are characterized by the following:

- An ammonia injection pressure increase from 500 bar to 2000 bar reduces ammonia induction period by 28% from 18° CAD to 13° CAD. Correspondingly, ammonia ignition delay after pilot diesel fuel combustion shortens from 6° CAD to 1° CAD. As a result, at injection pressure 500 bar, a transformation of double-phase combustion characteristic into single-phase at 800–2000 bar injection pressure leads to structural changes in the combustion cycle parameters and exhaust gas emissions.
- At 500 bar ammonia injection pressure, the long induction period and double-phase combustion characteristic results in a long ~60° CAD combustion duration (D100—90° CAD). Meanwhile, by gradually increasing the injection pressure of over 800 bar and approaching the single-phase combustion characteristic, the combustion duration is reduced to 20–30° CAD.
- Due to the long combustion duration at 500 bar ammonia injection pressure, 2.4% lower ITE was reached than compared to higher injection pressure due to higher heat balance losses through the cooling–exhaust system. At 500 bar, ITE = 41.6%, while at 800–2000 bar, ITE = 42.6–42.8%. Therefore, at ammonia injection pressure above 800 bar, 4.6% higher ITE was reached compared to the D100 combustion cycle.
- An injection pressure increase over 800 bar leads to a higher P_{max} and T_{max} than the D100 combustion cycle. P_{max} increases from 161 bar to 194 bar by increasing injection pressure from 800 bar to 2000 bar, while T_{max} increases from 1670 K to 1780 K. For comparison, D100 P_{max}–160 bar, and T_{max}–1565 K.

- GHG emissions decreased from 684 g/kWh at D100 to 490 g/kWh–195 g/kWh at ammonia injection pressure 500–2000 bar. GHG emissions from dual ammonia–diesel fuel combustion mainly did not depend on CO_2 as in the D100 case, but depended on N_2O, whose formation is associated with N atoms in fuel. The possibility to reduce N_2O emissions is associated with the increase of combustion chamber temperature.
- Unburnt NH_3 reached maximum of 8.95 g/kWh at 500 bar injection pressure, which indicated poor combustion efficiency. Gradually increasing injection pressure from 800 bar to 2000 bar reduced NH_3 emissions from 3.60 g/kWh to 0.65 g/kWh, respectively.
- NO_x emissions can be reduced from 4.81 g/kWh at D100 to 1.32 g/kWh at ammonia injection pressure 500 bar. However, at the same time, it can increase to 9.42 g/kWh at 2000 bar injection pressure. NO_x formation depends on N atoms in fuel, but at the same time depends on combustion chamber temperature. The lower NO_x levels at 500 bar ammonia injection pressure than D100 can be explained by the $deNO_x$ process. On the other hand, higher NO_x levels at high ammonia injection pressures are directly linked to thermal NO_x formation due to increased combustion chamber temperature.

In conclusion, when optimizing the combustion cycle of dual ammonia and diesel fuel, it is rational to achieve short, single-phase combustion, but within permissible P_{max} and T_{max} boundaries, ensuring the engine reliability factor compared to D100. In parallel, when selecting ammonia injection pressure, a compromise of GHG, unburnt NH_3, and NO_x emissions must be ensured to maintain optimal balance between these harmful substances. As a result, the optimal ammonia injection pressure for medium-speed, four-stroke Wartsila 6L46 marine diesel engine is 1000 bar. The next stage of ammonia combustion cycle optimization research will be associated with the optimization of pilot diesel and ammonia injection phases.

Author Contributions: Conceptualization, S.L.; Methodology, M.D.; Software, M.D.; Validation, M.D.; Formal analysis, M.D.; Data curation, S.L.; Writing – original draft, M.D.; Writing – review & editing, S.L.; Supervision, S.L. All authors have read and agreed to the published version of the manuscript.

Funding: The project was financed by Research Council of Lithuania and the Ministry of Education, Science and Sport Lithuania (Contract No. S-A-UEI-23-9).

Institutional Review Board Statement: Not applicable.

Informed Consent Statement: Not applicable.

Data Availability Statement: Data are contained within the article.

Acknowledgments: The authors thank AVL company for their long-term cooperation with Klaipeda University for providing AVL FIRE simulation software used in the PhD study program in maritime transport decarbonization research. Exceptional thanks to AVL FIRE personnel for sharing limited accessibility to the ammonia gas reaction mechanism, consultation, and professional advice.

Conflicts of Interest: The authors declare no conflict of interest.

Abbreviations

B/S	cylinder bore, stroke;
BSFC	specific fuel consumption;
CAD	crank angle degrees;
CH_4	methane;
CO_2	carbon dioxide;
D100	100% diesel fuel;
D5/A95	mixture of 5% diesel and 95% ammonia fuel;
DE	diesel engine;
EEDI	Energy Efficiency Design Index;
EEXI	Efficiency Existing Ship Index;
EU	European Union;

GHG	greenhouse gases;
GT	gross tonnage;
HPDF	high-pressure dual-fuel strategy;
HRC	heat release characteristic;
IMEP	indicated mean indicative pressure;
IMO	International Maritime Organization;
ITE	indicative thermal efficiency;
LNG	liquified natural gas;
LPDF	low-pressure dual-fuel strategy;
MM	mathematical model;
N_2O	dinitrogen oxide;
NH_3	ammonia;
NO_x	nitrous oxides;
PM	particulate matter;
SCR	selective catalytic reduction;
TDC	top dead center;

Symbols

H_u	fuel calorific value (kJ/kg);
m_f	hourly fuel consumption (kg/h);
n	cylinder number (-);
N	engine speed (RPM);
P_i	indicated power (kW);
P_{inj}	ammonia injection pressure (bar);
P_k	cylinder pressure reading (bar);
P_{max}	maximum cycle pressure (bar);
T_{max}	maximum cycle temperature (K);
T	temperature (K);
V_{cyl}	cylinder volume (m^3);
V_k	cylinder volume corresponding to P_k (m^3);
z	coefficient ($z = 1$ for 2-stroke engines, $z = 2$ for 4-stroke engines (-);
θ_0	crank angle degree value (-);
θ_f	crank angle degree value (-);

References

1. SEC (2021) 562 Final. Regulation of the European Parliament and of the Council on the Use of Renewable and Low-Carbon Fuels in Maritime Transport and Amending Directive 2009/16/EC. Available online: https://eur-lex.europa.eu/legal-content/EN/ALL/?uri=COM:2021:562:FIN (accessed on 3 January 2024).
2. International Maritime Organization. Energy Efficiency Measures. Available online: https://www.imo.org/en/OurWork/Environment/Pages/Technical-and-Operational-Measures.aspx (accessed on 3 January 2024).
3. International Maritime Organization. Rules on Ship Carbon Intensity and Rating System Enter into Force. Available online: https://www.imo.org/en/MediaCentre/PressBriefings/pages/CII-and-EEXI-entry-into-force.aspx (accessed on 3 January 2024).
4. International Maritime Organization. Annex 15, Resolution MEPC.377(80). 2023 Strategy on Reduction of GHG Emissions from Ships. Available online: https://wwwcdn.imo.org/localresources/en/OurWork/Environment/Documents/annex/MEPC%2080/Annex%2015.pdf (accessed on 5 January 2024).
5. DNV Report. Maritime Forecast to 2050. Available online: https://www.dnv.com/maritime/publications/maritime-forecast-2023/index.html (accessed on 5 January 2024).
6. UNCTAD. Review of Maritime Transport. Available online: https://unctad.org/system/files/official-document/rmt2021_en_0.pdf (accessed on 6 January 2024).
7. DNV Group. Maritime Forecast to 2050. Available online: https://www.dnv.com/maritime/publications/maritime-forecast-2022/download-the-report.html (accessed on 6 January 2024).
8. Aziz, M.; Wijayanta, A.T.; Nandiyanto, A.B.D. Ammonia as Effective Hydrogen Storage: A Review on Production, Storage and Utilization. *Energies* **2020**, *13*, 3062. [CrossRef]
9. Reiter, A.J.; Kong, S.C. Combustion and emissions characteristics of compression-ignition engine using dual ammonia-diesel fuel. *Fuel* **2011**, *90*, 87–97. [CrossRef]
10. Ryu, K.; Zacharakis-Jutz, G.E.; Kong, S.C. Performance characteristics of compression-ignition engine using high concentration of ammonia mixed with dimethyl ether. *Appl. Energy* **2014**, *113*, 488–499. [CrossRef]

11. Dimitriou, P.; Javaid, R. A review of ammonia as a compression ignition engine fuel. *Int. J. Hydrog. Energy* **2020**, *45*, 7098–7118. [CrossRef]
12. Nadimi, E.; Przybyła, G.; Lewandowski, M.T.; Adamczyk, W. Effects of ammonia on combustion, emissions, and performance of the ammonia/diesel dual-fuel compression ignition engine. *J. Energy Inst.* **2023**, *107*, 101158. [CrossRef]
13. Li, T.; Zhou, X.; Wang, N.; Wang, X.; Chen, R.; Li, S. A comparison between low and high-pressure injection dual-fuel modes of diesel-pilot-ignition ammonia combustion engines. *J. Energy Inst.* **2022**, *102*, 362–373. [CrossRef]
14. Scharl, V.; Sattelmayer, T. Ignition and combustion characteristics of diesel piloted ammonia injections. *Fuel Commun.* **2022**, *11*, 100068. [CrossRef]
15. Yousefi, A.; Guo, H.; Dev, S.; Liko, B.; Lafrance, S. Effects of ammonia energy fraction and diesel injection timing on combustion and emissions of an ammonia/diesel dual-fuel engine. *Fuel* **2022**, *314*, 122723. [CrossRef]
16. Li, T.; Zhou, X.; Wang, N.; Wang, X.; Chen, R.; Li, S. Pilot diesel-ignited ammonia dual fuel low-speed marine engines: A comparative analysis of ammonia premixed and high-pressure spray combustion modes with CFD simulation. *Renew. Sustain. Energy Rev.* **2023**, *173*, 113108.
17. Lyon, R.K. The chemistry of the thermal DeNOx process: A review of the thechnology's possible application to control of NOx from diesel engines. In Proceedings of the 7th Diesel Engine Emissions Reduction (DEER) Workshop 2001, Portsmouth, VA, USA, 5–9 August 2001.
18. Zinner, K. Results of real cycle calculations about the possibilities of influencing the efficiency of diesel engines. *MTZ* **1970**, *31*, 243–246.
19. Stechkin, B.S. On the efficiency of an ideal fast combustion cycle at a finite rate of heat release. Theory, design, calculation and testing of internal combustion engines. *Proc. Lab. Engines USSR Acad. Sci.* **1960**, 61–67.
20. Semenov, B.N.; Ivanchenko, N.N. Problems of increasing the fuel efficiency of diesel engines and ways to solve them. *Engine Engineering.* **1990**, *11*, 3–7.
21. Kruggel, O. Progress in the combustion technology of high performance diesel engines toward reduction of exhaust emissions without reduction of operation economy. In Proceedings of the Baden–Wurttemberg Technology Conference, Oslo, Norway, 8 February 1989; p. 14.
22. Schulte, H.; Duernholz, M.; Wuebbeke, K. The Contribution of the fuel injection system to meeting future demands on truck diesel engines. *SAE Trans.* **1990**, *99*, 157–162.
23. MTU, Deutsche Aerospace. *Series 396 Engines*; VTS 05076 (520 E); MTU, Deutsche Aerospace: Munich, Germany, 1990; 22p.
24. Shundoh, S.; Kakegawa, T.; Tsujimura, K.; Kobayashi, S. The effect of injection parameters and swirl on diesel combustion with high pressure fuel injection. *SAE Trans.* **1991**, *100*, 793–805.
25. Stas, M.; Wajand, J.; Bestimmung, d.V. Parameters for the two-phase combustion process in direct injection diesel engines. *MTZ Mot.* **1988**, *49*, 289–293.
26. Kruggel, O. Studies on nitrogen oxide reduction in high-speed large diesel engines. *MTZ Mot.* **1988**, *49*, 22–29.
27. Kakaee, A.H.; Rahnama, P.; Paykani, A. Influence of fuel composition on combustion and emissions characteristics of natural gas/diesel RCCI engine. *J. Nat. Gas Sci. Eng.* **2015**, *25*, 58–65. [CrossRef]
28. Yang, B.; Xi, C.; Wei, X.; Zeng, K.; Lai, M.C. Parametric investigation of natural gas port injection and diesel pilot injection on the combustion and emissions of a turbocharged common rail dual-fuel engine at low load. *Appl. Energy* **2015**, *143*, 130–137. [CrossRef]
29. Gatts, T.; Liu, S.; Liew, C.; Ralston, B.; Bell, C.; Li, H. An experimental investigation of incomplete combustion of gaseous fuels of a heavy-duty diesel engine supplemented with hydrogen and natural gas. *Int. J. Hydrog. Energy* **2012**, *37*, 7848–7859. [CrossRef]
30. Lebedevas, S.; Raslavičius, L.; Drazdauskas, M. Comprehensive correlation for the prediction of the heat release characteristics of diesel/CNG mixtures in a single-zone combustion model. *Sustainability* **2023**, *15*, 3722. [CrossRef]
31. Lebedevas, S.; Raslavičius, L.; Rapalis, P. Simulation of microalgae oil spray characteristics for mechanical fuel injection and CRDI systems. *Biomass Convers. Biorefinery* **2022**, 1–16. [CrossRef]
32. Lebedevas, S.; Čepaitis, T. Parametric analysis of the combustion cycle of a diesel engine for operation on natural gas. *Sustainability* **2021**, *13*, 2773. [CrossRef]
33. Lebedevas, S.; Čepaitis, T. Research of organic rankine cycle energy characteristics at operating modes of marine diesel engine. *J. Mar. Sci. Eng.* **2021**, *9*, 1049. [CrossRef]
34. Woodyard, D. Medium speed engines—Introduction. In *Marine Diesel Engines*; Elsevier: Amsterdam, The Netherlands, 2004; pp. 498–516.
35. Molland, A.F. Marine engines and auxiliary machinery. In *The Maritime Engineering Reference Book*; Elsevier: Amsterdam, The Netherlands, 2008; pp. 344–482.
36. Bergen B33:45L Marine Diesel Engine Specification. Available online: https://www.bergenengines.com/engines/b33-45l/ (accessed on 15 January 2024).
37. MAN L48/60CR Marine Diesel Engine Specification. Available online: https://www.man-es.com/marine/products/four-stroke-engines/propulsion (accessed on 15 January 2024).
38. Wartsila 46F Marine Diesel Engine Specification. Available online: https://www.wartsila.com/marine/products/engines-and-generating-sets/diesel-engines/wartsila-46f (accessed on 15 January 2024).

39. Glarborg, P.; Miller, J.A.; Ruscic, B.; Klippenstein, S.J. Modeling nitrogen chemistry in combustion. *Prog. Energy Combust. Sci.* **2018**, *67*, 31–68. [CrossRef]
40. Dukowicz, J.K. A Particle-fluid numerical model for liquid sprays. *J. Comput. Phys.* **1980**, *35*, 229–253. [CrossRef]
41. Schiller, L.; Naumann, A. A Drag Coefficient Correlation. *VDI Ztg.* **1933**, *77*, 318–320.
42. Abramzon, B.; Sirignano, W.A. Droplet vaporization model for spray combustion calculations. In Proceedings of the AIAA 26th Aerospace Sciences Meeting, Reno, NV, USA, 11–14 January 1988.
43. Javaherian, H.; Haskara, I.; Dagci, O.H. Technique for Calculating a High-Precision IMEP Using a Low-Resolution Coder and an Indirect Integration. Process Patent DE102011011485A1, 22 March 2012.
44. Mollenhauer, K.; Tschöke, H. *Handbook of Diesel Engines*; Springer: Berlin/Heidelberg, Germany, 2010; pp. 8–24.
45. Mondal, P.K.; Mandal, B.K. Effect of fuel injection pressure on the performances of a CI engine using water-emulsified diesel (WED) as a fuel. *Energy Sustain. Soc.* **2024**, *14*, 12. [CrossRef]
46. Geng, L.; Wang, Y.; Wang, Y.; Li, H. Effect of the injection pressure and orifice diameter on the spray characteristics of biodiesel. *J. Traffic Transp. Eng.* **2020**, *7*, 331–339. [CrossRef]
47. Hebbar, G.S. NOx from diesel engine emission and control strategies—A review. *Int. J. Mech. Eng. Eobotics Res.* **2014**, *3*, 471.

Journal of
Marine Science and Engineering

Article

Influence of Emission-Control Areas on the Eco-Shipbuilding Industry: A Perspective of the Synthetic Control Method

Lang Xu [1], Zeyuan Zou [1], Lin Liu [2,*] and Guangnian Xiao [3,*]

[1] College of Transport & Communications, Shanghai Maritime University, Shanghai 201306, China; xulang@shmtu.edu.cn (L.X.); zyzou2021@yeah.net (Z.Z.)

[2] Naval Medicine Center, Naval Medicine University, Shanghai 200433, China

[3] School of Economics & Management, Shanghai Maritime University, Shanghai 201306, China

* Correspondence: earnestliu@126.com (L.L.); gnxiao@shmtu.edu.cn (G.X.)

Abstract: Annex VI of the International Convention for the Prevention of Pollution from Ships (MARPOL Convention), adopted in October 2008, was dedicated to addressing environmental issues caused by ships, especially in ports, inland waterways, and some sea areas with concentrated routes and high navigational density. This study utilizes a regional-level ship dataset to assess the influences of emission-control areas (ECAs) on the ecological shipbuilding industry by fitting the policy utility through the synthetic control method and testing robustness via the difference-in-differences method. The outcomes of this study show that the cumulative new orders for eco-designed ships in China, The Netherlands, Republic of Korea, the UK, and the USA increased by 3401, 81, 234, 549, and -1435, respectively, after the implementation of ECAs. Compared to the implementation of ECAs, the increases were about 32%, 20%, 41%, 66%, and -83%, respectively.

Keywords: eco-designed ship; shipbuilding industry; synthetic control method; emission-control area

1. Introduction

As global trade accelerates, emission reduction from shipping is urgently needed. From 2012 to 2018, sulfur emissions from shipping increased from 10.8 million tons to 11.4 million tons, accounting for 13% of global sulfur emissions [1,2]. The International Maritime Organization (IMO) announced that if left unchecked, sulfur emissions from the sea would increase by 140% in 2012, which would hinder the achievement of environmental sustainability goals [3–5]. At the end of 2020, the IMO issued short-term emission reduction recommendations for ships, which set out requirements in terms of both technical energy efficiency indicators and operational SO_x and NO_x intensity rating mechanisms, of which the global eco-designed bulk cargo newbuilding orders for the period of 2000–2020 are shown in Figure 1.

At the same time, the shipping industry was ordered to adapt to stricter emission reduction regulations, which can facilitate the optimization of ship speed, the use of alternative fuels, and the accelerated application of technology in the shipping industry [6,7]. However, from a commercial aspect, the effect of increasingly stringent standards on the competitiveness of new and old ships cannot be underestimated. Therefore, ships must stand the test of the future, especially the designs of environmentally friendly ships [8,9]. To reduce SO_X and NO_X emissions, major shipping companies such as Maersk, COSCO, and MSC have acted to cope with increasingly stringent emission-control policies. For example, Maersk is gradually replacing traditional fuel with methanol and biofuels (maersk.com.cn/sustainability/reports-and-resources, accessed on 8 January 2024). Furthermore, as of 30 December 2019, COSCO has fully completed the conversion of high-sulfur oil to low-sulfur oil (lines.coscoshipping. com/home/About/socialResponsibility/sustainabilityReport, accessed on 8 January 2024). By limiting the upper sulfur content of marine fuel and gradually installing an exhaust gas purification system, MSC reduced sulfur emissions by 86% in 2020 compared to 2019

Citation: Xu, L.; Zou, Z.; Liu, L.; Xiao, G. Influence of Emission-Control Areas on the Eco-Shipbuilding Industry: A Perspective of the Synthetic Control Method. *J. Mar. Sci. Eng.* **2024**, *12*, 149. https://doi.org/10.3390/jmse12010149

Academic Editor: Claudio Ferrari

Received: 30 November 2023
Revised: 5 January 2024
Accepted: 7 January 2024
Published: 12 January 2024

(msccargo.cn/en/sustainability, accessed on 8 January 2024). The measures taken by major shipping companies in response to the ECAs are listed in Table 1. In summary, choosing an environmentally friendly ship design is crucial for reducing pollution emissions. The most available clean fuel is liquefied natural gas (LNG), and available environmental methods include SO_x Scrubber, electronic engines, and shore-to-ship electricity [2,10].

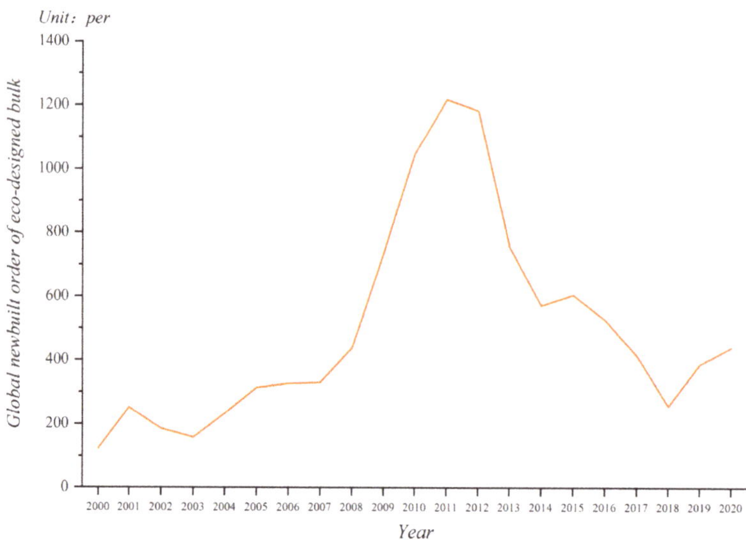

Figure 1. Global newly built orders of eco-designed bulk from 2000 to 2020 (Source: Clarkson SIN, www.clarksons.net.cn).

Table 1. Major shipping lines' actions on emission-control areas.

Shipping Line	Year	Actions
Maersk	2020	Install scrubbers to clean vessel exhaust
COSCO	2019	Promote green shipping to reduce ship emissions and pollution
CMA-CGM	2019	Implement a low-sulfur surcharge
MSC	2020	Exhaust gas cleaning systems, conventional low-sulfur fuel, and biofuels

It is worth noting that the most relevant research on the effectiveness of the MARPOL Annex mainly focuses on ship engineering and operational management [11–15]. In particular, Zhang et al. discussed the disposal of garbage by the IMO in international transportation, which promoted better governance of shipping pollution and a reduction of marine emissions [16]. In addition, Lion et al. and Sakib et al. also explored pollution from ship activities [17,18]. However, pollutants caused by ship navigations are caused by different fuel sources [19]. In particular, Halff et al. argued that the IMO has paradoxically slowed the rapid transition of the shipping market from traditional marine fuels to other fuels [20]. Beyond that, Kokosalakis et al. obtained the same result from cross-sectional analysis [21]. Meanwhile, Peng et al. estimated the detection efficiency and theoretical pollution diffusion of ship fuels at Yantian Port [22]. On this basis, Theocharis et al. evaluated the feasibility of the North Sea route for seasonal operations of finished oil tankers using alternative fuels under IMO restrictions [23].

Meanwhile, some scholars have considered the environmental effects of new shipbuilding orders from the perspectives of ship design, ship performance, and ship emissions [24–27]. As reported by the IPCC, Lindstad et al. discussed the probabilities of maximizing cargo-carrying capacity at the lowest shipbuilding costs to test the potential reductions by combining ship design and alternative power [28]. In particular, Li et al. investigated the economic and emission assessments in the Yangtze River inland waterway networks to propose the fuel consumption calculation formulas for LNG-fueled ships [29]. In addition, Fan et al. involved data preprocessing, sample size, model type, operating condition, and feature type in quantitatively predicting new-energy ship fuel consumption [30]. To our knowledge, most of the current research on shipbuilding is centered on ship emissions, ship fuel consumption, and ship pollution, but no scholars have been involved in considering the impact of ECA on new-energy shipbuilding from a macro perspective.

However, the above-mentioned scholars used regression models to conduct empirical analyses to examine whether ECA could improve the marine environment and how ship efficiency affects emissions. Based on annual shipbuilding data, the synthetic control method is used to quantify the effect of ECA on environmentally friendly ship construction. This study aims to provide recommendations for policymakers by examining the effects of ECA policy on environmentally friendly ship orders. The novelty of our research can be summarized as follows. Starting from the most recent data, the effects of ECAs on ship construction via the synthetic control method is quantified, which allows for the analysis of individual countries and further explores the effects of ECA policies in different countries. Next, the robustness of the synthetic control method via a placebo test and the difference-in-difference method is validated. In addition, the effects of ECAs are explored from the perspectives of shipowners and shipbuilders using textual analysis methods. Finally, our findings can explore the diversity of shipbuilding policies, as well as provide managerial and practical insights for the shipbuilding industry.

In this research, the impacts of emission-control areas on the shipbuilding industry are quantitatively evaluated. However, this research focuses on low-sulfur-emitting ships and ships powered by electricity, while other new energy sources, such as methanol, ammonia, and other new energy sources, are not taken into account due to the small sample size and insufficient laws. This paper compares the proportion of environmentally friendly ships by type, period, and building country via Clarkson's annual reports, as shown in Figure 2. Notice that the main reason for choosing bulk carriers is that they constitute the largest new orders for eco-designed ships and bear the burden of most of the global seaborne trade. New orders for environmentally friendly ships rose rapidly after the ECA policy came into effect in 2010. If ECA policies do not work for the shipbuilding industry, orders for eco-designed ships will be almost the same as in other years.

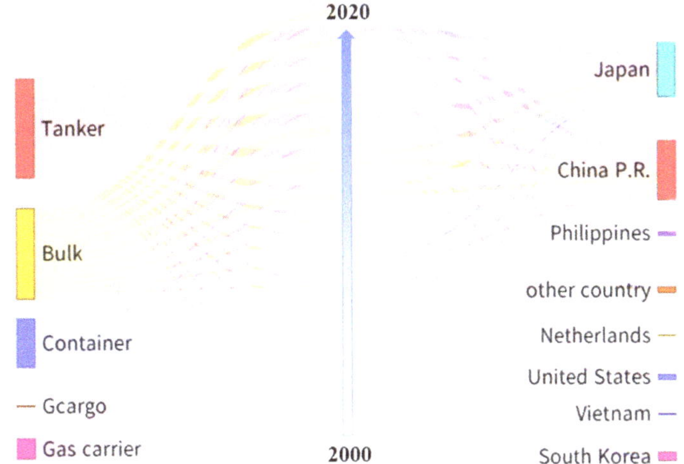

Figure 2. Main owner county of eco-designed bulk carriers from 2000 to 2020 (Source: Clarkson SIN, www.clarksons.net.cn).

The rest of our study is organized as follows. Section 2 presents the main methodology. We provide the original data and investigate the empirical results in Section 3. Finally, conclusions and managerial insights are given in Section 4.

2. Methodology

To explore whether a special policy will lead to a certain result, some scholars adopted the synthetic control method to address this issue [31,32]. The basic idea is to build a "counterfactual" control unit based on the available data and objectives, construct a virtual control unit by assigning different weights to a linear combination of several control groups similar to the treatment groups, and then compare the differences between the objectives and the control unit after the policy has been implemented, therefore evaluating the policy performance. Compared with other econometric methods, the advantage of the synthetic control method is the ability to assess the performance of individual countries and add the analysis of spatial variability to better derive the spatial–temporal effects of policies. On this basis, the implementation of ECA is considered to be a quasi-natural experiment, where the areas with and without implemented ECA are defined as experimental groups and control groups to ensure the optimal weight of linear combination in the control group through predictive variables. In addition, before implementing ECA, counterfactual composite control variables with similar characteristics are nested into the experimental group, and the impact of ECAs is discussed by comparing differences in new orders for ship construction.

In this study, it is supposed that the regions $i = 1, 2, \cdots, J + 1$ in the period $t \in [1, T_0]$ are collected, where T_0 $(1 \leq T_0 \leq T)$ is the setting time of emission-control areas. We define O_{itk}^{I} as the order of bulk in the i region at the t period, which is impacted by emission-control areas. Hence, when $t \in [1, T_0]$, there is $O_{it}^{I} = O_{it}^{N}$. After setting the emission-control area ($t > T_0$), $\theta_{it} = O_{it}^{I} - O_{it}^{N}$ is the effect of emission-control areas. For the regions without any emission-control area, O_{it}^{N} is a known value, whereas O_{it}^{I} is unknown where it needs to be synthesized. However, for the regions with emission-control areas, O_{it}^{N} is an unknown value, whereas O_{it}^{I} is known. According to previous studies, the data with the following equation is estimated to calculate the orders of ships in the i region at the t period as follows [1,31,33]:

$$O_{it}^{N} = \delta_t + \rho_t Z_i + \lambda_t \mu_i + \varepsilon_{it} \tag{1}$$

where δ_t is the fixed-time effect, ρ_t is a $1 \times F$-dimensional unknown vector of parameter, λ_t is a $1 \times F$-dimensional vector of unobserved regional fixed effect, and ε_{it} indicates the error term. Beyond that, Z_i represents a $F \times 1$-dimensional vector of control variables that is not influenced by the implementation of emission-control areas, whereas μ_i means a $F \times 1$ vector of unobservable regional fixed effect. Under this structure, if the i region sets emission-control areas and the rest of the regions without setting emission-control areas, an $F \times 1$-dimensional vector of weights $W = (w_2, w_3, \cdots, w_{k+1})$ is designed to ensure $w_i \geq 0$ and $\sum_{2}^{k+1} w_i = 1$. Hence, the specific value of each vector W represents a potential combination of synthetic control, i.e., a specific weight for J regions. For each control group, the value of the variable is weighted to be obtained as follows:

$$\sum_{i=2}^{J+1} w_i O_{it} = \delta_t + \rho_t \sum_{i=2}^{J+1} w_i Z_i + \lambda_t \sum_{i=2}^{J+1} w_i \mu_i + \sum_{i=2}^{J+1} w_i \varepsilon_{it} \tag{2}$$

From the above, there exists $(w_2^*, w_3^*, \cdots, w_{k+1}^*)$, which makes $\sum_{i=2}^{J+1} w_i^* O_{it} = O_{1t}$ and $\sum_{i=2}^{J+1} w_i^* Z_i = Z_1$ if $t \leq T_0$; thus

$$O_{it}^{N} - \sum_{i=2}^{J+1} w_i^* O_{it} = \sum_{i=2}^{J+1} w_i^* \sum_{s=1}^{T_0} \lambda_t \left(\sum_{n=1}^{T_0} \lambda_n' \lambda_n \right)^{-1} \lambda_t' (\varepsilon_{is} - \varepsilon_{is}) - \sum_{i=2}^{J+1} w_i^* (\varepsilon_{is} - \varepsilon_{1s}) \tag{3}$$

Under normal conditions, if the period before the policy is longer than the time range for the implementation of emission-control areas, then the mean value:

$$\sum_{i=2}^{J+1} w_i^* \sum_{s=1}^{T_0} \lambda_t \left(\sum_{n=1}^{T_0} \lambda_n^T \lambda_n \right)^{-1} \lambda_t' (\varepsilon_{is} - \varepsilon_{is}) - \sum_{i=2}^{J+1} w_i^* (\varepsilon_{is} - \varepsilon_{1s}) = 0 \tag{4}$$

Next, in the implementation period of the emission-control area, this study regards O_{it}^N as the value of unbiased estimation $\sum_{i=2}^{J+1} w_i^* O_{it}$; thus, the estimated value of policy effect is $\hat{\theta}_{it} = O_{it} - \sum_{i=2}^{J+1} w_i^* O_{it}$ and the average value of the policy effect is $\hat{\theta}_{i,mean} = \frac{1}{T-T_0} \sum_{t=T_0+1}^{T} \hat{\theta}_{it}$, where $t \in [T_0, T]$. We define $M = (m_1, m_2, \cdots, m_{T_0})$ as an F × 1-dimensional vector that results in linear combination before the policy implementation, e.g., $\overline{O}_{it} = \sum_{s=1}^{T_0} m_s O_{it}$. Thus, if $m_1 = m_2 = \cdots = m_{T_0-1} = 0$ and $m_{T_0} = 1$, the variable value is just a certain period before the policy implementation (e.g., $\overline{O}_{it} = O_{it}$); otherwise, $\overline{O}_{it} = \sum_{s=1}^{T_0} O_{it} / T_0$. Furthermore, we define X_1 as an F × 1-dimensional vector, which represents the data combination of control variables before the implementation policy, and X_2 as an F × J matrix, which reports the data combination of control variables. Thus, the optimal vector of weights W^* minimizes the distance:

$$\|X_1 - X_0 W\|_v = \sqrt{(X_1 - X_0 W^*)^T V (X_1 - X_0 W^*)} \tag{5}$$

where V can be regarded as an F × F symmetric and positive semidefinite matrix.

3. Empirical Results and Discussion

3.1. Data Source

Due to the requirements of the synthetic control methods, the time period before the policy intervention should be retained for a long period to make the degree-of-order curves between the actual group and the synthetic group fit [34]. Therefore, due to the absence of emission-control areas in our research before 2010, the monthly order data used for eco-designed ships from January 2000 to December 2020 are provided by the Clarkson Shipping Intelligence Network (https://sin.clarksons.net, accessed on 21 September 2021). In particular, the monthly order data for eco-designed ships from 20 regions (Argentina, Australia, Bangladesh, Brazil, Cyprus, India, Indonesia, Japan, Malaysia, Monaco, Philippines, Singapore, Thailand, UAE, and Vietnam) include the installation of SOx scrubbers and the design of eco-electronic engines, where five regions (China, The Netherlands, Republic of Korea, the UK, and the USA) have set emission-control areas as shown in Figure 3. Intuitively, after the gradual implementation of the ECA policy in 2010, the number of orders for environmentally friendly ships in both the treatment and control countries has increased significantly. This suggests that the treatment and control groups remain the same in terms of changes in the main variable of environmentally friendly ships, and those in the control group can simulate the changes in the orders of the countries in the treatment group better in the synthetic control method.

Since there are significant fluctuations in the monthly order data of eco-designed ships, it is not possible to visually display the differences between the actual group and the synthetic group. Hence, the cumulative order data of eco-designed ships is considered to display a stable trend. In addition, the reason these data are used is that some regions in the collected samples do not have new shipbuilding orders in the early stage, which can make it difficult to obtain the synthetic control subject through the weight of the control group; thus, these data are not employed. Here, as shown in Table 2, four predictive factors (oil consumption, GDP growth rate, goods import growth rate, and goods export growth rate) are considered to obtain the synthetic group.

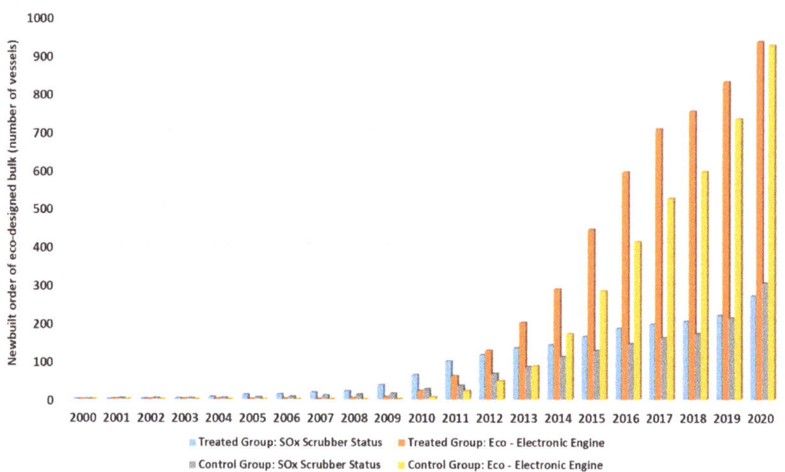

Figure 3. The cumulative eco-designed vessel orders between the treatment group and the control group (Source: Clarkson SIN, www.clarksons.net.cn).

Table 2. Data sources and selection basis of predictive factors.

Factors	Period	Unit	Data Sources	Selection Reason
Oil consumption	Yearly data 2000–2020	Percent	http://sin.clarksons.net (accessed on 21 September 2021)	Oil consumption affects the transport construction
GDP growth rate	Yearly data 2000–2020	Percent	https://www.kylc.com (accessed on 21 September 2021)	GDP growth measures economic scale and development level
Goods import growth rate	Yearly data 2000–2020	Percent	http://data.worldbank.org (accessed on 21 September 2021)	Goods import rate indicates the shipping industry level
Goods export growth rate	Yearly data 2000–2020	Percent	http://data.worldbank.org (accessed on 21 September 2021)	Goods export rate indicates the shipping industry level

3.2. Empirical Results

In this study, it is believed that all treatment groups use the default starting period to set emission-control areas. Because of the different regions setting emission-control areas at different time points, the treatment group is not analyzed as a mixed target according to existing research but instead constructs a synthetic group for different regions separately. Therefore, the effect of setting emission-control areas is measured by the difference between the monthly data of new shipbuilding orders in the treatment group and synthesis group. Taking the People's Republic of China as an example, the implementation time of emission-control areas was in 2015 and 2020, respectively. Therefore, this study introduces four predictive factors from 2000 to 2020 (oil consumption, GDP growth rate, good import growth rate, and good export growth rate) to fit the synthetic control variables. Next, the implementation influence of emission-control areas in China is reflected by the differences in shipbuilding orders between China and its synthetic group after 2015, with the weight-selection criterion minimizing the mean-square error of shipbuilding orders in the period prior to the implementation of emission-control areas. Stata 16 software was used in this study to perform synthetic control method regression on the collected data. Furthermore, the weight of synthetic control groups corresponding to all treatment groups is shown in Table 3.

From 2000 to 2020, the shipbuilding orders for eco-designed ships in the treatment group and corresponding synthesis group are shown in Figure 4, where the position of the vertical dashed line means the starting year of implementing emission-control areas in the region. Hence, the left side of the dashed line shows that the shipbuilding orders of eco-designed ships are very close to those of the synthetic group; otherwise, it gradually deviates on the right side of the dashed line. For example, China has implemented an emission-control area since 2015, where the shipbuilding orders for eco-designed ships from the actual and synthetic groups were very close before 2015. However, after the implementation of emission-control areas, the synthetic shipbuilding orders for eco-designed ships in China have been significantly lower than the actual order, where the gap between the above is the implementation effect of emission-control areas.

Table 3. The weight of the control groups corresponding to the treatment group.

	China	The Netherlands	Republic of Korea	UK	USA
Argentina	0	0.581	0	0.849	0
Australia	0	0	0.001	0	0
Bangladesh	0.001	0	0	0	0
Brazil	0	0.001	0	0	0
Cyprus	0	0.411	0	0	0.026
India	0	0.004	0	0	0
Indonesia	0	0	0.001	0	0
Japan	0.841	0	0.388	0.112	0.717
Malaysia	0	0	0	0.015	0
Monaco	0	0	0	0.015	0
Philippines	0	0	0	0.009	0
Singapore	0	0	0.431	0	0
Thailand	0	0.001	0	0	0
UAE	0	0.002	0	0	0.257
Vietnam	0.158	0	0.179	0	0

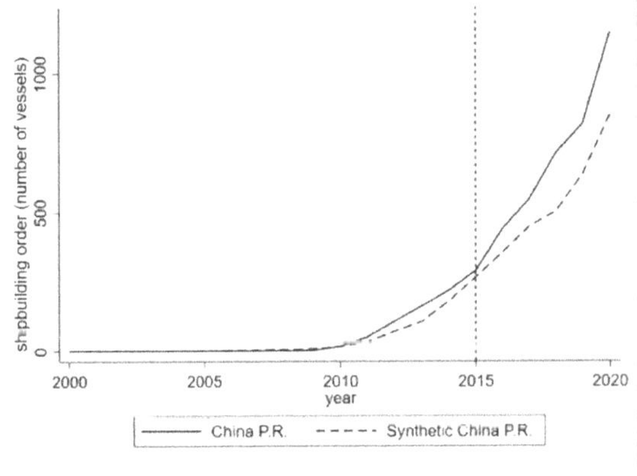

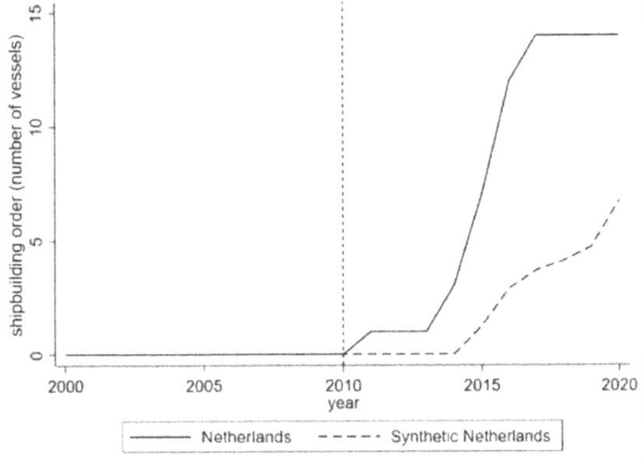

Figure 4. *Cont.*

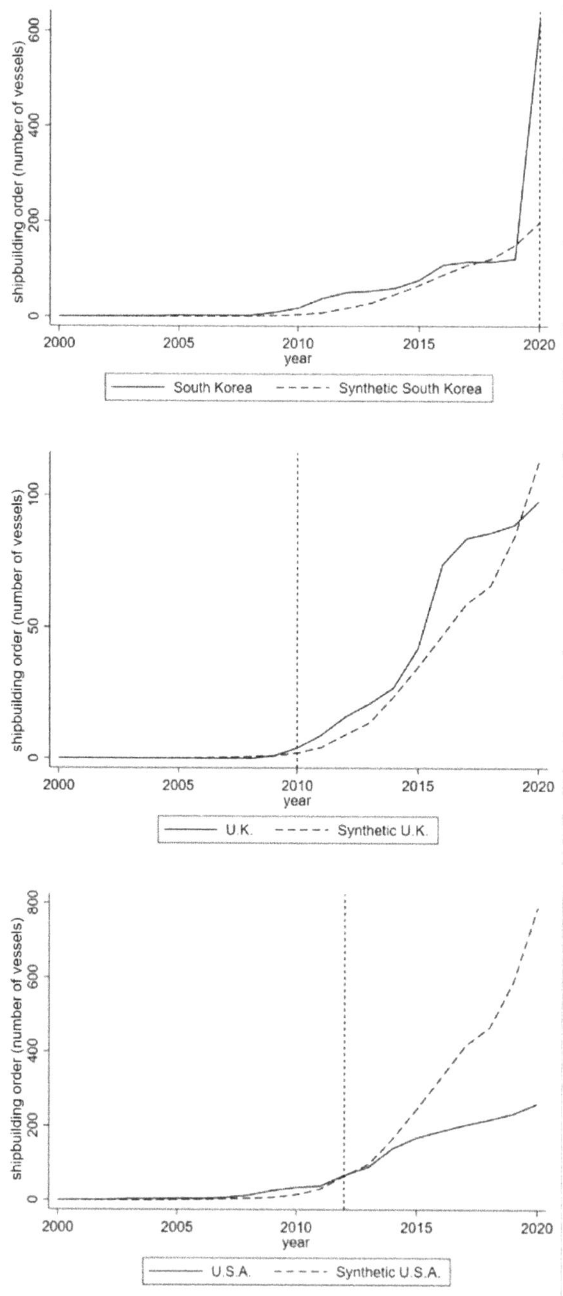

Figure 4. Trends in cumulative newbuilt orders of eco-designed vessels: treatment group vs. synthetic group in Stata 16. The vertical dashed line represents the year in which the treatment group country implemented ECA policies.

The results indicate that after the implementation of emission-control areas, the actual cumulative shipbuilding orders for eco-designed ships have significantly increased, while the synthetic cumulative shipbuilding orders are still steadily increasing, resulting in a huge gap. This also indicates that the implementation of emission-control areas has a positive impact on the shipbuilding orders of eco-designed ships and can promote the improvement of the marine environment. In addition, there are some special phenomena: (1) After 2010, the shipbuilding orders for eco-designed ships from China and Republic of Korea became more significant, which mainly contributed to the implementation of emission-control areas in Europe and North America. Since the beginning of international trade, shipbuilding orders have maintained a rapid growth trend to satisfy the needs of Europe and North America. Although China and Republic of Korea have not implemented emission-control areas, the rapid shipbuilding orders for eco-designed ships from the other regions have effectively promoted the improvement of the marine environment. (2) In 2020, due to the COVID-19 epidemic, the orders of the treatment group (The Netherlands, the UK, and the USA) for eco-designed ships declined significantly. Shipping companies must pay attention to the liquidity crisis in the shipping industry and the potential adverse effects of a short-term decrease in the orders for eco-designed ships. In particular, they must pay special attention to the direction of macroeconomic policy, seize opportunities for domestic demand-oriented reception, and try to replenish working capital as much as possible to overcome the crisis smoothly.

Additionally, as shown in Figure 4, there is a significant gap between the two curves. During 2010, the shipbuilding orders for eco-designed ships in the Netherlands and the UK increased rapidly, which may be the main reason. It is worth noting that the shipbuilding orders for eco-designed ships show a geometric doubling trend not only in The Netherlands and the UK but also in China, Republic of Korea, and the USA. This is mainly because of the development trend of world economic integration. Therefore, this not only helps to promote domestic order but also accelerates in other regions, which is beneficial for the global marine ecological environment.

However, since the early implementation of emission-control areas in North America and Europe, shipbuilding orders for eco-designed ships have been decreasing every year. In particular, with the slowdown in economic development and the adjustment of industrial layout, the growth of orders in the United States is not significant, even with the implementation of emission-control areas. In contrast, there has been a significant increase in annual orders from China and Republic of Korea, mainly due to the huge potential market. Please note that in 2020, the shipbuilding orders for eco-designed ships from China were the second-largest increase since the expansion of emission-control areas.

3.3. Robustness Test

From the above analysis, it can be seen that there are significant differences in the actual orders of eco-designed ships between the treatment group and the synthesis group. Hence, is this gap caused by the implementation of emission-control areas, or is it a coincidence? To assess the importance of our assumptions, two methods are used to test the robustness of the results.

3.3.1. Placebo Test

In this study, it is observed that the implementation of emission-control areas increased the shipbuilding orders for eco-designed ships. To further discuss the effectiveness and robustness of the variables based on the shipbuilding orders for eco-designed ships, the placebo test is first used to analyze the impact on the variables before the intervention, the basic idea of which is to utilize the same method for the regions where emission control is implemented, i.e., the test divides the pre-intervention period into an initial training period and a subsequent validation period. Therefore, in this study, the data from the starting date to the implementation of emission-control areas can be divided into initial training periods, whereas the data from the implementation of emission-control areas to December 2020 is divided into validation periods (Figure 5).

As shown in Figure 5, we present the placebo test results for estimating the impact of pseudo-interventions from the date of implementation of emission-control areas in each region. Here, the solid lines mean the difference between the five regions and their synthetic effect, whereas the solid lines represent the differences in the cumulative shipbuilding orders for eco-designed ships. Obviously, it can be seen that the MSPE (mean-square pure error) fluctuation between the treatment and control group countries before the policy occurred is insignificant, indicating that the treatment and control groups had similar development trends before the policy occurred, which ensures the placebo test is effective in validating the robustness of the results of the synthetic control method. Intuitively speaking, the curves of the treatment group and synthesis group show a similar trend and a stable trend before and after the expected pseudo-intervention in China and Republic of Korea. In addition, the results of the two exclusion regions correspond to the incomplete synthesis group in Figure 4.

Furthermore, the current gap between China and Republic of Korea is significantly larger than the two regions that did not implement the ECA policy. Taking China as an example, the curve is significantly higher than zero, indicating that the shipbuilding orders for eco-designed ships in 2015 were already higher than in the aforementioned years. On the other hand, it can be seen that Republic of Korea's efficiency in 2020 was significantly higher than that of other countries, meaning a significant increase in the shipbuilding orders for eco-designed ships from Republic of Korea after the implementation. However, the treatment effects after the implementation in the Netherlands, UK, and USA are not very significant. Although the treatment effect in the UK was greater in 2010 than in the control groups, it was lower after 2018 than in the control groups, which says that the implementation initially had a certain promoting effect on the shipbuilding orders for eco-designed ships but gradually disappeared.

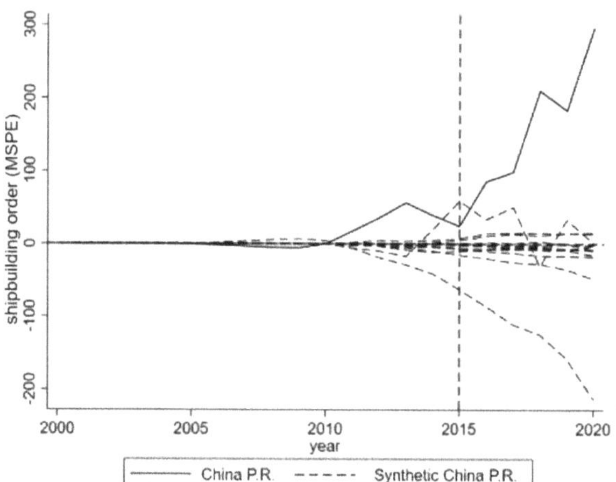

Figure 5. *Cont.*

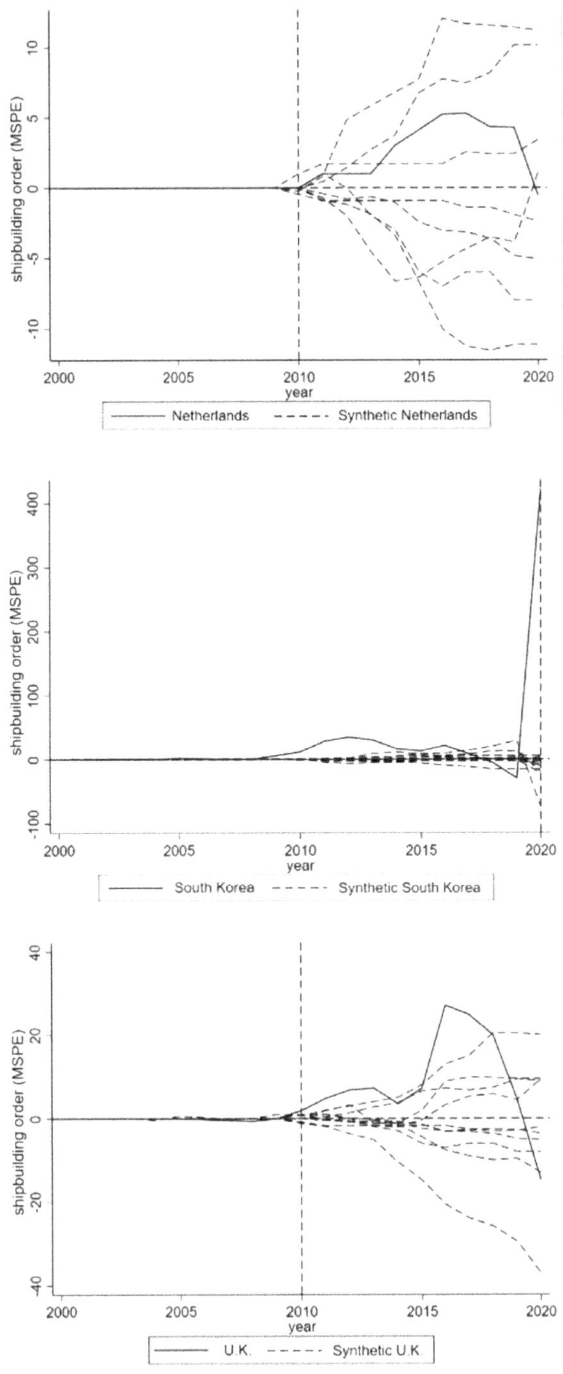

Figure 5. *Cont.*

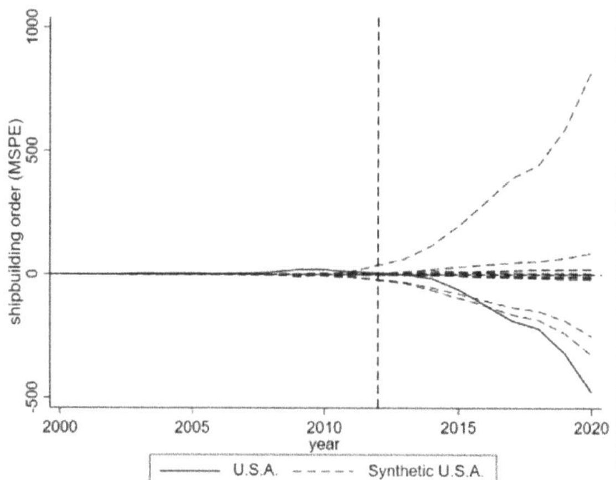

Figure 5. Placebo test of an eco-designed vessel: treatment group vs. synthetic group in Stata 16. The vertical dashed line represents the year in which the treatment group country implemented ECA policies.

3.3.2. Difference-in-Differences Test

Although the synthetic control method and difference-in-differences method are frequently used in comparative studies, the previous method introduces the variable presence that cannot be observed to influence the results over time in the latter method. Next, this study explores the effect of emission-control areas on the shipbuilding orders for eco-designed ships through the difference-in-differences method and then compares the outcome with that of the synthetic control method.

As shown in Table 4, the implementation of emission-control areas has had a positive impact on the shipbuilding orders for eco-designed ships in China, The Netherlands, Republic of Korea, the UK, and the USA. It is worth noting that for the difference-in-difference test, the implementation of emission-control areas has increased the shipbuilding orders for eco-designed ships in China, The Netherlands, Republic of Korea, the UK, and the USA by 3475, 23, 243, 1381, and 842, respectively, which is consistent with the verification results of robustness using the synthetic control method. However, due to the difference-in-differences method in testing, comparability is required between the treatment groups and control groups before the implementation of emission-control areas. Hence, it is difficult to observe the control group with the same related factors as the treatment group. On the other hand, the synthetic control method provides a better comparison of the treatment group using the weighted average of the control groups. Intuitively speaking, during the period before the implementation of emission-control areas, the trend of the shipbuilding orders for eco-designed ships from synthetic control methods is almost identical. On this basis, the estimation of the synthetic control method is more reasonable, which also provides monthly data to evaluate the impact of implementing emission-control areas.

Table 4. Comparative analysis between synthetic control and difference-in-differences method.

	China	The Netherlands	Republic of Korea	UK	USA
Synthetic control	4142	223	459	3030	1079
Difference-in-difference	3475	23	243	1381	842
$p > |t|$	0.17	0.37	0.31	0.69	0.68
R-square	0.48	0.06	0.06	0.14	0.17

4. Conclusions

To promote the sustainable development of the marine environment, IMO has adopted various policies. This research adopts a synthetic control method to quantitatively analyze the implementation effect of emission-control areas on the shipbuilding orders for eco-designed ships and further examine the robustness through the placebo test and difference-in-difference test. From the outcome, the implementation of emission-control areas has greatly increased the shipbuilding orders for eco-designed ships. Meanwhile, the shipbuilding orders for eco-designed ships from 2010 to 2020 in China, The Netherlands, Republic of Korea, the UK, and the USA increased by 3401, 81, 234, 549, and -1435, respectively. Compared with the absence of emission-control areas, the shipbuilding orders for eco-designed ships increased by 32%, 20%, 41%, 66%, and -83%. Except for the United States, the implementation effect is significantly improved.

Although the implementations of emission-control areas have positive impacts on the shipbuilding orders for eco-designed ships, governments have considered issuing guidance on supply guarantees and joint supervision of low-sulfur fuel oil, promoting the coordination and linkage mechanism of LNG clean-energy applications as a strengthening of the infrastructure construction of shore-to-ship power. Because the price of low-sulfur fuel oil is significantly higher than that of high-sulfur fuel oil, governments charge environmental tax on high-sulfur fuel oil and provide a certain number of subsidies for clean energy used by ships to reduce the investment. Beyond that, with a certain range, maritime authorities expand and upgrade the emission-control areas to increase shipbuilding orders for eco-designed ships. In particular, governments have expanded the scope of emission-control areas to cover key seaports and promote their transformation to reduce sulfur-oxide emissions. Expanding the scope of emission-control areas has motivated shipping companies to increase the orders of eco-designed ships, which is lower than the long-term investment of desulfurization transformation.

This study has some limitations. In this model, the implementation of emission-control areas on the shipbuilding orders for eco-designed ships is investigated. This may have a negative effect on other types of ships. In addition, relevant factors are not considered, which may drive the orders of eco-designed ships and have different impacts on the results. In terms of sample selections, five countries are selected to analyze the implementation of emission-control areas, where the sulfur limitation implemented by the IMO since 2010 is still a very short period of implementation, so the complete dataset has not been obtained yet. Finally, an additional limitation is that the aggregated data ignore detailed micro-insights into ships (tonnage, cost), which are necessary for the discussion of the impact of emission-control areas on shipbuilding orders for eco-designed ships.

Author Contributions: Methodology, L.X.; Software, L.L.; Validation, Z.Z.; Data curation, G.X. All authors have read and agreed to the published version of the manuscript.

Funding: National Natural Science Foundation of China (No. 52302393) and Shanghai Soft Science Research Program (No. 23692118800).

Institutional Review Board Statement: Not applicable.

Informed Consent Statement: Not applicable.

Data Availability Statement: All data have been used with permission and their sources have been indicated in the text.

Conflicts of Interest: The authors declare no conflict of interest.

References

1. Peng, J.; Xiao, J.; Zhang, L.; Wang, T. The impact of china's 'atmosphere ten articles' policy on total factor productivity of energy exploitation: Empirical evidence using synthetic control methods. *Resour. Policy* **2020**, *65*, 101544. [CrossRef]
2. Xu, L.; Di, Z.; Chen, J.H. Evolutionary game of inland shipping pollution control under government co-supervision. *Mar. Pollut. Bull.* **2021**, *171*, 112730. [CrossRef] [PubMed]

3. Ji, C.; El-Halwagi, M. A data-driven study of IMO compliant fuel emissions with consideration of black carbon aerosols. *Ocean Eng.* **2020**, *218*, 108241. [CrossRef]

4. Tran, N.K.; Lam, J.S.L.; Jia, H.; Adland, R. Emissions from container vessels in the port of Singapore. *Marit. Policy Manag.* **2022**, *49*, 306–322. [CrossRef]

5. Wan, Z.; Ji, S.; Liu, Y.; Zhang, Q.; Chen, J.; Wang, Q. Shipping emission inventories in China's Bohai Bay, Yangtze River Delta, and Pearl River Delta in 2018. *Mar. Pollut. Bull.* **2020**, *151*, 110882. [CrossRef] [PubMed]

6. Chen, J.; Xiong, W.; Xu, L.; Di, Z. Evolutionary game analysis on supply side of the implement shore-to-ship electricity. *Ocean Coast. Manag.* **2021**, *215*, 105926. [CrossRef]

7. Prehn, M. Climate strategy in the balance who decides? *Mar. Policy* **2021**, *131*, 104621. [CrossRef]

8. Eyring, V.; Köhler, H.W.; Lauer, A. Emissions from international shipping: Impact of future technologies on scenarios until 2050. *J. Geophys. Res. Atmos.* **2005**, *110*, D17306. [CrossRef]

9. Zhao, J.; Wei, Q.; Wang, S.; Ren, X. Progress of ship exhaust gas control technology. *Sci. Total Environ.* **2021**, *799*, 149437. [CrossRef]

10. Van, T.C.; Ramirez, J.; Rainey, T.; Ristovski, Z.; Brown, R.J. Global impacts of recent IMO regulations on marine fuel oil refining processes and ship emissions. *Transp. Res. Part D* **2019**, *70*, 123–134. [CrossRef]

11. Chen, J.; Zhang, W.; Song, L.; Wang, Y. The coupling effect between economic development and the urban ecological environment in Shanghai Port. *Sci. Total Environ.* **2022**, *841*, 156734. [CrossRef]

12. Shi, K.; Weng, J.; Li, G. Exploring the effectiveness of ECA policies in reducing pollutant emissions from merchant ships in Shanghai port waters. *Mar. Pollut. Bull.* **2020**, *155*, 111164. [CrossRef]

13. Wan, Z.; Zhou, X.; Zhang, Q.; Chen, J. Influence of sulfur emission control areas on particulate matter emission: A difference-in-differences analysis. *Mar. Policy* **2021**, *130*, 104584. [CrossRef]

14. Xu, L.; Huang, J.C.; Chen, J.H. How does the initiative of 21st century maritime silk road incentive logistics development in China's coastal region? *Ocean Coast. Manag.* **2023**, *239*, 106606. [CrossRef]

15. Xiao, G.; Chen, L.; Chen, X.; Jiang, C.; Ni, A.; Zhang, C.; Zong, F. A hybrid visualization model for knowledge mapping: Scientometrics, SAOM, and SAO. *IEEE Trans. Intell. Transp. Syst.* **2023**, 1–14. [CrossRef]

16. Zhang, S.; Chen, J.; Wan, Z.; Yu, M.; Shu, Y.; Tan, Z.; Liu, J. Challenges and countermeasures for international ship waste management: IMO, China, United States, and EU. *Ocean Coast. Manag.* **2021**, *213*, 105836. [CrossRef]

17. Lion, S.; Vlaskos, I.; Taccani, R. A review of emissions reduction technologies for low and medium speed marine Diesel engines and their potential for waste heat recovery. *Energy Convers. Manag.* **2020**, *207*, 112553. [CrossRef]

18. Sakib, N.; Gissi, E.; Backer, H. Addressing the pollution control potential of marine spatial planning for shipping activity. *Mar. Policy* **2021**, *132*, 104648. [CrossRef]

19. Van, T.C.; Zare, A.; Jafari, M.; Bodisco, T.A.; Surawski, N.; Verma, P.; Suara, K.; Ristovski, Z.; Rainey, T.; Stevanovic, S.; et al. Effect of cold start on engine performance and emissions from diesel engines using IMO-compliant distillate fuel. *Environ. Pollut.* **2019**, *255*, 113260. [CrossRef]

20. Halff, A.; Younes, L.; Boersma, T. The likely implications of the new IMO standards on the shipping industry. *Energy Policy* **2019**, *126*, 277–286. [CrossRef]

21. Kokosalakis, G.; Merika, A.; Merika, X. Environmental regulation on the energy-intensive container ship sector: A restraint or opportunity? *Mar. Policy* **2021**, *125*, 104278. [CrossRef]

22. Peng, X.; Huang, L.; Wu, L.; Zhou, C.; Wen, Y.; Chen, H.; Xiao, C. Remote detection sulfur content in fuel oil used by ships in emission control areas: A case study of the Yantian model in Shenzhen. *Ocean Eng.* **2021**, *237*, 109652. [CrossRef]

23. Theocharis, D.; Rodrigues, V.S.; Pettit, S.; Haider, J. Feasibility of the Northern Sea Route for seasonal transit navigation: The role of ship speed on ice and alternative fuel types for the oil product tanker market. *Transp. Res. Part A* **2021**, *151*, 259–283. [CrossRef]

24. Fan, L.; Li, Z.; Xie, J.; Yin, J. Container ship investment decisions—Newbuilding vs second-hand vessels. *Transp. Policy* **2023**, *143*, 1–9. [CrossRef]

25. Okay, O.S.; Karacık, B.; Güngördü, A.B.B.A.S.; Ozmen, M.; Yılmaz, A.; Koyunbaba, N.C.; Yakan, S.D.; Korkmaz, V.; Henkelmann, B.; Schramm, K.-W. Micro-organic pollutants and biological response of mussels in marinas and ship building/breaking yards in Turkey. *Sci. Total Environ.* **2014**, *496*, 165–178. [CrossRef]

26. Pettit, S.; Wells, P.; Haider, J.; Abouarghoub, W. Revisiting history: Can shipping achieve a second socio-technical transition for carbon emissions reduction? *Transp. Res. Part D* **2018**, *58*, 292–307. [CrossRef]

27. Zhu, Y.; Zhou, S.; Feng, Y.; Hu, Z.; Yuan, L. Influences of solar energy on the energy efficiency design index for new building ships. *Int. J. Hydrog. Energy* **2017**, *42*, 19389–19394. [CrossRef]

28. Lindstad, E.; Polić, D.; Rialland, A.; Sandaas, I.; Stokke, T. Decarbonizing bulk shipping combining ship design and alternative power. *Ocean Eng.* **2022**, *266*, 112798. [CrossRef]

29. Li, D.C.; Yang, H.L.; Xing, W.Y. Economic and emission assessment of LNG-fuelled ships for inland waterway transportation. *Ocean Coast. Manag.* **2023**, *246*, 106906. [CrossRef]

30. Fan, A.; Wang, Y.; Yang, L.; Tu, X.; Yang, J.; Shu, Y. Comprehensive evaluation of machine learning models for predicting ship energy consumption based on onboard sensor data. *Ocean Coast. Manag.* **2024**, *248*, 106946. [CrossRef]

31. Abadie, A.; Diamond, A.; Hainmueller, J. Synthetic control methods for comparative case studies: Estimating the effect of California's tobacco control program. *J. Am. Stat. Assoc.* **2010**, *105*, 493–505. [CrossRef]

32. Abadie, A.; Gardeazabal, J. The economic costs of conflict: A case study of the Basque country. *Am. Econ. Rev.* **2003**, *93*, 113–132. [CrossRef]
33. Li, K.; Wu, M.; Gu, X.; Yuen, K.F.; Xiao, Y. Determinants of ship operators' options for compliance with IMO 2020. *Transp. Res. Part D* **2020**, *86*, 102459. [CrossRef]
34. Chi, Y.; Wang, Y.; Xu, J. Estimating the impact of the license plate quota policy for ICEVs on new energy vehicle adoption by using synthetic control method. *Energy Policy* **2021**, *149*, 112022. [CrossRef]

Journal of
*Marine Science
and Engineering*

Article

A Bibliometric Analysis and Overall Review of the New Technology and Development of Unmanned Surface Vessels

Peijie Yang [1], Jie Xue [1,*] and Hao Hu [1,2]

[1] School of Naval Architecture, Ocean & Civil Engineering, Shanghai Jiao Tong University, Shanghai 200240, China; yangpeijie21@sjtu.edu.cn (P.Y.); hhu@sjtu.edu.cn (H.H.)
[2] State Key Laboratory of Ocean Engineering, Shanghai Jiao Tong University, Shanghai 200240, China
* Correspondence: j.xue@sjtu.edu.cn

Abstract: With the significant role that Unmanned Surface Vessels (USVs) could play in industry, the military and the transformation of ocean engineering, a growing research interest in USVs is attracted to their innovation, new technology and automation. Yet, there has been no comprehensive review grounded in bibliometric analysis, which concentrates on the most recent technological advancements and developments in USVs. To provide deeper insight into the relevant research trends, this study employs a bibliometric analysis to examine the basic features of the literature from 2000 to 2023, and identifies the key research hotspots and modeling techniques by reviewing their current statuses and the recent efforts made in these areas. Based on the analysis of the temporal and spatial trends, disciplines and journals' distribution, institutions, authors and citations, the publications relating to the new technology of USVs are assessed based on their keywords and the term analysis in the literature; six future research directions are proposed, including enhanced intelligence and autonomy, highly integrated sensor systems and multi-modal task execution, extended endurance and resilience, satellite communication and interconnectivity, eco-friendly and sustainable practices and safety and defense. The scientific literature is reviewed in a systematic way using a comparative analysis of existing tools, and the results greatly contribute to understanding the overall situation of new technology in USVs. This paper is enlightening to students, international scholars and institutions, as it can facilitate partnerships between industry and academia to allow for concerted efforts to be made in the domain of USVs.

Keywords: Unmanned Surface Vessel (USV); research trends; bibliometric analysis; research hotspots; VOSviewer

Citation: Yang, P.; Xue, J.; Hu, H. A Bibliometric Analysis and Overall Review of the New Technology and Development of Unmanned Surface Vessels. *J. Mar. Sci. Eng.* **2024**, *12*, 146. https://doi.org/10.3390/jmse12010146

Academic Editor: Sergei Chernyi

Received: 21 December 2023
Revised: 6 January 2024
Accepted: 8 January 2024
Published: 11 January 2024

1. Introduction

In recent years, unmanned vehicles have grown in popularity, with an ever-increasing number of applications in industry, the military and research within air, ground and marine domains [1]. This evolution has led to the emergence of sophisticated and self-sufficient watercrafts capable of undertaking complex tasks with minimal human intervention. USVs (Unmanned Surface Vessels), characterized by their autonomous navigation, operation and decision-making capabilities, represent a critical intersection between cutting-edge technologies and maritime operations [2]. By integrating advanced technologies such as Artificial Intelligence (AI), machine learning, computer vision and automation, USVs can not only optimize their operational efficiency but also enhance safety and reduce the risk of human errors. These vessels are equipped with a range of sensors, including radar, lidar and various imaging systems [3], enabling them to perceive and respond to dynamic marine environments effectively.

The application of USVs extends to diverse domains, including but not limited to, maritime surveillance, oceanographic research, marine resource exploration and environmental monitoring. For example, Tianyu Ma highlights the importance of USVs as weapons for various applications, and describes key technologies such as high-performance crafts, control systems, communication technology, collision avoidance and mission planning [4]. The collaborative system of Multiple Unmanned Surface Vehicles (MUSVs) has shown broad prospects in the field of civil and military applications [3]. To enhance the accuracy, a combined Nonlinear Model Predictive Control (NMPC) for the position and velocity tracking of under-actuated surface vessels, and the collision avoidance of static and dynamic objects into a single control scheme with side slip angle compensation and environmental disturbance counteraction, was presented and achieved high validity [5]. A novel path-planning methodologies for autonomous maritime vessels, including USVs, has been developed to operate efficiently in intricate and dynamic marine settings. This approach ensures rapid, reliable and definitive navigational solutions, even in scenarios characterized by multiple mobile maritime vessels and stationary hindrances, while also accommodating the variable and unforeseen maneuvers of surrounding vessels [6]. With the advent of machine learning, more research was focused on communication, remote sensing and connected automation, exemplified by a new autonomous data collection with dynamic goals and communication constraints for marine vehicles [7]; it was also focused on a powerful unmanned boat remote-control platform that can realize remote display sensing data, remote motion control functions and can ensure the unmanned boat under the safety requirements of various experiments [8], as well as a concurrent kinematic control tactic introduced for the landing of Unmanned Aerial Vehicles (UAVs) on USVs to mitigate the challenges of diminished precision and potential landing failures caused by the USVs' surface motion due to wave activity [9].

Despite increasingly growing interest in USVs, there is a lack of a systematic overview of solutions proposed in the scientific literature. Among exemplary review articles, there are a few focused on challenge analysis [2,10,11], collision avoidance methods [12–14], path planning [14–16] and control and autonomy [17,18]. However, according to the search results of WoS, there is a notable absence of a systematic literature review and featured bibliometric analysis regarding USVs in the context of the new technology applied in automation development, with a specific emphasis on assessing the development of the automation and future developments.

Therefore, this paper aims to provide a comprehensive overview of the literature relevant to USVs, with a focus on the technological developments, operational automation and the current state of research in the field. Through a meticulous analysis of existing scholarly works, this paper seeks to identify key trends, challenges and future prospects for the integration and proliferation of USVs in maritime applications. By examining the trajectory of the research and technological advancements in this domain, this paper aims to contribute to the collective understanding of the evolving field of USVs and their potential implications for the maritime industry and related fields.

In contrast with previous studies, this paper has three contributions. It is the first attempt at a bibliometric-based review of USVs, which comprehensively examines the basic features of the literature and identifies the current hotspots. Second, the focus is more concentrated on the new technologies and new methods utilized in USVs. Third, the future research directions and detailed analysis of different angles are highlighted to advance the knowledge in this area.

The paper is structured as follows: in Section 2, the methods used in the study are described. Section 3 presents the results showing, respectively, the bibliographic and comparative analyses. A discussion is provided in Section 4, while Section 5 summarizes and concludes the paper.

2. Data and Methodology

2.1. Research Framework

To offer more comprehensive insights into the advancement of cutting-edge technology in respect of USVs, this investigation conducted a systematic retrieval of scholarly articles focusing on USVs from the Web of Science (WoS) Core Collection database. Compared with other databases, WoS core database is more widely used and authoritative, and relevant published high-quality articles [19–23] have adopted WoS as their supporting database. Through the application of bibliographic analysis, this research delineated the prominent areas of investigation and the methodologies employed in the exploration of subjects related to USVs. The research framework is graphically represented in Figure 1 for reference.

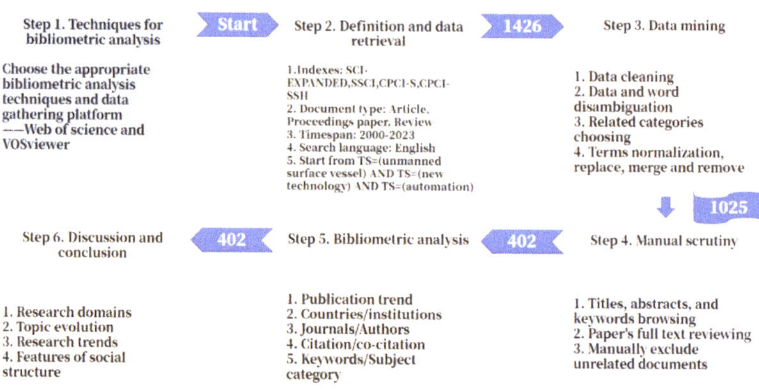

Figure 1. Research framework.

2.2. Bibliometric Methods and Visualization Tools Used in the Analysis

In this research, the utilization of bibliometric mapping methods and tools facilitated the visualization of scientific data through diverse procedures. Bibliometric mapping, as a research methodology, involves the application of quantitative techniques to visually depict the scientific literature derived from bibliographic data.

VOSviewer, a software tool developed by Van Eck and Waltman in 2010, specifically designed for the analysis and visualization of the scientific literature, was used in this present work. VOSviewer is a software widely used in the bibliometric field and in different disciplinary areas [24,25]. VOSviewer offers the capability to depict co-authorship networks among authors, institutions and geographic regions, co-citation networks of articles and journals, as well as co-occurrence networks of keywords and terms. This tool leverages clustering techniques for the visual detection of structural patterns within the research domain, and harnesses text-mining functionalities for recognizing trends and patterns in the addressed topics [26]. In this paper, the version 1.6.19 of VOSviewer was used for further research.

For a comprehensive understanding of the techniques and concepts underpinning bibliometric mapping, Li et al. [27] have provided an informative overview. Moreover, VOSviewer has gained widespread utilization in the realm of bibliometric analysis within the marine science field, as evidenced by various studies [20,28–40]. For a detailed examination of these applications, please refer to Li et al. [27].

Furthermore, two parameters, Impact Factor (IF) and H-index, were utilized in this paper to measure the impact of the journal or author. The IF and the H-index are two prominent metrics utilized to assess the significance and widespread influence of a publication [20]. They serve as valuable benchmarks when making informed decisions regarding the choice of journals and publications within a particular research domain. The ArcGIS, a comprehensive Geographic Information System (GIS) software that enables users to create, manage, analyze and visualize spatial data was utilized to facilitate the understanding of complex geospatial relationships of the publications, to aid in decision-making processes and to support the creation of detailed maps and visualizations for various applications.

Lastly, this study made use of the data analysis functionalities of WoS, in conjunction with VOSviewer, to conduct citation analysis and track the number of publications per year, with the overview methods shown in Figure 2.

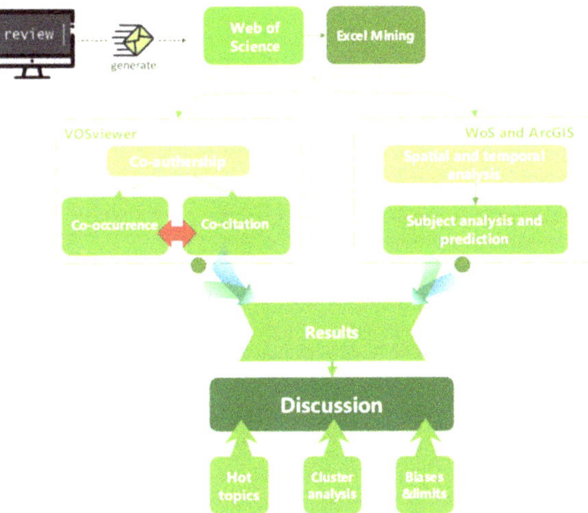

Figure 2. Flowchart for the methods applied in this research.

2.3. Data Extraction and Exclusion Criteria

The process of preparing the dataset comprised three distinct stages. First and foremost, the initial step (Stage 1) involved the formulation of a well-defined search strategy and the acquisition of relevant data. Subsequently, the acquired dataset underwent an initial screening (Stage 2) before proceeding to the final stage (Stage 3), where two distinct methods were employed for data refinement. The complete process of determining the dataset was visually depicted in Figure 3 and was elaborated upon in subsequent sections of this part.

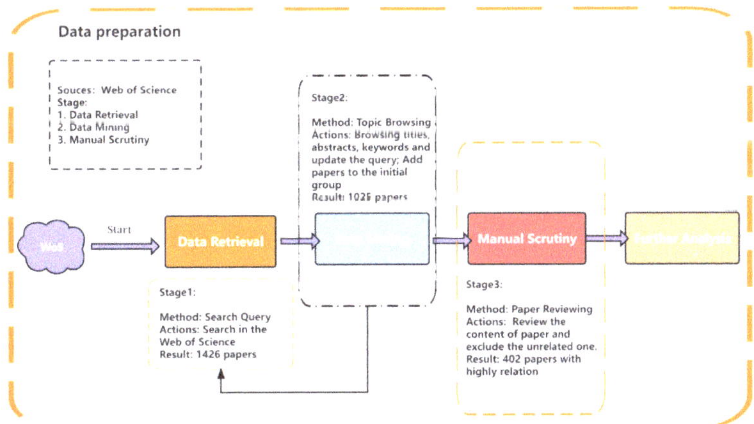

Figure 3. The process of determining the final data sample.

In the first stage of dataset compilation, a meticulously crafted search query was employed to gather the initial collection of documents. The query was solidly found in 2 themes: USVs and new technology. So that more insights into frontier developments would be taken, the choice of WoS as the primary data source was grounded in its recognition as a comprehensive and widely accepted repository of abstracts and citations from high-impact scientific publications [41]. The documents were sourced from the principal WoS Core Collections, namely the Science Citation Index Expanded (SCI-EXPANDED) and the Social Sciences Citation Index (SSCI).

As for the design of the search query, a methodology of vocabulary iteration was utilized to add more synonyms. For instance, the abbreviation of *USV* could be also interpreted as the *unmanned surface vessel, unmanned surface vehicle, autonomous surface vessel, automatic surface vessel*, et al. The main key 10 steps of query are shown in the Appendix A Table A1.

Furthermore, the field of USVs has likely seen rapid growth and evolution in the past two decades. The 2000s onwards was a period where significant regulatory and industry changes occurred, impacting USV deployment and integration into various sectors. Reviewing the literature from this period allowed us to track these changes and their implications. Hence, the timespan was restricted to a period from 2000 to 2023 to support the novelty. In this paper, documents preceding 2 November 2023 were obtained and downloaded.

The second phase of data selection entailed a rigorous evaluation of the documents obtained from WoS. This evaluation was primarily centered on scrutinizing the document titles, abstracts and associated keywords, including both Author Keywords and Keywords Plus, as the search query was accordingly updated over and over again until it had been searched for throughout the database. Documents meeting the initial validation criteria were identified as relevant and were earmarked for further analysis. The iteration of the search query is shown in the Appendix A. These qualified papers were included in the new dataset and were forwarded to the next step of filtering and determining the final sample (1025 papers).

Thence, during the third stage of dataset preparation, all documents underwent a thorough manual scrutiny to assess the content and title against the predetermined criteria by excluding unrelated documents. Papers meeting these criteria were then categorized into the final group, and, eventually, 402 scientific documents were included in the final data sample. The documents contained 225 articles, 170 proceeding papers and 7 review articles. The final data sample would be applied in bibliometric analysis.

3. The Results of the Bibliometric Analysis

3.1. Publication Trends

Figure 4 illustrates the annual number of papers with strong relevance to the new technology of USVs, and shows the predicted trends for the future. It is evident that the number of annually published papers has overall steadily increased since about 2014. According to the dataset retrieved from WoS, the earliest paper directly on USVs, written by Portuguese scholars in English, was published in September 2003 [42]. The number of articles in Chinese was relatively low before 2007; this period can be regarded as the initial stage of the internationalization of USV research in China. Additionally, Figure 4 shows that the total number of articles published worldwide has seen an approximately curvilinear growth since 2013. Therefore, 2014 can be considered a pivotal year in the internationalization of new technology in USV research. After that, the number of publications assumes a trend of growth, in companion with a polynomial simulating curve colored in orange, showing that it is likely to keep this tendency and to reach a summit in 2023 or years afterwards.

To obtain more insights into the concurrence of citations and publications, the time trends of publications in terms of publications and citations are also presented in Figure 4. The results show an overall increasing trend of research on USVs during the period from 2000 to 2023. This trend can be approximately divided into the following three stages: in (1) 2000–2013, the number of publications was minimal and increased slowly, producing fewer than 15 publications; in (2) 2014–2019, publications were driven by several special reports on USVs [43] and commitments using some new methods [44], and the research interest in considering USVs as promising auxiliary strategies had begun to increase. During this stage, the number of publications and citations increased from 11 to 57 and from 50 to 325, respectively; from (3) 2020–present, there is a stable stage in development, as the number of publications is always higher than the previous ones, along with a peak at 64 in 2022, during which time the average annual publication reached 40–60 documents.

Note that the soaring increase in citations clearly demonstrates the concentrated focus in this field and forecasts the popular direction of USVs.

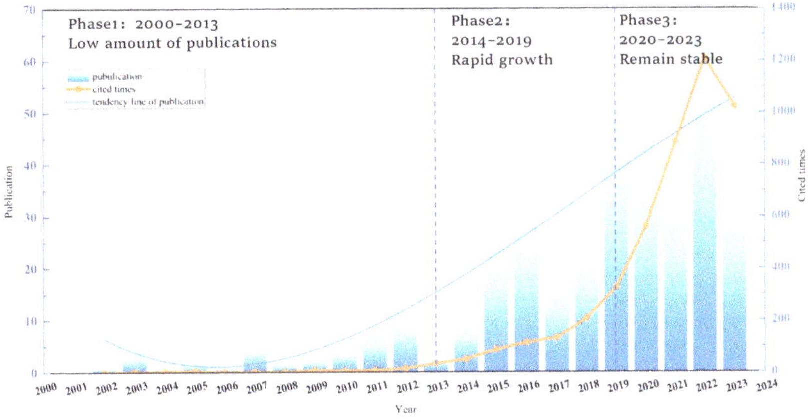

Figure 4. Total publications and citations on USV during 2000–2023.

3.2. Social Structure Analysis

3.2.1. Influential Authors Analysis

Scholars with substantially highly cited publications often dominate the conceptual and methodological trend of the pertinent research field and exert an influential impact on the development of the field [45]. Hence, it is essentially vital for us to identify the influential scholars so that we can gain more insights into the academic discourse.

To eliminate the ambiguous influence of these authors, we have made a detailed comparison among the top 10 productive authors and exhibit the information of their name, institution and some indicators of their contributions, as Table 1 shows. The results of the authors' productions show that the top 10 productive authors have, together, published 80 articles, accounting for nearly 20% of the total publications. As is listed in Table 1, the most productive and most cited author is Yuanchang Liu, from University College London, with 13 publications and 338 citations, followed by Kristan Matej, from University of Ljubljana, with 9 publications and Bucknall Richard from University College London with 8 publications. Moreover, Bucknall Richard was the second most cited author, with 251 citations. With 31.38 citations per paper, he also took the lead as the author with the highest average number of citations, demonstrating that his research received relatively significant attention from others. Note that Dalian Maritime University stood out with a larger number of prolific authors compared to other institutions, underscoring the significant contribution of this institution and its researchers within the USV research community.

Further, the analysis of an author's collaboration is essential and is carried out in lots of articles for the following reasons:

(1)　Revealing leading knowledge providers: the analysis of an author's collaboration network can identify the central figures or leading knowledge providers in a specific field; in this case, slip-and-fall incident research. These are the experts or authors who have made significant contributions to the body of knowledge in this domain.

(2)　Understanding social networks: beyond identifying individual experts, this analysis also delves into the social networks that exist among these authors. It reveals how these experts collaborate, communicate and share ideas, shedding light on the dynamics of knowledge creation and dissemination within the field.

(3)　Interest for early career researchers: early career researchers can benefit from such analyses when entering a new research domain. By identifying the key figures and their collaborative networks, they can find mentors, potential collaborators and resources to accelerate their research and integration into the field.

(4) Interest for external stakeholders: external stakeholders, such as industry professionals, policymakers or organizations interested in slip-and-fall incident research, can use this information to connect with world-class experts in the field. It enables them to seek advice, collaboration or expertise from those who are well-established in the domain.

Table 1. Top 10 most productive authors on USV.

Rank	Author	Institution	TP	TC	AC	H-Index
1	Yuanchang Liu	University College London	13	338	26.00	7
2	Kristan Matej	University of Ljubljana	9	231	25.67	6
3	Bucknall Richard	University College London National Engineering Research Center for Water Transport Safety	8	251	31.38	5
4	Yun Li	Shanghai Maritime University	8	62	7.75	5
5	Yan Peng	Shanghai University	8	31	3.88	3
6	Guofeng Wang	Dalian Maritime University	8	101	12.63	4
7	Dongdong Mu	Dalian Maritime University	7	101	14.43	4
8	Pers Janez	University of Ljubljana	7	209	29.86	5
9	Yunsheng Fan	Dalian Maritime University	6	82	13.67	3
10	Yuqing He	Chinese Academy of Sciences	6	40	6.67	3

Notes: TP = Total publications; TC = Times cited; AC = Average number of citations per publication.

Meanwhile, the selective inclusion of authors is needed to create a visually clear collaboration network, and it is common to limit the number of authors included. In this research, authors who have published more than two articles on the new technology of USVs were the target. This selective approach ensures that the collaboration network included individuals with a substantial body of work in this specific area, thereby emphasizing the most relevant and experienced contributors in Figure 5.

Figure 5. The network of author collaboration using analysis of the co-authorship.

In both visual representations, the size of the nodes corresponds to the number of publications attributed to each author [45]. The connecting lines between nodes depict instances of collaboration between these authors. In Figure 5, the color of a node signifies the specific clusters to which an author belongs. These clusters represent networks of authors whose collaborative work was evident through shared co-authorship relationships. The detailed setting for the parameters used for the figure was provided in Appendix A Table A2.

Authors can be categorized into distinct clusters based on the extent of their collaborative efforts, thereby revealing various sub-communities within the field of USV research. The authors analyzed were divided into five clusters. For example, Yuanchang Liu stood out as a pivotal figure within his group and had a substantial network of 22 collaboration links in the broader global research field. Similarly, authors such as Bo Wang (7), Yan Peng (10) and Yuqing He (12) took prominent roles within their respective groups. Further, at an individual level, with average publication year, it is noteworthy that the most recently active authors in USV included Bo Wang (2022), Peng Jiang (2021.5), Xinyu Zhang (2021) and Yulei Liao (2020). These scholars not only have exhibited high productivity but have also demonstrated significant activity, particularly in recent years.

More specifically, the authors of the red cluster mainly collaborated on system approach and validations of USV. The yellow cluster was more focused on the motion framework and navigation planning algorithms of USVs, while the purple cluster shared the early perception and navigation methods of USVs. Additionally, the blue cluster conducted cooperation in tracking control and neuro-adaptive control, and the green cluster worked on the detection and identity of USV.

3.2.2. Institutions Analysis

A total of 474 institutions were involved with the 402 documents in this study. Figure 6 illustrates the collaborative networks among institutions that have published more than two articles related to USV, with a total of 54 institutions meeting the threshold. In a manner similar to Section 3.1, Figure 6 delineates clusters of these institutions based on the strength of their collaborative ties. The size of the nodes in these network representations corresponds to the number of articles these institutions have contributed to the domain within our dataset. The lines connecting these nodes signify instances of collaboration among these institutions. The thicker the line, the more jointly authored articles the two connected institutions have published. Nodes sharing the same color indicate a higher level of collaboration in the new technology of USV research compared to others. The detailed setting for the parameters used for the figure was provided in Appendix A Table A2.

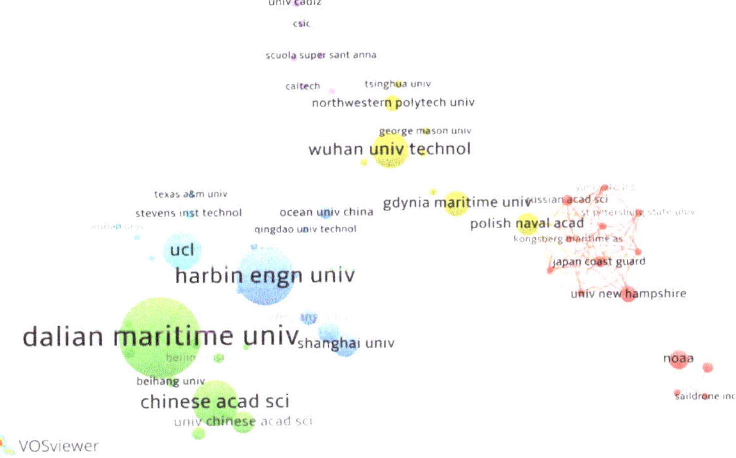

Figure 6. The network of institution collaboration using analysis of the co-authorship.

Figure 6 provides further insights into the collaborative research field, particularly highlighting the significant role of Dalian Maritime University. This institution stood out as the most prolific contributor, with a total of six collaborative links and 31 documents, primarily from Dalian Minzu University, UCL (University College London) and Southeast University. Notably, Dalian Maritime University also engaged in partnerships with universities in nearby northern cities, primarily Beijing and Harbin, thanks to their geographical proximity. Consequently, Dalian Maritime University served as the central hub within the green cluster due to these collaborative connections. Deduced from the light blue cluster, UCL, a university with the most productive author, acted as an important channel to conduct collaborative research with other institutions. As for the blue cluster, the institutions within this group were primarily situated near the ocean, encompassing areas like Zhejiang, Qingdao and Shanghai. In this cluster, Harbin Engineering University emerged as the focal point of collaborative activities. The yellow cluster, on the other hand, shows the strong link between Chinese institutions and European institutions, with Wuhan University of Technology playing a central role. Of particular interest was the fact that Polish Naval Academy had established connections with institutions in different regions, as it had 21 total link strengths, facilitated through its partnership with University of New Hampshire (UNH). Hence, UNH, became a crucial link in the red cluster, components of which were tightly intertwined and supported with each other and display a lasting and practical cooperation relationship. Lastly, the purple cluster consists of universities and research institutions in America, Singapore and Italy, etc., with National University of Singapore acting as the core of this cluster.

Among these institutions, Dalian Maritime University demonstrated the highest level of productivity, having authored 31 research papers. It was followed by Harbin Engineering University with 21 publications, Chinese Academy of Science with 15 and University College London with 13 publications. As evidenced in authors' analysis in Section 3.1, University College London had a dedicated team engaged in USV research, playing a central role in this specific domain. Furthermore, in accordance with Table 2, Shanghai Jiaotong University and Shanghai Maritime University were the top two institutions with the latest and highest average publication year, denoting that they were conducting most of the new technology research in recent years and paying sufficient attention to the burgeoning development of USVs.

Table 2. Bibliometric network and citation information of the top 16 most productive institutions.

Rank	Institution	Country	TP	TC	AC	TLS	APY
1	Dalian Maritime University	China	31	428	13.81	18	2020.2
2	Harbin Engineering University	China	21	160	7.62	17	2019.0
3	Chinese Academy of Science	China	15	91	5.69	14	2015.8
4	University College London	England	13	338	26.00	13	2019.2
5	Shenyang Institute of Autonomous CAS	China	12	57	4.75	12	2015.7
6	Wuhan University of technology	China	12	266	22.17	17	2019.4
7	University of Ljubljana	Slovenia	7	101	14.43	3	2018.7
8	Norwegian University of Science Technology NTNU	Norway	9	141	15.67	11	2020.2
9	Gdynia Maritime University	Poland	8	262	32.75	4	2014.1
10	Jiangsu University of science technology	China	8	5	0.63	8	2019.2
11	Korea Maritime Ocean University	Korea	8	100	12.50	14	2021.1
12	Shanghai University	China	8	53	6.63	11	2019.7
13	National Oceanic Atmospheric Admin NOAA USA	USA	7	48	6.86	12	2016.3
14	Shanghai Maritime University	China	7	29	4.14	9	2021.7
15	Harbin Institute of Technology	China	6	41	6.83	4	2019.8
16	Shanghai Jiaotong University	China	5	27	5.40	3	2021.8

Notes: TP = total publications; TC = times cited; AC = average number of citations per publication; TLS = total link strengths; APY = average publication year.

In spite of comparatively high total publications, institutions such as Shenyang Institute of Autonomous CAS, Gdynia Maritime University and Chinese Academy of Science embody an old APY of 2015.7, 2014.1 and 2015.8, respectively. The most cited or influential work of these institutions in the dataset is approaching a decade old. This could suggest that they made significant contributions to the field around that time. However, the older APY might also indicate a potential decrease in either the quantity or the impact of recent publications. It is importantly conduced that the institutions may have maintained their research momentums until 2014 or 2015, and then shifted its focus to other areas.

In addition, it should be noted that the table clearly demonstrates China's leading role in the field, as evidenced by the significant number of publications from the country.

Specifically, 10 out of the top 16 institutions are based in China, with the country dominating even more impressively among the top 6 institutions, 5 of which are affiliated with China.

3.2.3. Countries and International Cooperation Analysis

In terms of geographical distribution, a total of 55 countries and regions have participated in publishing 402 articles and reviews, while 30.9% of these countries contributed only one paper. Figure 7 illustrates the geographical distribution of publications from 2000 to 2023. The result shows that China took the lead with the largest number of publications at 177, followed by the United States with 68, England with 34, South Korea with 29 and Poland with 25.

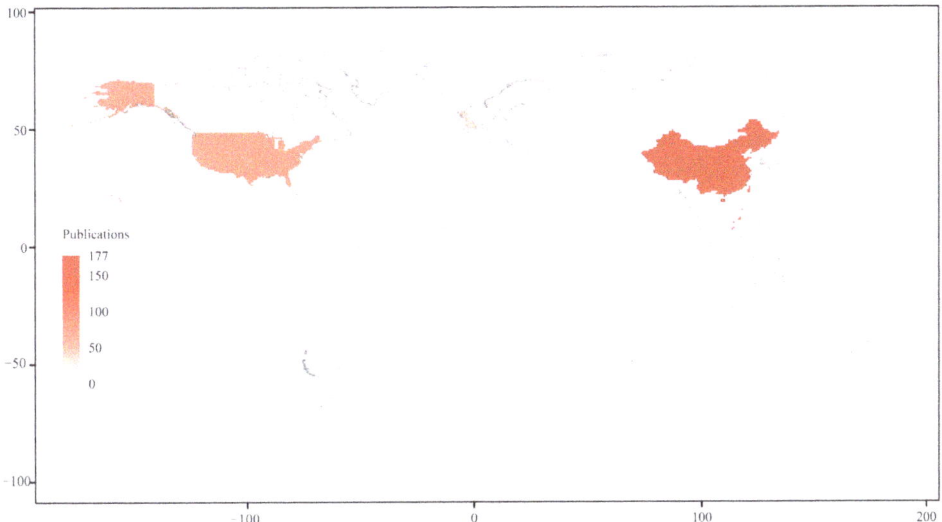

Figure 7. The distribution of publications arising from various countries.

When it comes to the total number of citations in Table 3, China (1655), the United States (1315), South Korea (702) and England (685) stood out, highlighting the substantial influence of these countries in this field.

Table 3. Total citation of the top ten countries.

Rank	Country	TC	AC
1	China	1655	8.40
2	USA	1315	19.39
3	South Korea	702	35.10
4	England	685	20.83
5	Bangladesh	413	413.00
6	Poland	332	17.47
7	Slovenia	231	23.10
8	Australia	223	27.88
9	Finland	157	31.40
10	Norway	146	12.17

Notes: TC = times cited; AC = average number of citations per publication.

Despite a smaller quantity of publications, Bangladesh and (413) Sweden (61.5) received a higher average number of citations than most other countries, indicating the exceptional quality of research conducted in these two nations. Overall, as per the statistics, the developing countries indeed had a limited global impact in the field of the USV in terms of publications and citations, as only two developing countries ranked within the top 15 productive countries.

To find the internal connection in the distribution, the international cooperation was analyzed by mapping countries based on the co-authorship. The detailed setting for the parameters used for the figure is provided in Appendix A Table A2.

It is well known that close research collaborations between countries drive technological and research advancements. Figure 8 illustrates the collaborative relationship among countries by examining co-authorship patterns. It is worth noting that the substantial number of publications on the new technology of USVs originated from China, as there was significant international cooperation between China and other countries, indicating that the research in this field factually requires a particular international cooperation. The network was constructed by using the full-counting method, with a minimum threshold of two documents per country, leading to the inclusion of 35 countries in the analysis. To normalize the data, association strength was applied. The color-coded overlay represents the year of publication, and the thickness of the links is proportional to the overall strength of the collaborations.

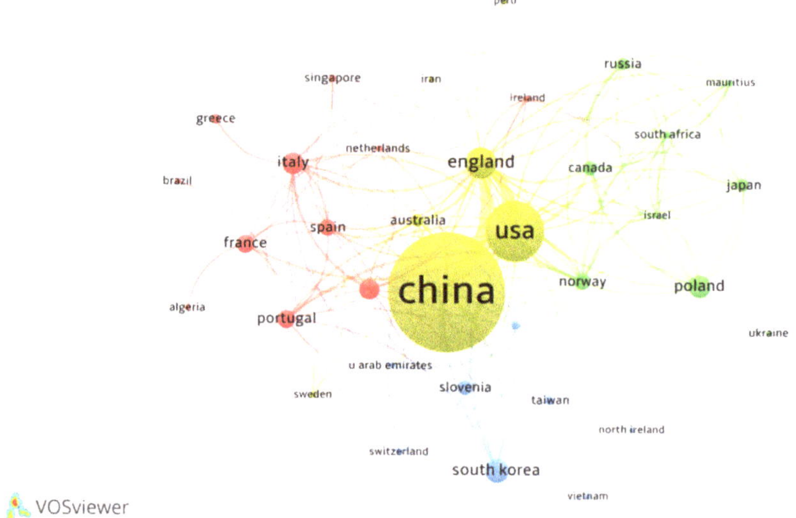

Figure 8. The network of international collaboration using analysis of co-authorship.

The international collaboration network reveals the presence of four primary clusters. Notably, China emerged as the most active collaborator, with the strongest link (10) established with England. It can also be inferred that Chinese researchers were currently responsible for the majority of document productions, as indicated by the color-coded data, and their collaborative network extended prominently at the international level. Additionally, note that a strong circle featured by the European region is shown in the red cluster, where Italy dominated the cooperative links in the red cluster, primarily conducting collaboration with England, Germany and Greece, along with some publications for a single time with other countries. South Korea and Poland took the privilege to engage in more contact with other countries in their clusters, respectively.

Figure 9 also presents information about the average publication year of each country, an indicator that tells us which country is more in tune with the era. It is palpably understood that France, Spain, Switzerland and the United Arab Emirates have the average latest publication year, showing that they have done much work in the past 3 years, whereas the USA (the United States), Canada and Japan, have published most of the papers 6 years ago, and still need to carry on researching new technology in the future.

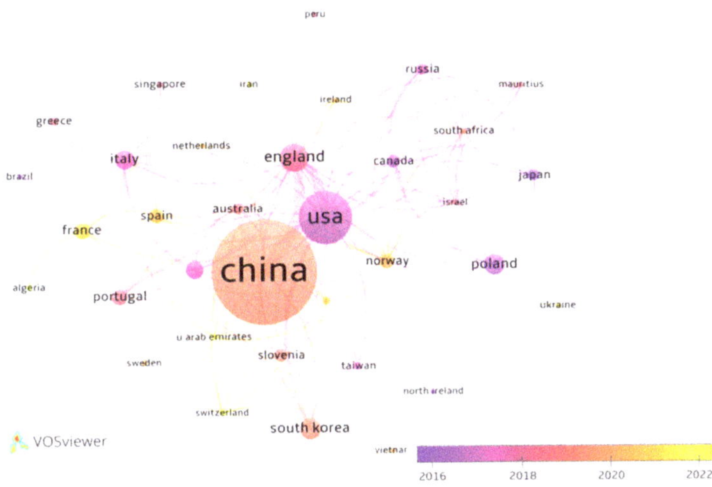

Figure 9. The network of international collaboration using analysis of co-authorship.

3.2.4. Disciplines and Subjects Analysis

The examination of the new technology in USVs as an interdisciplinary research field included a comprehensive spectrum of 77 distinct discipline categories, and disciplines with over 14 publications are visually represented in Figure 10. Within this diverse landscape of disciplines, it is notable that Engineering Electrical Electronic, Oceanography, Engineering Marine, Engineering Ocean and Automation Control Systems have prominently emerged as the dominant domains of study, as the publications related to these disciplines surpass 15%. To illustrate, the category of "Engineering Electrical Electronic" stood out as the most prolific with 108 publications, which account for a substantial 26.86% of the entire corpus of scrutinized documents. Following closely, "Oceanography" demonstrated its prominence with 103 publications, contributing significantly at 25.62%, while "Engineering Marine" occupied a vital role with 99 publications, representing 24.63% of the total.

Figure 10. Publication by subject category. Note: the number means the total number of publications by one subject. Source: Web of Science.

Within this interdisciplinary context, it is also intriguing to observe the presence of computer science disciplines among the top 10 most abundant categories. Disciplines such as "Engineering Electrical Electronic", "Robotics", "Computer Science Artificial Intelligence" and "Computer Science Information Systems" have secured their positions, underscoring the multifaceted and collaborative nature of new technology in USV research.

3.3. Citation and Co-Citation Network Analysis

3.3.1. Publications Citation and Co-Citation Analysis

Citation analysis is a way of measuring the influence and quality of a publication by counting the number of times that the publication has been cited by other publications [27]. The retrieved 402 papers were cited 3862 times by 3793 publications from 1270 publication sources. A total of 1881 out of 3793 publications were published by Chinese Institutions, which accounted for 49.60% of the citing publications. On top of that, out of 402 retrieved papers, 112 papers were never cited as of 1 October 2023; 235 papers, over half of the publications, were cited fewer than five times by any other publications.

Table 4 presents a compilation of the ten most extensively cited scholarly works in the domain of USV new technology authored by global researchers. Notably, the paper titled "6G Wireless Communication Systems: Applications, Requirements, Technologies, Challenges, and Research Directions" by Chowdhury et al. [46], was considered both the most cited publication and the one with the highest average citation count. This article, while focusing on the subject of 6G wireless communication, offered a comprehensive review of research developments pertaining to the application and prospects of 6G and reveals potential cultivation in USVs. The second most frequently cited article, written by Lyu et al. [6] and published in 2019, presented a real-time and deterministic path-planning method for USVs in complex and dynamic navigation environments. In this paper, a modified Artificial Potential Field (APF), which contained a new modified repulsion potential field function and the corresponding virtual forces, was developed to address the issue of Collision Avoidance (CA) with dynamic targets and static obstacles, including emergency situations [6]. Additionally, among the recent frequently cited literature, Hover et al. [47] developed and applied algorithms for the central navigation and planning problems on ship hulls to achieve a closed-loop control relative to features such as weld lines and biofouling, thus laying a cornerstone for operations on naval ships. Furthermore, it is worth mentioning that in the top 10 most highly cited publications, 6 of them were concentrated on the path-planning research of USVs, from the objective of collision avoidance to the real-time algorithms, showing tremendous attention from scholars to these aspects.

Table 4. Top 10 most highly cited publications on USV.

Article	Title	TC	ACY
Chowdhury et al., 2020 [46]	6G Wireless Communication Systems: Applications, Requirements, Technologies, Challenges, and Research Directions	415	103.75
Lyu et al., 2019 [6]	COLREGS-Constrained Real-time Path Planning for Autonomous Ships Using Modified Artificial Potential Fields	164	32.80
Hover et al., 2012 [47]	Advanced perception, navigation and planning for autonomous in-water ship hull inspection	157	13.08
Jahanbakht et al., 2021 [48]	Internet of Underwater Things and Big Marine Data Analytics-A Comprehensive Survey	129	43.00
Song et al., 2019 [49]	Smoothed A* algorithm for practical USV path planning	125	25.00
Lazarowska, A, 2015 [50]	Ship's Trajectory Planning for Collision Avoidance at Sea Based on Ant Colony Optimisation	116	12.89
Tsou et al., 2010 [51]	The Study of Ship Collision Avoidance Route Planning by Ant Colony Algorithm	116	8.29
Kahveci et al., 2013 [52]	Adaptive steering control for uncertain ship dynamics and stability analysis	99	9.00
Kristan et al., 2016 [53]	Fast Image-Based Obstacle Detection From USV	88	11.00
Kim et al., 2014 [54]	Angular rate-constrained path planning algorithm for USV	88	8.80

Notes: TC = times cited; ACY = average citations per year.

Co-citation, as originally articulated by Small [55] in 1973, refers to the frequency of instances where two documents are jointly referenced by other scholarly works. Employing co-citation analysis, a valuable method for assessing the resemblance between documents and revealing thematic structures within a research field, publications were organized into distinct clusters, as expounded upon by Li et al. [27] in 2020. Among the 10,108 references found within the 402 retrieved publications, a mere 1053 references received citations two or more times.

Figure 11 presents a graphical representation of the co-citation network, encompassing references that garnered citations exceeding seven occurrences within the retrieved publications of new technology in USV research. To conduct this analysis, VOSviewer was used as a powerful tool. The resultant network comprised 52 references that satisfied this citation frequency criterion. In this depiction, the sizes of the spheres correspond to the frequency of citations that each publication has received. Furthermore, the distinctive colors assigned to these spheres denote their affiliation with specific clusters, signifying overarching thematic patterns prevalent within the research domain. The detailed setting for the parameters used for the figure is provided in Appendix A Table A2.

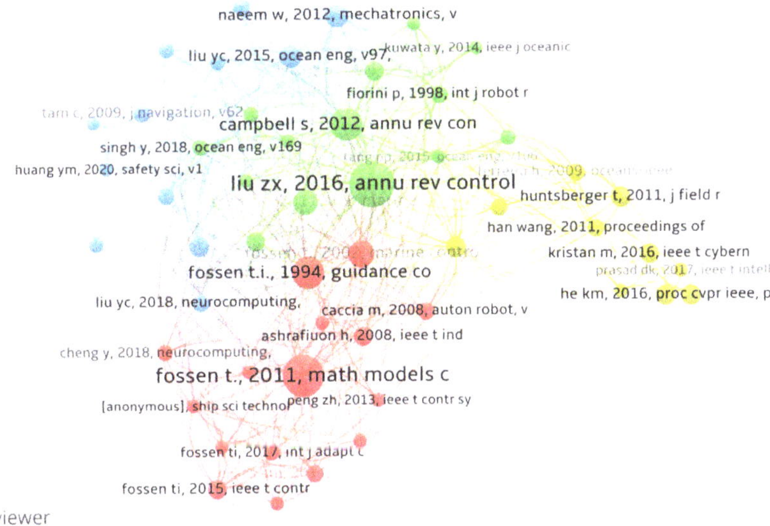

Figure 11. Co-citation network of references cited more than seven times in USVs.

Note that the total link strength within this network carries significant implications, as it reflects the cumulative intensity of connections between each unit. Notably, higher total link strength values signify closer inter-relationships with other units, underscoring their prominence within the co-citation framework.

As illustrated in Figure 11, these 52 references were effectively partitioned into four distinct clusters based on their co-citation associations. The demarcation of these clusters served as a valuable analytical approach for explaining the primary research themes characterizing each group. This identification process drew upon a comprehensive analysis of the titles and abstracts of the references comprising these clusters, thus shedding light on the fundamental narrative patterns underpinning the discourse within the research community.

The red cluster is the largest, with 16 references, primarily dedicated to the marine model control and control system. Within this cluster, the work titled "Handbook of Marine Craft Hydrodynamics and Motion Control", authored by Thor I. Fossen [56] in 2011, emerged as the most frequently cited publication, with 26 citations, and exhibited the highest total link strength of 76, thus establishing its pivotal role as the central reference in this thematic cluster. The green cluster is the second largest cluster, with 14 references. This cluster emphasizes the overview of USVs, robot automation and visual remote sensing. The article "Unmanned surface vehicles: An overview of developments and challenges" by Zhixiang Liu [2] in 2016 can be considered the core publication in this cluster, with the high citations (29) and total link strength (91), followed by the blue cluster with 12 references, primarily centered on the navigation and the neurocomputing application. Notably, the article titled "COLREGs-based collision avoidance strategies for unmanned surface vehicles", authored by W. Naeem et al. [57] in 2012, had the highest citation count at 13, and boasted the most substantial total link strength (78) within this cluster, thereby meriting recognition as the central reference shaping the discourse in this thematic group. Conversely, the smallest cluster, denoted by the color yellow and located at the right of the network, includes 10 references. The predominant themes in this cluster revolved around AI applications, such as deep learning, obstacle detection, object detection and autonomous visual perception. The article titled "Stereo Vision-Based Navigation for Autonomous Surface Vessels", authored by Terry Huntsberger et al. [58] in 2011, occupied the central position within this cluster, marked by the highest citation count of 11 and the most significant total link strength of 34.

3.3.2. Distribution and Co-Citation Analysis of Journal Sources

The final phase of the bibliometric analysis involved an investigation of the sources of the documents contained within the dataset. The analysis is needed to identify which journals are most frequently read and productive in the field of new technology in USV and to provide the scholars with tips for contributing to their manuscripts. The 402 retrieved articles were published from 295 unique publication sources. The sources with at least eight published documents are presented in Table 5, while mapping is depicted in Figure 12.

Table 5. The top six most productive journals on USV research.

Rank	Journal	TP	TC	AC	IF
1	*Ocean Engineering*	28	344	12.29	5.0
2	*Journal of Marine Science and Engineering*	16	89	5.56	2.9
3	*Sensors*	15	250	16.67	3.9
4	*IFAC-PapersOnLine*	14	80	5.71	/
5	*Applied Sciences-Basel*	9	100	11.11	2.7
6	*IEEE Access*	8	104	13.00	3.9

Notes: TP = total publications; TC = times cited; AC = average number of citations per publication; IF = influence factor.

The preeminent sources in this context are primarily represented by two distinguished journals: *Ocean Engineering* (OE) and *Journal of Marine Science and Engineering* (JMSE). The total amount of papers published through the two journals, numbering 21 and 15, respectively, constituted nearly 11% of all articles within the data sample. It is noteworthy, however, that a considerable discrepancy existed in the quantity of documents between the fourth-ranked source and the subsequent source in the collation, as elucidated in Table 5. *Sensors*, being in the third position, exhibited one publication fewer than that of JMSE, but demonstrates the highest average citation in the list, followed by *Ocean Engineering* with the second highest average citation (12.29). This is also consistent with the journal's impact factor ranking, where *Ocean Engineering* and *Sensors* have a relatively high impact factor among the top six most productive journals in the list. Additionally, there were also substantial groups of journals with either two or just one published paper, indicating that the papers in this field were not centralized but scattered in their quality, to some extent.

The journal co-citation analysis is a valuable method for categorizing journals by subject and pinpointing the key journals within each category. This is particularly beneficial for researchers seeking to identify the most pertinent and influential journals related to their specific research area. Hence, the sources were also analyzed in terms of how often they were cited together, a concept known as co-citation [26]. Herein, the relationship between the two publications is determined by the number of documents that reference both of them [26]. This bibliometric network method was employed to visualize the interconnections among the sources. A minimum threshold of 60 citations per source was set, resulting in 28 sources meeting this criterion. The strength of these connections was determined by the number of citations. Through fractional counting and fractionalization, four clusters were identified as a means of normalizing the data in Figure 12. The detailed setting for the parameters used for the figure is provided in Appendix A Table A2.

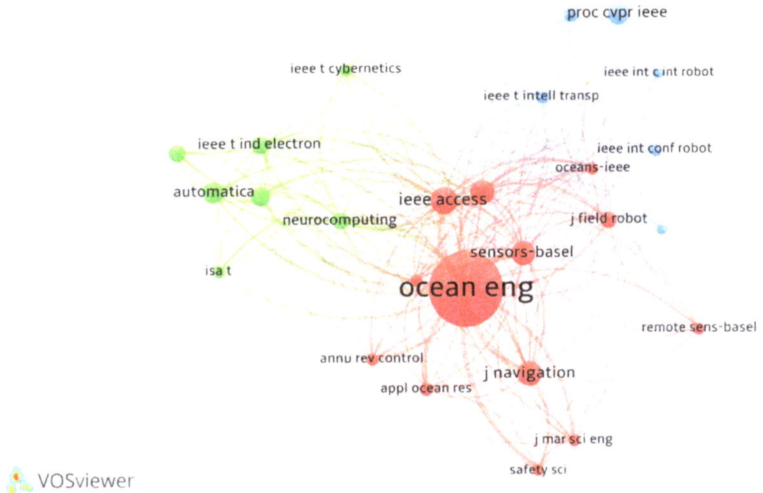

Figure 12. Co-citation network of journal sources.

The co-citation analysis reveals that *Ocean Engineering* and *IEEE Access* were the most significant sources. Interestingly, the cluster associated with *IEEE Conference on Computer Vision and Pattern Recognition* included several sources in the field of computing and robotics, such as *International Journal of Robotics Research*, *IEEE International Conference on Robotics* and *IEEE Transactions on Intelligent Transportation Systems*. These sources were connected to documents discussing regulations pertaining to maritime transportation design and maritime robotics innovation.

The second prominent cluster is centered around *Ocean Engineering*, a journal that encompassed sources related to marine engineering, ship dynamics, ship navigation and similar topics. The role of *Ocean Engineering* in the network is clear, acting as a bridge between safety and navigational-related sources on one hand, and sources focused more on mechanical engineering and hydrodynamics, such as *Journal of Field Robotics*, *Sensors*, *Remote Sensing* and *Applied Ocean Research* on the other hand. This suggests that the sources focusing on algorithms and models for path-planning and evasive maneuvers frequently cite documents chiefly originated from *Ocean Engineering*.

The third cluster primarily consists of automation control journals, strongly associated with publications like *Automatica*, *IEEE Transactions on Control Systems Technology*, *IEEE Transactions on Cybernetics* and *ISA Transactions*.

3.4. Keywords and Term Analysis

The content found within keywords, titles and abstracts serves as valuable data for deducing the primary subject matters and emerging patterns in a particular research domain. This is achieved through the application of text mining methodologies, as demonstrated by the study conducted by Li and colleagues in 2021 [27]. In the current research project, a co-occurrence analysis of keywords and a co-occurrence analysis of terms were carried out by VOSviewer software. The former one shows the research hotpots in a single special point while the latter one provides a more comprehensive elaboration of hot topics by involving the extraction of textual strings from the keywords, titles and abstracts of all publications contained within the database under investigation. The amalgamation of the two analyses can display a profound image of this field.

3.4.1. Keywords Analysis

Keywords usually contain valuable information regarding the intended focus of the author. The co-occurrence analysis of keywords is widely used to reveal the interconnection of keywords and to help readers gain insight into research hotspots and future trends [59]. In Figure 13, a visualization of the most significant keywords is presented, which included both the author's keywords and keywords plus.

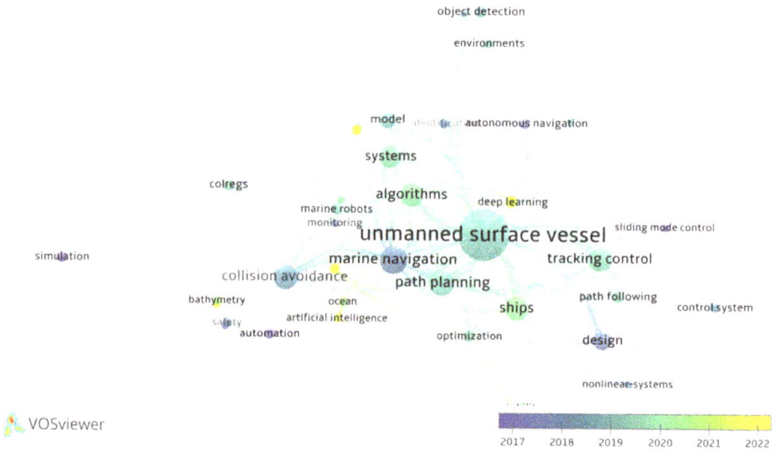

Figure 13. Co-occurrence network of USV keywords.

This visualization was based on the frequency of their shared occurrences in the documents under analysis. A total of 1734 keywords were identified by VOSviewer from the 402 retrieved articles. Among these, over 1500 terms appeared only once or twice, and 46 terms appeared more than five times, which is set as the criteria. A method of normalization was employed using association strength. The weights assigned to the keywords were calculated based on their frequencies, and the different clusters were distinguished by color-coding.

The analysis focused on new technology in USV development, leading to the identification of keywords predominantly associated with this domain. Notably, emerging trends were discernible within the dataset. High-frequency keywords highlighted the keen interest of researchers in new concepts such as path-planning algorithms, marine navigation, the design of systems and collision avoidance. Additionally, related keywords, though less frequent, included e-Navigation, marine robots, machine learning and sensor fusion. This heightened focus on collision-avoidance could be attributed to the advancements in e-Navigation and the issue of COLREGs.

Furthermore, the terms, i.e., control system, multi-agent systems, nonlinear-systems and tracking control indicated researchers' utilization of real-time or historical traffic data for developing novel systems grounded in AI. This shift could be linked to the progressive integration of AI in the shipping industry, as evidenced by relevant studies [60–70]. Additionally, the inclusion of environments, identification, remote sensing and object detection in the ranking suggested a growing inclination towards environmental perception in maritime decision-making process.

The research within the field of USVs has evolved significantly over the years. In 2017, the foundational aspects of simulation, autonomous navigation, and sensor technology formed the key areas of scholarly attention. Afterwards, the focus transitioned towards enhancing capabilities in monitoring, path planning and identification, laying the groundwork for more intricate system functionalities. By 2018, the emphasis shifted to the practical application of these technologies, with a heightened focus on collision avoidance, control systems, safety measures and environmental perception. The year 2019 marked a pivot towards the integration of sensor fusion and system identification, signaling a trend towards multifunctionality and the amalgamation of diverse technologies. The period from 2020 to 2022 witnessed a surge in AI-centered research, characterized by the incorporation of advanced AI algorithms, such as deep reinforcement learning, machine learning, and neural networks, and the utilization of satellite data and computer vision. These developments underscore a broader trend of enhancing communication and intelligent decision-making capabilities in USVs.

3.4.2. Term Analysis

A total of 11,270 terms, derived from keywords, titles and abstracts, were identified using the VOSviewer from the 402 retrieved articles. Among these, 9486 terms only appeared for one time and 448 terms appeared more than five times. To identify the hot topics in USVs, a restriction was set to include terms that appeared at least 10 times; a total of 109 terms, as 60% of the most relevant terms, were extracted, mapped and clustered in two-dimensional space.

Figure 14 illustrates the clustering of terms within new technology in the USV research domain, with an additional temporal dimension incorporated. In this depiction, the average publication year of each term's appearance within the research domain has been computed and assigned to each corresponding node on the map. Warmer, redder nodes signify more recent term appearances. A comparative analysis of the temporal factors in the three clusters revealed that terms within Cluster 1, encompassing multiple USVs, control strategy, external disturbance and stability, along with Cluster 2, associated with machine autonomy, numerical simulation and communication, as well as Cluster 3, featured by environmental identification and monitoring, were relatively recent in their emergence.

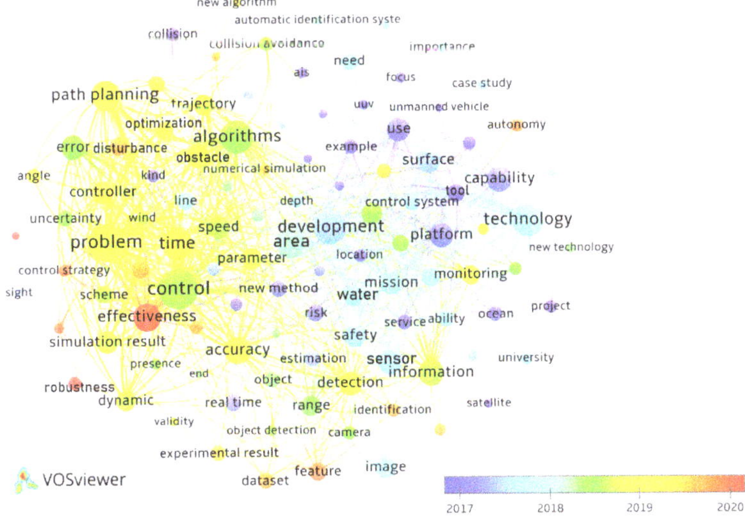

Figure 14. Term clusters of USV domains.

Comprehending the causes behind USVs constitutes a significant aspect of this research domain. The innovation of algorithms and the development of computer science have long been pivotal components of investigating the origins of new technology in USVs. This research area has historically scrutinized technological variables such as robustness, effectiveness, optimization, validity and accuracy. In recent years, a surge in research activity has been observed within Cluster 1, dedicated to the exploration of new vehicle forms with AI, and the new technology in control, communication and detection contributing to USVs.

4. Synthesis and Summary

4.1. Past and Current Trends

As is depicted in Figure 4, before 2008, there was a notable scarcity of English-language scholarly articles in the domain of USVs, perhaps due to the sluggish and outdated technology development. However, subsequent to this period, there has been a remarkable and approximately linear growth in the volume of such publications. Presently, China ranks first globally in terms of the total number of publications in this field. One plausible catalyst for this development could be traced to a series of advances in technology, such as the deep learning put forward by Hinton et al. [71] in 2006 and some national policies like Maritime Power Strategy (2012). With the development of AI and the accomplishments of unmanned underwater vessels, the focus has transferred from the underwater field to the surface field, thus forming an all-ranging application of unmanned vessels.

Currently, the publication in this field is predicted to keep soaring, and is likely to reach a new breakthrough, in line with the development of ChatGPT and 6G communication technology.

4.2. The Features of Social Structure

Considering the leading scholars in the field of new technology in USVs globally, Yuanchang Liu from University College London, Kristan Matej from University College London, Bucknall Richard from University College London and Yun Li from Shanghai Maritime University were the top four most productive authors, and Bucknall Richard had the highest average citations among all the authors, which testified to his powerful contribution and influence in this field.

As for the most productive institutions related to this realm, Dalian Maritime University, Harbin Engineering University, Chinese Academy of Science and University College London were the top four most productive institutions. It is intriguing to find that the four institutions were within or extremely close to the coastal regions, a compelling reality condition that provides them with abundant resources and scientific conditions. Moreover, Bo Wang, Peng Jiang, Xinyu Zhang and Yulei Liao not only exhibited high productivity but have also demonstrated significant activity, particularly in the last 3 years.

With regard to the international cooperation on new methods in USV research across the world, China, England, USA and South Korea were the alliances most frequently collaborating in USVs. International cooperation sometimes was confined to restricted international cooperation, since a noticeable link circle emerges with a majority of the European continent. On top of that, international cooperation was also embodied in institutions; Dalian Maritime University, UCL, Harbin Engineering University, Polish Nava Acad, National University of Singapore were the core institutions carrying out the collaboration between the regional institutes or international institutes, in their respective clusters, showing a strong capability to launch international cooperation and resource integration. There is telling evidence that the institutions in China have a strong mutual cooperation link with each other, in the domestic regard.

In terms of geographical distribution, China was the pioneer, with the largest number of publications at 177, followed by the United States with 68, England with 34, South Korea with 29 and Poland with 25. However, the most productive countries were located in economically developed coastal areas and areas with more universities in ocean engineering and computer science. It is worthwhile to point out that the developing countries had a limited global impact in the field of new technology in USVs in terms of publications and citations, as only two developing countries ranked within the top 15 productive countries, suggesting that new technology in USVs is well embraced and propelled in developed countries under most circumstances.

The information is likely to assist Chinese students with an inclination towards USVs in identifying appropriate educational institutions for advanced studies and suitable academic mentors for pursuing research degrees. Furthermore, the findings may serve as a valuable resource for international scholars or institutions seeking to identify high-performing establishments for collaborative endeavors. Additionally, they can facilitate partnerships between industry and academia for concerted efforts in the domain of USVs.

4.3. Citation and Co-Citation Network Summary

The citation analysis shows that, though China has the highest of citations, with reference to Table 6, the average citation rate of Chinese scholars in the USV field is comparatively low in the top five most cited countries, with nearly a third of the articles never having been cited. The USA has a total number of citations close to China, whereas it only has about the half of publications of China's, showing that the USA's new technology in USV research currently has a relatively powerful international impact, and that China should make efforts to enhance the quality of its articles.

Table 6. The top five countries with highest number of citations.

Rank	Country	TP	TC	AC	H-Index
1	China	172	1455	8.46	19
2	USA	68	1320	19.41	18
3	South Korea	20	704	35.2	8
4	England	29	605	20.86	10
5	Poland	19	334	17.58	8

Notes: TP = total publications; TC = times cited; AC = average number of citations per publication.

According to co-citation analysis, the new technology in USV research primarily focused on four aspects: control system, navigation technology, robot automation and visual remote sensing, as well as autonomous visual perception, such as path planning and object detection.

The primary journals that witnessed substantial contributions from international scholars and exhibited significant impact citations included *Journals of Ocean Engineering, Journal of Marine Science and Engineering, Sensors, IFAC-PapersOnLine* and *Applied Sciences-Basel.*

A co-citation analysis of these referenced journals reveals four principal thematic clusters where scholars, engaged in the field of USV, actively contribute: (i) models, algorithms and systems; (ii) navigation, sensing and safety; and (iii) automation and robots; (iv) AI, deep learning and intelligent machines.

4.4. Future Directions

The analysis of prevalent themes in new technology in USV research reveals a pronounced emphasis on path-planning algorithms, marine navigation, the design of systems and collision avoidance, marine robots, sensor fusion, e-Navigation, multi-agent systems and simulation. However, throughout the period from 2014 to 2018, the discernible observation shows the limited emergence of novel topics, always focused on simulation and navigation. The heated keywords featured with the latest version are centered on machine learning, AI and deep learning, which were actually initiated 10 years ago. Hence, the application of AI in USVs is a little hysteretic, and there is still a long way for us to go. While scholars have made substantial contributions to the new technology of USVs, evident in the extensive publication of English-language articles across diverse subjects, the scarcity of groundbreaking methodologies or research focus during this period is conspicuous, despite tremendous efforts put into planning algorithms. This suggests a potential gap in impact or innovation. Consequently, there exists considerable scope for enhancement within specific subdomains of the research field in USVs, necessitating efforts to align with and potentially surpass advancements achieved at the current time.

In light of these considerations, the prospect of shaping future research directions in the realm of USV is explored, drawing inspiration from the most recent advancements in global USV research. It is recommended that scholars should in future give further attention to the following research topics to achieve shipping safety, energy saving and emission reduction, and reducing operating costs.

4.4.1. Enhanced Intelligence and Autonomy

The intelligent waterway has been increasingly heated in recent years, spurring the technology of intelligence and autonomy. As the development of the Yangtze River's smart waterway infrastructure progresses, the deployment of USVs in its inland channels has seen a significant rise. For the autonomous navigation of the "Jinghai-I" USV within these inland waterways, a composite route determination technique has been formulated. This technique merges an enhanced A* algorithm with a refined model predictive control algorithm, demonstrating proficient autonomy and intelligent navigational capabilities [72]. To improve the autonomy level, a method combining the use of sparse random neighborhood graphs and constrained nonlinear Model Predictive Control (MPC), while implementing a feedback strategy that navigates through sparsely connected areas free of obstacles using a sequence of MPC policies, was proposed by Atasoy Simay [73]. New technology with intelligence was also used in the health management of USVs, exemplified by a hybrid neural network (HNN) prediction model, which integrates Convolutional Neural Networks (CNN), Bidirectional Long Short-Term Memory (BiLSTM) and Attention mechanisms, designed specifically for the prediction of Exhaust Gas Temperature (EGT) in marine diesel engines. This approach provides a new way of thinking for the research of fault early warning and the health management of marine diesel engines [74]. To speed up autonomy, the Unmanned Maritime Systems Program Office (PMS 406) is enhancing autonomy in unmanned maritime vehicles through the Unmanned Maritime Autonomy Architecture (UMAA) for software standards and the Rapid Autonomy Integration Lab (RAIL) for developing new capabilities [75]. Intelligence and autonomy likewise show their advantage in risk analysis [76].

In the future, it is predicted that USVs will become more intelligent and autonomous through the integration of AI, deep learning and machine learning technologies [77]. This will enable them to better perceive their environment, make autonomous decisions, and execute tasks. This advancement will encompass state-of-the-art sensing and obstacle avoidance systems, as well as robust decision support systems. Additionally, the level of autonomy in USVs will continue to rise, including autonomous path planning, autonomous cruising and autonomous maintenance. This will alleviate the burden on operators, enhance efficiency and reduce costs.

4.4.2. Highly Integrated Sensor Systems and Multi-Modal Task Execution

Recent advances in robotic design, autonomy and sensor integration create solutions for the exploration of deep-sea environments [78]. As marine exploration and exploitation continue to advance, coupled with the strides made in mechanical intelligence, there is a notable emphasis on the utilization of USVs and the intricate design of their guidance systems. A newly developed module in 2021, designated as the Three-Dimensional Perception Module (PMTD), employs a combination of camera and LiDAR technology to assimilate multi-dimensional environmental data. This module attains centimeter-level localization accuracy through the integration of Global Positioning System (GPS) and Inertial Measurement Unit (IMU) technologies [79]. A new integrated USV and unmanned underwater vehicle (UUV) platform connected via underwater cables capable of acquiring real-time underwater data and long-time operation was developed to gather more sensory data for decision making [80].

Additionally, new technology was presented in a novel recovery system: one a piece of hardware that ensures the attainment of unique attitude requirements and improves the effectiveness of stern chute recovery [81]. A new controller for dynamic positioning was then developed, combining model predictive control with predictions of short-term wave movements. This approach effectively compensated for wave effects, thereby significantly enhancing the operational capabilities of the vessel [82]. With more and more adjuncts devised, USVs will be expected to be equipped with a greater variety of and higher resolution sensors, such as high-resolution cameras, LiDAR, sonar, radar and more, to enhance their environmental perception capabilities [83]. This will contribute to a broader range of applications, including ocean surveys, marine conservation, military operations and emergency response.

In the meanwhile, USVs will be capable of executing a variety of tasks, including underwater exploration, subsea operations, cargo transportation, search and rescue, etc. This will necessitate advanced adaptive control systems and task-planning algorithms.

4.4.3. Extended Endurance and Resilience

For intelligent systems, the endurance and resilience are important in case of fatal failure. Effective human-AI interaction design is predicated on increased cross-disciplinary efforts, requiring reconciling productivity with resilience [84]. To investigate the fragile problems of the controller, a new non-fragile fault-tolerant scheme was provided against multiplicative and additive controller gain perturbations [85]. An adaptive error constraint line-of-sight guidance law was originally proposed with better transient and steady-state performance, and simplified the design of the controller by not relying on interference observers [86]. A new approach is suggested for the high-level processing of received data to evaluate their consistency, which is agnostic to the underlying technology of the individual sensory input [87]. This approach [87] demonstrates the feasibility of identifying anomalies in nominal operations when the navigation sensor is subjected to adversarial attacks or experiences malfunctions.

Furthermore, it is concluded that deploying multiple USVs as a formation fleet benefits great fault-tolerant resilience [88]. A new algorithm named the 'angle-guidance fast marching square' (AFMS) was presented, and a novel priority scheme based upon the distance to the closest point of approaching (DCPA) has been proposed and developed to efficiently and effectively navigate the USV formation [88]. A comprehensive training framework encompassing multiple tasks has been formulated for the management of formation control, tailored to the dynamic behavior of Unmanned Surface Vehicles and employing the leader–follower approach. Within this framework, the Soft Actor-Critic (SAC) reinforcement learning algorithm has been modified to facilitate the development of agent constructs [89].

Last but not least, the development of more efficient battery technology and the integration of renewable energy sources and power systems will also incessantly enable USVs to achieve extended endurance. This will allow them to carry out longer-duration missions, such as ocean monitoring and research [90]. A hybrid power system [91] comprising a solar array, an ocean wave energy converter, a fuel cell system, a diesel generator and a lithium ion battery pack was designed. The results show that optimization has been achieved with a 19.6% contribution from solar power during daylight hours and a 5.53% contribution of the wave energy harvester to meet the load demands [91]. And, an AutoNaut uncrewed surface vehicle powered entirely by renewable energy has been invented as the first USV to regularly run scientific missions off the coast of the UK [92].

Therefore, research related to fault-tolerant schemes and resilience amelioration by means of the USV formation has always been a heated discussion in the topic of USVs. And, in a more practical perspective, the industry will call for an increasing innovation of power systems in the future.

4.4.4. Satellite Communication and Interconnectivity

Nowadays, the majority of navigation systems employed by USVs utilize integrated approaches combining the global positioning system (GPS) and inertial navigation system(INS) methods to enhance overall navigation accuracy. To enhance and keep the performance in case of outages, interference and weakness, a continuous and accurate navigation solution was put forward via integrated GPS with micro-electro-mechanical (MEMS)-INS smartphone sensors and reduced-aided visual odometry (RAVO) using centralized Kalman filter (CKF) data fusion [93]. In case of insufficient satellites, the loose and tight integrated models of the USV navigation system were established with effective performance in a certain time range with relatively lower position errors [94]. A framework, comprised of an execution module and a multi-layered planner, was designed to enable the AUV to avoid collisions, maintain communication with the USV, and increase the sum of the rewards by reaching many of the discovered goals and demonstrate the efficiency of the approach to solve dynamic multi-goal motion-planning problems with communication constraints [7]. At the same time, some communication quality evaluation models were researched in this part, including the mobile communication quality evaluation model based on the multi-path effect, the Doppler effect and the ray beam method [95].

Hence, it is presumed that USVs will make wider use of satellite communication technology for remote telemetry and remote control. Additionally, they will be better equipped to engage in real-time communication and collaborative operations with other USV, vessels and terrestrial stations. It is hoped that the 6G communication [46] will be a recipe for the ground-breaking progress of USVs.

4.4.5. Eco-Friendly and Sustainable Practices

Climate change has become one of the top worldwide concerns and issues an urgent call for more sustainable developments in the near future [96]. A novel Remote Hyperspectral Imaging System (RHIS) was integrated into an Unmanned Aquatic Drone (UAD) for the purpose of detecting and identifying plastic debris in both coastal and freshwater ecosystems. The findings indicate that this newly implemented RHIS, in conjunction with the UAD, constitutes an environmentally friendly and effective method for the identification of plastic pollution in aquatic settings [97]. In the case that mobile sensors could not be deployed, the USV was taken to identify pollution sources, map the environmental impact and be an analysis tool for the further research of these ecosystems, promoting the green development of ecology [98].

The path-planning sector can also expedite the sustainable practices of USV, since an innovative path-planning algorithm, grounded in AI methodologies, was proposed to calculate viable navigation routes adhering to the Convention on the International Regulations for Preventing Collisions at Sea (COLREG). This algorithm uniquely incorporated considerations of energy efficiency, factoring in wind and sea current data to optimize energy consumption during maritime voyages [99]. Additionally, the USV was utilized in water quality monitoring [100,101], Arctic exploration [102] and persistent ocean observation [103], and was considered a catalyst for energy revolution [91], since some USVs are equipped with sustainable natural energy sources using solar arrays and ocean wave energy converters. An improved differential evolution particle swarm optimization algorithm (DePSO) was proposed, and demonstrates that it can effectively reduce the path intersection points, and thus greatly shorten the overall path length to reach sustainable development [104]. The integration of the chaotic quantum-behaved particle swarm algorithm (cQPSO) with the multi-population genetic algorithm (mGA) led to the development of a novel hybrid algorithm, termed cQPSO-mGA. This advanced algorithmic fusion demonstrated the effective resolution of the two-level scheduling model, as substantiated by empirical data. The results highlighted its proficiency in facilitating cost-effective transportation solutions and ensuring reliable path tracking [105].

Consequently, USV development will place greater emphasis on environmental sustainability, including emissions reduction, the adoption of renewable energy sources, and mitigating marine pollution through technological innovations, but the relevant research is blank at the moment.

4.4.6. Navigation Safety and Military Defense

The future of USVs holds great promise in enhancing security and defense capabilities, particularly in the military domain. This emphasis on safety and self-defense features underscores the growing importance of USVs in modern maritime operations.

In terms of navigation safety, object detection comes forth and trajectory planning follows in tune. The incorporation of a multi-scale feature extraction layer, encompassing dilation convolution and group convolution, into the baseline model of the Faster Region-based Convolutional Neural Network (Faster-RCNN), accompanied by modifications to the classification algorithm, was suggested. This enhancement aimed at augmenting the model's capability for obstacle detection, with a specific focus on bolstering its robustness and precision [106]. A fusion framework of field theoretical planning and a model predictive control (MPC) algorithm were proposed to obtain a realizable collision-free tracking trajectory to enhance safety, where the trajectory smoothness and collision avoidance constraints under a complex environment needed to be considered [107]. A new process of dynamic collision avoidance, combined with new attractive and repulsive potential field functions, was constructed to ensure the safety of USVs [108]. An innovative collision avoidance algorithm, predicated on Approximate Representation Reinforcement Learning (AR-RL), was developed to facilitate the collision avoidance capabilities of Maritime Autonomous Surface Ships (MASS) within a continuum state space. This algorithm was characterized by an interactive learning feature, mirroring the decision-making process of a human crew in navigational contexts. Its effectiveness has been established in significantly enhancing maritime safety, particularly in scenarios where mixed traffic consists of both manned vessels and MASS, promising substantial improvements in nautical safety in the foreseeable future [109]. In response to the critical requirement for assessing human error probability in terms of autonomous cargo ships with human–autonomy collaboration, a probabilistic model mixed with the Shore Control Centre (SCC), Technique for Human Error Rate Prediction (THERP)and Bayesian Networks was proposed, and was employed to evaluate the likelihood of human errors, specifically concentrating on the interaction between human operators and autonomous systems [110].

From the point of defense, the new technology of USVs mainly casts light on derivative techniques, combat theories and methods of arrangement in military scenarios. Multiple object tracking (MOT) in USV videos has many application scenarios in the military and civilian fields [111]. By projecting military power in a more affordable way, through the use of USVs, the exposure of human life to military risks should be significantly reduced [112]. A new attack strategy, USV combat based on wolves' attacks with weight, was testified to have conspicuous advantages in USV combat. In the future naval deployment of USVs, it is also proposed that, for the initial stages of exploration and investigation, the application of a queuing network theory was deemed more advantageous than relying on simulation-based analysis [112].

As is listed above, there are several hot research directions in this prediction, such as Enhanced Safety, Anti-Jamming Communication, Automatic Missile Defense Systems, Surveillance and Reconnaissance, Autonomous Swarming and Cost-Effective Alternatives.

4.4.7. Brief Summary

In summary, one of the primary areas of development will be the enhancement of intelligence and autonomy for USVs. This evolution is expected to be fueled by the integration of sophisticated AI, deep learning and machine learning technologies. These advancements will enable USVs to better perceive their environment, make autonomous decisions and execute complex tasks with greater efficiency and accuracy.

The focus will also be on developing advanced sensing and obstacle avoidance systems, as well as robust decision support systems, which will further enhance their operational capabilities in diverse environments. The progression in autonomy will likely include more sophisticated features such as autonomous path planning, cruising and maintenance, thereby reducing the reliance on human operators and enhancing operational efficiency.

Another significant area of development for USVs will be the integration of highly advanced sensor systems and the ability to execute multi-modal tasks. The advancement in sensor technology, including high-resolution cameras, LiDAR, sonar and radar, will significantly improve the environmental perception capabilities of USVs. This, in turn, will expand their applicability in various fields such as oceanographic research, marine conservation, military operations and emergency responses.

Furthermore, the enhancement in endurance and resilience through innovative power systems and fault-tolerant designs will enable USVs to undertake longer-duration missions and operate effectively in challenging conditions. The integration of renewable energy sources and eco-friendly practices in USV design will also contribute to sustainable maritime operations. In addition, the increasing use of satellite communication and improved interconnectivity will enhance remote operation capabilities, allowing for more coordinated and collaborative efforts in maritime missions.

Table 7 shows a brief and overall summary of this section.

Table 7. The summary of future directions.

Direction	New Trends	Effect
Enhanced Intelligence and Autonomy	Make autonomous decisions and execute tasks like autonomous cruising and autonomous maintenance	Alleviate the burden on operators, enhance efficiency and reduce costs
Highly Integrated Sensor Systems and Multi-Modal Task Execution	Marine conservation, military operations, underwater exploration, subsea operations, cargo transportation, search and rescue	Necessitate advanced adaptive control systems and task-planning algorithms
Extended Endurance and Resilience	More efficient battery technology and the integration of renewable energy sources	Allow them to carry out longer-duration missions
Satellite Communication and Interconnectivity	Real-time communication and collaborative operations with other USVs, vessels and terrestrial stations; 6G	A recipe for the ground-breaking progress for USV
Eco-Friendly and Sustainable Practices	Emissions reduction the adoption of renewable energy sources, mitigating marine pollution; new energy converter Intelligent feeding	More green to the world Cut down the cost
Safety and Defense	Enhanced Safety Anti-Jamming Communication Automatic Missile Defense Systems Surveillance and Reconnaissance Autonomous Swarming Cost-Effective Alternatives	Protect the lives of operators, countries and the sea
Eco-Friendly and Sustainable Practices	Emissions reduction The adoption of renewable energy sources, mitigating marine pollution new energy converter Intelligent feeding	More green to the world Cut down the cost

These developments, combined with the ongoing research in navigation safety and military defense applications, suggest that USVs will play an increasingly vital role in the future of maritime operations, offering both enhanced capabilities and cost-effective solutions.

4.5. Biases and Limitations

Despite the fact that this study has provided some valuable insights, it is essential to acknowledge its limitations, which stem from the publication retrieval process and the chosen analytical methodologies. Firstly, the exclusive utilization of the WoS database was necessitated by its adoption, attributed to variations in data standards. This constrained choice might have led to the inadvertent exclusion of pertinent papers concerning USV globally, potentially introducing biases in the outcomes and impeding a comprehensive elucidation of discernible patterns and developments. Moreover, this paper, based on existing research, primarily utilizes the WoS as the database for data retrieval. Future research could consider incorporating a wider range of databases to enrich the scope of the study and facilitate comparative analysis.

Secondly, to facilitate a broader understanding of the structural dynamics and evolutionary trends in USVs, this study focused exclusively on articles published in the English language, which is the predominant medium of academic discourse. Regrettably, a significant number of USV articles written by researchers from Non-English-speaking maritime powers, including countries like China, South Korea, Japan, Russia, etc., are published in their domestic language journals. Consequently, they are not included in the WoS database. The exclusion of these contributions could impact the comprehensiveness of the results.

Additionally, there are several common limitations that this paper may have encountered. These include potential publication bias, as certain types of articles or journals may be overrepresented or underrepresented in the chosen database, thus impacting the generalizability of the findings. The selected time frame for the analysis may influence the assessment of trends and patterns, as research topics and publication rates can evolve. The choice of bibliometric indicators, such as citation counts, can introduce sources of bias, as they may not capture the full impact or relevance of a paper. Moreover, the design of the search query, such as the confine of keywords and the combination of conjunctions, may also do a disservice to the final result.

Lastly, it is imperative to recognize that the scope of the database itself may not encompass all relevant articles, potentially resulting in the omission of some significant contributions to the field.

5. Conclusions

In this paper, to provide a detailed overview of the key contributions in this research domain, a comprehensive bibliometric analysis of USV research was conducted by utilizing 402 publications sourced from the WoS database. The analysis entailed the exploration of publication trends, the identification of influential authors, institutions, countries, notable articles, journals, disciplines, references and terms. Additionally, deeper insights were demonstrated of the international cooperation in USV research, explaining it from the perspectives of countries, institutions and authors. Furthermore, we used a dual-function word analysis to identify prevalent terms and keywords in USV research, thereby uncovering emerging research topics and trends. The evolution of focus areas in different periods was also analyzed, and future directions for the advancement of USV were put forth.

To facilitate a clearer understanding of the findings, this paper was supplemented with visual representations, including co-authorship networks of authors, countries and institutions, citation networks of publications and countries, co-citation networks of journals and density maps of focus topics, as well as the term networks. These visual aids serve to demonstrate the utility of bibliometric analysis in gaining insights into the development of USV research.

The implications of the results extend to various stakeholders. For scholars, the findings offer insights into the current state of research in the field, highlighting both strengths and weaknesses. Meanwhile, scholars can use the information to identify potential partners for collaborative educational and research endeavors in the realm of USVs. This collaboration has the potential to contribute not only to the sustainable development of sustainable marine development but also to related industrial activities worldwide.

Author Contributions: Conceptualization, J.X.; methodology, J.X. and P.Y.; software, J.X. and P.Y.; validation, J.X. and H.H.; resources, J.X. and H.H.; data curation, J.X. and P.Y.; writing—original draft preparation, J.X. and P.Y.; review and editing, J.X. and H.H.; visualization, J.X. and P.Y.; supervision, J.X. and H.H.; funding acquisition, J.X. All authors have read and agreed to the published version of the manuscript.

Funding: This research was funded by the National Natural Science Foundation of China, grant number 52201413, the Natural Science Foundation of Science and Technology Commission of Shanghai Municipality, grant number 23ZR1434400, the Opening Foundation of Key Laboratory of Safety and Risk Management on Transport Infrastructures for Ministry of Transport, grant number 2023KFKT015, the Oceanic Interdisciplinary Program of Shanghai Jiao Tong University, grant number SL2022PT107, and the Startup Fund for Young Faculty at Shanghai Jiao Tong University, grant number 22X010503469. The APC was funded by Shanghai Jiao Tong University.

Institutional Review Board Statement: Not applicable.

Informed Consent Statement: Not applicable.

Data Availability Statement: The data presented in this study are openly available in the Web of Science (WoS) Core Collection database (https://images.webofknowledge.com/WOKRS59B4/help/ja/WOS/hp_whatsnew_wos.html).

Conflicts of Interest: The authors declare no conflicts of interest.

Appendix A

Table A1. The iteration of query.

Time	Index Word Input	Number of Papers
1	TS = (unmanned surface vessel) AND TS = (new technology) AND TS = (automation)	1
2	TS = (driverless surface vessel or unmanned surface vessel or manless surface vessel) AND TS = (new technology or new science or new technique) AND TS = (automatic control or automation or automatization)	7
3	TS = (driverless surface vessel or unmanned surface vessel or manless surface vessel or unattended surface vessel or automatic surface vessel) AND TS = (new technology or new science or new technique or advanced technology) AND TS = (automatic control or automation or automatization or robotization or automate)	30
4	TS = (driverless surface vessel or unmanned surface vessel or manless surface vessel or unattended surface vessel or automatic surface vessel or unmanned surface ship or driverless surface ship) AND TS = (new technology or new science or new technique or advanced technology) AND TS = (automatic control or automation or automatization or robotization or automate)	30
5	TS = (driverless surface vessel or unmanned surface vessel or manless surface vessel or unattended surface vessel or automatic surface vessel or unmanned surface ship or driverless surface ship or unmanned surface ships driverless surface ships or unmanned surface marines or driverless surface marines) AND TS = (new technology or new science or new technique or advanced technology) AND TS = (automatic control or automation or automatization or robotization or automate)	32
6	TS = (driverless surface vessel or unmanned surface vessel or manless surface vessel or unattended surface vessel or automatic surface vessel or unmanned surface ship or driverless surface ship or unmanned surface ships or driverless surface ships or unmanned surface marines or driverless surface marines or unmanned surface vehicles or driverless surface vehicles or automatic surface vehicles) AND TS = (new technology or new science or new technique or advanced technology) AND TS = (automatic control or automation or automatization or robotization or automate)	91
7	TS = (driverless surface vessel or unmanned surface vessel or manless surface vessel or unattended surface vessel or automatic surface vessel or unmanned surface ship or driverless surface ship or unmanned surface ships or driverless surface ships or unmanned surface marines or driverless surface marines or unmanned surface vehicles or driverless surface vehicles or automatic surface vehicles) AND TS = (new technology or new science or new technique or advanced technology) AND TS = (automatic control or automation or automatization or robotization or automate or marine navigation)	108

Table A1. *Cont.*

Time	Index Word Input	Number of Papers
8	TS = (driverless surface vessel or unmanned surface vessel or manless surface vessel or unattended surface vessel or automatic surface vessel or unmanned surface ship or driverless surface ship or unmanned surface ships or driverless surface ships or unmanned surface marines or driverless surface marines or unmanned surface vehicles or driverless surface vehicles or automatic surface vehicles) AND TS = (new technology or new science or new technique or advanced technology or new technologies or new technological or new processes) AND TS = (automatic control or automation or automatization or robotization or automate or marine navigation)	145
9	TS = (driverless surface vessel or unmanned surface vessel or manless surface vessel or unattended surface vessel or automatic surface vessel or unmanned surface ship or driverless surface ship or unmanned surface ships or driverless surface ships or unmanned surface marines or driverless surface marines or unmanned surface vehicles or driverless surface vehicles or automatic surface vehicles or unmanned surface vessels or automatic surface vessels or driverless surface vessels) and TS = (new technology or new science or new technique or advanced technology or new technologies or new technological or new processes or new methods or new method or new model or new approach) and TS = (automatic control or automation or automatization or robotization or automate or marine navigation or automatic operation or automated or automations) OR TI = (driverless surface vessel or unmanned surface vessel or manless surface vessel or unattended surface vessel or automatic surface vessel or unmanned surface ship or driverless surface ship or unmanned surface ships or driverless surface ships or unmanned surface marines or driverless surface marines or unmanned surface vehicles or driverless surface vehicles or automatic surface vehicles unmanned surface vessels or automatic surface vessels or driverless surface vessels) and TI = (new technology or new science or new technique or advanced technology or new technologies or new technological or new processes or new methods) and TI = (automatic control or automation or automatization or robotization or automate or marine navigation or automatic operation or automated or automations)	216
10	TS = (USV or unmanned surface vehicle or driverless surface vessel or unmanned surface vessel or manless surface vessel or unattended surface vessel or automatic surface vessel or unmanned surface ship or driverless surface ship or unmanned surface ships or driverless surface ships or unmanned surface marines or driverless surface marines or unmanned surface vehicles or driverless surface vehicles or automatic surface vehicles or marine navigation or unmanned surface vessels or automatic surface vessels or driverless surface vessels) and TS = (new technology or new science or new technique or advanced technology new technologies or new technological or new processes or new methods or new model or new approach) and TS = (automatic control or automation or automatization or automate or robotization or automatic operation or automated or automations) OR TI = (USV or unmanned surface vehicle or driverless surface vessel or unmanned surface vessel or manless surface vessel or unattended surface vessel or automatic surface vessel or unmanned surface ship or driverless surface ship or unmanned surface ships or driverless surface ships or unmanned surface marines or driverless surface marines or unmanned surface vehicles or driverless surface vehicles or automatic surface vehicles or marine navigation or unmanned surface vessels or automatic surface vessels or driverless surface vessels) and TI = (new technology or new science or new technique or advanced technology new technologies or new technological or new processes or new methods or new model or new approach) OR AB = (USV or unmanned surface vehicle or driverless surface vessel or unmanned surface vessel or manless surface vessel or unattended surface vessel or automatic surface vessel or unmanned surface ship or driverless surface ship or unmanned surface ships or driverless surface ships or unmanned surface marines or driverless surface marines or unmanned surface vehicles or driverless surface vehicles or automatic surface vehicles or marine navigation or unmanned surface vessels or automatic surface vessels or driverless surface vessels) and AB = (new technology or new science or new technique or advanced technology new technologies or new technological or new processes or new methods or new model or new approach)	1025

Notes: TS = Topic; TI = Title; AB = Abstract.

The final search strings in this article.

TS = (USV or unmanned surface vehicle or driverless surface vessel or unmanned surface vessel or manless surface vessel or unattended surface vessel or automatic surface vessel or unmanned surface ship or driverless surface ship or unmanned surface ships or driverless surface ships or unmanned surface marines or driverless surface marines or unmanned surface vehicles or driverless surface vehicles or automatic surface vehicles or marine navigation or unmanned surface vessels or automatic surface vessels or driverless surface vessels) and TS = (new technology or new science or new technique or advanced technology new technologies or new technological or new processes or new methods or new model or new approach) and TS = (automatic control or automation or automatization or automate or robotization or automatic operation or automated or automations)

OR

TI = (USV or unmanned surface vehicle or driverless surface vessel or unmanned surface vessel or manless surface vessel or unattended surface vessel or automatic surface vessel or unmanned surface ship or driverless surface ship or unmanned surface ships or driverless surface ships or unmanned surface marines or driverless surface marines or unmanned surface vehicles or driverless surface vehicles or automatic surface vehicles or marine navigation or unmanned surface vessels or automatic surface vessels or driverless surface vessels) and TI = (new technology or new science or new technique or advanced technology new technologies or new technological or new processes or new methods or new model or new approach)

OR

AB = (USV or unmanned surface vehicle or driverless surface vessel or unmanned surface vessel or manless surface vessel or unattended surface vessel or automatic surface vessel or unmanned surface ship or driverless surface ship or unmanned surface ships or driverless surface ships or unmanned surface marines or driverless surface marines or unmanned surface vehicles or driverless surface vehicles or automatic surface vehicles or marine navigation or unmanned surface vessels or automatic surface vessels or driverless surface vessels) and AB = (new technology or new science or new technique or advanced technology new technologies or new technological or new processes or new methods or new model or new approach)

Table A2. Parameter setting of mapping.

Figure	Resolution	Min. Cluster Size	Attraction	Repulsion	Threshold
Figure 5	1	1	2	−3	2
Figure 6	1	1	2	−2	2
Figure 8	1	4	2	−2	2
Figure 9	1	4	2	−2	2
Figure 11	10	8	2	0	7
Figure 12	1	5	0	−1	60
Figure 13	10	6	2	−1	5
Figure 14	1	1	3	−1	10

References

1. Campbell, S.; Naeem, W.; Irwin, G.W. A review on improving the autonomy of unmanned surface vehicles through intelligent collision avoidance manoeuvres. *Annu. Rev. Control* **2012**, *36*, 267–283. [CrossRef]
2. Liu, Z.X.; Zhang, Y.M.; Yu, X.; Yuan, C. Unmanned surface vehicles.: An overview of developments and challenges. *Annu. Rev. Control* **2016**, *41*, 71–93. [CrossRef]
3. Liu, G.Q.; Wu, J.W.; Wen, N.F.; Zhang, R.B. A Review on Collaborative Planning of Multiple Unmanned Surface Vehicles. In Proceedings of the Chinese Automation Congress (CAC), Xi'an, China, 30 November–2 December 2018; pp. 2158–2163.
4. Ma, T.; Yang, S.; Wang, T.; Xin, L. Research status and development overview of multi-USV collaborative system. *Ship Sci. Technol.* **2014**, *36*, 7–13.
5. Abdelaal, M.; Fränzle, M.; Hahn, A. Nonlinear Model Predictive Control for trajectory tracking and collision avoidance of underactuated vessels with disturbances. *Ocean Eng.* **2018**, *160*, 168–180. [CrossRef]
6. Lyu, H.G.; Yin, Y. COLREGS-Constrained Real-time Path Planning for Autonomous Ships Using Modified Artificial Potential Fields. *J. Navig.* **2019**, *72*, 588–608. [CrossRef]
7. McMahon, J.; Plaku, E. Autonomous Data Collection With Dynamic Goals and Communication Constraints for Marine Vehicles. *IEEE Trans. Autom. Sci. Eng.* **2023**, *20*, 1607–1620. [CrossRef]
8. Sotelo-Torres, F.; Alvarez, L.V.; Roberts, R.C. An Unmanned Surface Vehicle (USV): Development of an Autonomous Boat with a Sensor Integration System for Bathymetric Surveys. *Sensors* **2023**, *23*, 4420. [CrossRef]
9. Li, W.; Ge, Y.; Guan, Z.; Ye, G. Synchronized Motion-Based UAV-USV Cooperative Autonomous Landing. *J. Mar. Sci. Eng.* **2022**, *10*, 1214. [CrossRef]

10. Wu, G.X.; Li, D.B.; Ding, H.; Shi, D.D.; Han, B. An overview of developments and challenges for unmanned surface vehicle autonomous berthing. *Complex Intell. Syst.* **2023**, *9*, 1–23. [CrossRef]

11. Barrera, C.; Padron, I.; Luis, F.S.; Llinas, O.; Marichal, G.N. Trends and Challenges in Unmanned Surface Vehicles (USV): From Survey to Shipping. *TransNav* **2021**, *15*, 135–142. [CrossRef]

12. Huang, Y.M.; Chen, L.Y.; Chen, P.F.; Negenborn, R.R.; van Gelder, P. Ship collision avoidance methods: State-of-the-art. *Saf. Sci.* **2020**, *121*, 451–473. [CrossRef]

13. Chen, L.Y.; Fu, Y.H.; Chen, P.F.; Mou, J.M. Survey on Cooperative Collision Avoidance Research for Ships. *IEEE Trans. Transp. Electrif.* **2023**, *9*, 3012–3025. [CrossRef]

14. Vagale, A.; Oucheikh, R.; Bye, R.T.; Osen, O.L.; Fossen, T.I. Path planning and collision avoidance for autonomous surface vehicles I: A review. *J. Mar. Sci. Technol.* **2021**, *26*, 1292–1306. [CrossRef]

15. Gu, N.; Wang, D.; Peng, Z.H.; Wang, J.; Han, Q.L. Advances in Line-of-Sight Guidance for Path Following of Autonomous Marine Vehicles: An Overview. *IEEE Trans. Syst. Man Cybern. Syst.* **2023**, *53*, 12–28. [CrossRef]

16. Xing, B.; Yu, M.; Liu, Z.; Tan, Y.; Sun, Y.; Li, B. A Review of Path Planning for Unmanned Surface Vehicles. *J. Mar. Sci. Eng.* **2023**, *11*, 1556. [CrossRef]

17. Steinberg, M. Intelligent autonomy for unmanned naval vehicles. In Proceedings of the Conference on Unmanned Systems Technology VIII, Kissimmee, FL, USA, 17–20 April 2006.

18. Farinha, A.; di Tria, J.; Zufferey, R.; Armanini, S.F.; Kovac, M. Challenges in Control and Autonomy of Unmanned Aerial-Aquatic Vehicles. In Proceedings of the 29th Mediterranean Conference on Control and Automation (MED), Puglia, Italy, 22–25 June 2021; pp. 937–942.

19. Gu, J.; Hu, M.; Gu, Z.; Yu, J.; Ji, Y.; Li, L.; Hu, C.; Wei, G.; Huo, J. Bibliometric Analysis Reveals a 20-Year Research Trend for Chemotherapy-Induced Peripheral Neuropathy. *Front. Neurol.* **2022**, *12*, 793663. [CrossRef]

20. Li, H.; Jiang, H.D.; Yang, B.; Liao, H. An analysis of research hotspots and modeling techniques on carbon capture and storage. *Sci. Total Environ.* **2019**, *687*, 687–701. [CrossRef] [PubMed]

21. Gil, M.; Wrobel, K.; Montewka, J.; Goerlandt, F. A bibliometric analysis and systematic review of shipboard Decision Support Systems for accident prevention. *Saf. Sci.* **2020**, *128*, 104717. [CrossRef]

22. Li, J.; Goerlandt, F.; Li, K.W. Slip and Fall Incidents at Work: A Visual Analytics Analysis of the Research Domain. *Int. J. Environ. Res. Public Health* **2019**, *16*, 4972. [CrossRef]

23. Wang, Q.; Wang, J.; Xue, M.; Zhang, X. Characteristics and Trends of Ocean Remote Sensing Research from 1990 to 2020: A Bibliometric Network Analysis and Its Implications. *J. Mar. Sci. Eng.* **2022**, *10*, 373. [CrossRef]

24. Bertocci, F.; Mannino, G. Can Agri-Food Waste Be a Sustainable Alternative in Aquaculture? A Bibliometric and Meta-Analytic Study on Growth Performance, Innate Immune System, and Antioxidant Defenses. *Foods* **2022**, *11*, 1861. [CrossRef]

25. Gizzi, F.T.; Potenza, M.R. The Scientific Landscape of November 23rd, 1980 Irpinia-Basilicata Earthquake: Taking Stock of (Almost) 40 Years of Studies. *Geosciences* **2020**, *10*, 482. [CrossRef]

26. Van Eck, N.J.; Waltman, L. VOS: A new method for visualizing similarities between objects. In Proceedings of the 30th Annual Conference of the German-Classification-Society, Berlin, Germany, 8–10 March 2006; pp. 299–305.

27. Li, J.; Goerlandt, F.; Reniers, G. An overview of scientometric mapping for the safety science community: Methods, tools, and framework. *Saf. Sci.* **2021**, *134*, 105093. [CrossRef]

28. Jin, R.; Zou, P.X.W.; Piroozfar, P.; Wood, H.; Yang, Y.; Yan, L.; Han, Y. A science mapping approach based review of construction safety research. *Saf. Sci.* **2019**, *113*, 285–297. [CrossRef]

29. Merigó, J.M.; Miranda, J.; Modak, N.M.; Boustras, G.; de la Sotta, C. Forty years of Safety Science: A bibliometric overview. *Saf. Sci.* **2019**, *115*, 66–88. [CrossRef]

30. Amin, M.T.; Khan, F.; Amyotte, P. A bibliometric review of process safety and risk analysis. *Process Saf. Environ. Prot.* **2019**, *126*, 366–381. [CrossRef]

31. Veiga-del-Bano, J.M.; Camara, M.A.; Oliva, J.; Hernandez-Cegarra, A.T.; Andreo-Martinez, P.; Motas, M. Mapping of emerging contaminants in coastal waters research: A bibliometric analysis of research output during 1986–2022. *Mar. Pollut. Bull.* **2023**, *194*, 115366. [CrossRef]

32. Demirci, S.M.E.; Elcicek, H. Scientific awareness of marine accidents in Europe: A bibliometric and correspondence analysis. *Accid. Anal. Prev.* **2023**, *190*, 107166. [CrossRef]

33. Duan, P.; Cao, Y.; Wang, Y.; Yin, P. Bibliometric Analysis of Coastal and Marine Tourism Research from 1990 to 2020. *J. Coast. Res.* **2022**, *38*, 229–240. [CrossRef]

34. Corsi, S.; Ruggeri, G.; Zamboni, A.; Bhakti, P.; Espen, L.; Ferrante, A.; Noseda, M.; Varanini, Z.; Scarafoni, A. A Bibliometric Analysis of the Scientific Literature on Biostimulants. *Agronomy* **2022**, *12*, 1257. [CrossRef]

35. Di Ciaccio, F.; Troisi, S. Monitoring marine environments with Autonomous Underwater Vehicles: A bibliometric analysis. *Results Eng.* **2021**, *9*, 100205. [CrossRef]

36. Bettencourt, S.; Costa, S.; Caeiro, S. Marine litter: A review of educative interventions. *Mar. Pollut. Bull.* **2021**, *168*, 112446. [CrossRef]

37. Vasconcelos, R.N.; Cunha Lima, A.T.; Lentini, C.A.D.; Miranda, G.V.; Mendonca, L.F.; Silva, M.A.; Cambui, E.C.B.; Lopes, J.M.; Porsani, M.J. Oil Spill Detection and Mapping: A 50-Year Bibliometric Analysis. *Remote Sens.* **2020**, *12*, 3647. [CrossRef]

38. Araujo, R.J.; Shideler, G. Celebrating 60 years of publication of the Bulletin of Marine Science: A bibliometric history (1951–2010). *Bull. Mar. Sci.* **2011**, *87*, 707–726. [CrossRef]
39. Costas, R.; Bordons, M. Development of a thematic filter for the bibliometric delimitation on interdisciplinary area: The case of Marine Science. *Rev. Esp. De Doc. Cient.* **2008**, *31*, 261–272. [CrossRef]
40. Erftemeijer, P.L.A.; Semesi, A.K.; Ochieng, C.A. Challenges for marine botanical research in East Africa: Results of a bibliometric survey. *S. Afr. J. Bot.* **2001**, *67*, 411–419. [CrossRef]
41. Li, J.; Hale, A. Identification of, and knowledge communication among core safety science journals. *Saf. Sci.* **2015**, *74*, 70–78. [CrossRef]
42. Oliveira, P.; Pascoal, A. On the design of multirate complementary filters for autonomous marine vehicle navigation. In Proceedings of the 1st IFAC Workshop on Guidance and Control of Underwater Vehicles, Newport, UK, 9–11 April 2003; pp. 151–156.
43. Papadopoulos, G.; Fallon, M.F.; Leonard, J.J.; Patrikalakis, N.M. Cooperative Localization of Marine Vehicles using Nonlinear State Estimation. In Proceedings of the IEEE/RSJ International Conference on Intelligent Robots and Systems, Taipei, Taiwan, 18–22 October 2010; pp. 4874–4879.
44. Xie, S.R.; Wu, P.; Liu, H.L.; Yan, P.; Li, X.M.; Luo, J.; Li, Q.M. A novel method of unmanned surface vehicle autonomous cruise. *Ind. Robot* **2016**, *43*, 121–130. [CrossRef]
45. Yang, Y.; Chen, G.; Reniers, G.; Goerlandt, F. A bibliometric analysis of process safety research in China: Understanding safety research progress as a basis for making China's chemical industry more sustainable. *J. Clean. Prod.* **2020**, *263*, 121433. [CrossRef]
46. Chowdhury, M.Z.; Shahjalal, M.; Ahmed, S.; Jang, Y.M. 6G Wireless Communication Systems: Applications, Requirements, Technologies, Challenges, and Research Directions. *IEEE Open J. Commun. Soc.* **2020**, *1*, 957–975. [CrossRef]
47. Hover, F.S.; Eustice, R.M.; Kim, A.; Englot, B.; Johannsson, H.; Kaess, M.; Leonard, J.J. Advanced perception, navigation and planning for autonomous in-water ship hull inspection. *Int. J. Robot. Res.* **2012**, *31*, 1445–1464. [CrossRef]
48. Jahanbakht, M.; Xiang, W.; Hanzo, L.; Azghadi, M.R. Internet of Underwater Things and Big Marine Data Analytics-A Comprehensive Survey. *IEEE Commun. Surv. Tutor.* **2021**, *23*, 904–956. [CrossRef]
49. Song, R.; Liu, Y.C.; Bucknall, R. Smoothed A* algorithm for practical unmanned surface vehicle path planning. *Appl. Ocean Res.* **2019**, *83*, 9–20. [CrossRef]
50. Lazarowska, A. Ship's Trajectory Planning for Collision Avoidance at Sea Based on Ant Colony Optimisation. *J. Navig.* **2015**, *68*, 291–307. [CrossRef]
51. Tsou, M.C.; Hsueh, C.K. The study of ship collision avoidance route planning by ant colony algorithm. *J. Mar. Sci. Technol.-Taiwan* **2010**, *18*, 746–756. [CrossRef]
52. Kahveci, N.E.; Ioannou, P.A. Adaptive steering control for uncertain ship dynamics and stability analysis. *Automatica* **2013**, *49*, 685–697. [CrossRef]
53. Kristan, M.; Kenk, V.S.; Kovacic, S.; Pers, J. Fast Image-Based Obstacle Detection From Unmanned Surface Vehicles. *IEEE Trans. Cybern.* **2016**, *46*, 641–654. [CrossRef]
54. Kim, H.; Kim, D.; Shin, J.U.; Kim, H.; Myung, H. Angular rate-constrained path planning algorithm for unmanned surface vehicles. *Ocean Eng.* **2014**, *84*, 37–44. [CrossRef]
55. Small, H. Co-citation in scientific literature—New measure of relationship between 2 documents. *J. Am. Soc. Inf. Sci.* **1973**, *24*, 265–269. [CrossRef]
56. Fossen, T.I. *Handbook of Marine Craft Hydrodynamics and Motion Control, 1. Aufl.*; Wiley: Hoboken, NJ, USA, 2011.
57. Naeem, W.; Irwin, G.W.; Yang, A. COLREGs-based collision avoidance strategies for unmanned surface vehicles. *Mechatronics* **2012**, *22*, 669–678. [CrossRef]
58. Huntsberger, T.; Aghazarian, H.; Howard, A.; Trotz, D.C. Stereo Vision-Based Navigation for Autonomous Surface Vessels. *J. Field Robot.* **2011**, *28*, 3–18. [CrossRef]
59. Mao, G.; Huang, N.; Chen, L.; Wang, H. Research on biomass energy and environment from the past to the future: A bibliometric analysis. *Sci. Total Environ.* **2018**, *635*, 1081–1090. [CrossRef] [PubMed]
60. Yang, X.; Han, Q. Improved reinforcement learning for collision-free local path planning of dynamic obstacle. *Ocean Eng.* **2023**, *283*, 115040. [CrossRef]
61. Yang, M.; Zhang, S.; Pan, Y.; Chang, X.; Xu, G. The Application Status and Development Trend of Artificial Intelligence in Ship Fields of China. In *International Conference on Marine Equipment & Technology and Sustainable Development*; Springer: Singapore, 2023; pp. 1132–1149.
62. Wang, K.; Wang, J.; Huang, L.; Yuan, Y.; Wu, G.; Xing, H.; Wang, Z.; Wang, Z.; Jiang, X. A comprehensive review on the prediction of ship energy consumption and pollution gas emissions. *Ocean Eng.* **2022**, *266*, 112826. [CrossRef]
63. Runnan, L.; Guangze, L.; Pengfei, H.; Xingzhi, L. Research on artificial intelligence safety prediction and intervention model based on ship driving habits. In *MATEC Web Conferences*; EDP Sciences: Les Ulis, France, 2022; Volume 355, pp. 03032–03038. [CrossRef]
64. Heubl, B. Fishing In Murky Waters Investigation-Sea Fish. *Eng. Technol.* **2021**, *16*, 48–50. [CrossRef]
65. Alexiou, K.; Pariotis, E.G.; Zannis, T.C.; Leligou, H.C. Prediction of a Ship's Operational Parameters Using Artificial Intelligence Techniques. *J. Mar. Sci. Eng.* **2021**, *9*, 681. [CrossRef]

66. Munim, Z.H.; Dushenko, M.; Jimenez, V.J.; Shakil, M.H.; Imset, M. Big data and artificial intelligence in the maritime industry: A bibliometric review and future research directions. *Marit. Policy Manag.* **2020**, *47*, 577–597. [CrossRef]

67. Mei, J.; Weigang, Z. Information Exchange Method of Ship Automatic Control Under Artificial Intelligence. In *Innovative Computing: IC 2020*; Springer: Singapore, 2020; pp. 1819–1826.

68. Yun, Z. The application of artificial intelligence in logistics and express delivery. *J. Phys. Conf. Ser.* **2019**, *1325*, 012085. [CrossRef]

69. Bekiros, S.; Loukeris, N.; Matsatsinis, N.; Bezzina, F. Customer Satisfaction Prediction in the Shipping Industry with Hybrid Meta-heuristic Approaches. *Comput. Econ.* **2019**, *54*, 647–667. [CrossRef]

70. Xue, J.; Wu, C.; Chen, Z.; Van Gelder, P.H.A.J.M.; Yan, X. Modeling human-like decision-making for inbound smart ships based on fuzzy decision trees. *Expert Syst. Appl.* **2019**, *115*, 172–188. [CrossRef]

71. Le Roux, N.; Bengio, Y. Representational power of restricted Boltzmann machines and deep belief networks. *Neural Comput.* **2008**, *20*, 1631–1649. [CrossRef] [PubMed]

72. Gao, P.C.; Xu, P.F.; Cheng, H.X.; Zhou, X.G.; Zhu, D.Q. Hybrid Path Planning for Unmanned Surface Vehicles in Inland Rivers Based on Collision Avoidance Regulations. *Sensors* **2023**, *23*, 8326. [CrossRef]

73. Atasoy, S.; Karagoz, O.K.; Ankarali, M.M. Trajectory-Free Motion Planning of an Unmanned Surface Vehicle Based on MPC and Sparse Neighborhood Graph. *IEEE Access* **2023**, *11*, 47690–47700. [CrossRef]

74. Ji, Z.G.; Gan, H.B.; Liu, B. A Deep Learning-Based Fault Warning Model for Exhaust Temperature Prediction and Fault Warning of Marine Diesel Engine. *J. Mar. Sci. Eng.* **2023**, *11*, 1509. [CrossRef]

75. Olena, J. Software Standardization and Infrastructure Development Efforts in Support of Unmanned Maritime Vehicle Autonomy. *Nav. Eng. J.* **2023**, *135*, 41–45.

76. Yu, Y.G.; Ahn, Y.J.; Lee, C.H. Using FRAM for Causal Analysis of Marine Risks in the Motor Vessel Milano Bridge Accident: Identifying Potential Solutions. *Appl. Sci.* **2023**, *13*, 8764. [CrossRef]

77. Wibisono, A.; Piran, M.J.; Song, H.-K.; Lee, B.M. A Survey on Unmanned Underwater Vehicles: Challenges, Enabling Technologies, and Future Research Directions. *Sensors* **2023**, *23*, 7321. [CrossRef]

78. Aguzzi, J.; Flögel, S.; Marini, S.; Thomsen, L.; Albiez, J.; Weiss, P.; Picardi, G.; Calisti, M.; Stefanni, S.; Mirimin, L.; et al. Developing technological synergies between deep-sea and space research. *Elementa-Sci. Anthrop.* **2022**, *10*, 19. [CrossRef]

79. Chen, Z.; Huang, T.; Xue, Z.F.; Zhu, Z.Z.; Xu, J.H.; Liu, Y. A Novel Unmanned Surface Vehicle with 2D-3D Fused Perception and Obstacle Avoidance Module. In Proceedings of the IEEE International Conference on Robotics and Biomimetics (IEEE ROBIO), Sanya, China, 27–31 December 2021; pp. 1804–1809.

80. Cho, H.J.; Jeong, S.K.; Ji, D.H.; Tran, N.H.; Vu, M.T.; Choi, H.S. Study on Control System of Integrated Unmanned Surface Vehicle and Underwater Vehicle. *Sensors* **2020**, *20*, 2633. [CrossRef]

81. Zhou, L.L.; Ye, X.M.; Huang, Z.H.; Xie, P.Z.; Song, Z.G.; Tong, Y.J. An Improved Genetic Algorithm for the Recovery System of USVs Based on Stern Ramp Considering the Influence of Currents. *Sensors* **2023**, *23*, 8075. [CrossRef]

82. Overaas, H.; Halvorsen, H.S.; Landstad, O.; Smines, V.; Johansen, T.A. Dynamic Positioning Using Model Predictive Control With Short-Term Wave Prediction. *IEEE J. Ocean. Eng.* **2023**, *13*, 1065–1077. [CrossRef]

83. Wang, N.; Gao, Y.; Weng, Y.; Zheng, Z.; Zhao, H. Implementation of an integrated navigation, guidance and control system for an unmanned surface vehicle. In Proceedings of the 2018 Tenth International Conference on Advanced Computational Intelligence, Xiamen, China, 29–31 March 2018; pp. 717–722.

84. Veitch, E.; Alsos, O.A. A systematic review of human-AI interaction in autonomous ship system. *Saf. Sci.* **2022**, *152*, 23. [CrossRef]

85. Xiong, J.; Li, J.N.; Du, P. A novel non-fragile H^∞ fault-tolerant course-keeping control for uncertain unmanned surface vehicles with rudder failures. *Ocean Eng.* **2023**, *280*, 11. [CrossRef]

86. Tong, H.Y. An adaptive error constraint line-of-sight guidance and finite-time backstepping control for unmanned surface vehicles. *Ocean Eng.* **2023**, *285*, 10. [CrossRef]

87. Dagdilelis, D.; Blanke, M.; Andersen, R.H.; Galeazzi, R. Cyber-resilience for marine navigation by information fusion and change detection. *Ocean Eng.* **2022**, *266*, 14. [CrossRef]

88. Liu, Y.C.; Bucknall, R. The angle guidance path planning algorithms for unmanned surface vehicle formations by using the fast marching method. *Appl. Ocean Res.* **2016**, *59*, 327–344. [CrossRef]

89. Jin, K.; Wang, J.; Wang, H.; Liang, X.; Guo, Y.; Wang, M.; Yi, H. Soft formation control for unmanned surface vehicles under environmental disturbance using multi-task reinforcement learning. *Ocean Eng.* **2022**, *260*, 112035. [CrossRef]

90. Suryanarayanan, S.; Cartes, D.; Sidley, R. *Energy Scavenging Modes from Renewable Sources for Unmanned Surface Vehicles: A Survey of Concepts*; SPIE: Bellingham, WA, USA, 2006; p. 62302K–62302K-7.

91. Khare, N.; Singh, P. Modeling and optimization of a hybrid power system for an unmanned surface vehicle. *J. Power Sources* **2012**, *198*, 368–377. [CrossRef]

92. Anonymous. Autonaut uncrewed surface vehicle selected by plymouth marine laboratory. *Ocean News Technol.* **2021**, *37*, 22.

93. Mostafa, M.Z.; Khater, H.A.; Rizk, M.R.; Bahasan, A.M. A novel GPS/RAVO/MEMS-INS smartphone-sensor-integrated method to enhance USV navigation systems during GPS outages. *MST* **2019**, *30*, 95103. [CrossRef]

94. Si, K.; Wei, J.; He, X.; Liu, D.; Li, P. Integrated Navigation Algorithm for USV with Insufficient Satellites. In Proceedings of the 2021 7th Annual International Conference on Network and Information Systems for Computers (ICNISC), Guiyang, China, 23–25 July 2021; pp. 743–748.

95. Lv, Z.; Zhang, J.; Jin, J.; Li, Q.; Liu, L.; Zhang, P.; Gao, B. Underwater Acoustic Communication Quality Evaluation Model Based on USV. *Shock. Vib.* **2018**, *2018*, 1–7. [CrossRef]

96. Dong, K.; Sun, R.; Dong, X. CO_2 emissions, natural gas and renewables, economic growth: Assessing the evidence from China. *Sci. Total Environ.* **2018**, *640–641*, 293–302. [CrossRef] [PubMed]

97. Alboody, A.; Vandenbroucke, N.; Porebski, A.; Sawan, R.; Viudes, F.; Doyen, P.; Amara, R. A New Remote Hyperspectral Imaging System Embedded on an Unmanned Aquatic Drone for the Detection and Identification of Floating Plastic Litter Using Machine Learning. *Remote Sens.* **2023**, *15*, 3455. [CrossRef]

98. Balanescu, M.; Suciu, G.; Badicu, A.; Birdici, A.; Pasat, A.; Poenaru, C.; Zatreanu, I. Study on Unmanned Surface Vehicles used for Environmental Monitoring in Fragile Ecosystems. In Proceedings of the IEEE 26th International Symposium for Design and Technology in Electronic Packaging (SIITME), Pitesti, Romania, 21–24 October 2020; pp. 94–97.

99. Barrera, C.; Maarouf, M.; Campuzano, F.; Llinas, O.; Marichal, G.N. A Comparison of Intelligent Models for Collision Avoidance Path Planning on Environmentally Propelled Unmanned Surface Vehicles. *J. Mar. Sci. Eng.* **2023**, *11*, 692. [CrossRef]

100. Jo, W.; Hoashi, Y.; Aguilar, L.L.P.; Postigo-Malaga, M.; Garcia-Bravo, J.M.; Min, B.C. A low-cost and small USV platform for water quality monitoring. *HardwareX* **2019**, *6*, 13. [CrossRef]

101. Siyang, S.; Kerdcharoen, T. Development of Unmanned Surface Vehicle for Smart Water Quality Inspector. In Proceedings of the 13th International Conference on Electrical Engineering/Electronics, Computer, Telecommunications and Information Technology (ECTI-CON), Chiang Mai, Thailand, 28 June–1 July 2016.

102. Cross, J.N.; Mordy, C.W.; Tabisola, H.M.; Meinig, C.; Cokelet, E.D.; Stabeno, P.J. Innovative Technology Development for Arctic Exploration. In Proceedings of the OCEANS MTS, Washington, DC, USA, 19–22 October 2015.

103. Daniel, T.; Manley, J.; Trenaman, N. The Wave Glider: Enabling a new approach to persistent ocean observation and research. *Ocean Dyn.* **2011**, *61*, 1509–1520. [CrossRef]

104. Gong, Y.H.; Zhang, S.J.; Luo, M.; Ma, S.A. A mutation operator self-adaptive differential evolution particle swarm optimization algorithm for USV navigation. *Front. Neurorobotics* **2022**, *16*, 11. [CrossRef] [PubMed]

105. Guo, X.H.; Ji, M.J.; Zhang, W.D. Research on a new two-level scheduling approach for unmanned surface vehicles transportation containers in automated terminals. *Comput. Ind. Eng.* **2023**, *175*, 16. [CrossRef]

106. Cai, J.H.; Du, S.; Lu, C.D.; Xiao, B.; Wu, M. Obstacle Detection of Unmanned Surface Vessel based on Faster RCNN. In Proceedings of the IEEE 6th International Conference on Industrial Cyber-Physical Systems (ICPS), Wuhan, China, 8–11 May 2023.

107. Peng, Y.; Li, Y. Autonomous Trajectory Tracking Integrated Control of Unmanned Surface Vessel. *J. Mar. Sci. Eng.* **2023**, *11*, 568. [CrossRef]

108. Liu, W.; Qiu, K.; Yang, X.F.; Wang, R.H.; Xiang, Z.R.; Wang, Y.; Xu, W.X. COLREGS-based collision avoidance algorithm for unmanned surface vehicles using modified artificial potential fields. *Phys. Commun.* **2023**, *57*, 13. [CrossRef]

109. Wang, C.B.; Zhang, X.Y.; Yang, Z.L.; Bashir, M.; Lee, K. Collision avoidance for autonomous ship using deep reinforcement learning and prior-knowledge-based approximate representation. *Front. Mar. Sci.* **2023**, *9*, 14. [CrossRef]

110. Zhang, M.Y.; Zhang, D.; Yao, H.J.; Zhang, K. A probabilistic model of human error assessment for autonomous cargo ships focusing on human-autonomy collaboration. *Saf. Sci.* **2020**, *130*, 12. [CrossRef]

111. Liang, Z.Q.; Xiao, G.; Hu, J.Q.; Wang, J.S.; Ding, C.S. MotionTrack: Rethinking the motion cue for multiple object tracking in USV videos. *Visual Comput.* **2023**, *39*, 1–13. [CrossRef]

112. Kouriampalis, N.; Pawling, R.J.; Andrews, D.J. Modelling the operational effects of deploying and retrieving a fleet of uninhabited vehicles on the design of dedicated naval surface ships. *Ocean Eng.* **2021**, *219*, 19. [CrossRef]

Journal of
Marine Science and Engineering

MDPI

Article

Research on Carbon Intensity Prediction Method for Ships Based on Sensors and Meteorological Data

Chunchang Zhang [1,2], **Tianye Lu** [3], **Zhihuan Wang** [2,4,*] **and Xiangming Zeng** [1,2]

[1] Merchant Marine College, Shanghai Maritime University, Shanghai 201306, China;
    cczhang@shmtu.edu.cn (C.Z.); xmzeng@shmtu.edu.cn (X.Z.)
[2] National Engineering Research Center for Special Equipment and Power Systems of Ships and Marine Engineering, Shanghai 201306, China
[3] Logistics Engineering College, Shanghai Maritime University, Shanghai 201306, China;
    202230210313@stu.shmtu.edu.cn
[4] Institute of Logistics Science and Engineering, Shanghai Maritime University, Shanghai 201306, China
*   Correspondence: zhwang@shmtu.edu.cn

Abstract: The Carbon Intensity Index (CII) exerts a substantial impact on the operations and valuation of international shipping vessels. Accurately predicting the CII of ships could help ship operators dynamically evaluate the possible CII grate of a ship at the end of the year and choose appropriate methods to improve its CII grade to meet the IMO requirement with minimum cost. This study developed and compared five CII predicting models with multiple data sources. It integrates diverse data sources, including Automatic Identification System (AIS) data, sensor data, meteorological data, and sea state data from 2022, and extracts 21 relevant features for the vessel CII prediction. Five machine learning methods, including Artificial Neural Network (ANN), Support Vector Regression (SVR), Least Absolute Shrinkage and Selection Operator (LASSO), Extreme Gradient Boosting (XG-Boost), and Random Forest (RF), are employed to construct the CII prediction model, which is then applied to a 2400 TEU container ship. Features such as the mean period of total swell, mean period of wind waves, and seawater temperature were considered for inclusion as inputs in the model. The results reveal significant correlations between cumulative carbon emissions intensity and features like cumulative distance, seawater temperature, wave period, and swell period. Among these, the strongest correlations are observed with cumulative distance and seawater temperature, having correlation coefficients of 0.45 and 0.34, respectively. Notably, the ANN model demonstrates the highest accuracy in CII prediction, with an average absolute error of 0.0336, whereas the LASSO model exhibits the highest error of 0.2817. Similarly, the ANN model provides more accurate annual CII ratings for the vessel. Consequently, the ANN model proves to be the most suitable choice for cumulative CII prediction.

Keywords: water transport; carbon intensity prediction of ship; machine learning; fuel consumption of ship; data fusion

Citation: Zhang, C.; Lu, T.; Wang, Z.; Zeng, X. Research on Carbon Intensity Prediction Method for Ships Based on Sensors and Meteorological Data. *J. Mar. Sci. Eng.* **2023**, *11*, 2249. https://doi.org/10.3390/jmse11122249

Academic Editor: Mihalis Golias

Received: 26 October 2023
Revised: 21 November 2023
Accepted: 25 November 2023
Published: 28 November 2023

1. Introduction

The IMO Marine Environment Protection Committee (MEPC 76), which took place in June 2021, introduced the Carbon Intensity Indicator (CII). This indicator is defined as the total amount of CO_2 emissions divided by the deadweight and the distance traveled per year. According to the requirements of the International Maritime Organization (IMO), starting from 1 January 2023, ships of 5000 GT or above on international voyages will be required to determine their operational carbon intensity indicator (CII) rating, and the CII rating requirement will be increased by 2% per year until 2026. After that time, the CII rating requirement is likely to be further increased to meet the requirements of the IMO 2023 international shipping greenhouse gas emission reduction strategy [1]. The CII ratings of ships are categorized into five classes from high to low: A, B, C, D, and E. Ships need to have a CII rating of at least C. If a ship is rated E, or if a ship has been rated D for three consecutive years, a corrective action plan needs to be developed as part of the ship's energy efficiency management plan [2].

In addition, as CII rating starts affecting the price and contract of ship chartering, a number of CII-related clauses have been added to the charter party. For example, the burden of CII is solely with the charterer [3]. If a chartered vessel does not have a CII rating of C, it may result in a breach of contract, leading to financial loss for the charterer. Increased waiting time in port, vessel idling, and charter suspension may have a negative impact on a ship's carbon intensity [4]. It is important for shipowner or operators to dynamically monitor and manage ship CII rating.

It is prudent for shipowners to monitor and assess the ship's actual CII in real time in order to determine how close it is to the required CII and to take appropriate measures and actions to avoid getting into a situation that would result in a lower CII rating [4]. Accurate prediction of the carbon intensity of a ship is an important basis for the future control of the CII rating of a ship. Previous studies are mainly focused on calculating and predicting ship fuel consumption, and very few studies investigate the prediction of ship CII prediction and upgrading.

Accurate prediction of the CII of a ship enables the assessment of the year-to-year change in CII grade of the ship. This foresight allows proactive measures to be taken, ensuring compliance with the annual carbon intensity qualification. Such proactive management not only aids in reducing carbon emissions and energy consumption but is also crucial in chartering contracts. Precise CII prediction prevents shipowners from incurring liquidated damage costs by ensuring the ship's carbon intensity level is maintained within the contractual requirements.

The main purpose of this work is to investigate the performance of five machine learning models for carbon intensity prediction using multiple data sources, including AIS data, fuel flow sensor data, meteorological data and sea state data, to identify the most suitable models for carbon intensity prediction. The best model will be utilized to further predict the CII rating at the end of the year. One container ship is taken as an example to illustrate the development process. This work first analyzes the change in carbon intensity of ships over a one-year period by using a carbon intensity assessment methodology. Second, the carbon intensity obtained from this assessment method is used as a prediction target to compare different machine learning models for carbon intensity prediction. This paper also analyzes the correlation between factors affecting fuel consumption and carbon intensity. The ultimate goal of this paper is to assess the accuracy of the carbon intensity classes obtained from the prediction using machine learning models.

This paper focuses on the urgent need for an annual rating of CII of ship operation. Taking a container ship with a capacity of about 2400 TEU as an example, this study uses the ship's AIS data, real-time fuel consumption data, sensor data, and meteorological data from 2022 to compare and analyze the accuracy of the different machine learning methods for predicting the CII of the ship to provide support for the prediction of CII for the ship and the ship's annual rating.

The remainder of this paper is organized as follows. Section 2 reviews related works on the prediction of fuel consumption and carbon intensity, while Section 3 introduces the main datasets used in our study and describes the processes and methods used to predict CII. Section 4 presents and discusses the results of the study. Finally, conclusions, limitations, and future work are presented in Section 5.

2. Related Work

Currently, related research in the same field consists of studies on the prediction of fuel consumption of ships, the prediction of resistance, and how to improve the carbon intensity level. However, there are fewer studies on the prediction of carbon intensity of ships, while most scholars have conducted studies on the prediction of fuel consumption and resistance of ships, and machine learning models have been frequently used for the prediction.

Wang carried out feature compression of the ship energy consumption data by means of the LASSO regression algorithm, and the results showed that the prediction effect of LASSO is better than neural network, support vector regression, and other models [5]. Jeon proposed a regression model based on an artificial neural network that adjusted the hyperparameters of the neural network, such as hidden layer, neuron, activation function and other hyperparameters. And the results showed that ANN has a higher prediction accuracy than the polynomial regression and support vector machine in fuel consumption prediction [6]. Ren compared the prediction results under different data sources using a ridge regression model based on AIS data, MRV data, and MRV-normalized data, respectively, and found that the model based on MRV report achieved the best results [7]. Li investigated the results of various prediction models with different combinations of data sources based on a variety of data sources, such as logbooks, meteorological data, and AIS data [8–10]. Uyanık proposed the methods of kernel ridge regression, Bayesian ridge regression, and Adaboost, and found that the ridge regression model had a higher accuracy [11]. In summary, different models are used to study the prediction of ship fuel consumption. The research methods are mainly based on neural networks, support vector machines, LASSO regression, and integrated learning models [12,13]. Under different scenarios and datasets, the best prediction models are different. In the area of resistance prediction, Yidiz presented a method for predicting the residual drag coefficient of a trimaran using artificial neural networks that had parameters such as lateral and longitudinal positions of the side hulls, longitudinal buoyancy centers, and Froude number [14]. Martić successfully applied artificial neural networks to the assessment of additional resistance on container ships [15].

In terms of carbon intensity, previous studies are mainly focused on how to upgrade ships to make them more efficient and attractive to charterers and to increase their competitiveness in the market [16]. Wang analyzes the effectiveness of four current CII metrics and considers designing an average CII calculation method for shipping companies rather than individual vessels [17]. Gianni considered a 180,000 GRT cruise ship as a case ship and designed seven scenarios to calculate the CII with reference to its power plant, then analyzed the impact of solid oxide fuel cells (SOFC) on the CII [18]. Hoffmann analyzed the connection between biofouling and ship CII and provided insights and strategies for improving hull performance related to the use of antifouling coatings [19]. During the gap period when the industry was waiting for alternative fuel solutions [20], Bayrakta designed seven different scenarios to test the EEXI and CII values of different vessels with different engine configurations and analyzed the impact of the utilization of two alternative fuels, LNG and methanol, on the EEXI and CII by calculating the EEXI and CII values of the vessels for the years 2019 to 2026 [21]. In addition, some studies have developed mathematical models to analyze the factors affecting carbon intensity. Elkafas found that carbon intensity values depend on the number of trips per year, the number of passengers carried, and the amount of fuel consumed, and that proper deceleration reduces the ship's emission rate [22]. Sun modeled the speed of a time-chartered vessel with CII penalties included and found that the larger the vessel, the more carbon emissions, and that carbon intensity and CII penalties are reduced when the charter speed is reduced for the same amount of time [23].

In general, compared with studies on fuel consumption, research on carbon intensity and CII grating are relatively rare; those that exist are mainly focused on factors that affect CII grading and how to upgrade CII level. Studies on how to predict CII and CII grating are still very limited. Although the fuel consumption of ships is closely related to the carbon intensity of a ship's operation, it is still rare to find studies that use the carbon intensity of ships as a direct prediction target.

3. Data and Methodology

3.1. Data

3.1.1. Case Ship Data

This study takes a container ship of 2400 TEU as an example. The information related to this container ship is listed in Table 1. The case ship has one main engine and two auxiliary engines. The main engine provides propulsion power, and the auxiliary engines provide electricity to the ship.

Table 1. Main information of the analyzed container ship.

Vessel Type	Container
Built	2019
Gross tonnage	26,771
Deadweight	35,337
Length	185
Breadth	32
Number of main engines	1
Main engine power	13,700 kW
Number of auxiliary engines	2
Auxiliary engine power	1370 kW, 1840 kW

The data sources for this study mainly include AIS data, sensor data, meteorological data, and sea state data.

3.1.2. AIS Data

AIS data provided by shipping companies. The ship's AIS data include dynamic and static data, and the update frequency differs according to the ship's speed and position. In addition to vessel identification and specific information (MMSI, Call Sign, Name, Draught, Length, Breadth), AIS data also contains specific navigational data, including Date, Longitude, Latitude, Speed, Course, ROT. AIS data useful for this paper are listed in Table 2.

Table 2. Main columns and sample data of the AIS data.

Date	Lon	Lat	Course (°)	Speed (kn)	ROT (°/s)	Draught (m)
2022-01-01 00:26:16	123.3059	30.9999	357.0	16.2	348.0	8.7
2022-01-01 00:28:21	123.3048	31.0051	345.0	16.1	340.0	8.7
2022-01-01 00:39:02	123.2830	31.0502	337.0	16.5	335.0	8.7
2022-01-01 00:43:50	123.2739	31.0711	342.0	16.6	339.0	8.7

The generation of missing data is necessary due to large amounts of missing information in some areas due to weather and location, which may make it impossible to fully calculate the distance traveled by the case vessel. This paper uses the linear interpolation method to interpolate the information in Table 2 with a time interval of 5 min. According to the change of latitude and longitude of neighboring AIS points, the sailing distance of the ship can be calculated. Data containing sailing distances are listed in Table 3.

Table 3. AIS data with 5 min sampling and sailing distances.

Date	Lon	Lat	Distance (m)
2022-01-01 00:25:00	123.3054	31.0025	294.5
2022-01-01 00:26:16	123.3059	30.9999	588.9
2022-01-01 00:28:21	123.3048	31.0051	2570.4
2022-01-01 00:30:00	123.2942	31.0263	2858.5

Since the frequency of the collected meteorological data and sea state data is hourly, to maintain a uniform time resolution, the ship's AIS data is aggregated according to the hour in this study [9]. In this paper, the sum of the distances traveled by the case ships and the average of the other data in the AIS data are calculated at intervals of one hour.

The speed and range data are distributed as shown in Figures 1 and 2 below.

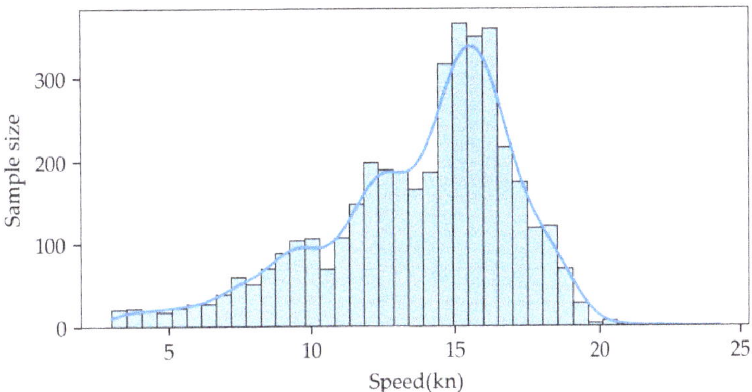

Figure 1. Distribution of the speed of the case ship.

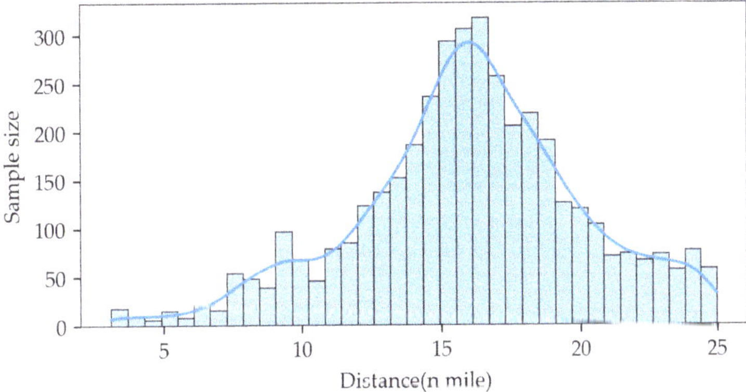

Figure 2. Distribution of the distance of the case ship.

From Figures 1 and 2, it can be found that the data processing ensures the consistency of the distribution of the speed and range data under one hour of data collection frequency. Because of the deviation between the voyage calculated by latitude and longitude in AIS data and the actual voyage, the speed and voyage data in the range of 20~25 do not correspond exactly.

3.1.3. Sensor Data

The case ship is equipped with various sensors to obtain real-time data about the ship. The main engine fuel consumption rate (MEActFOCons), generator fuel consumption rate (DGActFOCons), and boiler fuel consumption rate (BlrActFOCons) are recorded using mass flow meters. The main engine rotational speed (MERpm) and trim of the case ship (Trim) are recorded using the corresponding sensor data. The sensor data used in this paper are listed in Table 4.

Table 4. Main columns and sample data of sensor data.

PCDate	PCTime	MEActFOCons (kg/h)	DGActFOCons (kg/h)	BlrActFOCons (kg/h)	MERpm (r/min)	Trim (m)
2022-01-01	00:00:03	1256.8874	114.7474	0	81	3.97
2022-01-01	00:00:13	1251.0776	115.6824	0	81	3.06
2022-01-01	00:00:26	1239.1414	116.1036	0	82	3.1
2022-01-01	00:00:36	1238.5380	117.4958	0	82	3.69

The original unit of the fuel consumption of the main engine, auxiliary engine, and boiler is kg/h, which needs to be converted into kg according to the time interval of each data point. The converted fuel consumption data are listed in Table 5.

Table 5. Sensor data with converted fuel consumption data.

PCDate	PCTime	MEActFOCons (kg)	DGActFOCons (kg)	BlrActFOCons (kg)
2022-01-01	00:00:03	1.0474	0.0956	0
2022-01-01	00:00:13	3.4752	0.3213	0
2022-01-01	00:00:26	4.4746	0.4192	0
2022-01-01	00:00:36	3.4403	0.3263	0
2022-01-01	00:00:36	3.4403	0.3263	0

To maintain a uniform time resolution for data fusion, the data collected through the sensors also needs to be aggregated on an hourly basis. In this paper, the sum of the fuel consumption and the average of the other data in the sensor data are calculated at intervals of one hour. It was observed that the processed data was found to be less than 8760 (24×365) data, so there were also missing values in the sensor data. In this paper, linear interpolation is also taken for the data in the sensor to generate the missing data with a time interval of 1 h. On this basis, the total amount of heavy fuel and light fuel in each hour was calculated, and the total amount of carbon dioxide in each hour was calculated according to the emission factor. Carbon dioxide data are listed in Table 6.

Table 6. Carbon dioxide emissions at different times.

Date	CO_2 (kg)
2022-01-01 00:00:00	4180.33
2022-01-01 01:00:00	4208.82
2022-01-01 02:00:00	4172.76
2022-01-01 03:00:00	4137.75

3.1.4. Meteorological and Sea State Data and Processing

In this paper, meteorological and sea state data are obtained from the European Centre for Medium-Range Weather Forecasts (ECMWF) and the Copernicus Marine Service (Copernicus). The data from ECMWF cover a wide range of meteorological data sets from 1979 to present, including wind component at 10 m sea level, temperature, humidity, characteristic wave height, and cycle frequency. The scope of ECMWF data covers several meteorological datasets from 1979 to the present, including the wind component at 10 m above sea level, temperature, humidity, characteristic wave height, and cycle frequency, etc. The scope of Copernicus data covers several sea state datasets for each year, including seawater temperature, the current velocity component at different seawater depths, etc. The meteorological dataset used in this paper was collected at a frequency of 1 h, and the data are downloaded in the form of a grid divided according to latitude and longitude. The collected data information is recorded at each grid point, and there is a wide range of choices for the grid size; the minimum grid density is $0.125° \times 0.125°$, the maximum grid density is $1° \times 1°$, and the grid density selected in this paper is $0.25° \times 0.25°$. The sea state dataset used in this paper was also collected at a frequency of 1 h, and the current velocity component of seawater at a depth of 0.5 m was selected as the basis.

For the processing of meteorological data and sea state data, the latitude and longitude in the AIS data are first utilized to obtain the environmental data corresponding to the ship's position [8]. Eastward wind speed at 10 m above sea level (u10), northward wind speed at 10 m above sea level (v10), mean direction of total swell (mdts), mean direction of wind waves (mdww), mean period of total swell (mpts), mean period of wind waves (mpww), mean wave direction (mwd), mean wave period (mwp), sea surface temperature (sst), significant_height of combined wind waves and swell (swh), significant height of total swell (shts) and significant height of wind waves (shww) are recorded u ECMWF. Eastward sea water velocity (uo) and northward sea water velocity (vo) are recorded from Copernicus. The environmental data used in this paper are listed in Table 7.

Table 7. Meteorological data and sea state data from time–space matching acquisition.

Data Name	Sample 1	Sample 2
date	2022-01-01 00:00:00	2022-01-01 01:00:00
u10	−2.4207	−2.6627
v10	0.3611	0.5708
mdts	14.6175	14.2658
mdww	70.0711	91.5424
mpts	6.0399	6.0044
mpww	2.8725	2.1009
mwd	14.6768	14.4249
mwp	6.0288	5.9816
sst	289.3203	288.498
swh	0.9627	0.9532
shts	0.9616	0.9507
shww	0.0295	0.0578
vo	0.1896	0.1587
uo	−0.0190	−0.0428

Since the collected wind speed data and flow velocity data are east–west and north–south components, vector synthesis is needed to obtain the actual wind speed, wind direction, current speed, and current direction. The schematic of direction synthesis is shown in Figure 3.

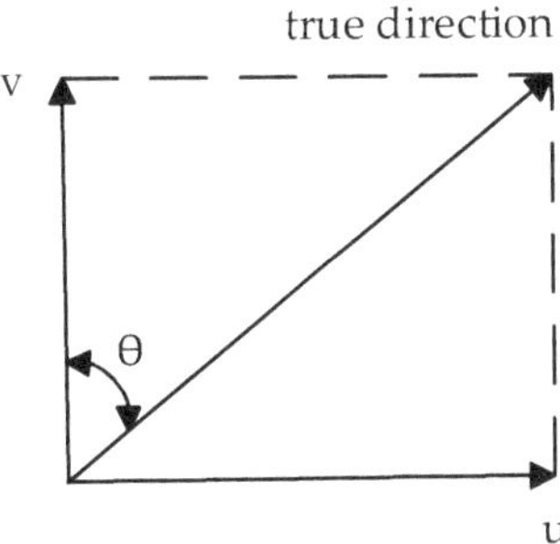

Figure 3. Schematic of vector synthesis.

Finally, it is necessary to convert the directions in the meteorological data into the relative directions of the ship in combination with the actual heading of the ship, and further fuse the preprocessed AIS data, sensor data, meteorological data, and sea state data according to the time [9,10].

3.1.5. Calculation of Cumulative Carbon Intensity of Ships

Finally, it is necessary to calculate the cumulative carbon intensity of the ships up to each point of the data, as well as the ratio of the total cumulative CO_2 mass of the ship at each moment in time to the total transport workload it carries out, by using the following formula [16]:

$$CII_t = \frac{M_t}{W_t} \tag{1}$$

where M_t is the total amount of carbon dioxide emission of the ship at the time, in kg and W_t is the total transportation workload accomplished by the ship at the time, in t·n mile.

The formula for calculating the total amount of carbon dioxide emission M_t at the time of the ship is as follows:

$$M_t = \sum FC_{jt} \times CF_j \tag{2}$$

where j is the fuel type; FC_{jt} is the total fuel consumption of the ship at the time, kg; and CF_j is the fuel mass to carbon dioxide mass conversion factor for fuel j.

The formula for calculating the total transportation workload of the ship at the time is as follows:

$$W_t = C \times D_t \tag{3}$$

where C is the DWT of the ship and D_t is the distance sailed by the ship at the time, n mile.

Equation (1) shows that CII could be calculated from carbon emission, deadweight tons, and sailing distance. Carbon emission could be directly computed by fuel consumption, which is collected through the mass flow meter on the case ship.

In addition, the variation of CII with time for the ships studied in this paper is shown in Figure 4.

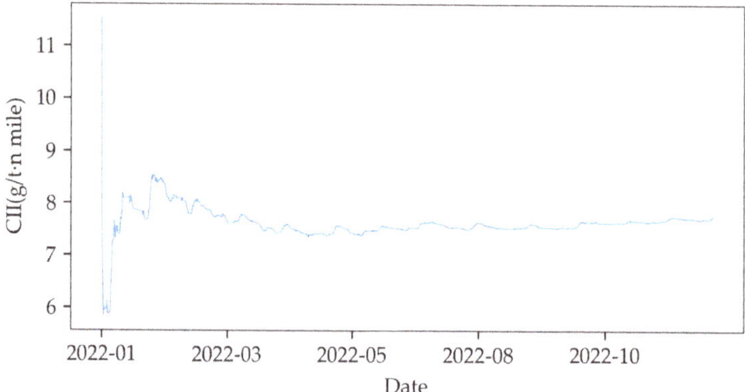

Figure 4. Time distribution of carbon intensity data.

Figure 4 shows that the fluctuations are large in January, but gradually stabilize in the range of 7 to 8 in the time that follows, which occurs because ships start with small voyages and relatively large carbon dioxide emissions, and as the cumulative value of the voyage increases, the calculation of the cumulative CII in conjunction with the deadweight tonnage leads to a certain degree of decrease and stabilization of its future value.

The cumulative CII of the ship mainly focuses on the situation when the ship is in sailing condition. In this paper, it is assumed that a ship is in sailing condition when its speed is more than 3 knots. Therefore, in this study, only data with speed greater than or equal to 3 knots are considered [24]. The total number of processed data is 4061, and some data samples are shown in Table 8.

Table 8. Sample data of the input for model development.

Data Name	Abbreviation	Data Sources	Unit	Sample 1
speed	speed	AIS	kn	16.4911
rate of turning	rot	AIS	$^\circ$/s	270.0357
wind speed	wind speed	ECMWF	m/s	2.4476
draught	draught	AIS	m	8.7000
distance	distance	AIS	n mile	10.2731
cumulative value of distance	distance sum	summation	n mile	10.2731
mean period of total swell	mpts	ECMWF	s	6.0399
mean period of wind waves	mpww	ECMWF	s	2.8726
mean period of wave	mwp	ECMWF	s	6.0288
height of combined wind waves and swell	shww	ECMWF	m	0.9628
height of total swell	swh	ECMWF	m	0.9617
height of wind waves	wwh	ECMWF	m	0.0295
current speed	current speed	Copernicus	m/s	0.1906
wind direction	wind direction	vector synthesis	$^\circ$	4.8576
current direction	current direction	vector synthesis	$^\circ$	80.6408
swell direction	swell direction	ECMWF	$^\circ$	100.9890
wind waves direction	dww	ECMWF	$^\circ$	156.4426
wave direction	wd	ECMWF	$^\circ$	101.0483
sea surface temperature	sst	ECMWF	$^\circ$C	16.1703
merpm	merpm	sensors	r/min	81.5403
trim	trim	sensors	m	3.4338
cumulative CII	CII	formula calculation	g/t·n mile	11.5154

3.2. Methodology

3.2.1. Research Framework

The main research framework could be divided into several steps. First, the ship AIS data, sensor data, and meteorological and sea state data are cleaned and preprocessed, and temporal and spatial fusion are performed. Second, relevant features are extracted, and the correlation between each feature is analyzed. Third, the data are divided into training and testing sets in the ratio of 0.75:0.25 (3046 trainset:1015 test set) according to the time series [25]. Fourth, five machine learning models are used to train and optimize the hyper-parameters of ANN, XGBoost, and RF models by random search and cross-validation, and the parameters of LASSO and SVR are optimized by grid search because of their fewer hyper-parameters. Figure 5 shows the general framework of ship CII prediction.

3.2.2. Model Performance Metrics

In this paper, mean absolute error (MAE), mean square error (MSE), root-mean-square-error (RMSE), and mean absolute percentage error (MAPE) are used to evaluate the performance of the model. The evaluation formulas are calculated as follows:

$$\text{MAE} = \frac{1}{m}\left(\sum_{i=1}^{m}\left|y_i - \hat{y}_i\right|\right) \tag{4}$$

$$\text{MSE} = \frac{1}{m}\sum_{i=1}^{m}\left(y_i - \hat{y}_i\right)^2 \tag{5}$$

$$\text{RMSE} = \sqrt{\frac{1}{m}\sum_{i=1}^{m}\left(y_i - \hat{y}_i\right)^2} \tag{6}$$

$$\text{MAPE} = \frac{1}{m}\sum_{i=1}^{m}\left|\frac{y_i - \hat{y}_i}{y_i}\right| \times 100\% \tag{7}$$

where m is the number of samples; y_i is the true value; and $\hat{y}_i$ is the predicted output value of the model.

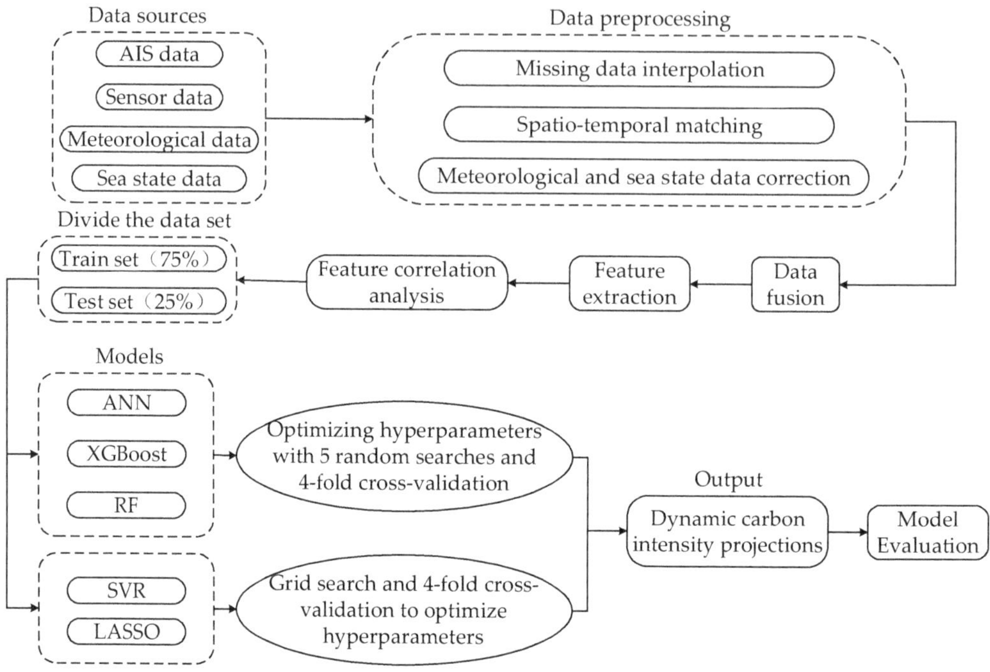

Figure 5. Ship carbon intensity prediction modeling framework.

3.2.3. CII Rating Calculation Model

The carbon intensity rating of a ship must be determined annually using the Carbon Intensity Index (Required CII) of the ship. This is based on the CII reference baseline for a specific ship type for the year 2019, and its rating boundaries are then determined based on the boundary parameters of the ship type. This allows its annual carbon intensity rating to be found. The specific calculation formula is as follows [16]:

$$\mathrm{CII}_{\mathrm{ref}} = a \cdot \mathrm{C}^{-c} \tag{8}$$

where $\mathrm{CII}_{\mathrm{ref}}$ is the reference baseline value of CII in 2019, g/(DWT·n mile) or g/(GT·n mile); C is the deadweight tonnage (DWT) of the ship; and a and c are the parameters for the different ship types. Since the research object of this paper is a container ship, $a = 1984$ and $c = 0.489$ [18].

The required CII is calculated as follows [26]:

$$\mathrm{CII}_{\mathrm{Req}} = \left(1 - \frac{Z}{100}\right) \times \mathrm{CII}_{\mathrm{ref}} \tag{9}$$

where Z is the discount factor of CII for different years; the year of study for this paper is 2022, which has the value of 3 [26]. For the convenience of rating, the rating mechanism is used to define four boundaries per year, which facilitates the division into five grades. Accordingly, the rating can be determined by comparing the annual CII of the ship with the boundary value, and the formula for calculating the boundary value B_i is as follows [27]:

$$B_i = \exp(d_i) \cdot \mathrm{CII}_{\mathrm{Req}} \tag{10}$$

where i = 1, 2, 3, 4, and $\exp(d_i)$ is the boundary parameter of the container ship. As shown in Table 9 [27].

Table 9. Carbon intensity rating boundary parameters for container ships.

Ship Type	Capacity	$\exp(d_1)$	$\exp(d_2)$	$\exp(d_3)$	$\exp(d_4)$
container ship	DWT	0.83	0.94	1.07	1.19

4. Results and Discussion

4.1. Feature Correlation Analysis

Since the carbon intensity of a ship is calculated based on the ship's fuel consumption, carbon emission factor, deadweight tonnage, and voyage, this paper extracts 21 relevant features that affect the fuel consumption as inputs to the cumulative carbon intensity prediction model based on the past research [8–10,25]. After normalizing all data, the correlation analysis is carried out, as shown in Figure 6. It was found that the correlation between the features, such as seawater temperature, wave period, and swell period, and the cumulative CII of the ship is high, which is different from the results of previous studies. In previous studies [12,28], it was found that the meteorological data, such as navigation speed, main engine speed, and wave height, have significant influence on the fuel consumption of the ship, and the CII of the ship needs to be further calculated according to the fuel consumption data to be obtained. However, Figure 6 shows that characteristics such as speed and engine speed do not show strong correlation with the cumulative CII of the ship. This probably occurs because this paper focuses on the prediction of the cumulative CII, which is calculated by the cumulative carbon emission and cumulative voyage, making it impossible to accurately understand the relationship between the cumulative carbon intensity and characteristics such as speed and engine speed.

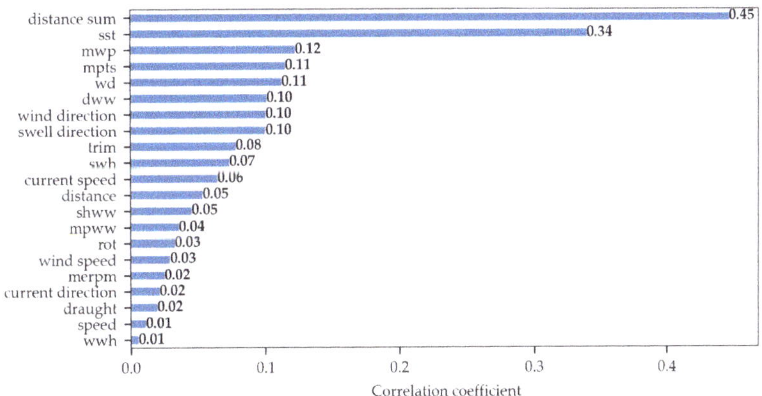

Figure 6. Cumulative carbon intensity and characteristic correlation analysis of ships.

4.2. Analysis of Carbon Intensity Prediction Models

In this paper, Windows 11, 12th Gen Intel(R) Core(TM) i7-12700H processor, 16 G RAM, and python version 3.9 were used as the experimental environment. The study takes the cumulative carbon intensity value of the ship as the prediction target and conduct experiments using five machine learning models in both the parameterized and unparameterized cases. In the case of tuning, this paper uses random search and grid search combined with 4-fold cross-validation to optimize the hyper-parameters of ANN, XGBoost, RF and LASSO, SVR models, respectively, in which the search range of the optimizer in the ANN model is (SGD, RMSprop, Adagrad, Adadelta, Adam, Adamax, Nadam), denoted by the abbreviation (S,N), and the search ranges of the remaining hyperparameters, the default values in the case of no parameter tuning, and the search results are shown in Table 10.

Table 10. Comparison of hyperparameter optimization results of different models.

Models	Hyperparameter Names	Default Values	Search Ranges	1	2	3	4	5
ANN	optimizer	Adam	(S,N)	RMSprop	Adamax	Adamax	RMSprop	Adamax
	neurons	100	(10,100)	40	80	30	100	80
	epochs	100	(10,200)	50	10	100	200	50
	batch_size	32	(10,100)	90	10	40	20	70
SVR	C	1	(0.01,10)				0.1	
	gamma	scale	(0.0001,10)				1	
LASSO	alpha	1	(0.00001,10)				10	
	max_iter	1000	(0,1000)				100	
XGBoost	subsample	None	(0.5,1)	0.7	0.6	0.9	0.7	0.8
	n_estimators	100	(0,300)	300	240	233	260	240
	min_child_weight	None	(0,10)	4	7	6	6	7
	max_depth	None	(2,10)	5	4	3	4	3
	learning_rate	None	(0.01,0.3)	0.21	0.3	0.21	0.1	0.21
	gamma	None	(0,10)	0	2	7	6	6
	colsample_bytree	None	(0.5,1)	0.9	0.8	0.9	0.6	0.6
RF	n_estimators	100	(0,300)	233	68	68	266	200
	min_samples_split	2	(1,20)	10	16	17	6	18
	max_depth	None	(2,10)	4	8	4	4	4

The evaluation indexes corresponding to the combination of hyperparameters for each optimization of each model were calculated by (4)–(7), as shown in Table 11.

Table 11. Comparison of predictive performance of models.

Models	Number of Optimizations	MAE	MSE	RMSE	MAPE
ANN	1	0.1179	0.0154	0.1243	1.5333
	2	0.0568	0.0045	0.0673	0.7432
	3 *	0.0476	0.0034	0.0591	0.6186
	4	0.1375	0.0204	0.1431	1.7895
	5	0.0524	0.0040	0.0634	0.6808
	None	0.2181	0.5513	0.7424	2.8392
SVR	1	0.1037	0.0123	0.1109	1.3489
	None	0.4203	0.1946	0.4412	5.4675
LASSO	1	0.2817	0.0864	0.2940	3.6654
	None	0.3687	0.1765	0.4201	3.9968
XGBoost	1	0.1050	0.0127	0.1131	1.3667
	2	0.1083	0.0144	0.1201	1.4091
	3	0.1384	0.0207	0.1439	1.8009
	4	0.0981	0.0112	0.1060	1.2770
	5 *	0.0968	0.0111	0.1055	1.2599
	None	0.1203	0.0161	0.1272	1.5658
RF	1	0.1262	0.0173	0.1317	1.6426
	2 *	0.1213	0.0162	0.1275	1.5777
	3	0.1266	0.0174	0.1322	1.6471
	4	0.1253	0.0171	0.1309	1.6304
	5	0.1262	0.0173	0.1317	1.6422
	None	0.1267	0.0177	0.1331	1.6487

* Indicates the parameter combination with the best performance of the results in the five replicated experiments.

The results show that, by repeating the experiments with random search, the ANN, XGBoost and RF models reach the optimum in the third, fifth, and second optimization, respectively, and for any kind of evaluation indexes, the error indexes obtained from the third optimization result of the ANN model are the smallest, and its prediction effect is better than that of the other models, while the performance of the LASSO is the poorest.

The experiments are carried out without setting the hyperparameters of the model, and the results are visualized and compared with the tuned models; the specific results are shown in Figure 7.

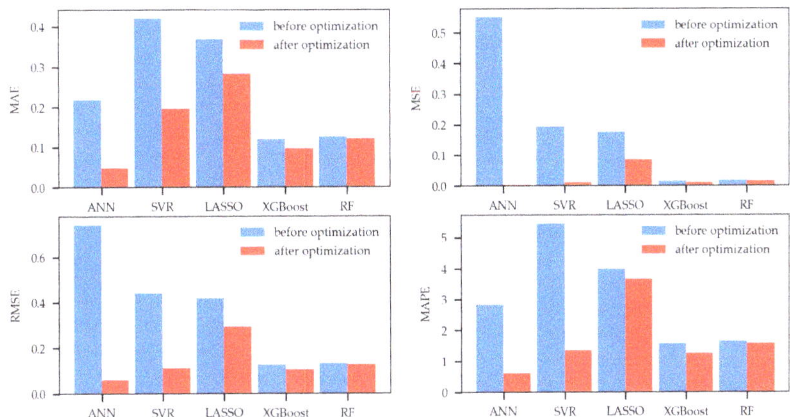

Figure 7. Performance comparison of different models before and after hyperparameter tuning.

Figure 7 shows that the error metrics of each model decreased after hyperparameter tuning compared to the model with default hyperparameters. Moreover, the tuning has the least enhancement for the XGboost and RF models.

In addition, in order to verify the influence of data dimensions on modeling [8–10,29,30] with dataset Set 1 (AIS, sensor, meteorological and sea state data), the above models are used to repeat the experiments five times in the datasets Set 2 (AIS and sensor data), Set 3 (AIS, sensor, meteorological data), and Set 4 (AIS, sensor and sea state data); the optimal parameter combination results are shown in Table 12. The comparison results of the optimal prediction performance of different models corresponding to each dataset are shown in Table 13.

Table 12. Comparison of model hyperparameter optimization results on different datasets.

Models	Hyperparameter Names	Set 1	Set 2	Set 3	Set 4
ANN	Optimizer	Adamax	Adamax	RMSprop	Nadam
	Neurons	30	40	80	10
	Epochs	100	200	150	200
	batch_size	40	40	90	40
SVR	C	0.1	10	0.1	0.1
	gamma	1	0.0001	1	1
LASSO	Alpha	10	0.0093	6.5793	0.0107
	max_iter	100	1000	100	300
XGBoost	subsample	0.8	0.7	0.9	0.8
	n_estimators	240	200	100	133
	min_child_weight	7	0	5	8
	max_depth	3	5	7	2
	learning_rate	0.21	0.11	0.11	0.21
	Gamma	6	7	1	1
	colsample_bytree	0.6	0.9	0.8	0.5
RF	n_estimators	68	266	233	266
	min_samples_split	16	3	13	13
	max_depth	8	6	6	5

Table 13. Comparison of best prediction performance of model on different datasets.

Models	Data Set	MAE	MSE	RMSE	MAPE
ANN	Set 1	0.0476	0.0034	0.0591	0.6186
	Set 2	0.0336	0.0015	0.0394	0.4377
	Set 3	0.0352	0.0016	0.0410	0.4601
	Set 4	0.0338	0.0017	0.0424	0.4402
SVR	Set 1	0.1037	0.0123	0.1109	1.3489
	Set 2	0.1694	0.0302	0.1739	2.2042
	Set 3	0.1037	0.0123	0.1109	1.3489
	Set 4	0.1037	0.0123	0.1109	1.3489
LASSO	Set 1	0.2817	0.0864	0.2940	3.6654
	Set 2	0.2655	0.0750	0.2738	3.4552
	Set 3	0.2677	0.0761	0.2759	3.4828
	Set 4	0.2652	0.0748	0.2735	3.4509
XGBoost	Set 1	0.0968	0.0111	0.1055	1.2599
	Set 2	0.1293	0.0183	0.1353	1.6828
	Set 3	0.0943	0.0104	0.1024	1.2273
	Set 4	0.1175	0.0155	0.1245	1.5286
RF	Set 1	0.1213	0.0162	0.1275	1.5777
	Set 2	0.1188	0.0157	0.1253	1.5452
	Set 3	0.1169	0.0152	0.1234	1.5207
	Set 4	0.1218	0.0163	0.1280	1.5844

Table 13 shows that, by validating different datasets, SVR is less affected by the datasets and has the same performance on Set 1, Set 3, and Set 4 datasets. The rest of the models perform better on the Set 3 dataset after dimensionality reduction, which occur because dimensionality reduction helps the model remove the redundant information, reduces the learning interference of the model, and improves the generalization ability of the model. In addition, compared with other models, the ANN model performs optimally in various situations.

According to the optimal hyper-parameter combinations obtained from each model experiment and input into the model again, the carbon intensity error curves of the ship under the five prediction models are depicted, as shown in Figure 8.

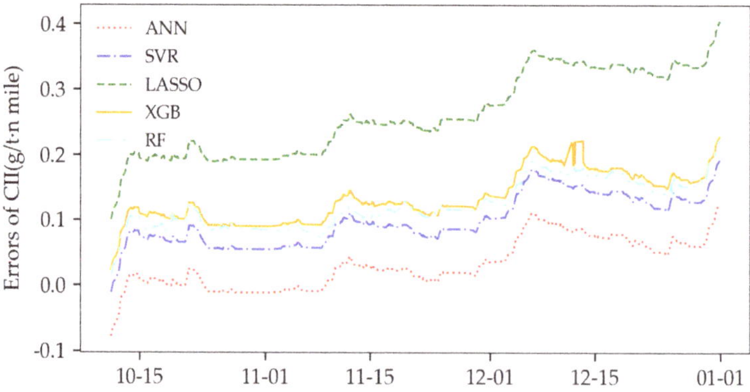

Figure 8. Comparison of the results of the cumulative carbon intensity prediction errors of the models.

It can be found through Figure 8 that, with each model performing as well as possible, the ANN model predicts the carbon intensity obtained with an error closest to 0, which is the smallest error among the five models.

4.3. Analysis of Carbon Intensity Rating Results

The carbon intensity level boundaries for the case ship can be calculated according to Equation (10), and the results are presented in Table 14.

Table 14. Carbon intensity rating boundary of container ships.

Boundary	Level
<9.5347	A
9.5347~10.7983	B
10.7983~12.2917	C
12.2917~13.6702	D
>13.6702	E

Since this paper divides the training set and test set according to the time series, the last value in the prediction result of each model is selected as the carbon intensity prediction result of the year and compared with the boundary parameters to determine the carbon intensity level achieved in the year, as shown in Table 15.

Table 15. Comparison of annual carbon intensity errors and ratings across models.

Models	Annual Carbon Intensity	Errors	Level
ANN	7.6330	0.1334	A
SVR	7.5734	0.1930	A
LASSO	7.3699	0.3965	A
XGBoost	7.5383	0.2281	A
RF	7.5431	0.2233	A
Real	7.7664	0	A

Table 14 show that the carbon intensity value of the container ship itself for the year 2022 is within the range of 9.5347, and the prediction error value of the model will not be too large; the result is that the predicted carbon intensity value of each model is within the range of 9.5347, and therefore, the rating results of each model are in line with the actual results. In addition, the annual carbon intensity value predicted by the ANN model has the smallest error with the actual value, so the ANN model is more suitable for predicting the carbon intensity of ships.

According to the results of the study, all models predicted carbon intensity within acceptable limits. In addition, since past studies did not analyze the factors affecting the carbon intensity of ships in depth, and the carbon intensity was calculated from fuel consumption, this study considers the relevant factors affecting the fuel consumption as the input features of the models in this paper. Since this paper considers cumulative carbon intensity as a prediction target, it is difficult to construct a strong correlation between fuel consumption characteristics and cumulative carbon intensity in terms of feature correlation.

In terms of carbon intensity prediction, five machine learning models commonly used in the past are selected for prediction in this paper. Since it is unknown under which combination of data the model achieves the highest accuracy, this paper divides the preprocessed data into four categories (Set 1: AIS, sensor, meteorological, and sea state data, Set 2: AIS and sensor data, Set 3: AIS, sensor, and meteorological data, and Set 4: AIS, sensor, and sea state data) and analyzes how the weather and sea state data affect the carbon intensity prediction model. In addition, this paper also gives specific instructions for setting and adjusting the hyperparameters of the model, and the hyperparameter tuning also proves to be effective for the carbon intensity prediction model.

This paper diverges from prior studies in several key aspects. First, it focuses on carbon intensity as the prediction target, in contrast to past studies that predominantly addressed fuel consumption. Second, in terms of feature selection, this paper incorporates factors such as the mean period of total swell, mean period of wind waves, mean period of waves, and seawater temperature; these features were not extensively explored in previous research. Third, in model comparison, the study evaluates the impact of different data combination scenarios on model performance, a consideration often overlooked in prior studies. Unlike previous research on carbon intensity, this paper introduces a cumulative carbon intensity assessment methodology. This methodology analyzes the change in carbon intensity for the selected vessels over a one-year period. Furthermore, the paper successfully applies a machine learning model to predict a ship's annual carbon intensity, providing insight into the carbon intensity level projected by the model.

This study could be beneficial to the shipping industry in several aspects. First, the proposed predicting method could be used by ship operators to dynamically calculate and monitor the carbon intensity of their ships. Second, carbon intensity prediction could further combine with history operation data to dynamically predict the CII grade of each ship at the end of each year. Third, providing reliable data support allows the ship company to adjust operation strategy or select decarbonization technologies that can upgrade the CII level to meet IMO requirements with minimum costs.

5. Conclusions

The CII grades of ships are crucial to international shipping companies. Failing to meet the CII requirements of IMO or charter contacts will cause dramatical market share and economic losses. This study investigates the performance of five different machine learning methods in predicting ship carbon intensity, including SVR, LASSO regression, XGBoost, and RF models, as well as the ANN model. Multiple data sources, such as AIS data, fuel flow sensor data, and meteorological and sea state data are considered to develop these models.

The results show that, compared with SVR, LASSO regression, XGBoost, and RF models, the ANN model performs the best, and the errors of all four evaluation indexes are minimized. Through several stochastic search optimization experiments, a better combination of hyperparameters can be found, which effectively improves the performance of the ship carbon intensity prediction model. By verifying the performance of the model on different datasets, appropriate dimensionality reduction can improve the accuracy of the ship carbon intensity prediction model. In the annual carbon intensity rating, the annual carbon intensity value predicted by the ANN model is the closest to the real value, while the error of the LASSO model is the largest, and the rating results are consistent with the actual results. The findings of this paper can help enterprises analyze the change of carbon intensity of a ship within a year and determine whether the annual carbon intensity value of a ship is within the qualified range or not. If the predicted carbon intensity grade of a ship is lower than C grade, the shipping company may have to take measures in advance to reduce the carbon intensity.

This is one of the first studies focus on predicting ship CII with multiple machine learning methods and high temporal and spatial data sources. The contribution of this study is as follows. First, it proposes a new carbon intensity assessment method to calculate the carbon intensity of a ship at different time points within a year, which takes new features such as swell and wind waves into consideration. Second, the ANN model was identified as the best carbon prediction method among those which are frequently utilized in studies on fuel consumption prediction. Third, while comparing the models, multiple data combination scenarios were considered (AIS data, sensor data, and meteorological data and sea state data). Fourth, this paper divides the training set and test set according to the time series, which can help to analyze the annual carbon intensity values of the model and, further, can determine the carbon intensity level.

However, there are some limitations in this study. First, only one sample ship is used, which may affect the generalization of the findings of this study. This study takes a relatively small-size container ship as an example, and larger ships are not considered, which may limit the application of the prediction model. Second, the accuracy of the prediction model probably could be further improved by advanced artificial intelligent models. This study mainly used machine learning models for its investigations, and the range of models considered is small. For carbon intensity prediction, models may exist with higher prediction accuracy. In this study, the technical condition of ship engines, systems, and mechanisms are not completely considered for the following reasons. First, our model actually included the rpm of the main engine and trim of ship dynamically to reflect the variance of main engine and ship technical conditions during the research period. Second, since the current analysis is limited to a single ship and the detailed change of its technical condition are difficult to obtain, this paper assumes that the mechanical condition throughout the study period is consistent and will not affect the ship's fuel consumption or Carbon Intensity Index (CII) rating. In future work, consideration could be given to selecting more ships, including larger-size ships, and applying more machine learning or deep learning models to improve the accuracy of CII prediction.

Author Contributions: Conceptualization, Z.W.; methodology, Z.W.; validation, T.L.; formal analysis, T.L. and Z.W.; investigation, C.Z.; resources, C.Z., Z.W. and X.Z.; data curation, T.L.; writing—original draft preparation, T.L.; writing—review and editing, C.Z., T.L., Z.W. and X.Z.; visualization, C.Z.; supervision, C.Z.; project administration, C.Z.; funding acquisition, C.Z. All authors have read and agreed to the published version of the manuscript.

Funding: This research was funded by National Key R&D Program of China (2022YFB4301403).

Institutional Review Board Statement: Not applicable.

Informed Consent Statement: Not applicable.

Data Availability Statement: Data is contained within the article.

Conflicts of Interest: The authors declare no conflict of interest.

References

1. MEPC.328(76). Amendments to the Annex of the Protocol of 1997 to Amend the International Convention for the Prevention of Pollution from Ships, 1973, as Modified by the Protocol of 1978 Relating Thereto. 2021 Revisez MARPOL Annex VI. 2021. Available online: https://wwwcdn.imo.org/localresources/en/KnowledgeCentre/IndexofIMOResolutions/MEPCDocuments/MEPC.328(76).pdf (accessed on 8 December 2022).
2. Chuah, L.F.; Mokhtar, K.; Ruslan, S.M.M.; Abu Bakar, A.; Abdullah, M.A.; Osman, N.H.; Bokhari, A.; Mubashir, M.; Show, P.L. Implementation of the energy efficiency existing ship index and carbon intensity indicator on domestic ship for marine environmental protection. *Environ. Res.* **2023**, *222*, 115348. [CrossRef] [PubMed]
3. Oldendorff. CII Is Not the Answer, What Do We Do Now? 2022. Available online: https://oldendorff-website-assets.s3.amazonaws.com/assets/downloads/Oldendorff-EMISSIONS.pdf (accessed on 8 January 2023).
4. BIMCO. Cii Operations Clause for Time Charter Parties. 2022. Available online: https://www.bimco.org/contracts-and-clauses/bimco-clauses/current/cii-operations-clause-2022 (accessed on 30 December 2022).
5. Wang, S.; Ji, B.; Zhao, J.; Liu, W.; Xu, T. Predicting ship fuel consumption based on LASSO regression. *Transp. Res. Part D Transp. Environ.* **2018**, *65*, 817–824. [CrossRef]
6. Jeon, M.; Noh, Y.; Shin, Y.; Lim, O.K.; Lee, I.; Cho, D. Prediction of ship fuel consumption by using an artificial neural network. *J. Mech. Sci. Technol.* **2018**, *32*, 5785–5796. [CrossRef]
7. Ren, F.; Wang, S.; Liu, Y.; Han, Y. Container Ship Carbon and Fuel Estimation in Voyages Utilizing Meteorological Data with Data Fusion and Machine Learning Techniques. *Math. Probl. Eng.* **2022**, *2022*, 4773395. [CrossRef]
8. Li, X.; Du, Y.; Chen, Y.; Nguyen, S.; Zhang, W.; Schönborn, A.; Sun, Z. Data fusion and machine learning for ship fuel efficiency modeling: Part I—Voyage report data and meteorological data. *Commun. Transp. Res.* **2022**, *2*, 100074. [CrossRef]
9. Du, Y.; Chen, Y.; Li, X.; Schönborn, A.; Sun, Z. Data fusion and machine learning for ship fuel efficiency modeling: Part II—Voyage report data, AIS data and meteorological data. *Commun. Transp. Res.* **2022**, *2*, 100073. [CrossRef]
10. Du, Y.; Chen, Y.; Li, X.; Schönborn, A.; Sun, Z. Data fusion and machine learning for ship fuel efficiency modeling: Part III—Sensor data and meteorological data. *Commun. Transp. Res.* **2022**, *2*, 100072. [CrossRef]
11. Uyanık, T.; Karatuğ, Ç.; Arslanoğlu, Y. Machine learning approach to ship fuel consumption: A case of container vessel. *Transp. Res. Part D Transp. Environ.* **2020**, *84*, 102389. [CrossRef]
12. Onur, Y.; Murat, B.; Mustafa, S. Comparative study of machine learning techniques to predict fuel consumption of a marine diesel engine. *Ocean Eng.* **2023**, *286*, 115505.
13. Yuan, Z.; Liu, J.; Zhang, Q.; Liu, Y.; Yuan, Y.; Li, Z. Prediction and optimisation of fuel consumption for inland ships considering real-time status and environmental factors. *Ocean Eng.* **2021**, *221*, 108530. [CrossRef]

14. Yildiz, B. Prediction of residual resistance of a trimaran vessel by using an artificial neural network. *Brodogr. Teor. Praksa Brodogr. Pomor. Teh.* **2022**, *73*, 127–140. [CrossRef]
15. Martić, I.; Degiuli, N.; Grlj, C.G. Prediction of Added Resistance of Container Ships in Regular Head Waves Using an Artificial Neural Network. *J. Mar. Sci. Eng.* **2023**, *11*, 1293. [CrossRef]
16. Rauca, L.; Batrinca, G. Impact of Carbon Intensity Indicator on the Vessels' Operation and Analysis of Onboard Operational Measures. *Sustainability* **2023**, *15*, 11387. [CrossRef]
17. Wang, S.; Psaraftis, H.N.; Qi, J. Paradox of international maritime organization's carbon intensity indicator. *Commun. Transp. Res.* **2021**, *1*, 100005. [CrossRef]
18. Gianni, M.; Pietra, A.; Coraddu, A.; Taccani, R. Impact of SOFC Power Generation Plant on Carbon Intensity Index (CII) Calculation for Cruise Ships. *J. Mar. Sci. Eng.* **2022**, *10*, 1478. [CrossRef]
19. Hoffmann, M. The Impact of 'Fouling Idling' on Ship Performance and Carbon Intensity Indicator (CII). 2022. Available online: https://selektope.com/wp-content/uploads/2022/06/HullPIC-2022_ITech-conference-paper-.pdf (accessed on 4 January 2023).
20. Kalajdžić, M.; Vasilev, M.; Momčilović, N. Power reduction considerations for bulk carriers with respect to novel energy efficiency regulations. *Brodogr. Teor. Praksa Brodogr. Pomor. Teh.* **2022**, *73*, 79–92. [CrossRef]
21. Bayraktar, M.; Yuksel, O. A scenario-based assessment of the energy efficiency existing ship index (EEXI) and carbon intensity indicator (CII) regulations. *Ocean Eng.* **2023**, *278*, 114295. [CrossRef]
22. Elkafas, A.G.; Rivarolo, M.; Massardo, A.F. Environmental economic analysis of speed reduction measure onboard container ships. *Environ. Sci. Pollut. Res.* **2023**, *30*, 59645–59659. [CrossRef]
23. Sun, L.; Wang, X.; Lu, Y.; Hu, Z. Assessment of ship speed, operational carbon intensity indicator penalty and charterer profit of time charter ships. *Heliyon* **2023**, *9*, e20719. [CrossRef] [PubMed]
24. Li, X.Y.; Zuo, Y.; Jiang, J.H. Application of Regression Analysis Using Broad Learning System for Time-Series Forecast of Ship Fuel Consumption. *Sustainability* **2023**, *15*, 380. [CrossRef]
25. Yuan, Z.; Liu, J.; Liu, Y.; Yuan, Y.; Zhang, Q.; Li, Z. Fitting analysis of inland ship fuel consumption considering navigation status and environmental factors. *IEEE Access* **2020**, *8*, 187441–187454. [CrossRef]
26. MEPC.338(76). Guidelines on the Operational Carbon Intensity Reduction Factors Relative to Reference Lines (CII Reduction Factors Guidelines, G3). 2021. Available online: https://www.ccs.org.cn/ccswzen/specialDetail?id=202206220276449608 (accessed on 15 December 2022).
27. MEPC.354(78). Guidelines on the Operational Carbon Intensity Rating of Ships (CII Rating Guidelines, G4). 2022. Available online: https://wwwcdn.imo.org/localresources/en/OurWork/Environment/Documents/Air%20pollution/MEPC.339(76).pdf (accessed on 8 December 2022).
28. Su, M.; Su, Z.Q.; Cao, S.L.; Park, K.S.; Bae, S.H. Fuel Consumption Prediction and Optimization Model for Pure Car/Truck Transport Ships. *J. Mar. Sci. Eng.* **2023**, *11*, 1231. [CrossRef]
29. Chen, Z.S.; Lam, J.S.L.; Xiao, Z. Prediction of harbour vessel fuel consumption based on machine learning approach. *Ocean Eng.* **2023**, *278*, 114483. [CrossRef]
30. Xie, X.; Sun, B.; Li, X.; Olsson, T.; Maleki, N.; Ahlgren, F. Fuel Consumption Prediction Models Based on Machine Learning and Mathematical Methods. *J. Mar. Sci. Eng.* **2023**, *11*, 738. [CrossRef]

Journal of
*Marine Science
and Engineering*

Article

Exploring Drivers Shaping the Choice of Alternative-Fueled New Vessels

Shun Chen [1], Xingjian Wang [1], Shiyuan Zheng [2,*] and Yuantao Chen [1]

[1]   School of Economics and Management, Shanghai Maritime University, Shanghai 201306, China; shunchen@shmtu.edu.cn (S.C.)

[2]   College of Transport and Communications, Shanghai Maritime University, Shanghai 201306, China

*   Correspondence: syzheng@shmtu.edu.cn; Tel.: +86-(0)21-38282455

Abstract: The urgent imperative for maritime decarbonization has driven shipowners to embrace alternative marine fuels. Using a robust orderbook dataset spanning from January 2020 to July 2023 (encompassing 4712 vessels, 281 shipyards, and 967 shipping companies), four distinct multinomial logit models were developed. These models, comprising a full-sample model and specialized ones for container vessels, dry bulk carriers, and tankers, aim to identify the key determinants influencing shipowners' choices of alternative fuels when ordering new vessels. It is interesting to find that alternative fuels (e.g., liquefied natural gas) are the most attractive choice for gas ships and ro-ro carriers; others prefer to use conventional fuels. Furthermore, this study reveals that shipowners' choices of new fuels significantly correlate with their nationality. While it is well-established that economic factors influence shipowners' choices for new ship fuel solutions, the impacts of bunker costs, freight rates, and CO_2 emission allowance prices remain relatively limited. It is evident that the policies of the International Maritime Organization (IMO) to reduce carbon emissions have increased the demand for building new energy ships. This research contributes to bridging research gaps by shedding light on the intricate interplay of factors that influence shipowners' preferences for alternative marine fuels amidst global regulatory shifts. It also offers valuable insights for policymakers aiming to incentivize shipowners to transition towards sustainable energy sources.

Keywords: alternative fuels; multinomial logit model; orderbook; new vessels

Citation: Chen, S.; Wang, X.; Zheng, S.; Chen, Y. Exploring Drivers Shaping the Choice of Alternative-Fueled New Vessels. *J. Mar. Sci. Eng.* **2023**, *11*, 1896. https://doi.org/10.3390/jmse11101896

Academic Editor: Mihalis Golias

Received: 1 September 2023
Revised: 22 September 2023
Accepted: 25 September 2023
Published: 29 September 2023

1. Introduction

International shipping, responsible for moving over 90% of globally traded goods, plays a pivotal role in the world economy. Currently, it contributes about 2–3% of global emissions, a figure projected to rise to 17% by 2050 [1]. The IMO has set crucial milestones for the industry, targeting a minimum 40% reduction in carbon intensity by 2030 and 70% by 2050, compared to 2008 levels. Additionally, a 50% reduction in total annual greenhouse gas emissions from shipping by 2050 is aimed, all underpinned by the adoption of low-carbon fuels to accommodate the sector's growth [2].

The maritime sector's intricate nature, characterized by high capital requirements, risks, and specialization, makes ordering new vessels a complex decision. Shipowners must assess market conditions before investment, considering shipbuilding capacity, compliance with conventions, and market competitiveness. Adhering to new emissions standards, however, necessitates significant adjustments that could cost the container shipping industry up to USD 10 billion [3], exerting a substantial influence on shipping companies' revenue. Approximately 47% of voyage costs in the maritime sector are attributed to bunker costs, contingent upon fuel prices and vessel specifications [4].

Given the shipping sector's magnitude and reliance on fuel expenditures, even incremental energy efficiency improvements can yield significant outcomes [5]. To align with IMO goals, ships need to transition to low-carbon alternative fuels like liquefied natural gas (LNG), methanol, liquefied petroleum gas (LPG), and biofuels, with future adoption of even more environmentally friendly options like hydrogen and ammonia anticipated [6]. While existing research mainly focuses on emissions and fuel performance, a gap exists between academic findings on alternative fuels and shipowners' practical responses to changing regulations. Meeting the IMO's 2050 carbon intensity targets prompts shipowners to assess when and whether to invest in vessels powered by alternative fuels.

This study employs the multinomial logit (MNL) model to identify key factors influencing shipowners' decisions to invest in such vessels. These factors include ship-related, shipowner-related, market-related, and regulation-related ones. This research draws on a robust dataset derived from a vessel orderbook spanning from January 2020 to July 2023, encompassing 4712 vessels, 281 shipyards, and 967 shipping companies.

This research bears three significant contributions. Firstly, most existing studies in the literature have primarily focused on assessing the commercial, operational, and technical viability of alternative marine fuels, as well as their potential for reducing carbon emissions on an experimental basis. Little attention has been given to the practical applications of these alternative fuels in new vessels and the factors influencing shipowners' decisions to adopt them. Our analysis in this paper fills this gap and enhances our understanding of shipowners' choices in alternative marine fuels when constructing new ships. Secondly, our findings reveal that vessel type, shipowner nationality, and IMO policies significantly influence fuel choices, whereas the effects of bunker costs, freight rates, and CO_2 emission allowance prices remain relatively limited. These findings hold important policy implications. They underscore that economic incentives alone may not be sufficient to drive the industry's adoption of new environmentally friendly fuels. Prompt technological advancements, government policy support, regulatory requirements, and a well-developed supply chain and infrastructure are crucial catalysts for this transition. Thirdly, the methodology proposed in this study can be effectively applied and extended to analyze behaviors of navigating the transition towards sustainable energy sources by various stakeholders.

The remainder of this paper is structured as follows: Section 2 reviews the literature on shipowners' choices of alternative marine fuels. Section 3 outlines the methodology. Section 4 offers the descriptions and analysis of orderbook data for new ships. Section 5 demonstrates the regression results of multinomial logit models and discussions. The conclusions drawn from this study are presented in Section 6.

2. Literature Review

2.1. Review of Alternative Marine Fuels and Shipowners' Choices

The urgent need for maritime decarbonization has motivated shipowners to adopt various emission abatement solutions, including improving energy efficiency, slow steaming, using innovative power plants, and renewable fuels [7–12]. For newbuilding vessels, adopting alternative fuels could be one of the most important solutions, especially when considering the more stringent carbon emission regulations set to be stipulated by the IMO in the future. Thus, research into alternative marine fuels has gained extensive attention. Previous research has heavily focused on LNG, with growing interest in methanol, ammonia, and hydrogen due to their potential to lower or have zero net carbon emissions [13,14]. LNG served as an interim solution [15,16], while e-fuel, methanol and ammonia will be the source of future fuels [14]. The competitiveness of methanol, compared with conventional fuels, depends mainly on ship productivity and the price difference between methanol and marine diesel oil (MDO) [17,18]. The marginal abatement costs and greenhouse gas (GHG) abatement potential of alternative marine fuels including methanol, ammonia, liquid hydrogen, LNG, LPG and bio-diesel for a newbuilding vessel depend on the cost of carbon capture and storage, electricity cost, and shipping route [19].

Key drivers behind shipowners' decisions to invest in emission abatement solutions, including alternative marine fuels, have been investigated. Previous research has showed that financial factors, such as investment costs, operational costs, and government support and regulations, all have significant impacts on shipowner decisions [20–22]. Furthermore, freight rate index, ship type, and shipowner nationality are all highly correlated with shipowners' emission abatement solutions [23]. Some authors have argued that for Norwegian shipowners, long-term profitability, company strategy, and financial and intellectual resources serve as significant factors affecting their adoption of alternative fuels [24]. IMO policies have accelerated shipping decarbonization, but some measures still remain uncertain and discrepancies exist between the IMO's incentives and industry perspectives [25]. Practicality and short-term returns matter in shipowners' preferences for emission abatement solutions. When fuel oil prices are relatively high, regulations can stimulate shipowners to complete the fuel transition in a more aggressive direction [26].

2.2. Review on Multinomial Logit Model

The well-established multinomial logit model serves as a valuable tool for estimating choice probabilities and discerning influential factors. It has been applied in various contexts to shed light on critical decision-making processes within the maritime industry.

It was used to investigate cruise lines' compliance decisions with the 2020 sulfur cap, revealing that fuel price fluctuations and government support had a minimal impact, while new vessel orders favored alternative fuels like LNG [27]. It was also applied to assess scrapping probabilities, considering vessel characteristics, market factors, and deviations from average freight rates [28]. Some authors explored shipowners' vessel selection and size preferences using a multinomial logit model, and synthesized factors including internal company traits, market conditions, and competitor performance [29]. In addition, the model proved instrumental in estimating port-to-port cargo flow. By employing multinomial logit models, one study analyzed the effect of trade volume on ship size choice. It was found that trade volume, voyage distance, and dry bulk shipping index all impacted ship size preferences [30].

The existing literature predominantly focuses on assessing the commercial, operational, and technical viability of alternative marine fuels, as well as their potential for carbon emission reductions on an experimental basis. Notably, little research investigates the practical applications of these alternative fuels in the construction of new vessels. Furthermore, limited attention has been given to investigating the influential factors that shape shipowners' decisions regarding the adoption of alternative fuels, and the temporal evolution of such choices. This study bridges the gap between academic research on energy choices in building new vessels and the actual responses of shipowners to regulations. It enriches the understanding of emission compliance decisions made by shipowners when ordering new vessels.

3. Methodology

3.1. Conceptual Framework

The orderbook data underwent preprocessing before being used in the study. Empirical analysis was conducted using multinomial logit models, including the full-sample model, dry bulk model, container model, and tanker model. This process is illustrated in Figure 1, which outlines the conceptual framework consisting of five steps.

(1) Data collection and cleaning: orderbook data collected from the Clarksons, spanning from January 2020 to July 2023, cover 4712 vessels, 281 shipyards, and 967 shipping companies. Records with errors or missing data were excluded. (2) Variable description and analysis: this study defines and describes both the explained and explanatory variables. The explained variables relate to different fuels used in new vessels, while the explanatory variables encompass ship-related, shipowner-related, market-related, and regulation-related factors. The descriptive analysis is presented. (3) Model estimation and fitting: four distinct models were developed, namely the full-sample model along with specialized models for container vessels, dry bulk carriers, and tankers. These latter three models focus on dissecting shipowners' decision-making processes within each respective category. Model fit was assessed using likelihood ratio tests within the context of the multinomial logit model. (4) Discussion: the findings are discussed with theoretical explanations and the

practical implications of the observed findings. (5) Conclusions and limitations: in the final section, we will present our conclusions and highlight any limitations of the study.

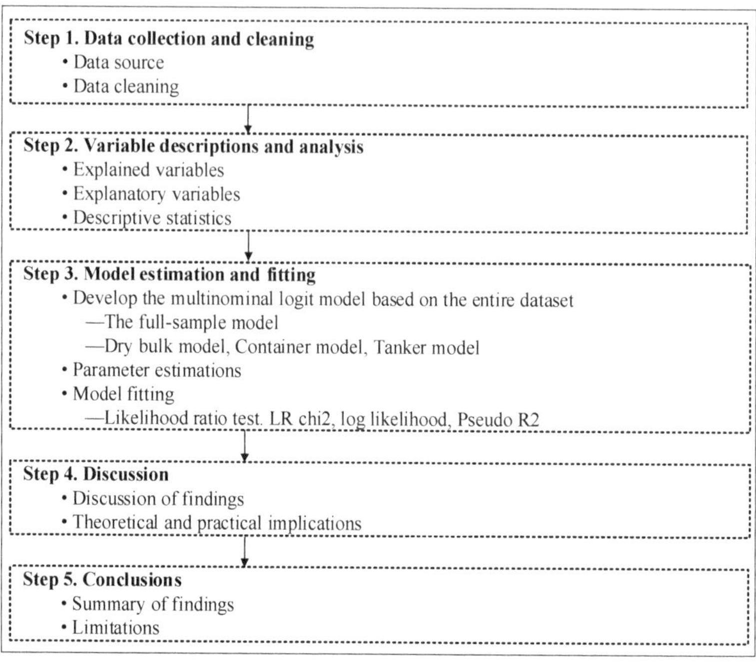

Figure 1. Conceptual framework of this research.

3.2. Model Building

A shipowner's choice of fuel for a new vessel is a complex interplay of individual circumstances, market dynamics, and environmental concerns. Discrete choice models, treating decision-makers as utility maximizers, offer a robust framework to explain and predict selections from multiple alternatives [31]. The multinomial logit model, commonly employed in such analyses, is suitable for scenarios with various choices. It comprises binomial logit models, each catering to specific choice behaviors, and encompasses diverse explanatory variables.

The discrete choice model, grounded in random utility theory, captures preference as a utility value, combining observable and unobservable random variables. Observable attributes and personal traits constitute the observable aspect, while other influences are encapsulated in an unobservable error term. Given the impact of random errors on precise utility prediction, choice probability represents the decision-maker's utility. Thus, the utility function guiding a shipowner's selection of an alternative-fueled vessel can be expressed as:

$$U_{ij} = V_{ij} + \varepsilon_{ij} \tag{1}$$

where the utility value (U_{ij}) of alternative fueled vessel j for shipowner i is determined by two components: the observable component (V_{ij}) and the unobservable component (ε_{ij}). V_{ij} can often be approximated as a linear function of a set of explanatory variables. This linear approximation can be expressed as:

$$V_{ij} = \beta_1 X_{i1} + \beta_2 X_{i2} + \cdots \beta_k X_{ik} \tag{2}$$

where $X_{i1}, X_{i2}, \cdots X_{ik}$ are the explanatory variables affecting the shipowner's decision to order an alternative-fueled vessel, and $\beta_1, \beta_2, \cdots \beta_k$ are the estimated corresponding parameters for the explanatory variables.

The probability of shipowner i choosing alternative fueled vessel j is given by:

$$P_i(j) = \Pr(U_{ij} > U_{ik}), \; for k \neq j, j = 1, 2, 3, 4, 5, 6. \tag{3}$$

where U_{ij} represents the maximum utility that shipowner i obtains when selecting alternative fueled vessel j. It is assumed that all ε_{ij} are independently and identically distributed (i.i.d.) following a Gumbel distribution with a mean value $\eta = 0$ and a scalar value u. Consequently, the probability of shipowner i selecting an alternative fueled vessel j can be expressed as follows:

$$P_i(j) = \frac{e^{V_{ij}}}{\sum_{k \in J} e^{V_{ik}}} = \frac{\exp\left(X'\beta_j\right)}{\sum_{k \in J} \exp\left(X'\beta_k\right)} \tag{4}$$

$$\sum_{j=1}^{J} P_i(j) = 1 \tag{5}$$

where X' is the set of attributes, β_j is the parameter vector of attributes to be estimated, and j denotes the selection set. Selection probabilities are then calculated using collected data [32]. Equation (5) ensures probabilities sum to one. Different parameter sets are estimated for each alternative-fueled vessel. The β_j for a new energy vessel type is set to 1 as the baseline, with coefficients of other options explained relative to this baseline.

The probability of the alternative fueled vessel scenario j can be expressed as:

$$P_i(j) = \frac{\exp\left(X'\beta_j\right)}{1 + \sum_{j=2}^{J-1} \exp\left(X'\beta_j\right)} \tag{6}$$

Similarly, the probability of the baseline option can be represented as:

$$P_i(j) = \frac{1}{1 + \sum_{j=2}^{J-1} \exp\left(X'\beta_j\right)} \tag{7}$$

4. Descriptions and Analysis of Variables

4.1. Explained and Explanatory Variable

4.1.1. Explained Variables

The explained variables in this study are different fuels of new vessels, which are classified as follows: conventional fuel, LNG capable, methanol, ready (including LNG ready, ammonia ready, methanol ready, LPG ready), and "other fuels" (battery, LPG, ethane, and other blended fuels). "LNG ready" refers to a type of vessel that currently operates using conventional fuels but is designed and built with the necessary infrastructure and adaptations to be easily converted to use LNG as fuel in the future. This concept applies similarly to ammonia ready, methanol ready, and LPG ready.

LNG is widely considered an option, with emissions about 15% lower than conventional fuel after accounting for leakage. However, LNG faces storage challenges due to its low temperature and requires specialized infrastructure [33–35]. Methanol, possessing a low sulfur content and igniting easily, presents competitiveness and can store surplus power through carbon capture [36,37]. DNV data show that container vessels using methanol fuel slightly exceed capital costs but are one-third of the cost of LNG counterparts [38]. Ammonia, a hydrogen carrier, emits no CO_2 but necessitates emission reduction measures for nitrogen oxides [38,39]. LPG fuels have good environmental performance, but their long-term decarbonization efficacy is limited. Batteries and hydrogen fuel cells offer potential, with the latter needing infrastructure enhancements. Hydrogen is promising, but it is the least mature among several fuels, facing obstacles in production, transportation, and storage [40–43].

None of these alternative green, zero-carbon, or low-carbon fuels currently have a globally available or cost-effective infrastructure to support the global shipping fleet. The shipping industry has yet to determine which fuel is the best choice. We define conventional fuel as the baseline category, and the options include LNG capable, methanol "ready", and "other fuels".

4.1.2. Explanatory Variables

The study's explanatory variables encompass ship-related, shipowner-related, market-related, and regulation-related variables.

Research indicates that vessel size and type significantly influence shipowners' newbuilding order decisions [23]. For instance, Gas carriers tend to prefer LNG, while smaller vessels often opt for conventional fuel [44].

Shipowner nationality also plays a pivotal role [23]. Distinct development levels and cultural factors across countries lead to varying incentives and policies. Research has established a symbiotic relationship between the freight and shipbuilding markets [45,46]. However, the impact of freight rates on shipowners' newbuilding decisions remains unclear. The ClarkSea Index gauges global freight performance, now covering 80% of fleet capacity, including LNG and chemical vessels since early 2022. In parallel, Clarkson Average Earnings have been adopted for dry bulkers, container vessels, and tankers, respectively.

Fuel expenses, constituting a substantial portion of voyage costs, significantly influence shipowners' responses to emission policies [47]. Among the leading alternatives, the cost of LNG stands as a pivotal consideration. Shipowners' sensitivity may be heightened by the interplay between LNG bunker prices and those of very low sulfur fuel. Notably, capital-intensive shipping operations remain susceptible to fluctuations in interest rates [48]. A marked transition from the London Interbank Offered Rate (LIBOR) to risk-free rates, such as the Secured Overnight Financing Rate (SOFR), Sterling Overnight Index Average (SONIA), and Swiss Average Rate Overnight (SARON), has been observed. SOFR, intricately linked to US Treasury repurchase rates, has replaced LIBOR and serves as a benchmark for gauging the capital costs of ships.

The idle rate, representing the ratio of idle vessels to the total fleet, serves as an indicator of how prevailing freight market conditions influence shipowners' decisions. The global carbon emissions trading system exerts an influence on the shipping industry, fostering the construction of energy-efficient vessels [48]. Notably, the European Union Emissions Trading System relies upon the CO_2 EUA price as a pivotal parameter. The utilization of CO_2 EUA price data, procured from Europe's primary carbon trading market, assumes a crucial role in scrutinizing its impact on shipowners' investments in vessels utilizing alternative fuels. This strategic approach aims at attaining the objectives of carbon emission reduction.

The growing alignment with the ambitious emission targets set by the IMO is steering the trajectory of new vessel orders towards alternative fuels [44]. The transition from a goal of reducing emissions by 50% to a more profound objective of completely eliminating greenhouse gases by 2050 underscores the need to scrutinize the influence of policies released by the IMO on shipowners' preference for alternative fuels.

All these explanatory datasets are matched with each individual vessel order, ensuring a comprehensive analysis for each distinct case. A comprehensive overview of the explanatory variables is presented in Table 1.

4.2. Data and Variable Analysis

4.2.1. Data Collection

This study employs vessel orderbook data from the Clarkson database spanning January 2020 to July 2023. The "Alternative Fuel Types" column specifies the vessel fuel type, with blanks indicating conventional fuel use. Order details include status, builder, contract date, gross tonnage (GT), DWT, vessel type, construction date, owner, etc. Earnings, LNG/FO, SOFR, idle rate, CO_2 price and policy are matched with contract dates. Data cleaning eliminated orders lacking DWT info. The final dataset comprises 4712 vessels from 281 shipyards and 967 companies. Table 2 summarizes vessel distribution by type.

Table 1. Detailed description of the explanatory variables.

	Explanatory Variables	**Description**
Ship-related	DWT	Deadweight tonnage of the vessel
	Type	Ship type in the orderbook; 1 if ship type is dry bulker; 2 if container vessel; 3 if tanker; 4 if multipurpose; 5 if gas carrier; 6 if ro-ro; 7 if general cargo vessel
Shipowner-related	Nation	China, Japan, Greece, Singapore, Republic of Korea, Germany, Norway, France etc. The selected countries account for over 75% of all orders
Market-related	Earnings (USD/day)	Monthly value of ClarkSea Index for full sample model (composite index of freight market performance). For the dry bulk model, container model, and tanker model, Clarkson Average Earnings are used for dry bulks, containers, and tankers, respectively.
	LNG/FO	The ratio of monthly LNG bunker prices over very low sulfur fuel oil prices
	SOFR (%)	Secured overnight financing rate
	Idle rate (%)	The ratio of the idle fleet over the total
	CO_2 price (USD/day)	CO_2 European Union Allowances price
Regulation-related	Policy	Dummy variables of decarbonization policies

Table 2. The number of merchant vessels categorized according to the vessel type.

Vessel Type	Number of Orders	Proportion
Bulk	1467	31.13%
Container	1152	24.45%
Tanker	890	18.89%
Multi-purpose	108	2.29%
Gas carrier	545	11.56%
Ro-ro	185	3.92%
General cargo	365	7.74%
Total	4712	100%

Source: compiled by authors.

4.2.2. Descriptive Statistics

(1)   Alternative fuels

Alternative fuels are categorized into five groups (Table 3) across 4712 vessels. Conventional fuel is used by 67.98% of vessels, while LNG capable, methanol, ready, and "other fuels" are chosen by 16.36%, 2.99%, 6.92%, and 5.75%, respectively. Notably, shipowners favor low-sulfur fuel oil and scrubbers, with about 15.35% opting for scrubbers with conventional fuel, which is consistent with the results of other studies on in-service fleets [44]. LNG remains the prime alternative fuel due to improved infrastructure.

Table 3. Fuel selection distribution statistics.

Alternative Fuels	Number of Orders	Proportion	Tonnage (Million)	Proportion
Conventional fuel	3203	67.98%	179.24	56.47%
LNG capable	771	16.36%	78.5	24.76%
Methanol	141	2.99%	16.9	5.33%
Ready	326	6.92%	32.8	10.35%
Other fuels	271	5.75%	9.74	3.07%
Total	4712	100%	317.18	100%

Source: compiled by authors.

"Ready vessels" such as LNG ready, methanol ready, ammonia ready, and LPG ready vessels follow LNG as viable options. These vessels provide flexibility and adaptability, allowing shipowners to transition to cleaner fuels gradually as market conditions and regulations evolve. This approach acknowledges the uncertainty and challenges associated with the rapid adoption of new fuels while still positioning companies to embrace sustainability initiatives.

Methanol is gaining attention as a potential alternative marine fuel due to its relatively lower emissions and wider availability. The "other" category, representing 5.75% of vessels, likely includes vessels exploring less common or emerging alternative fuels such as biofuels, ethane, battery propulsion, and LPG. This choice suggests a willingness to experiment with newer technologies and fuels that might not yet be widely adopted due to factors like technological maturity, availability, or cost-effectiveness.

It is significant to highlight that both ammonia and hydrogen fuel cells are currently in developmental stages and have not yet been deployed for maritime operations based on the observations of the sample vessels.

(2)   Vessel type analysis

Table 4 shows the percentage distribution of each alternative fuel based on vessel type. Notably, over 60% of bulk, container, tanker, gas carrier, and general cargo vessels favor conventional fuels. Bulk carriers, in particular, exhibit the highest preference at 92.2%, as their uncertain routes per voyage and the prolonged downturn in the market lead to a greater reliance on the stable returns provided by conventional fuels [49].

Table 4. The percentage distribution of each alternative fuel based on vessel type.

Vessel Type	Conventional	LNG Capable	Methanol	Ready	Other Fuels
Bulk	92.2%	4.9%	0.5%	2.2%	0.1%
Container	60.3%	18.1%	9.5%	11.3%	0.7%
Tanker	70.3%	8.7%	2.5%	13.5%	5.1%
Multi-purpose	83.3%	2.8%	0.0%	9.3%	4.6%
Gas carrier	14.7%	53.4%	0.0%	1.5%	30.5%
Ro-ro	13.0%	59.5%	1.1%	13.5%	13.0%
General cargo	66.2%	1.8%	27.9%	0.0%	4.2%

Source: compiled by authors.

Container shipowners, accounting for over 24% of the orderbook, display a greater inclination towards alternative fuels, with 18.1% opting for LNG capable vessels. Tankers also prefer conventional fuels (70.3%), with a notable 13.5% showing interest in "ready" vessels. Gas carriers (53.4%) and ro-ro vessels (59.5%) predominantly utilize LNG due to ample onboard storage. Container and general cargo vessels exhibit higher methanol usage. Remarkably, container, tanker, and ro-ro vessels exhibit a higher percentage in the "ready" category. Additionally, gas carriers and ro-ro vessels display an elevated preference for other fuels such as batteries, LPG, ethane, and blends.

(3)   DWT analysis

Table 5 underscores the trend that vessels employing LNG, methanol fuels, and ready fuels exhibit notably higher average DWT (deadweight tonnage) compared to those relying on conventional fuels. The data demonstrate that shipowners show a greater inclination toward alternative fuels as ship size increases. Remarkably, ships using methanol fuel possess the largest average and minimum DWT.

It is worth noting that other fuels such as battery, biofuels, and ethane predominantly find application in small- and medium-sized vessels with substantially lower DWT due to their ranges and technical limitations.

(4)   Policy analysis

Considering that our dataset encompasses the period from January 2020 to July 2023, our primary focus lies on decarbonization policies within this timeframe (Table 6). Our key objective is to assess whether these policies had a significant impact on shipowners' decision-making.

Table 5. Statistical description of newbuilding vessels' DWT according to alternative fuels.

Alternative Fuel	DWT				
	Mean	Std. Dev.	Max.	Min.	Median
Conventional fuel	55,962.04	51,921.87	319,202	72	49,000
LNG capable	101,803.74	67,689.98	321,020	2500	96,000
Methanol	119,532.93	68,809.73	225,000	4000	140,000
Ready	100,661.84	89,198.77	320,000	110	63,598
Other fuels	35,936.34	22,722.50	64,012	900	30,108
Total	67,305.98	61,627.50	321,020	72	55,077

Source: compiled by authors.

Table 6. IMO actions to reduce GHG emissions from ships.

Year	Milestone Actions
2020.11	Encouraging member states to develop and submit voluntary National Action Plans to address GHG emissions from ships
2021.06	Three additional measures adopted including a mandatory Carbon Intensity Indicator (CII), an Energy Efficiency Existing Ship Index (EEXI), and a strengthening SEEMP
2021.12	Initiating the revision of the Initial IMO Strategy on Reduction of GHG Emissions from Ships
2022.06	A series of 10 technical Guidelines adopted to support the implementation of the short-term GHG reduction measure
2022.12	Amendments adopted to MARPOL Annex VI to revise the data collection system for fuel oil consumption for the implementation of the EEXI and the CII framework
2023.07	Resolution adopted on Guidelines on lifecycle GHG intensity of marine fuels (LCA guidelines)

Source: compiled from the IMO website.

To ensure a thorough examination, we chose two crucial IMO policies: one issued in November 2020 and another in June 2021. It is worth noting that the policy introduced in June 2022 primarily involves a revision of the Initial IMO Strategy, while the one from December 2022 mainly pertains to a data collection system for fuel oil consumption, with seemingly minimal impact on alternative fuels. Additionally, the policy introduced in July 2023 holds significant relevance to marine fuels, but due to limited available data, its impact remains inconclusive.

In November 2020, the IMO encouraged member states to develop voluntary National Action Plans to address GHG emissions from ships. These efforts were paralleled by amendments to the MARPOL Annex VI, aimed to enhance and strengthen the Energy Efficiency Existing Ship Index (EEXI) Phase 3 requirements for three specific ship types, including container, general cargo, and LNG carriers. In June 2021, the IMO adopted amendments to MARPOL Annex VI introducing short-term GHG reduction measures containing a technical EEXI, an operational CII, and an enhanced SEEMP, and approved a work plan to advance the development of mid- and long-term GHG reduction measures in alignment with the Initial IMO Strategy.

These IMO policies incentivize shipowners to invest in ordering new vessels utilizing alternative fuels. Figure 2 showcases the orderbook distribution across different fuels from Q1 2020 to Q2 2023. The impact of the pandemic in 2020 had a noticeable effect on the orderbook, with percentages indicating the proportion of new vessels opting for conventional fuel. Notably, this percentage rose to 86.9% in Q3 2020 and then steadily decreased. It dropped to 66.9% in Q4 2022 and further declined to 44.2% by the end of

2022. Although there was a slight increase in the first three quarters of 2023, the trend of ordering more new energy vessels becomes evident. Meanwhile, there was an increase in the ordering of LNG capable vessels since 2020 since LNG is favored by shipowners for meeting the 2030 mid-term emissions reduction target, and the ratio of LNG capable vessels to the whole fleet reached 34.8% in Q3 2022. Methanol exhibited upward trends post Q3 2021. Since the end of 2022, there was a tendency toward the adoption of other fuels, such as biofuels, batteries, ethane, and blends. The maturation of technologies surrounding new alternative fuel production, storage, and utilization is driving shipowners to explore options beyond LNG.

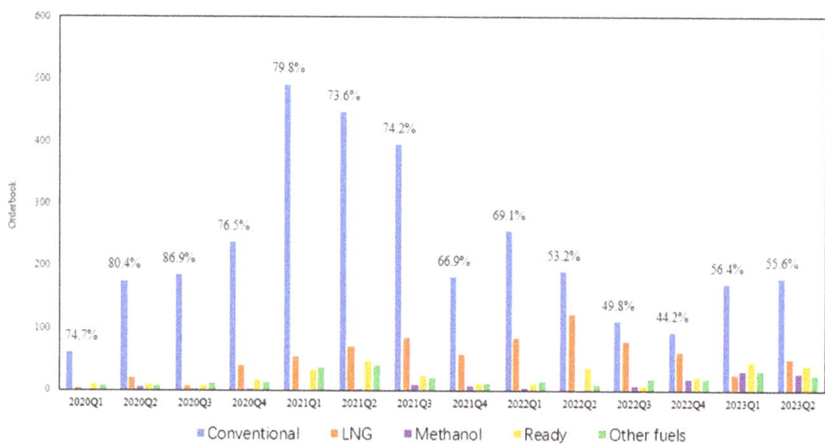

Figure 2. Orderbook distribution across different fuels from Q1 2020 to Q2 2023.

(5)   Shipowner analysis

Table 7 sheds light on the leading shipowners based on their orderbook volumes, revealing an interesting trend in their fuel category utilization. Noteworthy entities like CDB Leasing, Nisshin, Wan Hai, and SITC predominantly opt for conventional fuel, suggesting a cautious approach, possibly awaiting industry advancements before fully committing to alternative fuels.

Table 7. Top ten shipowners in terms of orderbook volume.

Rank	Shipowners	Conventional Fuel	LNG Capable	Methanol	Ready	Other Fuels	Total
1	Eastern Pacific Shpg	35	36	0	2	34	107
2	CMA CGM	16	55	24	0	0	95
3	CDB Leasing	81	0	0	0	0	81
4	MSC	12	60	0	3	0	75
5	Evergreen Marine	47	0	24	0	0	71
6	Seaspan Corporation	40	25	0	0	0	65
7	BoCom Leasing	44	18	0	0	0	62
8	Nisshin Shipping	50	0	0	0	0	50
9	Wan Hai Lines	48	0	0	0	0	48
10	SITC	47	0	0	0	0	47

Notes: CDB Leasing, known as China Development Bank Financial Leasing Co., Ltd., Shenzheng China. BoCom Leasing, known as Bank of Communications Financial Leasing Co., Ltd., Shanghai China. Source: Compiled by authors.

Conversely, Eastern Pacific Shipping, CMA CGM, MSC, Seaspan, and BoCom prominently emphasize LNG-fueled orders, indicative of recognizing LNG's transitional role in emission reduction. Additionally, CMA CGM and Evergreen display a notable volume of methanol-fueled orders, while Eastern Pacific Shipping features significant orders across various alternative fuels, including six ethane-fueled ones. While LNG adoption is relatively widespread, the lesser prevalence of methanol adoption might imply an ongoing industry exploration of its feasibility and benefits as a marine fuel.

The varying degrees of adoption suggest that the industry is in a phase of transition, with some companies leading the way and others assessing the best path forward.

In terms of shipowner nationalities, this study covers shipowners from 83 different countries. However, our focus has been on the top seven countries (Table 8), which collectively contribute to over 75% of all orders in the maritime industry. It is noteworthy that Singapore, the Republic of Korea, and Norway exhibit relatively higher proportions of orders (each exceeding 40%) which involve the utilization of alternative fuels. In particular, Korean shipowners have undergone a noticeable shift from their prior inclination towards scrubber installations on conventionally fueled vessels to actively adopting alternative fuels as a strategy for emissions reduction, as documented by Kim and Seo (2019) [20].

Table 8. Top seven countries in terms of orderbook volume.

Rank	Top 7	Conventional Fuel	LNG Capable	Methanol	Ready	Other Fuels	Total
1	China	1133	111	16	50	38	1348
2	Japan	746	100	12	22	36	916
3	Greece	312	76	0	70	19	477
4	Singapore	113	55	19	6	39	232
5	Republic of Korea	86	70	10	31	21	218
6	Germany	120	19	2	35	11	187
7	Norway	49	58	0	11	22	140

Source: compiled by authors.

In contrast, shipowners originating from major shipbuilding nations such as China and Japan continue to exhibit a comparatively lower inclination towards embracing alternative fuel options in their new orders. This intriguing dynamic reflects the intricate interplay between various factors influencing shipowners' decisions, ranging from environmental considerations to economic viability, and underscores the evolving nature of the maritime industry's response to sustainability challenges.

5. Empirical Analysis

5.1. Model Estimation

This study encompasses four distinct models, namely the full-sample model along with specialized models for container vessels, dry bulk carriers, and tankers. The latter three models concentrate on dissecting the decision-making processes of shipowners within each respective category. The logit regression model for the complete dataset is expressed by the following equation:

$$\ln \frac{P_i(j)}{1-P_i(j)} = \beta_0 + \beta_1 DWT_i + \beta_2 NATION_i + \beta_3 TYPE_i + \beta_4 Earnings_i + \beta_5 LNG/FO_i \\ + \beta_6 SOFR_i + \beta_7 IdleRate_i + \beta_8 CO_2 Price_i + \beta_9 Policy_{1i} + \beta_{10} Policy_{2i} \tag{8}$$

where $P_i(j)$ denotes the probability of shipowner i choosing alternative fuels j; β_1, β_2, $\beta_3 \cdots \beta_{10}$ are the estimated corresponding parameters for the explanatory variables. For dry bulk ships, container ships and tankers, the variable $TYPE_i$ is not included in the regression model. $Policy_{1i}$ is the dummy variable assigned a value of one after November 2020 and zero before that. $Policy_{2i}$ is the dummy variable assigned a value of one after June 2021 and zero before that.

We employ two types of tests to assess the significance of individual independent variables. The likelihood ratio test evaluates the overall relationship between the dependent variable and a series of independent variables in the model. Conversely, the Wald test scrutinizes whether a specific independent variable holds statistical significance in distinguishing between two alternative groups. The results of the likelihood ratio tests for the four models are presented in Table 9.

Table 9. Likelihood ratio test of explanatory variables.

Explanatory Variable	Full		Dry Bulk		Container		Tanker	
	Chi-Square	Sig	Chi-Square	Sig	Chi-Square	Sig	Chi-Square	Sig
Nation	491.77	0.000	67.58	0.000	501.27	0.000	177.55	0.000
Type	2097.35	0.000	(-)	(-)	(-)	(-)	(-)	(-)
DWT	460.81	0.000	25.65	0.000	263.23	0.000	93.54	0.000
Idle	29.34	0.000	16.67	0.000	54.59	0.000	7.93	0.09
SOFR	95.88	0.000	6.68	0.03	27.18	0.000	10.76	0.03
LNG/FO	22.23	0.000	0.16	0.92	37.02	0.000	7.29	0.12
CO_2 price	26.40	0.000	2.01	0.37	35.53	0.000	10.57	0.03
Earnings	39.72	0.000	9.61	0.000	14.99	0.000	6.71	0.15
Policy1	5.46	0.24	9.60	0.000	11.16	0.02	23.50	0.000
Policy2	53.32	0.000	6.19	0.04	33.37	0.000	10.26	0.03

Notes: (-) in the table represents that the variable "TYPE" is not applicable in the Likelihood ratio test.

In the full sample model, we observe that nearly all the explanatory variables, with the exception of Policy1, exhibit significance at the 1% significance level. In the container model, Policy2 demonstrates significance at the 5% level, while all other variables are significant at the 1% level. Turning to the dry bulk model, both LNG/FO and CO_2 prices are deemed insignificant, and others are all significant at the 5% level. Finally, in the tanker model, LNG/FO and earnings are not found to be significant, while the remaining variables hold significance at the 10% level.

To explore whether the explanatory variables exhibit varying effects on different groups, a Wald test is conducted on the model. One of the key advantages of the MNL model is its capability to discern the specific impact of individual explanatory variables on each group. The Wald test quantifies the significance of particular explanatory variables by testing the null hypothesis that their estimates are equal to zero. Conventional fuel is established as the baseline group and serves as a basis for comparison with other alternative fuels. The parameter estimates and results of the Wald test are illustrated in Tables 10–13.

The coefficient represents the estimated value in the MNL model, whereas the relative risk ratio (RRR) reflects the modification in the odds ratio for each explanatory variable concerning the reference group. The RRR is computed by raising the estimated coefficient to a certain power. Economically, interpreting the relative risk coefficient entails observing the alteration in the log odds of selecting a specific category relative to the reference group. An RRR exceeding 1 indicates that an increase in the explanatory variable amplifies the log odds of choosing that particular option group. Conversely, an RRR lower than 1 suggests that an increase in the explanatory variable diminishes the log odds of selecting that option group. An RRR of 1 signifies that a change in the explanatory variable has no discernible impact on the log odds of the group.

5.2. Model Fitting

The likelihood ratio test is a commonly used method for evaluating model fit in the context of MNL models [50], and the results are presented in Table 14. The LR chi-squared test statistic serves as an indicator of the overall goodness of fit of the model, testing the joint significance of all variables except the constant. The p-values are all 0, indicating that, compared to the model containing only the constant term, the comprehensive model provides a significantly better fit. Thus, this set of explanatory variables has a substantial and statistically significant effect on the explained variables. The log likelihood is calculated for both the null model containing only a constant variable and the full model containing all explanatory variables, facilitating the comparison of nested models in terms of their explanatory power.

Pseudo R^2, also known as McFadden's R^2, is a likelihood ratio index used to compare the relative size of log likelihood values between models incorporating only the constant term and those incorporating all explanatory variables. The higher the value, the better the fit of the model. An R^2 value between 0.2 and 0.4 is considered "very satisfactory" [51]. The full sample model has an R^2 value of 0.40, while the dry bulk, container vessels, and tanker models present R^2 values of 0.21, 0.51, and 0.23, respectively. These findings collectively indicate that all four models achieve a high level of goodness of fit.

Table 10. Parameter estimates and Wald test results for alternative fuels in full sample model.

	LNG Capable		Methanol		Ready		Other Fuels	
	Parameters	RRR	Parameters	RRR	Parameters	RRR	Parameters	RRR
Nation (China)	−1.105 *** (0.172)	0.331	−0.928 *** (0.333)	0.395	0.819 *** (0.193)	0.441	1.488 *** (0.246)	0.226
Nation (Japan)	−1.172 *** (0.197)	0.310	−0.704 * (0.379)	0.495	−1.096 *** (0.258)	0.334	−1.829 *** (0.270)	0.161
Nation (Greece)	−1.162 *** (0.214)	0.313	−16.39 (678.3)	0	0.0404 (0.197)	1.041	−1.935 *** (0.331)	0.144
Nation (Singapore)	0.489 ** (0.217)	1.631	0.821 ** (0.343)	2.273	−1.298 *** (0.441)	0.273	1.136 *** (0.275)	3.114
Nation (Republic of Korea)	−0.185 (0.262)	0.831	0.717 (0.449)	2.048	0.697 *** (0.255)	2.008	−0.750 ** (0.357)	0.472
Nation (Germany)	0.356 (0.296)	1.428	−0.574 (0.779)	0.563	1.222 *** (0.253)	3.394	−0.239 (0.447)	0.787
Nation (Norway)	1.151 *** (0.287)	3.161	−13.94 (1037)	0	0.945 ** (0.384)	2.573	0.856 ** (0.352)	2.354
Nation (France)	2.100 *** (0.287)	8.166	1.472 *** (0.430)	4.358	−19.62 (7574)	0	0.102 (0.769)	1.107
TYPE (Dry Bulk)	−2.393 *** (0.420)	0.091	12.18 (729.0)	1.94×10^5	14.42 (836.5)	1.83×10^6	−3.977 *** (0.772)	0.019
TYPE (Container)	−1.098 *** (0.411)	0.334	15.70 (729.0)	6.58×10^6	16.47 (836.5)	1.42×10^7	−2.377 *** (0.472)	0.093
TYPE (Tanker)	−1.086 ** (0.425)	0.338	14.04 (729.0)	1.25×10^6	16.49 (836.5)	1.45×10^7	−0.367 (0.326)	0.693
TYPE (Multipurpose)	−1.586 ** (0.711)	0.205	−0.685 (1615)	0.504	15.79 (836.5)	7.20×10^6	−0.698 (0.547)	0.498
TYPE (Gas carrier)	2.675 *** (0.412)	14.512	−1.481 (1854)	0.242	16.04 (836.5)	9.25×10^6	3.679 *** (0.318)	39.607
TYPE (Ro-ro)	3.915 *** (0.438)	50.149	15.04 (729.0)	3.40×10^6	18.79 (836.5)	1.45×10^8	2.397 *** (0.401)	10.990
DWT [1]	1.355 *** (0.0794)	1.014	0.681 *** (0.129)	1.007	0.440 *** (0.0699)	1.004	0.110 (0.0743)	1.001
Earnings [1]	1.599 *** (0.378)	1.016	−1.428 (0.883)	0.986	1.926 *** (0.425)	1.0169	1.327 *** (0.496)	1.013
LNG/FO	0.153 (0.0933)	1.165	−0.0936 (0.171)	0.911	−0.468 *** (0.141)	0.626	0.194 (0.127)	1.214
SOFR	0.0612 (0.0509)	1.063	0.630 *** (0.099)	1.878	0.388 *** (0.0604)	1.474	0.348 *** (0.0714)	1.416
Idle Rate	−1.107 *** (0.249)	0.331	−0.252 (0.448)	0.777	−0.905 *** (0.313)	0.405	0.123 (0.351)	1.131
CO_2 Price [1]	−1.180 *** (0.396)	0.988	−3.880 *** (0.951)	0.962	−0.974 ** (0.461)	0.990	−1.224 ** (0.551)	0.988
Policy1	−0.207 (0.327)	0.813	−0.388 (0.984)	0.678	−0.104 (0.400)	0.901	0.904 * (0.470)	2.469
Policy2	0.896 *** (0.303)	2.450	5.694 *** (1.090)	297.08	0.0585 (0.338)	1.060	0.772 * (0.426)	2.164
Constant	−23.02 *** (3.598)		2.239 (729.1)		−34.24 (836.5)		−12.89 *** (4.718)	
Traditional_fuel (based outcome)								
Observations	4712		4712		4712		4712	

Notes: [1] means variables are transformed into logarithms in order to make sense in explanations, *** $p < 0.01$, ** $p < 0.05$, * $p < 0.1$. RRR, known as relative risk ratio, is calculated by taking the exponential of the parameter estimate. Shipowner nations were selected for the analysis based on a single criterion: accounting for over 70% of orderbook volume, combined.

5.3. Discussion

New orders from top-ranking nations such as China, Japan, and Greece predominantly favor conventional fuels in the full sample model. Similar trends are observed in both the dry bulk and tanker models. Conversely, Singapore, Norway, and France exhibit a propensity for LNG capable-fueled vessels with odds ratios of alternative fuels over conventional fuels significantly surpassing 1. Shipowners from Singapore and France are more likely to choose methanol. Furthermore, "ready" vessels are favored by shipowners from the Republic of Korea and Germany, while Singapore and Norway display an inclination towards exploring "other fuels".

Table 11. Parameter estimates and Wald test results for alternative fuels in dry bulk model.

	LNG Capable		Other Alternative Fuels	
	Parameters	**RRR**	**Parameters**	**RRR**
Nation (China)	−0.959 *** (0.312)	0.383	−2.312 *** (0.517)	0.099
Nation (Japan)	−1.363 *** (0.338)	0.256	−2.090 *** (0.512)	0.124
Nation (Greece)	−0.872 * (0.475)	0.418	−15.99 (578.9)	0
DWT [1]	0.481 ** (0.189)	1.004	1.411 *** (0.335)	1.014
Earnings [1]	−1.902 *** (0.622)	0.981	−1.008 (0.936)	0.989
LNG/VLSFO	0.00925 (0.209)	1.010	0.0866 (0.214)	1.090
SOFR	−0.401 ** (0.183)	0.670	0.318 (0.247)	1.374
Idle Rate	−1.591 *** (0.434)	0.234	0.267 (0.408)	1.306
CO_2 Price [1]	1.026 (0.895)	1.010	−1.267 (1.610)	0.987
Policy1	2.266 *** (0.760)	9.641	0.276 (1.397)	1.318
Policy2	−0.795 (0.503)	0.452	2.147 * (1.248)	8.559
Constant	13.44 ** (6.315)		−6.943 (10.08)	
Traditional_fuel (based outcome)				
Observations	1467		1467	

Notes: [1] means variables are transformed into logarithms in order to make sense in explanations, *** $p < 0.01$, ** $p < 0.05$, * $p < 0.1$. RRR, known as relative risk ratio, is calculated by taking the exponential of the parameter estimate. Top three shipowner nations make up over 70% of total orderbook volume in terms of dry bulk vessels.

In the container model, shipowners from France, Singapore, Italy, and Germany are more inclined to LNG capable with probabilities of roughly 96 times, 21 times, 28 times, and 11 times that of conventional fuels, respectively. Additionally, shipowners from Singapore, France, and the Republic of Korea demonstrate a preference for methanol usage. The tendency for the Republic of Korea and Germany to favor "ready" vessels is also evidenced within the container vessel category. For dry bulk vessels and tankers, shipowners from top-ranking nations in terms of orderbook volume tend to favor conventional fuel vessels.

Different national policies and shipping power dynamics drive varied preferences, suggesting emission reduction progress varies by nation.

Despite the significant correlation observed between the independent variable "DWT" and the dependent variable across all models, the RRRs associated with "DWT" for each alternative fuel option remain remarkably close to 1. In other words, even though distinct ship sizes are associated with each alternative fuel choice, the influence of ship size on the decision-making process of shipowners is limited. This same pattern holds true for dry bulkers, container ships, and tankers. Similar findings have been noted in prior studies investigating the determinants influencing emission abatement strategies within operational fleets [44].

When analyzing ship preferences by type, it becomes evident that conventional fuels are more favored by dry bulkers, container ships, tankers, and multipurpose vessels compared to LNG. In contrast, gas carriers and ro-ro vessels tend to prefer LNG, with probabilities approximately 14 times and 50 times higher than those of conventional fuels. For gas carriers, this preference can likely be attributed to their ability to leverage existing LNG infrastructure. Additionally, the unit newbuilding prices of gas carriers and ro-ro vessels are higher than those of conventional ships, making it more feasible for them to better manage the increased expenses related to constructing and operating new energy vessels.

Table 12. Parameter estimates and Wald test results for alternative fuels in container model.

	LNG Capable		Methanol		Ready		Other Fuels	
	Parameters	RRR	Parameters	RRR	Parameters	RRR	Parameters	RRR
Nation (China)	−0.802 * (0.418)	0.448	−1.700 * (0.901)	0.183	−1.207 *** (0.327)	0.299	−2.727 *** (0.526)	0.065
Nation (Greece)	−1.435 *** (0.361)	0.238	−16.96 (1667)	0	−0.667 ** (0.271)	0.513	−16.52 (1086)	0
Nation (Japan)	−1.663 *** (0.586)	0.190	−1.480 (0.981)	0.228	−17.84 (1053)	0	−3.771 *** (0.875)	0.023
Nation (Singapore)	0.0807 (0.407)	1.084	0.912 (0.708)	2.489	−1.936 *** (0.551)	0.144	−1.146 * (0.644)	0.318
Nation (Republic of Korea)	−18.19 (1833)	0	−16.58 (3897)	0	−1.153 ** (0.470)	0.316	−17.40 (1715)	0
DWT [1]	0.864 *** (0.141)	1.009	−0.373 (0.232)	1.003	0.252 *** (0.0957)	1.003	−0.924 *** (0.176)	0.991
Earnings [1]	0.525 * (0.302)	1.005	−4.804 ** (2.342)	0.953	−6.5e−04 (0.252)	1.000	−0.483 (0.450)	1.005
LNG/VLSFO	0.303 (0.289)	1.354	2.216 ** (0.959)	9.171	−0.416 (0.320)	1.516	0.503 * (0.282)	1.654
SOFR	0.0208 (0.168)	1.021	3.090 ** (1.249)	21.977	0.162 (0.148)	1.176	0.596 *** (0.223)	1.815
Idle Rate	−0.995 ** (0.399)	0.369	−3.879 ** (1.837)	0.021	−0.0348 (0.304)	0.966	−0.422 (0.519)	0.656
CO_2 Price [1]	−2.771 *** (1.004)	0.973	4.359 (3.325)	1.045	−0.953 (0.834)	0.991	−1.692 (1.603)	0.983
Policy1	3.792 *** (0.920)	44.345	−2.997 (2.096)	0.050	1.353 * (0.698)	3.869	−0.724 (1.508)	2.063
Policy2	−0.742 (0.815)	0.476	−3.482 (3.357)	0.031	−0.137 (0.708)	0.872	3.325 ** (1.420)	28.560
Constant	−11.50 ** (5.503)		41.47 (28.05)		−0.0912 (4.549)		17.38 ** (8.821)	
Traditional_fuel (based outcome)								
Observations	890		890		890		890	

Notes: [1] means variables are transformed into logarithms in order to make sense in explanations, *** $p < 0.01$, ** $p < 0.05$, * $p < 0.1$. RRR, known as relative risk ratio, is calculated by taking the exponential of the parameter estimate. Top five shipowner nations make up over 70% of total orderbook volume in terms of container vessels.

The independent variable "earnings" exhibits a significant relationship with the dependent variable in the full sample model for various fuel options. For instance, the probability ratio of opting for LNG compared to conventional fuel is 1.011. Likewise, the probability ratios of choosing "ready" and "other fuels" relative to conventional fuels are 1.001 and 1.003, respectively. Consequently, higher levels of earnings correlate with an elevated inclination among shipowners to contemplate alternative fuels when placing orders for new ships.

However, the odds ratios, which are in proximity to 1, indicate that the impact of "earnings" is relatively modest in all models. The extent of the "earnings" variable does not play a significant role in steering shipowners toward selecting alternative fuels. Economic factors hold limited sway over shipowners' decisions, consistent with prior research [23]. Fuel maturity, ease of refueling, and compliance with IMO policies drive decisions.

The price spread between LNG bunker prices and very low sulfur fuel oil prices is not significant in most cases. In the full sample model, shipowners are more likely to choose conventional vessels compared to "Ready" ones when the price spread is larger, and the finding is the same for container vessels. For dry bulkers, the price spread does not have an impact on shipowners' decisions on choosing alternative fuels. For tankers, when LNG bunker prices are less competitive, shipowners prefer to order new vessels powered by methanol or other fuels.

It is worth highlighting that as the SOFR level rises, there is an observable increase in the likelihood of selecting methanol, "ready", and "other fuels" in the full sample model. This finding is consistent with the observation that, under elevated interest rates, shipowners tend to favor investments in LNG capable, methanol, or "ready" options for container vessels and lean towards methanol and "other fuels" for tankers. In a higher-interest-rate financial environment, conventional-fuel-powered vessels might be attractive in the short term. However, new energy vessels could potentially offer advantages in terms of reduced operational costs and environmental benefits over the long term. For dry bulkers, the interest rate appears to have no significant impact on shipowners' decision-making processes.

Table 13. Parameter estimates and Wald test results for alternative fuels in tanker model.

	LNG Capable		Methanol		Ready	
	Parameters	RRR	Parameters	RRR	Parameters	RRR
Nation (China)	−3.981 *** (0.805)	0.019	−1.698 *** (0.614)	0.183	0.406 (0.351)	1.501
Nation (France)	4.572 *** (0.472)	96.737	3.309 *** (0.574)	27.358	−16.61 (4926)	0
Nation (Singapore)	3.059 *** (0.448)	21.306	2.690 *** (0.553)	14.732	0.470 (0.777)	1.600
Nation (Italy)	3.364 *** (0.487)	28.904	−17.00 (2428)	0	1.032 (0.710)	2.807
Nation (Republic of Korea)	0.821 (0.670)	2.273	1.449 ** (0.593)	4.259	2.636 *** (0.392)	13.957
Nation (Greece)	−0.787 (0.794)	2.197	−16.76 (1645)	0	2.427 *** (0.379)	11.325
Nation (Germany)	2.467 *** (0.572)	11.787	1.405 (0.858)	4.076	3.358 *** (0.409)	28.732
DWT [1]	2.979 *** (0.274)	1.030	2.008 *** (0.268)	1.020	0.869 *** (0.146)	1.009
Earnings [1]	3.338 *** (0.980)	1.034	0.984 (1.052)	1.010	1.552 ** (0.756)	1.016
LNG/VLSFO	0.00525 (0.301)	1.005	0.259 (0.308)	1.296	−1.890 *** (0.378)	0.151
SOFR	1.134 *** (0.283)	3.108	1.268 *** (0.291)	3.554	0.429 ** (0.215)	1.536
Idle Rate	−1.053 *** (0.339)	0.349	−0.710 ** (0.341)	0.492	1.401 *** (0.281)	4.059
CO_2 Price [1]	−7.390 *** (1.374)	0.929	−6.028 *** (1.630)	0.942	−1.048 (1.083)	0.989
Policy1	2.704 *** (0.835)	14.939	2.428 (1.985)	11.336	0.356 (0.972)	1.428
Policy2	2.300 *** (0.881)	9.974	3.55 *** (1.91)	34.81	1.691 *** (0.577)	5.425
Constant	−42.90 *** (8.031)		−30.60 (1731)		−27.03 *** (6.089)	
Observations	1144		1144		1144	

Notes: [1] means variables are transformed into logarithms in order to make sense in explanations, *** $p < 0.01$, ** $p < 0.05$. RRR, known as relative risk ratio, is calculated by taking the exponential of the parameter estimate. Top seven shipowner nations make up over 70% of total orderbook volume in terms of tankers.

Table 14. Likelihood ratio test for the MNL model.

Regression Model	LR chi2	Prob > chi2	−2 Log Likelihood	Pseudo R^2
Full sample model	3854.57	0.00	5688.57	0.40
Dry bulk model	185.82	0.00	7765.62	0.21
Container model	1264.04	0.00	1219.92	0.51
Tanker model	399.53	0.00	1330.23	0.23

Fleet idle rates exert a substantial impact on LNG-fueled vessel orders across all models. Elevated idle rates diminish shipowners' proclivity towards LNG-fueled vessels, reflecting concerns about market oversupply. Notably, new energy vessels typically entail higher costs than conventional counterparts. In the case of container vessels, shipowners' preference for methanol would also be reduced to approximately 0.492 times that for conventional vessels, while their interest in "ready" vessels would increase. For tankers, a higher idle rate would likely lead shipowners to be less inclined towards ordering methanol-powered vessels.

The price of CO_2 EUAs significantly influences the options of alternative fuels. As the CO_2 EUA price escalates, the likelihood of shipowners placing orders for LNG capable-, LNG ready-, and methanol-powered vessels experiences a slight decrease. It is worth noting that both LNG and methanol fuels emit certain greenhouse gases and may not fully align with the current carbon-neutral policy of the IMO. However, due to the odds ratio being in proximity to 1, the impact of the CO_2 EUA price is relatively muted. This pattern holds true for container ships and tankers. For dry bulkers, the CO_2 price holds no relevance in explaining shipowners' preferences for alternative fuel options.

Regarding the regulation-related variable, our findings highlight the noteworthy influence of the "policy" variable on shipowners' decisions. In our investigation, we delved into the impact of the IMO policies released in both November 2020 and June 2021. "Policy1" was designed to assess the effect of the policy unveiled in November 2020 on shipowners' inclination towards new energy vessels. Notably, this variable demonstrates statistical significance at a level of $p < 0.01$ for the "other fuels" option, though it does not yield significance for the other alternatives in the full sample model. In the context of dry bulk, container, and tanker vessels, "Policy1" acts as a catalyst, prompting shipowners to increase their investments in LNG capable vessels. Specifically, the odds ratios stand at 9.641, 14.939, and 44.345, respectively. This underscores a substantial shift in shipowners' preferences towards LNG capable ships influenced by the policy implemented in November 2020.

The dummy variable "Policy2" was employed to investigate the effects of the policy introduced in June 2021 on the selection of new energy vessels. In the full sample model, the RRRs for the "Policy2" variable are 2.45, 297, and 2.164 for the LNG capable, methanol, and "other fuels" options, respectively. This observation signifies that the policy endorsed by the IMO in June 2021 carries notable influence over shipowners' inclinations to invest in alternative fuels.

In the case of dry bulkers and tankers, following the release of this policy, shipowners exhibit an increased likelihood of ordering vessels propelled by alternative fuels other than those initially considered. For container vessels, shipowners are more prone to selecting LNG, methanol, and "ready" vessels. This attests to the policy's impact on reshaping shipowners' preferences in favor of various new energy options.

6. Conclusions

Implementing more stringent environmental regulations has placed significant pressure on the maritime industry. Meeting the IMO 2050 targets necessitates a profound transformation within global fleets. One part of the effort to achieve low or zero carbon shipping is to diversify marine fuels away from fossil fuels. This paper may be among the first to present an empirical analysis of drivers shaping shipowners' preferences for alternative fuels when ordering new vessels. We propose an MNL model and employ worldwide newbuilding ship data spanning from January 2020 to July 2023 (encompassing 4712 vessels, 281 shipyards, and 967 shipping companies) for this analysis. Several noteworthy findings have emerged.

First, shipowners' choices exhibit a significant correlation with their nationality. For example, shipowners from France, Singapore, Italy, and Germany display a greater inclination toward LNG capable, while shipowners from Singapore, France, and the Republic of Korea demonstrate a preference for methanol usage. "Ready" vessels are favored by shipowners from the Republic of Korea and Germany, while Singapore and Norway display an inclination towards exploring "other fuels" such as biofuel, battery, and ethane.

Second, vessel type emerges as a significant factor in shipowners' selection of alternative fuels, while the size of the ship wields limited influence. Gas ships and ro-ro carriers tend to favor LNG, while others exhibit a higher probability for conventional fuels.

Third, the impact of freight market conditions is relatively modest, although a more favorable freight market may induce the adoption of alternative fuels when procuring new ships. But higher fleet idle rates diminish shipowners' adoption of LNG and methanol. Notably, the ratio of LNG-to-fuel prices appears to exert no influence on shipowners' inclination toward alternative fuels.

Fourthly, interest rates play a significant role in shaping shipowners' decisions regarding various alternative marine fuels, primarily due to their association with financing costs.

Furthermore, the carbon emissions trading system acts as a catalyst for shipowners' leanings towards wholly zero-carbon fuels, underlining the potency of policy mechanisms in shaping industry behaviors. It is also interesting to find that IMO policies and regulations have spurred demand for the construction of new energy-efficient ships.

These findings underscore that economic incentives alone may not be sufficient to motivate the maritime industry to embrace environmentally friendly fuels. Technological improvements and national policies wield substantial influence over these decisions. As the shipping industry charts its course towards sustainability, the shifting landscape necessitates that shipowners carefully assess the impact of various drivers on their investments. An in-depth understanding of these dynamics can lead to a balanced approach that aligns economic viability, environmental responsibility, and compliance with evolving regulations.

This research primarily focuses on examining the impacts of two IMO policies on shipowners' choices, with the potential to expand our analysis to include other policy conditions such as carbon taxes or emission trading systems. Moreover, our study relies on a dataset spanning only the most recent three years and seven months, suggesting that future investigations could benefit from a more extensive and precise newbuilding orderbook dataset for a deeper analysis.

Author Contributions: Conceptualization, S.C. and S.Z.; methodology, X.W.; software, X.W. and Y.C.; validation, S.C., X.W., S.Z. and Y.C.; formal analysis, S.C.; investigation, S.C.; data curation, X.W.; writing—original draft preparation, S.C. and Y.C.; writing—review and editing, S.Z.; supervision, S.Z.; project administration, S.Z.; funding acquisition, S.C. All authors have read and agreed to the published version of the manuscript.

Funding: This work is supported by China National Social Science Fund Post-Funding Project (22FGLB113).

Institutional Review Board Statement: Not applicable.

Informed Consent Statement: Not applicable.

Data Availability Statement: Readers can access our data by sending an email to the corresponding author Shiyuan Zheng.

Acknowledgments: The authors are grateful to the anonymous reviewers and the editor for their valuable comments.

Conflicts of Interest: The authors declare no conflict of interest.

References

1. Zis, T.; Psaraftis, H.N. Operational Measures to Mitigate and Reverse the Potential Modal Shifts Due to Environmental Legislation. *Marit. Policy Manag.* **2019**, *46*, 117–132. [CrossRef]
2. Psaraftis, H.N. Decarbonization of Maritime Transport: To Be or Not to Be? *Marit. Econ. Logist.* **2019**, *21*, 353–371. [CrossRef]
3. Hoffmann, P.N.; Eide, M.S.; Endresen, O. Effect of Proposed Co2 Emission Reduction Scenarios on Capital Expenditure. *Marit. Policy Manag.* **2012**, *39*, 443–460. [CrossRef]
4. Kim, K.; Lim, S.; Lee, C.H.; Lee, W.J.; Jeon, H.; Jung, J.W.; Jung, D.H. Forecasting Liquefied Natural Gas Bunker Prices Using Artificial Neural Network for Procurement Management. *J. Mar. Sci. Eng.* **2022**, *10*, 1814. [CrossRef]
5. Vilhelmsen, C.; Lusby, R.; Larsen, J. Tramp Ship Routing and Scheduling with Integrated Bunker Optimization. *EURO J. Transp. Logist.* **2013**, *32*, 143–175. [CrossRef]
6. Xing, H.; Stuart, C.; Spence, S.; Chen, H. Alternative Fuel Options for Low Carbon Maritime Transportation: Pathways to 2050. *J. Clean. Prod.* **2021**, *297*, 126651. [CrossRef]
7. Halim, R.A.; Kirstein, L.; Merk, O.; Martinez, L.M. Decarbonization Pathways for International Maritime Transport: A Model-Based Policy Impact Assessment. *Sustainability* **2018**, *10*, 2243. [CrossRef]

8.   Balcombe, P.; Brierley, J.; Lewis, C.; Skatvedt, L.; Speirs, J.; Hawkes, A.; Staffell, I. How to Decarbonise International Shipping: Options for Fuels, Technologies and Policies. *Energy Convers. Manag.* **2019**, *182*, 72–88. [CrossRef]
9.   Serra, P.; Fancello, G. Towards the Imo's Ghg Goals: A Critical Overview of the Perspectives and Challenges of the Main Options for Decarbonizing International Shipping. *Sustainability* **2020**, *12*, 3220. [CrossRef]
10.   Romano, A.; Yang, Z. Decarbonisation of Shipping: A State of the Art Survey for 2000–2020. *Ocean. Coast. Manag.* **2021**, *214*, 105936. [CrossRef]
11.   Mallouppas, G.; Yfantis, E.A. Decarbonization in Shipping Industry: A Review of Research, Technology Development, and Innovation Proposals. *J. Mar. Sci. Eng.* **2021**, *9*, 415. [CrossRef]
12.   Bouman, E.A.; Lindstad, E.; Rialland, A.I.; Strømman, A.H. State-of-the-Art Technologies, Measures, and Potential for Reducing Ghg Emissions from Shipping—A Review. *Transp. Res. Part D Transp. Environ.* **2017**, *52*, 408–421. [CrossRef]
13.   Ampah, J.D.; Yusuf, A.A.; Afrane, S.; Jin, C.; Liu, H.F. Reviewing Two Decades of Cleaner Alternative Marine Fuels: Towards Imo's Decarbonization of the Maritime Transport Sector. *J. Clean. Prod.* **2021**, *320*, 128871. [CrossRef]
14.   Tadros, M.; Ventura, M.; Soares, C.G. Review of Current Regulations, Available Technologies, and Future Trends in the Green Shipping Industry. *Ocean. Eng.* **2023**, *280*, 114670. [CrossRef]
15.   Moshiul, A.M.; Mohammad, R.; Hira, F.A.; Maarop, N. Alternative Marine Fuel Research Advances and Future Trends: A Bibliometric Knowledge Mapping Approach. *Sustainability* **2022**, *14*, 4947. [CrossRef]
16.   Le Fevre, C. *A Review of Demand Prospects for Lng as a Marine Transport Fuel*; Oxford Institute for Energy Studies: Oxford, UK, 2018.
17.   Hansson, J.; Brynolf, S.; Fridell, E.; Lehtveer, M. The Potential Role of Ammonia as Marine Fuel-Based on Energy Systems Modeling and Multi-Criteria Decision Analysis. *Sustainability* **2020**, *12*, 3265. [CrossRef]
18.   Priyanto, E.M.; Olcer, A.I.; Dalaklis, D.; Ballini, F. The Potential of Methanol as an Alternative Marine Fuel for Indonesian Domestic Shipping. *Int. J. Marit. Eng.* **2020**, *162*, 115–129. [CrossRef]
19.   Lagouvardou, S.; Lagemann, B.; Psaraftis, H.N.; Lindstad, E.; Erikstad, S.O. Marginal Abatement Cost of Alternative Marine Fuels and the Role of Market-Based Measures. *Nat. Energy* **2023**, 1–12. [CrossRef]
20.   Kim, A.R.; Seo, Y.J. The Reduction of Sox Emissions in the Shipping Industry: The Case of Korean Companies. *Mar. Policy* **2019**, *100*, 98–106. [CrossRef]
21.   Stalmokaite, I.; Yliskyla-Peuralahti, J. Sustainability Transitions in Baltic Sea Shipping: Exploring the Responses of Firms to Regulatory Changes. *Sustainability* **2019**, *11*, 1916. [CrossRef]
22.   Hansson, J.; Mansson, S.; Brynolf, S.; Grahn, M. Alternative Marine Fuels: Prospects Based on Multi-Criteria Decision Analysis Involving Swedish Stakeholders. *Biomass Bioenergy* **2019**, *126*, 159–173. [CrossRef]
23.   Zhang, X.; Bao, Z.H.; Ge, Y.E. Investigating the Determinants of Shipowners' Emission Abatement Solutions for Newbuilding Vessels. *Transp. Res. Part D Transp. Environ.* **2021**, *99*, 102989. [CrossRef]
24.   Makitie, T.; Steen, M.; Saether, E.A.; Bjorgum, O.; Poulsen, R.T. Norwegian Ship-Owners' Adoption of Alternative Fuels. *Energy Policy* **2022**, *163*, 112869. [CrossRef]
25.   Chen, S.; Zheng, S.Y.; Sys, C. Policies Focusing on Market-Based Measures Towards Shipping Decarbonization: Designs, Impacts and Avenues for Future Research. *Transp. Policy* **2023**, *137*, 109–124. [CrossRef]
26.   Kaya, A.Y.; Erginer, K.E. An Analysis of Decision-Making Process of Shipowners for Implementing Energy Efficiency Measures on Existing Ships. The Case of Turkish Maritime Industry. *Ocean. Eng.* **2021**, *241*, 110001. [CrossRef]
27.   Bao, Z.H.; Zhang, X.; Fu, G.Y. Factors Influencing Decision to Sulphur Oxide Emission Abatement for Cruise Shipping Companies. *Int. J. Logist. Res. Appl.* **2022**, 1–20. [CrossRef]
28.   Alizadeh, A.H.; Strandenes, S.P.; Thanopoulou, H. Capacity Retirement in the Dry Bulk Market: A Vessel Based Logit Model. *Transp. Res. Part E-Logist. Transp. Rev.* **2016**, *92*, 28–42. [CrossRef]
29.   Fan, L.X.; Xie, J.Q. Identify Determinants of Container Ship Size Investment Choice. *Marit. Policy Manag.* **2023**, *50*, 219–234. [CrossRef]
30.   Kanamoto, K.; Liwen, M.R.; Nakashima, M.; Shibasaki, R. Can Maritime Big Data Be Applied to Shipping Industry Analysis? Focussing on Commodities and Vessel Sizes of Dry Bulk Carriers. *Marit. Econ. Logist.* **2021**, *23*, 211–236. [CrossRef]
31.   Talluri, K.; van Ryzin, G. Revenue Management under a General Discrete Choice Model of Consumer Behavior. *Manag. Sci.* **2004**, *50*, 15–33. [CrossRef]
32.   Louviere, J.J.; Hensher, D.A.; Swait, J.D. *Stated Choice Methods: Analysis and Applications*; Cambridge University Press: Cambridge, UK, 2000.
33.   Balcombe, P.; Staffell, I.; Kerdan, I.G.; Speirs, J.F.; Brandon, N.P.; Hawkes, A.D. How Can Lng-Fuelled Ships Meet Decarbonisation Targets? An Environmental and Economic Analysis. *Energy* **2021**, *227*, 120462. [CrossRef]
34.   Lindstad, E.; Eskeland, G.S.; Rialland, A.; Valland, A. Decarbonizing Maritime Transport: The Importance of Engine Technology and Regulations for Lng to Serve as a Transition Fuel. *Sustainability* **2020**, *12*, 8793. [CrossRef]
35.   Wang, S.Y.; Notteboom, T. The Adoption of Liquefied Natural Gas as a Ship Fuel: A Systematic Review of Perspectives and Challenges. *Transp. Rev.* **2014**, *34*, 749–774. [CrossRef]
36.   Lee, B.; Lee, H.; Lim, D.; Brigljevic, B.; Cho, W.; Cho, H.S.; Kim, C.H.; Lim, H. Renewable Methanol Synthesis from Renewable H-2 and Captured Co2: How Can Power-to-Liquid Technology Be Economically Feasible? *Appl. Energy* **2020**, *279*, 115827. [CrossRef]
37.   Verhelst, S.; Turner, J.W.G.; Sileghem, L.; Vancoillie, J. Methanol as a Fuel for Internal Combustion Engines. *Prog. Energy Combust. Sci.* **2019**, *70*, 43–88. [CrossRef]

38. Inal, O.B.; Zincir, B.; Deniz, C. Investigation on the Decarbonization of Shipping: An Approach to Hydrogen and Ammonia. *Int. J. Hydrogen Energy* **2022**, *47*, 19888–19900. [CrossRef]
39. Zincir, B. Environmental and Economic Evaluation of Ammonia as a Fuel for Short-Sea Shipping: A Case Study. *Int. J. Hydrogen Energy* **2022**, *47*, 18148–18168. [CrossRef]
40. Foretich, A.; Zaimes, G.G.; Hawkins, T.R.; Newes, E. Challenges and Opportunities for Alternative Fuels in the Maritime Sector. *Marit. Transp. Res.* **2021**, *2*, 100033. [CrossRef]
41. Lindstad, E.; Lagemann, B.; Rialland, A.; Gamlem, G.M.; Valland, A. Reduction of Maritime Ghg Emissions and the Potential Role of E-Fuels. *Transp. Res. Part D Transp. Environ.* **2021**, *101*, 103075. [CrossRef]
42. Solakivi, T.; Paimander, A.; Ojala, L. Cost Competitiveness of Alternative Maritime Fuels in the New Regulatory Framework. *Transp. Res. Part D Transp. Environ.* **2022**, *113*, 103500. [CrossRef]
43. Steen, M.; Bach, H.; Bjørgum, Ø.; Hansen, T.; Kenzhegaliyeva, A. *Greening the Fleet: A Technological Innovation System (Tis) Analysis of Hydrogen, Battery Electric, Liquefied Biogas, and Biodiesel in the Maritime Sector*; SINTEF: Trondheim, Norway, 2019.
44. Li, K.; Wu, M.; Gu, X.H.; Yuen, K.F.; Xiao, Y. Determinants of Ship Operators' Options for Compliance with Imo 2020. *Transp. Res. Part D Transp. Environ.* **2020**, *86*, 102459. [CrossRef]
45. Beenstock, M. A Theory of Ship Prices. *Marit. Policy Manag.* **1985**, *12*, 215–225. [CrossRef]
46. Xu, J.J.; Yip, T.L. Ship Investment at a Standstill? An Analysis of Shipbuilding Activities and Policies. *Appl. Econ. Lett.* **2012**, *19*, 269–275. [CrossRef]
47. Jiang, L.; Kronbak, J.; Christensen, L.P. The Costs and Benefits of Sulphur Reduction Measures: Sulphur Scrubbers Versus Marine Gas Oil. *Transp. Res. Part D Transp. Environ.* **2014**, *28*, 19–27. [CrossRef]
48. Zhu, M.; Yuen, K.F.; Ge, J.W.; Li, K.X. Impact of Maritime Emissions Trading System on Fleet Deployment and Mitigation of Co2 Emission. *Transp. Res. Part D Transp. Environ.* **2018**, *62*, 474–488. [CrossRef]
49. Yang, J.L.; Zhang, X.; Ge, Y.E. Measuring Risk Spillover Effects on Dry Bulk Shipping Market: A Value-at-Risk Approach. *Marit. Policy Manag.* **2022**, *49*, 558–576. [CrossRef]
50. Mazzanti, M. Discrete Choice Models and Valuation Experiments. *J. Econ. Stud.* **2003**, *30*, 584–604. [CrossRef]
51. Law, P.K. A Theory of Reasoned Action Model of Accounting Students' Career Choice in Public Accounting Practices in the Post-Enron. *J. Appl. Account. Res.* **2010**, *11*, 58–73. [CrossRef]

Journal of
*Marine Science
and Engineering*

MDPI

Article

Evaluating the Readiness of Ships and Ports to Bunker and Use Alternative Fuels: A Case Study from Brazil

Huang Wei [1], Eduardo Müller-Casseres [1], Carlos R. P. Belchior [2] and Alexandre Szklo [1,*]

[1] Centre for Energy and Environmental Economics (CENERGIA), Centro de Tecnologia, Sala C-211, Cidade Universitária, Ilha do Fundão, Rio de Janeiro 21941-972, RJ, Brazil; huangkenwei@ppe.ufrj.br (H.W.); ecasseres@ppe.ufrj.br (E.M.-C.)

[2] Ocean Engineering Program—COPPE/UFRJ, Centro de Tecnologia, Sala C-203, Cidade Universitária, Ilha do Fundão, Rio de Janeiro 21945-970, RJ, Brazil; belchior@oceanica.ufrj.br

* Correspondence: szklo@ppe.ufrj.br; Tel.: +55-21-3938-8760

Abstract: The International Maritime Organization (IMO) has recently revised its strategy for shipping decarbonization, deepening the ambition to reduce annual greenhouse gas emissions by 2050. The accomplishment of this strategy requires the large-scale deployment of alternative maritime fuels, whose diversity and technical characteristics impose transition challenges. While several studies address the production of these fuels, a notable gap lies in the analysis of the required adaptations in vessels and ports for their usage. This study aims to fill this gap with a comprehensive review of material compatibility, storage in ports/vessels, and bunkering technology. First, we analyze key aspects of port/vessel adaptation: physical and chemical properties; energy conversion for propulsion; fuel feeding and storage; and bunkering procedures. Then, we perform a maturity assessment, placing each studied fuel on the technological readiness scale, revealing the most promising options regarding infrastructure adaptability. Finally, we develop a case study from Brazil, whose economy is grounded on maritime exports. The findings indicate that multi-product ports may have the potential to serve as multi-fuel hubs, while the remaining ports are inclined to specific fuels. In terms of vessel categories, we find that oil tankers, chemical ships, and gas carriers are most ready for conversion in the short term.

Keywords: alternative fuels; port; ship; bunker; biofuels; LNG; ammonia; methanol

Citation: Wei, H.; Müller-Casseres, E.; Belchior, C.R.P.; Szklo, A. Evaluating the Readiness of Ships and Ports to Bunker and Use Alternative Fuels. A Case Study from Brazil. *J. Mar. Sci. Eng.* **2023**, *11*, 1856. https://doi.org/10.3390/jmse11101856

Academic Editors: Jian Wu and Xianhua Wu

Received: 28 August 2023
Revised: 19 September 2023
Accepted: 22 September 2023
Published: 24 September 2023

1. Introduction

Maritime Transport is a key sector of the global economy, accounting for approximately 90% of the global trade in mass basis [1,2]. Shipping is a fundamental mode of trade for consuming less fuel per mass transported and distance covered compared with alternative modes. According to the Fourth IMO (International Maritime Organization) GHG (greenhouse gas) Study [3], the shipping world fleet consumed 13.6 exajoules (EJ) in 2018 and emitted 1.056 billion tonnes of carbon dioxide equivalent (CO_2eq), being responsible for nearly 3% of global greenhouse gas emissions. International shipping was responsible for 87% of the total emissions. Smith et al. [4] suggest that in the absence of measures to reduce greenhouse gas emissions, these emissions could increase by 250% by the year 2050. Among the available strategies to mitigate such emissions is to set speed, power, and fuel consumption limits [5]. Conversely, the vast diversity of ship types, with its associated challenges in construction and operation, has been a great barrier to standardization [6], in addition to the long lifetime of long-distance ships. Several studies have evidenced that the implementation of measures and technologies targeting a reduction in greenhouse gases (GHG) holds the potential to curtail emissions by up to 75% of the current levels [7–9].

In 2023, IMO established a goal of achieving net zero GHG emissions[1] by 2050, accounting for the life cycle emissions of fuels, while a medium-term goal entails achieving a minimum 20% reduction in GHG emissions from international shipping by 2030, as compared with emissions levels recorded in 2008 [11]. This new strategy exhibits a greater degree of firmness when contrasted with IMO's initial and ambitious approach, which primarily focused on a reduction in shipping direct GHG emissions by a minimum of 50% in relation to 2008 levels [12]. Smith et al.'s [4] estimation indicated that the shipping sector emitted 921 million tons of carbon dioxide (CO_2) by 2008. According to DNV GL [13], to achieve previous IMO 2050 goals, it was imperative that 40% of the energy supplied to the shipping fleet was derived from fuels characterized by net zero emissions in ships. Faber et al. [3] predicted that without intervention, emissions could escalate to over 1300 million tons of CO_2 by 2030 and surpass 2300 million tons by 2050. Consequently, in comparison with a scenario with no actions to lessen the emissions, a decrease of more than 560 million tons of CO_2 emitted would be necessary by 2030.

To mitigate GHG emissions [14], several measures can be used, but the utilization of fuels with lower emissions levels or net zero emissions throughout their life cycle will be required [15]. The 2023 IMO guidelines on the removal of regulatory barriers concerning the blend of marine fuels with up to 30% of alternative fuels, specifically biofuels or synthetic fuels, encompass a fundamental factor in promoting the entrance of these alternative fuels into the shipping market. The blends with alternative fuels are to be treated on par with regular fuels, implying that they can be utilized as long as they comply with NOx emission limits [16,17].

Therefore, the investigation of alternative fuels for maritime transport has earned significant interest from both the academic and professional community. Recently, there has been a substantial number of studies delving into the subject of the production and consumption of biofuels [18–22], hydrogen and ammonia [23–27], liquefied natural gas (LNG) [28–30], and methanol [31–34] for shipping. While a significant share of these studies focuses on the technical aspects of production [35–41], emissions mitigation [7,42–44], and their consumption in marine engines [45–47], few have given due attention to the necessary adaptations required in ships and ports to the operation of these alternative fuels. Actually, the implementation of alternative fuels in the maritime sector drives various adjustments within ships. These modifications encompass alterations in fuel tanks and engine locations, utilization of distinct materials for storage tanks and pipelines, reinforcement of pipe structures, enhancement in ventilation systems to mitigate potential gas leakage [48], and changes in port infrastructure.

Therefore, the primary aim of this study is to assess the current progress of adjusting ships and ports to effectively use selected alternative fuels, with a particular emphasis on their applicability to long-haul cargo shipping, mostly characterized by large vessels, which significantly contributes to the sector's overall energy demand and GHG emissions [3]. By doing so, this analysis seeks to determine the technological readiness for the conversion of ports and ships to the storage, bunkering, and use of the chosen fuels. Some of the highlighted fuels can have their production based on both fossil and sustainable sources. For instance, LNG, methanol, ammonia, and hydrogen can be produced from fossil fuels and biomass or with electrolysis and carbon capture and storage (CCS) and direct air capture (DAC), known as e-fuels [49]. Both bio and e-fuel alternatives possess the potential to reduce GHG emissions when compared with fossil-based fuels [43]. Given that the primary goal of this analysis is to assess the compatibility of alternative fuel handling, storage, and usage, our discussion does not encompass an evaluation of the GHG emissions of these fuels from their production to their consumption. This has been completed in several works, such as Muller-Casseres et al. [35] and Brynolf et al. [39].

In addition, since this study addresses long-distance freight transportation based on large vessels, energy carriers and storage options, such as hydrogen and batteries, are not evaluated given their low suitability for deep-sea large ships, as shown by Gray et al. [50] and Xing et al. [40]. Indeed, these alternatives lead to a substantial spatial allocation loss in comparison with conventional fuels, making them impractical for long-distance shipping [51].

Then, to validate and illustrate the assessment conducted in this study, a case study was carried out to assess the capacity of the Brazilian fleet and port infrastructure to adopt alternative fuels. The Brazilian case is emblematic since the country's economy heavily relies on marine routes [52] for exporting goods and sustaining its economic activities [53]. Additionally, Brazil has an impressive potential for alternative fuel production, particularly biofuels, given its abundant availability of biomass resources and established expertise in biofuels production [54]. For instance, according to Carvalho et al. [37], the comparative analysis encompassing Brazil, Europe, South Africa, and the USA illustrates that "biomass concentration in Brazil makes it the region with highest biobunker potential, which are mostly close to coastal areas and surpasses regional demand".

The next section outlines the methods and materials used in the evaluation. In Section 3, the results of the analysis are presented, focusing on determining and comparing the readiness of each alternative fuel. Section 4 delves into a comprehensive discussion of previous findings by applying them to a specific case study. Lastly, Section 5 provides the conclusions, along with recommendations and barriers identified in this study.

2. Materials and Methods

The primary objective of this study is to analyse the necessary adaptations in large deep-sea ships and ports for the proper storage, transfer, and utilization of alternative marine fuels. As such, it does not encompass fuels that can be classified as fully drop-in [49], such as Fischer–Tropsch liquids [38,55] from biomass and electric-derived hydrogen and CO_2. The deployment of these drop-in fuels can rely on existing ships and bunkering infrastructure, thereby enabling a direct replacement or blend with conventional fuels [56]. In contrast, most candidate alternative marine fuels require some level of adaptation in ships and ports. Some of them can be seen as partially drop-in, meaning that they only require minor adjustments and specific attention compared with conventional fuels to be used in the existing infrastructure. On the other hand, a second group (non-drop-in fuels) require substantial changes and investments in vessel technology and bunkering infrastructure. As this paper will discuss further, some of these non-drop-in fuels already have an established infrastructure in several ports (for example, the case of tradeable ammonia and methanol) and have a relevant usage record in dual engines (LNG and, to a lesser extent, methanol). However, the categorization here considers that more than 95% of large ships are still based on diesel engines and the ports associated with their routes are mostly single hubs to store and bunker petroleum-derived fuels for them [4]. This study focuses on the assessment of specific fuels encompassed by these two categories, as listed in Table 1. As mentioned before, it is worth noting that ammonia, LNG, and methanol can be produced from fossil, bio, and synthetic feedstocks. Our focus here is not on their production but on their handling and usage.

Table 1. Fuel grouping.

Partially Drop-In [1]	Non-Drop-In [1]
Biodiesel	Ammonia
Hydrotreated pyrolysis oil (HPO)	Liquefied natural gas (LNG)
Hydrotreated vegetable oil (HVO)	Methanol
Straight vegetable oil (SVO)	

[1] [19,35,57].

A comprehensive and thorough review of the technical literature was conducted, with a specific emphasis on the essential properties to be taken into consideration for achieving a successful adaptation in retrofitting both ships and ports to enable proper storage, transfer, and utilization of alternative fuels. Figure 1 provides a summary of the steps undertaken in this study. This analytical study first examined various aspects pertaining to selected alternative fuels. As a second step, considering the existing ships and bunkering infrastructure globally, along with regulatory frameworks and tests designed to assess fuel performance on ships, the analysed fuels were categorized into those that are partially or non-drop-in. This categorization was succeeded by an assessment of technological readiness based on the guidelines provided by the US Department of Energy [58].

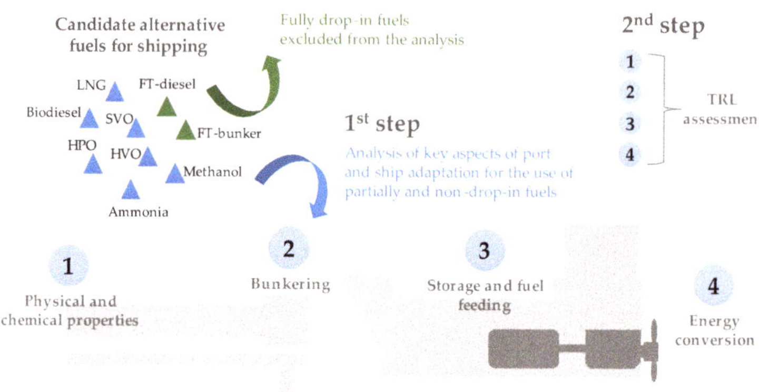

Figure 1. Methodological procedure.

As Figure 1 displays, the first step in the analysis encompasses the key aspects of port and ship conversion for the proper utilization of the selected alternative fuels. The first step was split into four main aspects, namely, physical, and chemical characteristic properties, bunkering procedures, storage and fuel feeding systems, and energy conversion systems. Table 2 displays the main aspects analysed for each of the aforementioned segments.

Table 2. Main aspects analysed for each section concerning the adaptation of ships and ports to the use of partial and non-drop-in fuels.

Segment	Analysed Aspects
Physical and chemical properties	Heating value Volumetric density Energy density Kinematic viscosity Acidity Flash point Self ignition temperature CCAI Other properties
Bunkering	Pressurization Liquefaction Tank shape Inertisation Ventilation Maintenance
Storage and fuel feeding	Pressurization Liquefaction Tank location Tank volume Inertisation Ventilation reinforcement Maintenance Need for double-wall Materials Drainage Preheating Filtering
Energy conversion	Converter type Need for pilot fuel Engine adjustments

As Table 2 illustrates, the initial analysis includes a review of the main properties of fuels in comparison with conventional fossil bunker fuels. Heating value and volumetric density are both linked to energetic density, which represents the amount of energy per cubic meter. In shipping, greater energetic density is preferable as it allows for increased autonomy due to the higher energy demand for fuels (e.g., Ref. [59]), as well as smaller losses of freight space [50]. High levels of kinematic viscosity directly impact the spray and flow characteristics of fuel [60]. Acidity is associated with the content of free fatty acids in fuel. A high content of free fatty acids can result in engine deterioration, as well as degradation of engine feed [61]. Flash point refers to the minimum temperature at which gases ignite when exposed to a flame [62]. Hence, low-flash point fuels are undesirable for shipping. Ellis and Tanneberger [31] underscored that low flash points trigger additional safety measures in order to prevent the fuel from being exposed to ignition sources. A high self-ignition temperature leads to obstacles in achieving auto-ignition, a characteristic considered unfavourable for use in diesel engines [63]. The aromaticity index, measured with the calculated carbon aromaticity index (CCAI), is used to assess fuel quality based on ignition delay. CCAI is calculated with an evaluation of density and viscosity. For marine engines, a CCAI below 870 is recommended [64]. Viscosity and CCAI values for LNG and ammonia are not evaluated in the literature since they are equivalent to or lower than those of traditional fuels. As a result, these factors were not considered in this study, nor were the acidity levels in LNG, methanol, ammonia, and HVO. Other properties, such as oxygen and water content, play a pivotal role in determining the requisite adjustments for utilizing these fuels in the current infrastructure.

Having addressed the main properties of fuels, this study evaluated the necessary adjustments to bunkering infrastructure to accommodate the usage of each selected fuel. As indicated in Table 2, certain aspects were examined, including the requirements for pressurization, liquefaction, different tank shapes, inertisation, ventilation reinforcement, and an increase in maintenance. This evaluation encompassed not only the bunkering process but also storage at ports.

Then, this study revised the challenges related to storage and fuel feeding in ships. The analysis carried out addressed significant modifications resulting from distinct properties of the chosen fuels, as opposed to conventional fossil bunker fuels. Aspects such as demands for pressurization and liquefaction during storage, different shapes, locations, and volumes of tanks, double walls, and filtering were highlighted.

Finally, the energy conversion analysis addressed the available choices of energy converters for each fuel, with a specific emphasis on a potential pilot fuel demand and adjustments in engines for the proper use of the fuels. The analysed options for energy converters are diesel engines, dual-fuel engines, and fuel cells. According to the Fourth GHG IMO Study [3], conventional fossil bunker fuels, namely, heavy fuel oil (HFO) and marine diesel oil (MDO), are the two primary fuels commonly used in the marine industry, representing 66.0% and 30.5% of the world's consumption, respectively. Additionally, LNG accounted for roughly 3.4% of the world's consumption, whereas methanol represented a mere 0.05% of the overall shipping consumption. As a result, the predominant energy converter to propulsion in the vessel fleet is the two-stroke diesel engine. In 2018, low, medium, and high diesel engines accounted for over 98% of the global marine fleet, while dual-fuel LNG engines were installed in less than 0.5% of ships, and engines adapted to methanol were reported in less than 0.15% of the fleet [3]. Diesel engines designed for marine applications are available in two configurations: two- and four-stroke variants. Larger ships typically opt for two-stroke engines due to their ability to achieve lower propulsion speeds effectively. In contrast, medium- and high-speed engines predominantly use four-stroke cycles to optimize the operation of these vessels [65].

In relation to the conversion of diesel engines to dual-fuel engines, Tiwari [66] reported that the dual-fuel engine is essentially a diesel engine equipped with supplementary devices that enable the utilization of fuels such as LNG. Bhavani and Murugesan [67] further pointed out that the conversion from diesel to a dual-fuel mode solely necessitates external modifications to the engine, while the internal components remain unchanged. Furthermore, the authors emphasized that the conversion process involves the addition of a set of retrofit components, including fuel supply systems, pilot and supplemental fuel inlet controllers, air and gas mixers, engine cooling systems, flameproof kits, and gas detectors. Another viable energy converter option is the use of fuel cells, which are currently in the developmental phase for marine applications. Nevertheless, fuel cells present superior efficiency and emit fewer pollutants during tank-to-wake, namely, the use in ships, when compared with internal ignition and gas engines. In addition, a steam reformer can be incorporated into vessels to enable the use of hydrocarbons as an energy vector. Although this process generates carbon dioxide emissions, they are significantly lower than those produced by conventional engines utilizing fossil fuels, and the emissions of other pollutants remain nearly negligible [68]. Xing et al. [51] stated that recent research and demonstration projects have validated the technical feasibility of fuel cells for maritime applications regarding power capacity, safety, durability, and operational terms. These developments contribute to promoting the adoption of fuel cells in vessel fleets in the future. However, it is important to note that despite these advancements, commercial viability remains a challenge [46], and their suitability for long-haul shipping is still limited [51].

Having addressed all segments of the first step, the evaluation of TRL for each fuel is thus complete. Figure 2 summarizes the assessment approach.

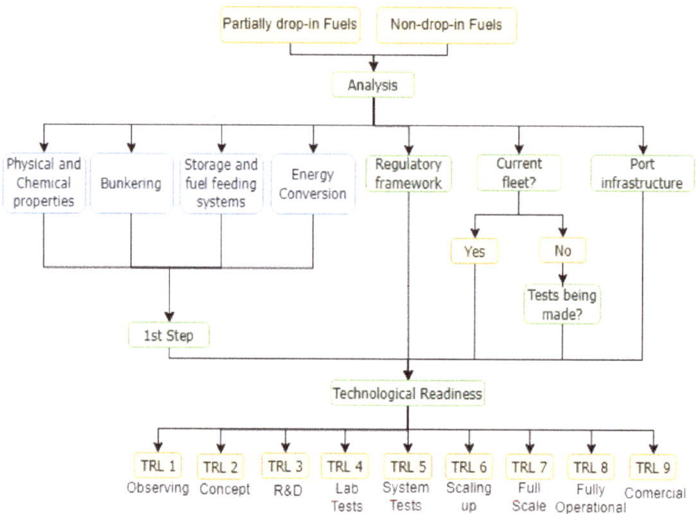

Figure 2. TRL evaluation of each fuel type.

As Figure 2 illustrates, the determination of TRL for each fuel resulted from the analysis performed, also considering the current regulatory and port infrastructure. A detailed exploratory review was performed to identify the established standards, guidelines, and whitepapers conducting procedural aspects associated with the utilization of each designated fuel, thus enabling an assessment of the regulatory framework. Current infrastructure evaluation was also performed by compiling data pertaining to vessels that already adopted alternative bunker fuels. In the absence of ships using fuel, a review encompassing not only vessels but also other modes of transportation were conducted. Furthermore, an evaluation of port infrastructure was conducted to identify existing port facilities offering bunkering services for each fuel. The required adjustment to each fuel for use in maritime infrastructure facilities leads to the estimation of TRL. This ranges from observation of technology (TRL 1) to conceptualization (TRL 2), research and development or R&D (TRL 3), laboratory tests (TRL 4), systems tests in real conditions (TRL 5), scaling up in real conditions tests (TRL 6), full scale in real conditions tests (TRL 7), and fully operational functioning (TRL 8) to reach commercial status (TRL 9) [58].

Finally, after conducting a comprehensive assessment of the obstacles and complexities involved in adapting the existing maritime infrastructure to accommodate alternative fuels, this study applied it to a case study as a representative example. The case study was based in Brazil, given its high economic dependency on maritime routes, from cabotage to national trade and long-haul distances for exportation [52,53], as well as its notable potential as a major future biobunker producer [37]. The case study examined the current state of the Brazilian shipping sector, including high-priority ports, given their cargo movement and initiatives to bunkering of alternative fuels; an analysis of potential multi-fuel hubs; the progress and challenges in converting ships for alternative fuels; the initiatives assumed by local governments and companies linked to the maritime sector to achieve decarbonization of Brazil's maritime transport; thermal stability of fuels in maritime routes; and the problem of loss in cargo space. The primary objective was to develop a coherent framework that would evaluate the potential of introducing alternative fuels in the country. This framework can serve as the first roadmap for assessing the feasibility of applying alternative fuel solutions in Brazil and potentially extrapolate these findings to other countries and regions with similar characteristics.

3. Results

3.1. Physical and Chemical Properties

Table 3 lists the main properties of the selected alternative fuels.

Table 3. Properties of marine fuels.

Fuel Property	Heating Value	Volumetric Density	Energy Density	Viscosity at 40 °C	Acidity	Flash Point	Self-Ignition Temperature	Aromaticity Index (CCAI)
Unit	MJ/kg	kg/m^3	MJ/m^3	mm/s^2	Mg KOH/g	°C	°C	-
HFO	40.0 [a]	991 [a]	39,640	380 [i]	2.5 [i]	>60 [i]	407 [p]	856.5 [u]
MGO	42.0 [a]	890 [a]	37,380	3.5 [i]	0.5 [i]	>60 [i]	257 [q]	808.1 [u]
LNG	50.0 [b]	415 [b]	20,750	-	-	−188 [b]	537 [o]	-
Biodiesel	37.1 [c]	885 [c]	32,833.5	4–6 [j]	0.052–0.295 [m]	>93 [c]	374–449 [r]	822.6 [u]
SVO	37–39.62 [a]	900–930 [a]	33,300–36,847	14–40 [k]	0.02–20 [n]	>400 [k]	405 [s]	836.6–878.7 [u]
HVO	44.1 [d]	780 [d]	34,398	3 [d]	-	99 [d]	204 [o]	738.4 [u]
HPO	28.9 [e]	1150 [h]	33,235	9 [h]	21.3–76.1 [h]	53–101 [h]	340 [t]	1076 [u]
Ammonia	18.6 [g]	758 [g]	14,101	-	-	132 [o]	630 [o]	-
Methanol	20.1 [f]	798 [f]	16,040	0.58 [l]	-	12 [f]	470 [o]	837.6 [u]

a—[62]; b—[69]; c—[18]; d—[70]; e—[71]; f—[72]; g—[73]; h—[74]; i—[75]; j—[76]; k—[77]; l—[31]; m—[78]; n—[79]; o—[57]; p—[80]; q—[81]; r—[82]; s—[83]; t—[84]; u—[85].

In the comparison among fossil fuels, LNG stands as the option for the mitigation of sulphur oxides, nitrogenous oxides, and particulate matter emissions [86]. It is predominantly composed of methane, accompanied by minor proportions of other hydrocarbons such as ethane, propane, and butane [29]. Under atmospheric temperature and pressure, LNG is in the gaseous phase and has low density. In order to optimize storage, natural gas is liquefied at a temperature of −162 °C and atmospheric pressure, thereby reducing the required volume for storage [69].

The properties of biofuels vary depending on the feedstock used in production. Biodiesel, SVO, HVO, and HPO have energy density levels close to HFO and MGO compared with the other assessed fuels, suggesting that those fuels have greater potential to provide increased autonomy or reduced storage space requirements. SVO is a biofuel that entails a straightforward production process in comparison with other fuels. The production steps involve biomass collection, low-temperature seed pressing, and filtration to remove sludge. The quality of the fuel is heavily influenced by the quality of the feedstock and the conditions during production and processing [87]. When contrasted with traditional marine fuels, SVO has a slightly lower energy density and a higher flash point, viscosity, and acidity. These characteristics can potentially result in corrosion of engine feed pipelines [62]. Biodiesel (or FAME), widely regarded as one of the most promising biofuels, is repeatedly stated as a potential blend component for diesel in the road transport sector [88].

HVO consists of straight chains of paraffinic hydrocarbons, which undergo additional production steps in comparison with SVO. These steps include catalytic saturation (hydrogenation), hydrodeoxygenation, hydrodecarboxylation, and isomerization. HVO is distinguished by its exceedingly low sulphur content and minimal emission factors [70]. As a paraffinic compound, HVO exhibits a high cetane number, typically ranging from 75 to 95 [89].

Hydrotreated pyrolysis oil, also known as bio-oil or HPO [90], is derived from biomass, which undergoes a high-temperature process in the absence of oxygen. The biomass is subjected to a temperature of 500 °C for a brief duration [21]. Hydrogenation is the final step, transforming the pyrolysis oil into hydrotreated pyrolysis oil. Depending on the pyrolysis process, the water content in bio-oil can reach up to 30%, which is sufficient to induce phase separation when stored at ambient temperature for six months [20]. Treatment of bio-oil can result in a compound with a significant reduction in oxygen content and an increase in light aromatic compounds.

In relation to viscosity, SVO and HPO have elevated levels, necessitating appropriate measures to viscosity decrease such as preheating. Moreover, these fuels are also notable for their high acidity levels. Biodiesel has a viscosity greater than traditional diesel yet not as high as SVO and HPO; therefore, preheating is advisable [22]. HPO has a high and unstable viscosity, posing a challenge for both its use as a fuel and storage [91]. Notably, the low flash point of biodiesel limits its practical utilization in low air temperature conditions [45]. HVO has a flash point higher than traditional fuels [89].

The acidity level of SVO, as is the case for biodiesel, is associated with its specific feedstock, which is also the case for biodiesel. While certain vegetable oils may present higher acidity levels compared with HFO, others exhibit relatively low acid values, as exemplified by rapeseed oil, which has an acidity level below 2.5 mg KOH/g [87]. Despite undergoing a reduction of approximately 70% in acidity after treatment, the resultant HPO acidity level remains notably higher when compared with traditional marine fuels [74].

The majority of the discussed fuels exhibit an aromaticity index below the recommended limit. However, depending on the feedstock used, the aromaticity index of SVO may exceed the suggested limit, as is the case for HPO. Ellis and Tanneberger [31] draw attention to the possibility of utilizing a lubricant oil to address the issue of low lubricity. In comparison with traditional fuels, biodiesel has superior lubricity and lower toxicity levels. However, it possesses a high oxygen content, typically ranging between 10 and 11%, and a low pour point [45,47,62]. To mitigate the risk of corrosion, the usage of a corrosion inhibitor known as tert-butylamine is advisable, with a recommended concentration of 250 ppm [47].

Methanol [92] and ammonia [93] are widely used as feedstocks in the chemical industry. Given their high toxicity, it is essential to implement safety measures to prevent leaks and human exposure to these substances, such as gas detectors. As stated by Kay et al. [94], ammonia leakage not only into the air, but also into the sea, can lead to critical damage, and lethality can be greatly reduced if the release duration is shortened. Overall, the authors found that a 30 s leakage is 70% less lethal than a 60 s leakage. Safety measures must be targeted to mitigate toxicity, especially at potential sources of leakage such as inlet and outlet manifolds for hose connection. According to Hansson et al. [27], the presence of high concentrations of ammonia poses health risks and can prove lethal within certain concentrations and exposure durations.

Ammonia has been proposed as a potential sustainable energy carrier of hydrogen due to its composition of three hydrogen atoms per ammonia molecule (NH_3) [95]. In addition, the storage of liquid hydrogen requires extremely low temperatures, specifically, -253 °C [96]. Hydrogen is recognized as a promising marine fuel, with ongoing tests aimed at advancing its utilization in the shipping industry. However, as reported by ABS [97], hydrogen currently offers a very limited power output, associated with substantial costs and limited production. Additionally, hydrogen storage in vessels addresses significant problems that marine communities have yet to overcome. Kim et al. [73] also highlight that ammonia possesses 1.7 times higher energy content compared with hydrogen, along with a 50% greater hydrogen content by volume [26], leading to a reduced volume requirement of fuel storage. Alongside LNG, ammonia also necessitates lower temperatures and pressurization to maintain its liquid state during storage. Ammonia can be stored at 25 °C when pressurized at 10 bar, whereas under atmospheric pressure, the required storage temperature is -33.4 °C [73]. For ammonia, refrigerated storage is preferable due to its better effectiveness in reducing operational risk [94]. Methanol and LNG are low-flash point fuels, making them highly flammable. Methanol is flammable and exhibits lower lubricity compared with conventional marine fuels [31]. The flammability of ammonia [98], methanol [31], and LNG necessitates safety protocols to prevent the risk of leaks and spills, particularly in areas where ignition sources are present [99]. Regardless of its high flash point, ammonia has lower flame velocity compared with conventional fuels [93].

3.2. Bunkering

The bunkering of conventional fuels can be carried out using tank trucks (truck-to-ship-transfer or TTS), bunker vessels (ship-to-ship or STS), as well as shore tanks or pipelines (shore tank-to-ship or TPS) [100]. Regarding alternative fuels, the three aforementioned methods can be applied for bunkering, with specific protocols designed for each fuel type based on its distinct characteristics.

When using LNG bunkering, a security protocol must be followed to avoid leakages of the fuel under cryogenic conditions. If materials such as steel come into contact with LNG, they tend to become fragile and may experience cracking. The procedures for leak prevention are as follows: checking the connection of the supply pipeline, inertization of the pipeline with nitrogen gas, cleaning the interior of the pipeline with vapour from liquefied natural gas at cryogenic temperatures, bunkering, cleaning the remaining LNG inside the pipeline with vapour from natural gas at cryogenic temperatures, inertization of the pipeline with nitrogen, and disconnection of the supply pipeline [101]. Aneziris et al. [99] asserted that the utilization of low-temperature pipelines, loading arms, and hoses is mandatory for LNG bunkering. Furthermore, the authors also highlighted that it is essential to acknowledge that extremely low LNG temperatures may pose a significant hazard, impacting not only the structural integrity of materials used by causing potential cracks but also the safety of individuals in proximity to the LNG due to the risk of frostbite.

To ensure the appropriate bunkering of biofuels, it is imperative to modify storage tanks in accordance with specific fuel properties [19]. Ideally, the tanks should possess a narrow shape, aiming to reduce the retention of oil and fats during the cleaning process. Furthermore, the tank bottoms should be tapered to facilitate effective drainage [102]. The fuelling processes for SVO [62] and HVO [103] are comparable to those already established for HFO and marine diesel, respectively. However, as Kesieme et al. [62] stated, certain adjustments are necessary to safeguard against corrosion and water contamination. Additionally, the authors recommended that maintenance procedures should be reinforced to ensure prolonged use. Regarding HPO, the complete supply chain must be developed, including the development of suitable bunkering infrastructure to accommodate the unique fuelling requirements of bio-oil [104].

All fuelling methods applicable to LNG can be used for ammonia as well. However, additional requirements must be met, specifically, the filling station must be equipped with appropriate ventilation, either using natural means or machinery. Additionally, the piping system must be self-draining and composed of inert materials [105]. Those adjustments are demanded due to the toxicity and flammability issues of ammonia, as previously addressed in Section 3.1 and extensively examined by Fan and Enshaei [98] and Kay et al. [94]. The latter study also recommended the use of multiple hoses with lower flow rates instead of a single hose with a higher flow rate, thereby resulting in a reduction in bunkering time and increased safety conditions. In order to ensure the appropriate bunkering of ammonia and prevent the release of the substance, it is imperative, as emphasized by Duong et al. [106], to develop a comprehensive strategy aimed at minimizing ammonia leakage during the bunkering process.

3.3. Storage and Fuel Feeding

Due to the low temperatures observed during storage, specific tanks become necessary when utilizing LNG in ships. Several options for storage tanks are available: IMO type A, which resembles the ones commonly used for standard marine fuel [107], IMO type C, designed as pressure vessels, and membrane tanks. Additionally, there exists a category called type B, encompassing all tanks that are neither type A, type C, nor membrane tanks. Among the mentioned tank types, type A and type B are the most suitable for larger vessels due to their generally prismatic shape [108]. However, an obstacle to the effective utilization of LNG as fuel is the occurrence of methane slip [86], which involves gas leakage during both storage and engine operation. This issue can be mitigated if the leaked gas is reclaimed and reused by other ship machinery, such as in gas combustion units [107]. Regarding engine fuel supply, in order to enhance LNG safety procedures, ABS [109] recommends the utilization of gas detection systems for instant shutdown, double-wall piping with at least 30 air changes per hour, a maximum 10 bar pressure limit, nitrogen-based inertization for emergencies, and independent pumps and compressors from other circuits.

The utilization of biofuels, such as biodiesel, necessitates the use of appropriate materials for tanks and pipelines. It is recommended that stainless steel, as a material, be used for this purpose. However, when the blends comprise no more than 20% biodiesel in the overall volume, conventional materials can be used if adequately coated with zinc. The construction of feed pipelines using mild steel is permissible, provided that filters are installed to ensure the smooth operation of the system [110]. Moreover, to maintain the integrity of the biofuel infrastructure and prevent any potential water contamination, regular and careful inspections, maintenance activities, and constant cleaning of tanks and piping are essential [102].

The coexistence of water within a fuel blend poses a significant risk of degrading fuel filter cartridges, potentially leading to cavitation [62]. To mitigate such hazards, the use of stainless steel is recommended as the material of choice for constructing pipelines and tanks to ensure optimal safety. Alternatively, mild steel can be considered for tank and pipeline construction if suitably coated with an inert material. However, it is imperative to conduct regular inspections of the tanks to assess the condition of the coatings and ensure their integrity is preserved. Furthermore, it is of utmost importance that all materials utilized in tanks and auxiliary machinery, including heating units, must be inert to vegetable oils [102].

HVO exhibits a great level of resemblance to conventional diesel-based fuels, rendering it compatible with the materials already used in marine infrastructure for pipelines, tanks, feed systems, and engines. Nevertheless, it is recommended to take on maintenance and cleaning procedures for storage tanks before fuelling to ensure optimal performance. Additionally, strict supervision is advised to prevent any contact between HVO and water within the tanks and feed system as this could lead to detrimental effects. Remarkably, HVO sets itself apart from other biofuels by causing minimal corrosion of the materials commonly utilized in the naval industry's infrastructure. This exceptional property contributes to enhanced durability and safety in marine operations involving HVO usage [103].

The high viscosity characteristic of HPO leads to an increase in engine deposits, which subsequently requires more energy for pumping and results in accelerated wear on fuel pump components and injectors. To mitigate these effects, preheating the fuel is essential as it effectively reduces the viscosity level. For the engine feed system, it is imperative to construct it using corrosion-resistant materials to withstand the high acidity of the oil. Copper can be considered as a viable material option for tank storage and pipelines; however, it is recommended to utilize stainless steel for tanks and pipes. The high acidity of HPO poses limitations on the use of carbon steel in pumps, fuel lines, and burners. These components must be made of materials that can resist the corrosive nature of the fuel. Furthermore, due to the presence of solid particles with high energy density, filtering them is not considered desirable. Nevertheless, the careful design of fuel supply piping is essential to prevent any blockages resulting from solid particle materials. Moreover, both pumping and atomizing processes should be equipped with suitable filtration mechanisms to ensure smooth and efficient operation [104].

When utilizing fuels with high acidity and/or flammability in ships, it is imperative that fuel storage tanks in ships adhere to a double-walled construction for enhanced safety measures. These tanks can be positioned either at the main deck, offering a more economical and less complex installation, or at lower decks as long as they are sufficiently distanced and detached from accommodation and machinery spaces. To minimize the risk of gas leakages, stringent preventive measures must be used. These include the implementation of inert systems, reinforcement of ventilation, and the utilization of specialized materials such as aluminium or, preferably, stainless steel for storage, feed, and engine components [29]. Furthermore, it is crucial to ensure that the pressure within the feed system does not exceed 10 bar [109] to maintain operational safety. By sticking to these guidelines, the potential hazards associated with fuel storage and usage in ships can be effectively mitigated.

In 2020, IMO [105] issued a comprehensive set of guidelines regarding the utilization of methanol and ethanol in vessels, encompassing fuelling procedures and safety practices. Some of these practices were already disseminated by DNV GL. The recommended safety measures include the implementation of double-walled feed pipelines and storage tanks constructed from stainless steel or austenitic steel, the incorporation of inert gas purging devices to facilitate the controlled release of gas, the installation of service tanks with the capacity to power operational loads for a minimum of eight hours, and the use of high-pressure pumps with a minimum pressure of 10 bar to facilitate the fuel feed to engines [111]. It is preferable to position the service tank on the main deck, while the pilot fuel tank may be situated in the engine room [34]. Due to the highly toxic nature of methanol, all areas containing pipelines or tanks are required to have adequate ventilation reinforcement. Specifically, normal spaces require a minimum of 15 air renovations per hour, while spaces more susceptible to fuel leakage require 30 air renovations per hour [31].

Regarding the utilization of ammonia in vessels, the required tanks for storing ammonia should be pressurized, with a minimum pressure of 8.6 bar, while the recommended pressure level stands at 17 bar [112]. For optimal cost-effectiveness, the type C tank has demonstrated its superiority and versatility, as it can be conveniently installed on the main deck and seamlessly integrated into the majority of existing ships [113]. To ensure the safe handling of ammonia, the feed pipelines must be constructed using durable materials such as carbon and stainless steel [25]. These pipelines should be displayed in a double-walled configuration to mitigate the risks of leakage [111]. Additionally, it is mandatory to equip all spaces associated with the fuel storage system with a comprehensive ventilation system. This measure is indispensable in preventing any potential ammonia leakages [112], thereby enhancing overall safety and minimizing associated hazards. Additionally, with respect to fuel feeding, it is recommended to avoid corrosive materials such as copper, high-nickel alloys, and plastic. To prevent corrosion, it is advisable to use Teflon in engine seals instead of rubber and plastic [25]. A system for emergency ventilation must be installed and operated in accordance with either of the following principles: a reduction in ammonia concentration to below 10 ppm with dilution or the capture of excessive ammonia [112]. In order to reduce the potential risks associated with ammonia leakage in the engine room, it is advisable to install both a tank and feed system on the deck, coupled with its connection to the engine using dual-walled piping. Another alternative is to place the feed system and tanks within the engine room if an airlock system to prevent ammonia dispersion on-site is installed [113]. DNV GL [111] suggested a mandatory provision of secondary enclosures for all fuel piping to securely contain any potential leaks. Furthermore, an arrangement involving the infusion of nitrogen into the secondary enclosure, coupled with ongoing pressure monitoring, can also be an alternative solution to ensure safety.

3.4. Energy Converters

The analysed fuels are applicable for one or more of the three energy converters considered in this study. With appropriate adjustments to adapt feed and combustion requirements, all fuels can be effectively applied in existing marine engines. Biodiesel [45], SVO [62], HVO [70], and HPO [91] demand relatively minor modifications to existing marine diesel engines and feed infrastructure. On the other hand, methanol and LNG, due to their high ignition temperature and consequently low cetane number, face ignition complications. To tackle this issue, dual-fuel engines can be used, in which a pilot fuel, such as marine diesel, is injected to start ignition [29].

Regarding SVO, to achieve the desirable viscosity levels, fuel preheating is imperative. The recommended heating temperature is within the range of 67 to 78 °C, which is comparatively lower than the temperatures required to preheat HFO [61]. Similarly, for the proper use of HPO in diesel engines, preheating within the temperature range of 40 to 80 °C is required [104]. It is crucial to be cautious of potential impurities in vegetable oils, since their presence may lead to engine failure or damage when used as marine fuel [62]. The combustion properties of HVO are similar to those of conventional fuels, such as marine diesel, although it has a lower density. Therefore, it is advisable to make adjustments in order to enable longer fuel injections for engine optimization, thereby increasing efficiency and fuel savings [70].

According to Dincer and Siddiqui [114], the use of ammonia in diesel engines presents drawbacks, notably, its limited flammability range, low kinetic rate, and high self-ignition temperature. Ammonia's combustion properties demand modifications to conventional combustion engines, as well as blending with fuels exhibiting superior combustion properties [115] and using dual-fuel engines. Burning ammonia may have the potential to produce more NOx emissions than regular fuels [116], potentially also releasing N_2O, a much stronger greenhouse gas than CO_2 [117]. Another alternative for ammonia is the use of fuel cells [114]. Kim et al. [73] compared the use of a polymer electrolyte membrane fuel cell (PEMFC), a low-cost alternative, and a solid-oxide fuel cell (SOFC) for ammonia chemical energy conversion. The results indicated that the latter is a simpler and more optimized operation for an ammonia-fuelled 2500 TEU container ship. SOFC used 12% less fuel on a volumetric basis.

3.5. Technological Readiness

The analysis by El-Gohary [118] demonstrated that the utilization of LNG as the primary fuel instead of conventional marine fuels has the potential to reach a notable reduction in annual expenses associated with fuel and maintenance, ranging from 30% to 40%. The implementation of LNG as the primary fuel for ships is rapidly becoming a reality. As of July 2023, a substantial portion of the global fleet, specifically 403 ships, has already adopted the use of LNG as fuel, and 275 terminals worldwide have equipped bunkering facilities for these vessels [119]. Consequently, the infrastructure for LNG bunkering has been firmly established, and all requisite fuel procedures have been meticulously documented by classification societies [109]. This thorough development and documentation have led to the classification of LNG's practical use as commercially available, indicated by a TRL of 9.

Among all the analysed fuels in this study, only biodiesel was mentioned in standards until 2022, allowing its use in marine fuel blends. Specifically, ISO 8217:2017 enables the utilization of up to 7% v/v of biofuel in such blends [120]. Mohd et al. [45] demonstrated that the direct use of biofuel in ships could potentially compromise current power supply systems, decrease efficiency and, consequently, increase specific consumption. However, Mohd et al. also pointed out that certain engine manufacturers, such as MAN, Wärtsilä and Caterpillar, have conducted tests showing satisfactory performance without necessitating modifications if the blend contains up to 30% v/v of biofuel. Additionally, Ogunkunle and Ahmed [121] reported that blends containing 30% biofuel (B30) and diesel do not result in engine alterations, although there is an increase in specific consumption. Countless marine engine manufacturers have undertaken research and testing to enhance the implementation of biofuels in vessels. Despite this progress, the biofuel bunkering process in ships still requires further development, even though minor adjustments may be necessary [18]. Consequently, as a marine fuel, biofuel is still in the full-scale testing phase, awaiting validation under real operating conditions, characterized as TRL 7.

Kesieme et al. [62] asserted that although SVO and HFO share some similarities, it is improbable that a blend of these two types of fuels would be compatible. Consequently, the most practical and viable solution would be a complete replacement of HFO with SVO. The usage of SVO in marine applications is still under research, both as a drop-in replacement and as a blend with traditional fuels. It has been observed that if a blend contains no more than 20% v/v of SVO with diesel, no changes in the fuel feeding systems of engines are necessary [122]. Furthermore, No [123] reported that a blend containing 20% v/v of SVO and diesel does not require any alterations to the marine engine systems. Additionally, it was found that pre-heating SVO at temperatures ranging from 55 to 85 °C allows for an increase in the percentage of SVO in the blend by 30% to 60% v/v without requiring changes in engine structures. Blin et al. [79] proposed that for drop-in usage of SVO in ships, a dual injection system should be used, where diesel would be injected at the start of the engine, and once it warmed up, SVO would be injected. The implementation of SVO as a marine fuel demands the development of a bunkering infrastructure [62], as well as further testing and refinement, leading to an assumed TRL regarding the use of the biofuel of 5.

HVO exhibits the potential to serve as a viable substitute for marine diesel, owing to its similar characteristics and compatibility with conventional ignition engines [123]. Currently, HVO is undergoing tests in the transport sector. Notably, numerous experiments have been conducted involving trucks and cars utilizing HVO either as a drop-in fuel or as a component in the fuel blend. These tests have been carried out in diverse countries, including Germany, Canada, the United States, Finland, and Sweden. One particularly significant test took place in the city of Alberta, Canada, demonstrating HVO's capability to function efficiently even in extremely cold temperatures reaching as low as -44 °C. However, despite investigations in road transportation, there was no documented record of HVO being tested in ships until the year 2022 [103]. Therefore, HVO emerges as the alternative marine fuel in this study, imposing the least modifications for its implementation in existing fleet and bunkering infrastructure. However, there exist certain barriers to the widespread adoption of HVO in the maritime sector, such as limited production capacity and high pricing, along with competition from the road and air sector [21]. To overcome these challenges and establish HVO as a viable marine fuel, further comprehensive studies and research are vital to assuring an assumed TRL of 5.

Concerning its utilization in marine engines, Chong and Bridgewater [90] stated that the blend of HPO with diesel and alcohol should not exceed 40% v/v. There is an emerging prospect that HPO may serve as a replacement for heavy oil in the future. However, its widespread adoption requires further research and comprehensive testing [104]. As a result of its early stage of development, HPO has been classified as having a low maturity level, specifically, a TRL of 2.

In July 2023, methanol became the fuel for 25 ships worldwide, and 127 terminals were successfully supplying ships with this fuel [119]. As previously mentioned, the technologies and procedures for using methanol as a marine fuel and for bunkering applications were established and are regulated by the IMO and classification societies. According to the report from the ABS [32], methanol-burning engines utilizing high-pressure diesel combustion processes have been made available by the manufacturers MAN and Wärtsilä. Moreover, methanol has been transported in chemical carriers for several decades and is also utilized by offshore support vessels (OSVs) and platform supply vessels (PSVs) for the offshore industry [32], facilitating its widespread adoption as a marine fuel. Due to these favourable factors and the potential for rapid integration into the marine fleet, methanol was estimated to possess high potential for widespread use in the short term. Nevertheless, the major source of methanol production is fossil-based (coal or natural gas) [124], presenting an obstacle to the widespread adoption of renewable methanol for maritime transport applications. As a result, the technological readiness level assigned to methanol as a marine fuel is TRL 7, indicating an advanced stage of technological development and readiness for practical implementation, yet renewable production still demands further expansion.

Ammonia currently benefits from an established supply chain network primarily catered to its use in the chemical industry [73] with efficient transportation via ships worldwide. The MAN dual-fuel engine, originally designed to operate with methanol and diesel, can also be adapted to use ammonia as an alternative fuel, provided certain modifications are made to the feed system's pressure [23]. As a result, the technologies, materials, and procedures necessary for its application are well-known within the industry. Nonetheless, further adaptation and development are required to utilize ammonia as a marine fuel [113]. The use of this fuel would face competition from the chemical sector and encounter challenges such as high toxicity and the technology's premature stage for integration into engines and fuel cells. Consequently, in order for ammonia to attain full commercial viability in the long term, further technological advancement is required, and as a result, the assumed TRL for ammonia is 5.

3.6. Summary of Results

In Table 4, a comparison between fuels is summarized by topics: energy density compared with HFO, bunkering readiness, material compatibility, storage tanks, engine feed, engine option, safety, and TRL.

Table 4. Summary of the comparison between fuels.

Criteria	LNG	Biodiesel	SVO	HVO	HPO	Methanol	Ammonia
Energy density HFO/fuel	1.91	1.21	1.19–1.08	1.15	1.19	2.47	2.81
Bunkering readiness	Already worldwide established	Adaptation to biodiesel properties, narrow shaped tanks, constant cleaning	Procedures are similar to HFO bunkering	Procedures are similar to MDO bunkering	Urge of development all bunkering process	Under establishment, ventilation reinforcement	Ammonia bunkering is already performed in the chemical industry
Material compatibility	Aluminium and stainless steel	Stainless steel or zinc reinforcement	Stainless or mild steel if coated with zinc silicate	No changes are needed	Stainless steel	Stainless or austenitic manganese steel	Stainless steel
Storage tanks	Double-walled, cryogenic storage (−162°), 10 bar pressure, inert	Isolated from machinery	Isolated from machinery, coated with vegetable oil inert material	Constant maintenance to avoid water contamination	Isolated from machinery, coated with biomass oil inert material	Double-walled, detection system to leakages	Double-walled, isolated from machinery, pressure of 8.6 bar
Engine feed	Double-walled, Ventilation reinforcement, 10 bar feed pressure	Filtering, constant maintenance	Pre-heating (67 to 78 °C), filtering, constant maintenance	No changes are needed	Pre-heating, piping designed to not block solid particles, filtering	Double-walled, ventilation reinforcement, pressure of 10 bar	Double-walled, ventilation reinforcement
Engine option	Dual fuel	Diesel engine	Diesel engine	Diesel engine	Diesel engine	Dual fuel	Fuel cell (dual-fuel is also an option)
Safety	Cryogenic and flammable	Low temperature use restricted due to low pour point, low toxicity	Low toxicity	Low toxicity	Low toxicity	Highly toxic and flammable	Highly toxic and flammable
TRL	9	7	5	5	2	7	5

4. Case Study

The Brazilian maritime sector has a fleet of approximately 2700 vessels [52] and more than 380 ports or terminals [125]. According to ANTAQ (Agência Nacional de Transportes Aquaviários) [52], long-haul navigation accounts for the highest cargo and travel movement, indicating the significant flow of Brazilian trade goods with foreign countries. Cabotage has some heavily travelled routes, such as Santos to Pecém, which is mainly focused on container transportation. However, this type of freight represents roughly one-third of the cargo and travel compared with deep-sea navigation. Concerning the energy transition of the maritime sector, the Brazilian Ministry of Mines and Energy (MME) initiated a program in 2012 aimed at the promotion of sustainable technologies applicable to all modes of transportation, particularly marine transport [126].

4.1. Main Port Profiles and Future Hubs

Brazilian port facilities exhibiting higher activity rates, as determined using the 2021 cargo movement data, namely, Ponta da Madeira, Santos, Tubarão, Angra dos Reis, São Sebastião, Paranaguá, Açu, Itaguaí, Itaqui, and Ilha da Guaíba [52], can be identified as primary hotspots for the transition of the Brazilian maritime transportation sector. Furthermore, ports and terminals with registered bunkering or movement of alternative fuels as cargo, meaning there is infrastructure in place to handle the loading or unloading of selected fuels, should also be accounted for. Finally, there are also ports that exhibit planned implementation of infrastructure dedicated to the bunkering of alternative fuels. Figure 3 summarizes Brazilian port information, classified according to the previously mentioned criteria.

Figure 3. High-activity and alternative fuels available, available handling infrastructure, and planned ports and terminals.

Regarding bunkering, in July 2023, an agreement was finalized with ports and companies within the Brazilian maritime sector, with the primary objective of promoting the utilization of alternative fuels in ships [127]. Given the limited number of Brazilian ports equipped with the necessary infrastructure for bunkering non-conventional fuels, such initiatives are of utmost importance in stimulating the transformation of Brazil's maritime infrastructure. As exposed in Figure 3, notably, the ports located in Santos, Rio Grande, Paranaguá and Salvador possess infrastructure for ammonia bunkering, whereas the facilities in Santos and Paranaguá are additionally equipped for methanol bunkering [119].

Figure 3 also shows ports and terminals that have the infrastructure to handle SVO and biodiesel. Since 2013, biodiesel has been transported by ships departing from various ports in Brazil, namely, Belém, Itacoatiara, Itaituba, Manaus, Paranaguá, Porto Velho, and Rio Grande [52]. Additionally, ANTAQ [52] completes the transportation of vegetable oils (specifically, palm and soybean) using specific Brazilian ports, including Barcarena, Belém, Manaus, Paranaguá, Porto Velho, Santos, Recife, Rio de Janeiro, Rio Grande, and Santarém. This indicates the existence of adequate infrastructure to handle the bunkering of vegetable oils and their derivatives at major ports throughout Brazil.

Furthermore, with regard to forthcoming adaptations, Paranaguá port has undertaken plans to construct infrastructure to facilitate LNG bunkering, with the projected beginning of operations in 2025 [128]. Simultaneously, the port is also actively investigating the implementation of a biodigester plant dedicated to the production of biomethane, which can be liquefied and turned into a green alternative to LNG [129]. Parallelly, in 2021, Pecém port created a proposal for the establishment of a hydrogen hub in its facilities [130]. This strategic move holds the potential to equip the ports with a dedicated infrastructure for the transportation and handling of hydrogen. As outlined earlier, hydrogen handling demands liquefaction and pressurization to optimize storage, along with precise conditions for loading and unloading operations [131]. Consequently, the procedures governing the handling of hydrogen closely mirror those already used for LNG and ammonia, rendering the port susceptible to the bunkering procedures of the aforementioned fuels.

The port of Açu also has plans to enable the bunkering of not only hydrogen but also ammonia. In partnership with the oil company Shell, the port authority is arranging the establishment of a facility dedicated to the production of the aforementioned fuels, along with the development of the necessary supply infrastructure [132]. Similarly, the port of Suape is also engaged in ongoing projects for the production of green hydrogen and ammonia [133].

The selected ports were also examined in terms of cargo movement, main products handled, and destinations. Table 5 displays their main compiled data.

Table 5. Total cargo movement (in millions of metric tonnes) in 2021, main products, and destinations departing from each analysed port.

Port	Cargo Movement (10^6 Metric-Ton)	Main Products	Main Destinations
Açu	39.0	Oil and derivatives, containers, cooper, iron and steel	Suape, Madre de Deus, Santos, Rio de Janeiro, Vitória
Angra dos Reis	29.3	Iron and steel, oil and derivatives	Alexandria (Egipt), Mersin (Turkey), Kabil (Indonesia), Qingdao (China), Aratu
Belém	2.6	Containers, oil and derivatives, corn, general cargo	Manaus, Barcarena, Fortaleza, Madre de Deus, Santarém
Guaíba	26.3	Iron ore, wood, cellulose pulp	Rio de Janeiro, Rio Grande, Port Talbot (Wales), Ijmuiden and Rotterdam (the Netherlands)
Itacoatiara	7.0	Soy, soy oil, ethanol, fossil fuels, oil and derivatives	Fortaleza, Manaus, Itaqui
Itaguaí	46.9	Containers	Santos, Imbituba, Suape, Callao (Peru), Rotterdam (the Netherlands)
Itaituba	6.1	Oil and derivatives, corn, soy	Belém, Manaus, Porto Velho, Santarém, Santana
Itaqui	20.3	Oil and derivatives, containers, ethanol, chemical products	Belém, Aratu, Fortaleza, Santos, Suape
Manaus	6.0	Oil and derivatives, containers, general cargo	Belém, Fortaleza, Santos, Suape, Itacoatiara

Table 5. *Cont.*

Port	Cargo Movement (10^6 Metric-Ton)	Main Products	Main Destinations
Paranaguá	32.6	Containers, oil and derivatives, chemical products, wheat	Belém, Fortaleza, Santos, Suape, Itaguaí
Pecém	10.4	Containers, iron and steel, oil and derivatives, manganese	Los Angeles (USA), Manaus, Cubatão, Brownsville (USA), Santos
Ponta da Madeira	186.6	Iron ore	Qingdao (China), Labuan (Malaysia), Kwangyang (Korea), Sohrar (Oman), Pecém
Porto Velho	14.2	Soy, corn, containers, general cargo	Santarém, Itacoatiara, Belém, Long Beach (USA), Montoir De Bretagne (France)
Recife	0.3	Sugar, salt, oil and derivatives, fossil fuels	Dubai (UAE), Fernando de Noronha, Baltimore (USA), Barra Do Riacho, Douala (Cameroon)
Rio Grande	20.0	Soy, containers, wood, fertilizers	Tanger (Morocco), Pecém, Antwerpen (Belgiun), Porto Alegre, Dafeng (China)
SãoSebastião	12.6	Oil and derivatives, sugar	Singapore, Qingdao (China), Manaus, Itaqui, Itacoatiara
Salvador	4.5	Oil and derivatives, cellulose pulp, containers	Vila do Conde, Belém, São Sebastião, Changshu (China), Santos
Santarém	6.5	Oil and derivatives, soy, corn, fertilizers	Itaituba, Algete and Barcelona (Spain), Belém, Rotterdam (the Netherlands)
Santos	99.1	Soy, oil and derivatives, soy oil, containers	Anshan, Koh Sichang (China), Bandar Khomeini (Iran), Singapore, São Sebastião
Suape	11.8	Oil and derivatives, containers, sugar, ethanol	Singapore, Manaus, Fortaleza, Itaqui, Santos
Tubarão	62.7	Iron ore, soy	Tangshan, Qingdao and Rizhao (China), Labuan (Malaysia), Rio de Janeiro

Data from ANTAQ [52].

One important outlook of the analysis of main Brazilian ports is that shipping is focused on bulk and container products. Routes are diverse, yet most of the cargo movements are concentrated in international destinations, confirming the importance of long-haul navigation to Brazil's economy. China is the busiest destination for Brazilian exports, mainly due to iron ore, soy, corn, oil, and containers [52]. Another output is the high activity in the Brazilian north region, mostly in the Legal Amazon Area. Ports such as Ponta da Madeira, Manaus, Belém, Porto Velho, and Santarém heavily contribute to local shipping.

Considering cargo movement and the potential conversion of ports for the bunkering of alternative fuels, it can be concluded that ports characterized by high cargo movement—herein presumed to be ports sustaining an annual cargo movement greater than 10 million tonnes—alongside a diverse product flow, encompassing a minimum of four distinct products categories, and consequently having a varied array of types of ships docked, are more acceptable for implementation as multi-fuel hubs. The ports satisfying these criteria, as listed in Table 5, encompass Açu, Itaqui, Paranaguá, Porto Velho, Rio Grande, Santos, and Suape.

Additionally, ports that envision the integration of infrastructure designed to enable the provision of two or more alternative fuel bunkering exhibit heightened precedence in relation to the establishment of multi-fuel hubs. Ports that have handled any of the analysed fuels as cargo also meet this criterion. Specifically, as Figure 3show, these ports are Açu, Manaus, Paranaguá, Porto Velho, Santos, Suape, and Rio Grande.

Taking into account the two abovementioned criteria, our analysis delineates the following ports as possessing the potential to serve as a multi-fuel hub: Açu, Paranaguá, Porto Velho, Rio Grande, Santos, and Suape.

Conversely, ports such as Ponta da Madeira, Itaguaí, and Tubarão, distinguished by substantial cargo movement, although with a concentrated product range, are assessed to be prone to experiencing a more restricted bunkering of alternative fuels. In other words, these ports are better suited to the bunkering of a particular alternative fuel, considering factors such as the final destinations of the product's fuel availability, and even the local production disposal of alternative fuels.

4.2. Fleet and Cargo Profile: Challenges and Progress in Conversion to Alternative Fuel Use

In 2023, the Brazilian ship fleet recorded an average age of approximately 19.5 years. Support vessels, despite being smaller, stand out due to their significant quantity, representing 90% of the fleet. Port support vessels account for 73% of this total, while maritime support vessels represent 27% [52]. Among the ships with the highest gross tonnage, bulk carriers and container ships are highlighted. Based on ANTAQ [52], Table 6 displays the products transported, age, and average deadweight tonnage (DWT), along with the number of ships, for the types of vessels with the highest average DWT in the Brazilian fleet.

Table 6. Products transported, average age, deadweight tonnage, and number of ships of the main Brazilian ship types.

Ship Type	Products Transported	Average Age (2023)	Average DWT	Fleet Size
Tanker	Crude oil and derivatives	10	89,054	54
Bulk	Dry bulk	15	57,007	21
Container	Container	13	45,009	33
Chemical tanker	Chemical products	18	26,234	8
Pipe laying support vessel (PLSV)	Offshore pipes	9	10,661	8
Subsea equipment support vessel	Subsea equipment	15	7570	2
LPG tanker	Liquefied petroleum gas	11	5481	8
Liquefied gas tanker	Liquefied gases	13	5455	11

[52].

Given that the typical lifespan of a ship is 30 years [50], it can be concluded that the highlighted types of vessels exhibit a residual lifespan of no less than 12 years, a scenario particularly applicable to the chemical tanker fleet. Therefore, replacement of the existing fleet due to the end of its lifetime remains an impractical course of action for a short period. In this regard, a priority arises to optimize the ship retrofits required for the adoption of alternative fuels.

LPG and liquefied gas tankers are notably suited to embrace the utilization of liquefied and pressurized fuels, namely, LNG, ammonia, and methanol. This advantage stems from the existing infrastructure designed for the storage and management of these fuels, which leads to a simplified conversion than other vessels.

Chemical tankers are also more suitable for ammonia and methanol. These fuels are flammable, demanding ships to be meticulously constructed and operated with intensified attention to potential incidents concerning the cargo [134]. This condition particularly applies to chemical ships, easing the adaptation to the use of the aforesaid fuels.

Tanker ships also exhibit a notable advantage in terms of adaptability due to their operation with fuel as cargo. However, changes in the entire infrastructure, encompassing storage tanks, fuel feeding and engines, are imperative. Given their intrinsic lack of operational experience with liquefaction and extreme pressurization, these vessels are better suited to undergo conversion for the utilization of other fuels, preferably having higher readiness levels, such as biodiesel, SVO and HVO. An analogous circumstance applies to the remaining selected types of vessels, given their inherent limitation of lacking experience in the handling of fuel as cargo.

Concerning the current stage of fuel usage, in 2022, Bunker One, a Danish bunkering company actively engaged in operations along the Brazilian coast, entered into a collaborative partnership with the Federal University of Rio Grande do Norte to conduct experimental trials on a fuel blend composed of HFO and 7% *v/v* biodiesel. These trials are specifically focused on tugboats operating within the area of the Port of Rio de Janeiro, with the aim of gathering valuable data on the performance and suitability of this mixture in the maritime context [135]. Petrobras has undertaken the implementation of a fuel blend consisting of 90% HFO and 10% biodiesel in an LPG tanker, with the primary objective of conducting a comprehensive analysis of its performance characteristics and identifying any potential logistics challenges that may arise. The dedicated Research Laboratories at Petrobras conducted testing and assessment of this fuel mixture in January 2023, observing that its integration necessitates no modifications to the existing maritime infrastructure [136]. In July 2023, the company made an announcement regarding its plans to conduct additional tests on vessels using a blend of 24% v/v of biodiesel [137]. Additionally, the company is actively investing in and establishing the development of large-scale production of HVO within its refineries [138].

As previously mentioned, companies linked to the maritime and energy sectors have taken the lead in the effort to introduce alternative fuels into vessels. Apart from these companies, governmental and regulatory bodies must be prepared to assume a pivotal role in facilitating the transition of the maritime sector [7]. Their contribution encompasses measures targeted not only at facilitating fuel production but also at proposing the conversion of marine fleet and port infrastructure. The actions of governments, such as those in Norway, range from setting more ambitious targets relative to those defined by IMO, directing mandatory percentages of biofuels within maritime fuel blends, to instituting fiscal incentives for enterprises that champion the utilization of alternative fuels [139], present examples that Brazil could consider to follow.

4.3. Thermal Stability of Fuels in the Main Routes

In terms of the thermal stability of the selected fuels, as highlighted in Sections 3.1 and 3.6, biodiesel exhibits a low pour point compared to traditional marine fuels and other alternative fuels. This particular property restricts its widespread usage in regions characterized by low temperatures or during cold seasons [45]. Given the routes departing from the main Brazilian ports, displayed in Table 4, and global historical average temperatures across various regions [140], it can be concluded that international routes transiting through South Africa, Europe, the United States, and North Asia demand the use of distinct fuels from biodiesel during periods of low temperature.

4.4. Fleet Profile: Loss of Cargo Space

Shipping companies, particularly those specializing in long-haul navigation, are continuously in search of strategies to optimize the allocation of cargo freight, aiming to maximize its utilization during a voyage. This pursuit explains the quest for achieving economies of scale in bulk shipping [141], whose vessels have progressively larger cargo capacities. For instance, standard dry bulk carriers have reached a capacity of 400,000 DWT with the deployment of Valemax vessels, the regular ships for the Ponta da Madeira to Qingdao iron ore route [142]. As clarified in Section 3, the adoption of alternative fuels brings a consequential requirement for increased storage tank volume due to the relatively lower energy density in contrast to conventional fuels. This decrease in space availability, particularly seen in the cases of LNG, ammonia, and methanol, is set to decrease the allocation of cargo space [46]. Given the substantial reliance on bulk shipping in the Brazilian context, this loss of cargo space emerges as a considerable barrier to the effective use of alternative fuels. In response to this challenge, Lindstad et al. [143] proposed some initiatives aimed at mitigating the loss of cargo space, including the increase in maximum draught and length of vessels. In the short term, however, this loss of space tends to be solved with more ships [144].

5. Conclusions

This study reviewed and summarized the major changes required for ports and long-distance large cargo ships to store, feed, and use alternative fuels. Considering the focus on fuel usage, this work did not encompass aspects related to the production chain, such as feedstock diversity. Therefore, no distinctions were made between fossil, bio, and e-fuels.

The handling, bunkering, and usage of alternative fuels must deal with: (i) the low energy density of fuels compared with HFO, particularly, LNG, ammonia, and methanol, leading to a loss in cargo space; (ii) the need for liquefaction (LNG) and/or pressurization (ammonia and methanol) of fuels to optimize storage or enable proper fuel feeding; (iii) the use of different materials such as stainless and mild steel in storage tanks and fuel feeding systems; (iv) the requirement for double-walled storage tanks and fuel feeding systems, as observed in the cases of LNG, ammonia, and methanol; (v) the need for enhanced precautions to prevent water contamination, particularly to biofuels usage; (vi) the high toxicity of fuels, notably, ammonia and methanol, which require extra ventilation inside ships; (vii) thermal stability issues impacting biodiesel utilization, particularly in extreme low temperatures; and (ix) modifications in engine fuel feeding and ignition (biofuels), adjustments for dual-fuel (LNG and methanol), and substitution for fuel cells (ammonia).

It is worth noting that although economic factors are not discussed in this study, they represent a challenge for the marine and academic communities, as evidenced by the research conducted by DNV GL [57], Xing et al. [40], UMAS [145], Bilgili [46], and Carvalho et al. [49]. Alternative fuels are still costlier than fossil fuels due to their more expensive production and capital and the operational cost of vessels, especially ammonia and hydrogen [46]. Economic competitiveness will be unreachable without actions from stakeholders to enhance alternative fuel usage, such as incentives, national and regional policies, and carbon pricing [9].

While the demand for alternative fuels is increasing, further advancement is necessary to significantly broaden the array of options. While fuels like LNG and methanol are already in use on specific vessels, fuels like HPO and SVO are still in the experimental stage. This posed challenges when reviewing technical and scientific literature related to these emerging fuel options.

The conducted case study underscored the feasibility of single or multi-fuel bunkering within the main Brazilian ports by indicating the main products and routes and the prospective development of alternative bunkering infrastructure within each port studied. Ports such as Açu, Paranaguá, Porto Velho, Rio Grande, Santos, and Suape exhibit potential for accommodating multi-fuel bunkering, while Ponta da Madeira, Itaguaí, and Tubarão tend to accommodate single-fuel bunkering.

Concerning the Brazilian fleet, given the limited number of alternative fuel trials within the country, the analysis int this study was conducted by evaluating vessel types requiring fewer adaptations for the utilization of alternative fuels. Due to the operational characteristics of ships, LPG and liquefied gas tankers are leading the way in terms of conversion to utilize fuels like LNG, ammonia, and methanol. A similar trend is observed for chemical vessels, which are more suitable for conversion to ammonia and methanol use, as well as tanker ships, which hold potential for the use of fuels such as biodiesel, SVO, and HVO. In the pursuit of establishing a fleet powered by alternative fuels, stakeholders may adopt diverse strategies, including the establishment of more ambitious targets, mandatory incorporation of biofuels in blends, and fiscal incentives promoting the integration of alternative fuels in their fleets. The analysis of these different strategies should be deepened in further studies. Further studies could also widen our analysis to other types of ships, for example, those that are more appropriate for hydrogen and electric batteries, such as ferryboats, offshore support vessels, etc. Additionally, the implementation of alternative fuel bunkering in a specific port can be a factor in reducing port congestion. Shipping companies struggle to avoid queuing in port areas and search for alternatives to avoid it, such as adapting shipping routes and destination ports in order to diminish cost losses [146]. The impact of a wide network of alternative fuel bunkering ports can be evaluated in future studies targeting not only port congestion reduction but also energy inflation reduction, which are intrinsically related [98].

Author Contributions: Conceptualization, H.W., C.R.P.B. and A.S.; methodology, H.W.; formal analysis, H.W., E.M.-C. and A.S.; investigation, H.W.; writing—original draft preparation, H.W.; writing—review and editing, E.M.-C. and A.S. All authors have read and agreed to the published version of the manuscript.

Funding: This research received no external funding.

Institutional Review Board Statement: Not applicable.

Informed Consent Statement: Not applicable.

Data Availability Statement: Not applicable.

Acknowledgments: We thank the Brazilian agencies CNPq and CAPES for their support on the early stages of this study.

Conflicts of Interest: The authors declare no conflict of interest.

Notes

1. Net zero emissions are achieved when human caused GHG emissions are balanced globally by human induced removals of CO_2 on a global scale during a defined period [10].

References

1. Longarela-Ares, Á.; Calvo-Silvosa, A.; Pérez-López, J.B. The Influence of Economic Barriers and Drivers on Energy Efficiency Investments in Maritime Shipping, from the Perspective of the Principal-Agent Problem. *Sustainability* **2020**, *12*, 7943. [CrossRef]
2. United Nations. *The First Global Integrated Marine Assessment*; Cambridge University Press: Cambridge, UK, 2017; ISBN 9781316510018.
3. Faber, J.; Hanayama, S.; Zhang, S.; Pereda, P.; Comer, B.; Hauerhof, E.; van der Loeff, W.S.; Smith, T.; Zhang, Y.; Kosaka, H.; et al. *Fourth IMO GHG Study 2020*; IMO: London, UK, 2020.
4. Smith, T.W.P.; Jalkanen, J.P.; Anderson, B.A.; Corbett, J.J.; Faber, J.; Hanayama, S.; O'Keeffe, E.; Parker, S.; Johansson, L.; Aldous, L.; et al. *Third IMO GHG Study 2014*; IMO: London, UK, 2014.
5. Nepomuceno de Oliveira, M.A.; Szklo, A.; Castelo Branco, D.A. Implementation of Maritime Transport Mitigation Measures According to Their Marginal Abatement Costs and Their Mitigation Potentials. *Energy Policy* **2022**, *160*, 112699. [CrossRef]
6. Lagouvardou, S.; Psaraftis, H.N.; Zis, T. A Literature Survey on Market-Based Measures for the Decarbonization of Shipping. *Sustainability* **2020**, *12*, 3953. [CrossRef]
7. Bouman, E.A.; Lindstad, E.; Rialland, A.I.; Strømman, A.H. State-of-the-Art Technologies, Measures, and Potential for Reducing GHG Emissions from Shipping—A Review. *Transp. Res. Part D Transp. Environ.* **2017**, *52*, 408–421. [CrossRef]
8. Maritime Knowledge Centre; TNO; TU Delft. *Final Report—Framework CO_2 Reduction in Shipping*; Maritime Knowledge Centre: Wageningen, The Netherlands, 2017.
9. Serra, P.; Fancello, G. Towards the IMO's GHG Goals: A Critical Overview of the Perspectives and Challenges of the Main Options for Decarbonizing International Shipping. *Sustainability* **2020**, *12*, 3220. [CrossRef]
10. Intergovernmental Panel on Climate Change. *Global Warming of 1.5 °C*; Intergovernmental Panel on Climate Change: Geneva, Switzerland, 2018; ISBN 9781009157940.
11. IMO. 2023 IMO Strategy on Reduction of GHG Emissions from Ships. Available online: https://wwwcdn.imo.org/localresources/en/OurWork/Environment/Documents/annex/2023IMOStrategyonReductionofGHGEmissionsfromShips.pdf (accessed on 1 August 2023).
12. IMO. *Adoption of the Initial IMO Strategy on Reduction of GHG Emissions from Ships and Existing IMO Activity Related to Reducing GHG Emissions in the Shipping Sector*; IMO: London, UK, 2018.
13. DNV. *GL Maritime Forecast To 2050*; DNV: Bayrum, Norway, 2019.
14. IPCC. *Climate Change 2014—Synthesys Report*; IPCC: Geneva, Switzerland, 2015; ISBN 978-92-9169-143-2.
15. Bengtsson, S.; Andersson, K.; Fridell, E. A Comparative Life Cycle Assessment of Marine Fuels: Liquefied Natural Gas and Three Other Fossil Fuels. *Proc. Inst. Mech. Eng. Part M J. Eng. Marit. Environ.* **2011**, *225*, 97–110. [CrossRef]
16. IMO. *MEPC.1/Circ.795/Rev.6: Unified Interpretations to MARPOL Annex VI*; IMO: London, UK, 2022.
17. Mærsk Mc-Kinney Møller Center. *Using Bio-Diesel Onboard Vessels: An Overview of Fuel Handling and Emission Management Considerations*; Mærsk Mc-Kinney Møller Center: København, Denmark, 2023.
18. ABS. *Biofuels as Marine Fuel*; ABS: Houston, TX, USA, 2021.
19. Florentinus, A.; Hamelinck, C.; van den Bos, A.; Winkel, R.; Maarten, C. *Potential of Biofuels for Shipping*; Ecofys: Utrecht, The Netherlands, 2012.
20. IEA. Biofuels for the Marine Shipping Sector: An Overview and Analysis of Sector Infrastructure, Fuel Technologies and Regulations. Available online: https://www.ieabioenergy.com/wp-content/uploads/2018/02/Marine-biofuel-report-final-Oct-2017.pdf (accessed on 25 November 2020).

21. Winnes, H.; Fridell, E.; Hansson, J.; Jivén, K. *Biofuels for Low Carbon Shipping*; IVL Swedish Environmental Research Institute: Stockholm, Sweden, 2019.
22. Laursen, R.; Barcarolo, D.; Patel, H.; Dowling, M.; Penfold, M.; Faber, J.; Király, J.; van der Ven, R.; Pang, E.; van Grinsven, A. *Update on Potential of Biofuels in Shipping*; EMSA: Lisbon, Portugal, 2022.
23. ABS. *Ammonia as Marine Fuel*; ABS: Houston, TX, USA, 2020.
24. ABS. *Hydrogen as Marine Fuel*; ABS: Houston, TX, USA, 2021.
25. Ash, N.; Scarbrough, T. *Sailing on Solar. Could Green Ammonia Decarbonise International Shipping?* Environmental Defense Fund: London, UK, 2019.
26. Earl, T.; Ambel, C.C.; Hemmings, B.; Gilliam, L.; Abbasov, F.; Officer, S. *Roadmap to Decarbonising European Shipping*; Transport & Environment: Brussels, Belgium, 2018.
27. Hansson, J.; Brynolf, S.; Fridell, E.; Lehtveer, M. The Potential Role of Ammonia as Marine Fuel-Based on Energy Systems Modeling and Multi-Criteria Decision Analysis. *Sustainability* **2020**, *12*, 3265. [CrossRef]
28. ABS. *LNG as Marine Fuel*; ABS: Houston, TX, USA, 2021.
29. Boulougouris, E.K.; Chrysinas, L.E. *LNG Fueled Vessels Design Training*; University of Strathclyde: Glasgow, UK, 2015.
30. Ge, J.; Wang, X. Techno-Economic Study of LNG Diesel Power (Dual Fuel) Ship. *WMU J. Marit. Aff.* **2017**, *16*, 233–245. [CrossRef]
31. Ellis, J.; Tanneberger, K. *Study on the Use of Ethyl and Methyl Alcohol as Alternative Fuels in Shipping*; EMSA: Lisbon, Portugal, 2015.
32. ABS. *Methanol as Marine Fuel*; ABS: Houston, TX, USA, 2021.
33. MAN Diesel & Turbo. *Using Methanol Fuel in the MAN B&W ME-LGI Series*; MAN Energy Solutions: Augsburg, Germany, 2014.
34. Rachow, M.; Loest, S.; Bramastha, A.D. Analysis of the Requirement for the Ships Using Methanol as Fuel. *Int. J. Mar. Eng. Innov. Res.* **2018**, *3*, 58–68. [CrossRef]
35. Müller-Casseres, E.; Carvalho, F.; Nogueira, T.; Fonte, C.; Império, M.; Poggio, M.; Wei, H.K.; Portugal-Pereira, J.; Rochedo, P.R.R.; Szklo, A.; et al. Production of Alternative Marine Fuels in Brazil: An Integrated Assessment Perspective. *Energy* **2021**, *219*, 119444. [CrossRef]
36. Tanzer, S.E.; Posada, J.; Geraedts, S.; Ramírez, A. Lignocellulosic Marine Biofuel: Technoeconomic and Environmental Assessment for Production in Brazil and Sweden. *J. Clean. Prod.* **2019**, *239*, 117845. [CrossRef]
37. Carvalho, F.; Portugal-pereira, J.; Junginger, M.; Szklo, A. Biofuels for Maritime Transportation: A Spatial, Techno-economic, and Logistic Analysis in Brazil, Europe, South Africa, and the Usa. *Energies* **2021**, *14*, 4980. [CrossRef]
38. Brynolf, S.; Taljegard, M.; Grahn, M.; Hansson, J. Electrofuels for the Transport Sector: A Review of Production Costs. *Renew. Sustain. Energy Rev.* **2018**, *81*, 1887–1905. [CrossRef]
39. Brynolf, S.; Fridell, E.; Andersson, K. Environmental Assessment of Marine Fuels: Liquefied Natural Gas, Liquefied Biogas, Methanol and Bio-Methanol. *J. Clean. Prod.* **2014**, *74*, 86–95. [CrossRef]
40. Xing, H.; Stuart, C.; Spence, S.; Chen, H. Alternative Fuel Options for Low Carbon Maritime Transportation: Pathways to 2050. *J. Clean. Prod.* **2021**, *297*, 126651. [CrossRef]
41. Harahap, F.; Nurdiawati, A.; Conti, D.; Leduc, S.; Urban, F. Renewable Marine Fuel Production for Decarbonised Maritime Shipping: Pathways, Policy Measures and Transition Dynamics. *J. Clean. Prod.* **2023**, *415*, 137906. [CrossRef]
42. Bengtsson, S.; Fridell, E.; Andersson, K. Environmental Assessment of Two Pathways towards the Use of Biofuels in Shipping. *Energy Policy* **2012**, *44*, 451–463. [CrossRef]
43. Gilbert, P.; Walsh, C.; Traut, M.; Kesieme, U.; Pazouki, K.; Murphy, A. Assessment of Full Life-Cycle Air Emissions of Alternative Shipping Fuels. *J. Clean. Prod.* **2018**, *172*, 855–866. [CrossRef]
44. Huang, J.; Fan, H.; Xu, X.; Liu, Z. Life Cycle Greenhouse Gas Emission Assessment for Using Alternative Marine Fuels: A Very Large Crude Carrier (VLCC) Case Study. *J. Mar. Sci. Eng.* **2022**, *10*, 1969. [CrossRef]
45. Mohd Noor, C.W.; Noor, M.M.; Mamat, R. Biodiesel as Alternative Fuel for Marine Diesel Engine Applications: A Review. *Renew. Sustain. Energy Rev.* **2018**, *94*, 127–142. [CrossRef]
46. Bilgili, L. A Systematic Review on the Acceptance of Alternative Marine Fuels. *Renew. Sustain. Energy Rev.* **2023**, *182*, 113367. [CrossRef]
47. Paulauskiene, T.; Bucas, M.; Laukinaite, A. Alternative Fuels for Marine Applications: Biomethanol-Biodiesel-Diesel Blends. *Fuel* **2019**, *248*, 161–167. [CrossRef]
48. Deniz, C.; Zincir, B. Environmental and Economical Assessment of Alternative Marine Fuels. *J. Clean. Prod.* **2016**, *113*, 438–449. [CrossRef]
49. Carvalho, F.; Müller-Casseres, E.; Poggio, M.; Nogueira, T.; Fonte, C.; Wei, H.K.; Portugal-Pereira, J.; Rochedo, P.R.R.; Szklo, A.; Schaeffer, R. Prospects for Carbon-Neutral Maritime Fuels Production in Brazil. *J. Clean. Prod.* **2021**, *326*, 129385. [CrossRef]
50. Gray, N.; McDonagh, S.; O'Shea, R.; Smyth, B.; Murphy, J.D. Decarbonising Ships, Planes and Trucks: An Analysis of Suitable Low-Carbon Fuels for the Maritime, Aviation and Haulage Sectors. *Adv. Appl. Energy* **2021**, *1*, 100008. [CrossRef]
51. Xing, H.; Stuart, C.; Spence, S.; Chen, H. Fuel Cell Power Systems for Maritime Applications: Progress and Perspectives. *Sustainability* **2021**, *13*, 1213. [CrossRef]
52. ANTAQ Anuário ANTAQ. Available online: http://web.antaq.gov.br/ANUARIO/ (accessed on 24 June 2021).
53. Ministério da Indústria, C.E.eS. Exportação e Importação Geral. Available online: http://comexstat.mdic.gov.br/pt/geral (accessed on 10 May 2023).

54. Tagomori, I.S.; Rochedo, P.R.R.; Szklo, A. Techno-Economic and Georeferenced Analysis of Forestry Residues-Based Fischer-Tropsch Diesel with Carbon Capture in Brazil. *Biomass Bioenergy* **2019**, *123*, 134–148. [CrossRef]
55. He, Z.; Cui, M.; Qian, Q.; Zhang, J.; Liu, H.; Han, B. Synthesis of Liquid Fuel via Direct Hydrogenation of CO_2. *Proc. Natl. Acad. Sci. USA* **2019**, *116*, 12654–12659. [CrossRef] [PubMed]
56. Carvalho, F.; Müller-Casseres, E.; Portugal-Pereira, J.; Junginger, M.; Szklo, A. Lignocellulosic Biofuels Use in the International Shipping: The Case of Soybean Trade from Brazil and the U.S. to China. *Clean. Prod. Lett.* **2023**, *4*, 100028. [CrossRef]
57. DNV. *GL Comparison of Alternative Marine Fuels*; DNV: Bayrum, Norway, 2019.
58. US Department of Energy. *Technology Readiness Assessment Guide*; US Department of Energy: Washington, DC, USA, 2011.
59. Mestemaker, I.B.T.W.; Van Biert, L.; Visser, I.K. The Maritime Energy Transition from a Shipbuilder's Perspective. In Proceedings of the International Naval Engineering Conference (iNEC 2020), Delft, The Netherlands, 6–8 October 2020.
60. Russo, D.; Dassisti, M.; Lawlor, V.; Olabi, A.G. State of the Art of Biofuels from Pure Plant Oil. *Renew. Sustain. Energy Rev.* **2012**, *16*, 4056–4070. [CrossRef]
61. Jiménez Espadafor, F.; Torres García, M.; Becerra Villanueva, J.; Moreno Gutiérrez, J. The Viability of Pure Vegetable Oil as an Alternative Fuel for Large Ships. *Transp. Res. Part D Transp. Environ.* **2009**, *14*, 461–469. [CrossRef]
62. Kesieme, U.; Pazouki, K.; Murphy, A.; Chrysanthou, A. Biofuel as an Alternative Shipping Fuel: Technological, Environmental and Economic Assessment. *Sustain. Energy Fuels* **2019**, *3*, 899–909. [CrossRef]
63. Mallouppas, G.; Yfantis, E.A. Decarbonization in Shipping Industry: A Review of Research, Technology Development, and Innovation Proposals. *J. Mar. Sci. Eng.* **2021**, *9*, 415. [CrossRef]
64. *British Standard ISO/FDIS 8217*; Petroleum Products—Fuels (Class F)—Specifications of Marine Fuels. ISO: Geneva, Switzerland, 2012; p. 42.
65. Taylor, D.A. *Introduction to Marine Engineering*, 2nd ed.; Elsevier, Ed.; Elsevier: Oxford, UK, 1996; ISBN 0-7506-2530-9.
66. Tiwari, A. Converting a Diesel Engine to Dual-Fuel Engine Using Natural Gas. *Int. J. Energy Sci. Eng.* **2015**, *1*, 163–169. [CrossRef]
67. Bhavani, K.; Murugesan, S. Diesel to Dual Fuel Conversion Process Development. *Int. J. Eng. Technol.* **2018**, *7*, 306–310. [CrossRef]
68. De-Troya, J.J.; Álvarez, C.; Fernández-Garrido, C.; Carral, L. Analysing the Possibilities of Using Fuel Cells in Ships. *Int. J. Hydrogen Energy* **2016**, *41*, 2853–2866. [CrossRef]
69. Burel, F.; Taccani, R.; Zuliani, N. Improving Sustainability of Maritime Transport through Utilization of Liquefied Natural Gas (LNG) for Propulsion. *Energy* **2013**, *57*, 412–420. [CrossRef]
70. Ushakov, S.; Lefebvre, N. Assessment of Hydrotreated Vegetable Oil (HVO) Applicability as an Alternative Marine Fuel Based on Its Performance and Emissions Characteristics. *SAE Int. J. Fuels Lubr.* **2019**, *12*, 4–12. [CrossRef]
71. Mendes, F.L.; da Silva, V.T.; Pacheco, M.E.; Toniolo, F.S.; Henriques, C.A. Bio-Oil Hydrotreating Using Nickel Phosphides Supported on Carbon-Covered Alumina. *Fuel* **2019**, *241*, 686–694. [CrossRef]
72. Ammar, N.R. An Environmental and Economic Analysis of Methanol Fuel for a Cellular Container Ship. *Transp. Res. Part D Transp. Environ.* **2019**, *69*, 66–76. [CrossRef]
73. Kim, K.; Roh, G.; Kim, W.; Chun, K. A Preliminary Study on an Alternative Ship Propulsion System Fueled by Ammonia: Environmental and Economic Assessments. *J. Mar. Sci. Eng.* **2020**, *8*, 183. [CrossRef]
74. Veses, A.; Martínez, J.D.; Callén, M.S.; Murillo, R.; García, T. Application of Upgraded Drop-in Fuel Obtained from Biomass Pyrolysis in a Spark Ignition Engine. *Energies* **2020**, *13*, 89. [CrossRef]
75. PETROBRAS. *Combustíveis Marítimos*; PETROBRAS: Rio de Janeiro, Brazil, 2021.
76. Kass, M.; Abdullah, Z.; Biddy, M.; Drennan, C.; Hawkins, T.; Jones, S.; Holladay, J.; Longman, D.; Newes, E.; Theiss, T.; et al. *Understanding the Opportunities of Biofuels for Marine Shipping*; Oak Ridge National Laboratory: Springfield, VA, USA, 2018.
77. Bart, J.C.J.; Palmeri, N.; Cavallaro, S. Vegetable Oil Formulations for Utilisation as Biofuels. *Biodiesel Sci. Technol.* **2010**, *4*, 114–129. [CrossRef]
78. Nguyen, T.A. *Sustainability Assessment of Inedible Vegetable Oil-Based Biodiesel for Cruise Ship Operation in Ha Long Bay, Vietnam*; Osaka Prefecture University: Gakuencho, Japan, 2017.
79. Blin, J.; Brunschwig, C.; Chapuis, A.; Changotade, O.; Sidibe, S.S.; Noumi, E.S.; Girard, P. Characteristics of Vegetable Oils for Use as Fuel in Stationary Diesel Engines—Towards Specifications for a Standard in West Africa. *Renew. Sustain. Energy Rev.* **2013**, *22*, 580–597. [CrossRef]
80. Ilo Bunker Fuel. Available online: https://www.inchem.org/documents/icsc/icsc/eics1774.htm (accessed on 14 September 2023).
81. Marathon Petroleum. *Marine Gas Oil: Safety Data Sheet*; Marathon Petroleum: Findlay, OH, USA, 2017.
82. Marathon Petroleum. *Biodiesel B100: Safety Data Sheet*; Marathon Petroleum: Findlay, OH, USA, 2016.
83. Suhartono; Suharto; Ahyati, A.E. The Properties of Vegetable Cooking Oil as a Fuel and Its Utilization in a Modified Pressurized Cooking Stove. *IOP Conf. Ser. Earth Environ. Sci.* **2018**, *105*, 012047. [CrossRef]
84. Chevron Phillips. *Light Pyrolysis Oil: Safety Data Sheet*; Chevron Phillips: Woodland, TX, USA, 2020.
85. Dan-Bunkering CCAI Calculator. Available online: https://dan-bunkering.com/Pages/Solutions/Tools-and-specs/Tool-ccai.aspx (accessed on 15 June 2020).
86. Schuller, O.; Whitehouse, S.; Poulsen, J.; Stoffregen, A.; Hengstler, J.; Kupferschmid, S. *Life Cycle GHG Emission Study on the Use of LNG as Marine Fuel*; Thinkstep AG: Stuttgart, Germany, 2019.
87. Torres-García, M.; García-Martín, J.F.; Jiménez-Espadafor Aguilar, F.J.; Barbin, D.F.; Álvarez-Mateos, P. Vegetable Oils as Renewable Fuels for Power Plants Based on Low and Medium Speed Diesel Engines. *J. Energy Inst.* **2020**, *93*, 953–961. [CrossRef]

88. Lin, C.Y. Strategies for Promoting Biodiesel Use in Marine Vessels. *Mar. Policy* **2013**, *40*, 84–90. [CrossRef]
89. Engman, M.A.; Hartikka, T.; Honkanen, M.; Kiiski, U.; Kuronen, M.; Mikkonen, S.; Oyj, N.O. *Hydrotreated Vegetable Oil (HVO)—Premium Renewable Biofuel for Diesel Engines*; Neste: Espoo, Finland, 2014.
90. Chong, K.J.; Bridgwater, A.V. Fast Pyrolysis Oil Fuel Blend for Marine Vessels. *Environ. Prog. Sustain. Energy* **2014**, *36*, 677–684. [CrossRef]
91. Veses, A.; López, J.M.; García, T.; Callén, M.S. Prediction of Elemental Composition, Water Content and Heating Value of Upgraded Biofuel from the Catalytic Cracking of Pyrolysis Bio-Oil Vapors by Infrared Spectroscopy and Partial Least Square Regression Models. *J. Anal. Appl. Pyrolysis* **2018**, *132*, 102–110. [CrossRef]
92. Huang, Y. *Conversion of a Pilot Boat to Operation on Methanol*; Chalmers University of Technology: Göteborg, Sweden, 2015.
93. Hansson, J.; Fridell, E.; Brynolf, S. *On the Potential of Ammonia as Fuel for Shipping—A Synthesis of Knowledge*; Lighthouse: Göteborg, Sweden, 2019.
94. Kay, C.; Ng, L.; Liu, M.; Siu, J.; Lam, L.; Yang, M. Accidental Release of Ammonia during Ammonia Bunkering: Dispersion Behaviour under the Influence of Operational and Weather Conditions in Singapore. *J. Hazard. Mater.* **2023**, *452*, 131281. [CrossRef]
95. Lewis, J. *Fuels Without Carbon: Prospects and the Pathway Forward for Zero-Carbon Hydrogen and Ammonia Fuel*; Clean Air Task Force: Boston, MA, USA, 2018.
96. Sheriff, A.M.; Tall, A. *Assessment of Ammonia Ignition as a Maritime Fuel, Using Engine Experiments and Chemical Kinetic Simulations*; World Maritime University: Malmo, Sweden, 2019.
97. ABS. *Low Carbon Shipping*; ABS: Houston, TX, USA, 2019.
98. Fan, H.; Enshaei, H.; Jayasinghe, S.G.; Tan, S.H.; Zhang, C. Quantitative Risk Assessment for Ammonia Ship-to-Ship Bunkering Based on Bayesian Network. *Process Saf. Prog.* **2022**, *41*, 395–410. [CrossRef]
99. Aneziris, O.; Koromila, I.; Nivolianitou, Z. A Systematic Literature Review on LNG Safety at Ports. *Saf. Sci.* **2020**, *124*, 104595. [CrossRef]
100. IMO. *Studies on the Feasability and Use of LNG as a Fuel for Shipping*; IMO: London, UK, 2016.
101. American Bureau of Shipping. *LNG Bunkering: Technical and Operational Advisory*; American Bureau of Shipping: Houston, Texas, USA, 2014.
102. FAO. *Code of Practice for the Storage and Transport of Edible Fats and Oils in Bulk*; FAO: Rome, Italy, 2015.
103. Neste. *Neste Renewable Diesel Handbook*; Neste: Espoo, Finland, 2020.
104. Lehto, J.; Oasmaa, A.; Solantausta, Y.; Kytö, M.; Chiaramonti, D. *Fuel Oil Quality and Combustion of Fast Pyrolysis Bio-Oils*; VTT Technical Research Centre of Finland P.O.: Espoo, Finland, 2013; ISBN 978-951-38-7930-3.
105. IMO. *MSC.1-Circ.1621: Interim Guidelines for the Safety of Ships Using Methyl/Ethyl Alcohol as Fuel*; IMO: London, UK, 2020.
106. Duong, P.A.; Ryu, B.R.; Song, M.K.; Van Nguyen, H.; Nam, D.; Kang, H. Safety Assessment of the Ammonia Bunkering Process in the Maritime Sector: A Review. *Energies* **2023**, *16*, 4019. [CrossRef]
107. Mokhatab, S.; Mak, J.Y.; Valappil, J.V.; Wood, D.A. *Handbook of Liquefied Natural Gas*; Elsevier: Oxford, UK, 2013; ISBN 9780124046450.
108. Mærsk Mc-Kinney Møller Center. *Preparing Container Vessels for Conversion to Green Fuels*; Mærsk Mc-Kinney Møller Center: København, Denmark, 2022.
109. American Bureau of Shipping. *Propulsion and Auxiliary Systems for Gas Fuelled Ships*; American Bureau of Shipping: Houston, Texas, USA, 2011.
110. Mathew, B.C.; Thangaraja, J. Material Compatibility of Fatty Acid Methyl Esters on Fuel Supply System of CI Engines. *Mater. Today Proc.* **2018**, *5*, 11678–11685. [CrossRef]
111. DNV. *GL Part 6 Additional Class Notations Chapter 2 Propulsion, Power Generation and Auxiliary Systems*; DNV: Bayrum, Norway, 2020.
112. Man Energy Solutions. *Engineering the Future Future in the: Two-Stroke Green-Ammonia Engine*; MAN Energy Solutions: Augsburg, Germany, 2019.
113. Laval, A.; Hafnia, H.T.; Vestas, S.G. Ammonfuel-an Industrial View of Ammonia as a Marine Fuel. *Hafnia* **2020**, *7*, 32–59.
114. Dincer, I.; Siddiqui, O. *Ammonia Fuel Cells*; Elsevier: Amsterdam, The Netherlands, 2020; Volume 1, ISBN 978-0-12-822825-8.
115. Erdemir, D.; Dincer, I. A Perspective on the Use of Ammonia as a Clean Fuel: Challenges and Solutions. *Int. J. Energy Res.* **2021**, *45*, 4827–4834. [CrossRef]
116. Kobayashi, H.; Hayakawa, A.; Somarathne, K.D.K.A.; Okafor, E.C. Science and Technology of Ammonia Combustion. *Proc. Combust. Inst.* **2019**, *37*, 109–133. [CrossRef]
117. Wolfram, P.; Kyle, P.; Zhang, X.; Gkantonas, S.; Smith, S. Using Ammonia as a Shipping Fuel Could Disturb the Nitrogen Cycle. *Nat. Energy* **2022**, *7*, 1112–1114. [CrossRef]
118. El-Gohary, M.M. The Future of Natural Gas as a Fuel in Marine Gas Turbine for LNG Carriers. *Proc. Inst. Mech. Eng. Part M J. Eng. Marit. Environ.* **2012**, *226*, 371–377. [CrossRef]
119. DNV. GL Alternative Fuels Insight. Available online: https://afi.dnvgl.com/ (accessed on 3 June 2023).
120. World Fuel Services. *ISO 8217 2017: Fuel Standard for Marine Distillate Fuels*; World Fuel Services: Miami, FL, USA, 2019.
121. Ogunkunle, O.; Ahmed, N.A. Exhaust Emissions and Engine Performance Analysis of a Marine Diesel Engine Fuelledwith Parinari Polyandra Biodiesel–Diesel Blends. *Energy Reports* **2020**, *6*, 2999–3007. [CrossRef]
122. Van Uy, D.; The Nam, T. Fuel Continuous Mixer—An Approach Solution to Use Straight Vegetable Oil for Marine Diesel Engines. *TransNav, Int. J. Mar. Navig. Saf. Sea Transp.* **2018**, *12*, 151–157. [CrossRef]

123. No, S.Y. Application of Hydrotreated Vegetable Oil from Triglyceride Based Biomass to CI Engines—A Review. *Fuel* **2014**, *115*, 88–96. [CrossRef]

124. Tabibian, S.S.; Sharifzadeh, M. Statistical and Analytical Investigation of Methanol Applications, Production Technologies, Value-Chain and Economy with a Special Focus on Renewable Methanol. *Renew. Sustain. Energy Rev.* **2023**, *179*, 113281. [CrossRef]

125. BNDES Portos. Available online: https://hubdeprojetos.bndes.gov.br/pt/setores/Portos#1 (accessed on 22 June 2021).

126. MME Programa Combustível Do Futuro. Available online: https://www.gov.br/mme/pt-br/programa-combustivel-do-futuro (accessed on 5 July 2023).

127. Rede Brasil Pacto Global Da ONU No Brasil Lança GT de Negócios Oceânicos Para Impulsionar a Descarbonização de Portos e Transportes Marítimos. Available online: https://www.pactoglobal.org.br/noticia/676/pacto-global-da-onu-no-brasil-lanca-gt-de-negocios-oceanicos-para-impulsionar-a-descarbonizacao-de-portos-e-transportes-maritimos (accessed on 30 July 2023).

128. Reis, R.G. Nimofast Signed a Partnership with Kanfer Shipping to Sell and Deliver LNG via Small-Scale LNG. Available online: https://www.nimofast.com/post/nimofast-signed-a-partnership-with-kanfer-shipping-to-sell-and-deliver-lng-via-small-scale-lng (accessed on 10 January 2023).

129. Portos do Paraná. *Relatório de 2021: Sustentabilidade*; Portos do Paraná: Paranaguá, Brazil, 2021.

130. Pecém Hub de Hidrogênio Verde No Complexo de Pecém. Available online: https://www.complexodopecem.com.br/hubh2v/ (accessed on 20 August 2023).

131. Smith, T.; Krantz, R.; Mouftier, L.; Christiansen, E.S. *Insight Briefing Series: Hydrogen as a Cargo*; Getting to Zero Coallition: Copenhagen K, Denmark, 2021.

132. Porto do Açu. *Relatório de Sustentabilidade 2021*; Porto do Açu: São João da Barra, Brazil, 2021.

133. Engie Porto de Suape Quer Construir Planta de Hidrogênio Verde Estimada Em US$ 3.5 Bi. Available online: https://www.alemdaenergia.engie.com.br/porto-de-suape-quer-construir-planta-de-hidrogenio-verde-estimada-em-us-35-bi/ (accessed on 20 August 2023).

134. Carlton, J. Ship Types, Duties, and General Characteristics. *Encycl. Marit. Offshore Eng.* **2018**, *01*, 1–20. [CrossRef]

135. UDOP. Bunker One Testará Biodiesel Em Embarcações No Brasil Em 2022. Available online: https://www.udop.com.br/noticia/2021/12/21/bunker-one-testara-biodiesel-em-embarcacoes-no-brasil-em-2022.html (accessed on 10 March 2023).

136. Petrobras Navio Da Transpetro Recebe Primeiro Abastecimento de Bunker Com Conteúdo Renovável. Available online: https://transpetro.com.br/transpetro-institucional/noticias/navio-da-transpetro-recebe-primeiro-abastecimento-de-bunker-com-conteudo-renovavel.htm (accessed on 11 March 2022).

137. EPBR. Petrobras Testa Combustível Marítimo Com 24% de Biodiesel. Available online: https://epbr.com.br/petrobras-testa-combustivel-maritimo-com-24-de-biodiesel/ (accessed on 15 July 2023).

138. Petrobras Vamos Ampliar Capacidade de Produção de Diesel Com Conteúdo Renovável Ainda Em 2023. Available online: https://petrobras.com.br/fatos-e-dados/vamos-ampliar-capacidade-de-producao-de-diesel-com-conteudo-renovavel-ainda-em-2023.htm (accessed on 20 July 2023).

139. Norwegian Government. *The Government's Action Plan for Green Shipping*; Norwegian Government: Oslo, Norway, 2019.

140. NASA NASA Data Access Viewer. Available online: https://power.larc.nasa.gov/data-access-viewer/ (accessed on 8 August 2023).

141. Goulielmos, A.M. The "Kondratieff Cycles" in Shipping Economy since 1741 and till 2016. *Mod. Econ.* **2017**, *8*, 308–332. [CrossRef]

142. VALE Terminal Marítimo Ponta Da Madeira Completa 35 Anos Com Novo Patamar de Embarque. Available online: https://www.vale.com/pt/w/terminal-maritimo-ponta-da-madeira-completa-35-anos-com-novo-patamar-de-embarque (accessed on 5 August 2023).

143. Lindstad, E.; Polić, D.; Rialland, A.; Sandaas, I.; Stokke, T. Decarbonizing Bulk Shipping Combining Ship Design and Alternative Power. *Ocean Eng.* **2022**, *266*, 112798. [CrossRef]

144. Law, L.C.; Mastorakos, E.; Evans, S. Estimates of the Decarbonization Potential of Alternative Fuels for Shipping as a Function of Vessel Type, Cargo, and Voyage. *Energies* **2022**, *15*, 7468. [CrossRef]

145. Lloyd's Register, UMAS. *Techno-Economic Assessment of Zero-Carbon Fuels*; Lloyd's Register: London, UK, 2020.

146. Meng, L.; Ge, H.; Wang, X.; Yan, W.; Han, C. Optimization of Ship Routing and Allocation in a Container Transport Network Considering Port Congestion: A Variational Inequality Model. *Ocean Coast. Manag.* **2023**, *244*, 106798. [CrossRef]

Journal of
Marine Science and Engineering

Article

Optimal Ship Deployment and Sailing Speed under Alternative Fuels

Haoqing Wang [1,†], Yuan Liu [2,*,†], Shuaian Wang [1,†] and Lu Zhen [3,†]

1   Department of Logistics and Maritime Studies, Faculty of Business, The Hong Kong Polytechnic University, Hong Kong, China
2   School of Economics and Management, Wuhan University, Wuhan 430072, China
3   School of Management, Shanghai University, Shanghai 200444, China
*   Correspondence: 2020301052156@whu.edu.cn
†   These authors contributed equally to this work.

Abstract: The European Union (EU) has implemented a sub-quota of 2% for renewable marine fuels to be utilized by vessels operating within its jurisdiction, effective starting from 2034. This progressive policy signifies a significant leap towards reducing carbon emissions and promoting sustainable development. However, it also presents notable challenges for shipping companies, particularly in terms of fuel costs. In order to support shipping companies in devising optimal strategies within the framework of this new policy, this study proposes a mixed-integer linear programming model. This model aims to determine the optimal decisions for fuel choice, sailing speed and the number of vessels on various routes. Furthermore, we showcase the adaptability of our model in response to fluctuations in fuel prices, relevant vessel costs, and the total fleet size of vessels. Through its innovative insights, this research provides invaluable guidance for optimal decision-making processes within shipping companies operating under the new EU policy, enabling them to minimize their total costs effectively.

Keywords: sustainable maritime transportation; green shipping; energy efficiency; sailing speed

Citation: Wang, H.; Liu, Y.; Wang, S.; Zhen, L. Optimal Ship Deployment and Sailing Speed under Alternative Fuels. *J. Mar. Sci. Eng.* **2023**, *11*, 1809. https://doi.org/10.3390/jmse11091809

Academic Editor: Claudio Ferrari

Received: 22 August 2023
Revised: 12 September 2023
Accepted: 13 September 2023
Published: 16 September 2023

1. Introduction

The global shipping industry plays a vital role in ensuring the movement of goods and commodities across the world [1–3]. However, traditional fuels used in this sector, primarily fossil fuels, have raised significant concerns due to their adverse environmental impacts, including carbon emissions and air pollution [4]. As societies worldwide strive to combat climate change and transition towards sustainable practices, the exploration and integration of renewable fuels have emerged as a pivotal solution for the maritime industry [5–7]. Renewable fuels, also known as biofuels or alternative fuels, are derived from organic matter such as plants, algae, and waste materials. Unlike traditional fossil fuels, these sources are considered sustainable as they can be replenished through natural processes, reducing dependency on finite resources [8]. The integration of renewable fuels into the maritime sector holds tremendous potential to greatly mitigate the environmental footprint of shipping operations while promoting sustainable development. Furthermore, the utilization of renewable energy sources aligns with international efforts to achieve the objectives of the Paris Agreement and the International Maritime Organization's (IMO) related emission reduction targets [9,10]. Understanding the significance of renewable fuels in maritime transportation is significant for guiding policy decisions and driving sustainable practices within the shipping industry.

According to [11], the EU has established a sub-quota of 2% for renewable marine fuels to be utilized by vessels operating within its jurisdiction, effective from the year 2034. This means that starting from 2034, at least 2% of the fuel used by vessels during voyages within the EU must be derived from renewable sources. Note that for these voyages linking the EU and non-EU areas, half of the fuel consumption generated is attributed to the EU. Undoubtedly, this exerts a crucial impact on the optimal decision-making of shipping companies.

In this paper, we put forth mathematical frameworks aimed at facilitating the attainment of optimal decision-making in the pursuit of cost minimization for shipping companies, while simultaneously ensuring compliance with the renewable fuel policy recently introduced by the EU. To the utmost extent of our understanding, our study represents a groundbreaking inquiry, uniquely incorporating this freshly established EU policy. Specifically, our investigation addresses the ensuing research inquiries:

1. What are the optimal fuel choices within different types of areas that result in the minimal overall expenditure while adhering to the renewable fuel target stipulated by EU policy?
2. What are the optimal sailing speeds, taking into account both traditional and renewable fuels, within the EU and non-EU areas, to minimize the aggregate expenses incurred by shipping companies in accordance with the recently proposed EU renewable fuel regulations?
3. What is the optimal number of deployed vessels in diverse shipping routes, including considerations for the chartering in or chartering out ships, that will culminate in the lowest total costs while simultaneously meeting the EU's 2% renewable fuel target?
4. How do the optimal fuel choices within the EU and non-EU areas, sailing speeds associated with different fuel types, as well as the optimal number of vessels to be equipped in each shipping route, including the consideration of vessels chartering in or chartering out, vary in response to fluctuations in fuel prices, relevant vessel costs, and the total fleet size of vessels?

To tackle the four research inquiries mentioned above, we initially introduce a sophisticated mixed integer nonlinear optimization (MINLP) model characterized by its intricate nature and challenging problem-solving complexity. Subsequently, we employ advanced mathematical techniques to convert the MINLP model into a mixed-integer linear programming (MILP) formulation. This transformation enables the utilization of readily available optimization solvers for solving the MILP model. Lastly, we conduct a series of comprehensive experiments and thorough sensitivity analyses to assess the model's performance in response to various parameter variations.

1.1. Literature Review

We review two streams of literature closely related to our study: (i) the renewable fuel in shipping; (ii) the optimal decisions in shipping.

1.1.1. The Renewable Fuel in Shipping

In the level of policy, many countries and organizations have shed light on the transformative potential of renewable fuels and their role in shaping the future of sustainable shipping [12]. In detail, at the IMO, there are ongoing deliberations regarding the implementation of Market-Based Measures (MBMs) to enhance the economic desirability of low- and zero-carbon fuels compared to fossil fuels. The proposed MBMs encompass a range of approaches, including the imposition of global levies on marine fuels, the establishment of an Emissions Trading System (ETS), and the exploration of other hybrid mechanisms [13]. Moreover, the FuelEU Maritime Directive exemplifies a joint endeavor to decrease the greenhouse gas (GHG) of energy employed aboard maritime vessels, elucidating a profound ambition of achieving a 75% reduction by 2050 [14]. This multipronged objective can be actualized by means of actively advocating and extensively embracing renewable and low-carbon fuels [14]. In addition, a comprehensive overhaul of the European Energy Taxation Directive (EU ETD) looms on the horizon, enacting thresholds for the minimum taxation rates on bunker fuel. Noteworthy is the fact that fossil fuels bear the brun of the highest minimum tax rate, valiantly standing at EUR 10.75/GJ, while renewable fuels are subject to the lowest rate, valiantly striking at EUR 0.15/GJ [15]. In addition, the revised renewable energy directive (RED II) propounds an exacting target of a minimum 13% reduction in GHG intensity within the transport sector by 2030 [15]. Moreover, it establishes sub-targets, tailored specifically to advance biofuels and non-biological renewable fuels [16].

In the aspect of renewable fuel utilization, various studies have explored the potential of different alternatives. Hydrogen fuel, due to its notable efficiency and environmental advantages, stands as a paramount sustainable fuel option. In practical applications, methanol is commonly employed to generate hydrogen. The alcohol–hydrogen fuel mixed by a series of high-temperature catalytic reactions can be used in the shipping industry [17]. Moreover, biodiesel emerges as a sustainable energy source that exhibits exceptional biodegradability and possesses low toxicity, making it an excellent alternative to fossil fuels across various sectors [18]. Additionally, a comprehensive study [19] focused on evaluating the environmental and economic aspects of using Liquefied Natural Gas (LNG) as a ship fuel. The results from this extensive life-cycle analysis and cost assessment demonstrated that LNG generates lower greenhouse gas emissions, up to 28% less than heavy fuel oil, while producing slightly higher nitrogen oxide emissions. From the prospective of technology, LNG, Liquefied Petroleum Gas (LPG), and methanol are considered more mature technologies, along with biodiesel, hydrogen, and ammonia fuels, which exhibit greater potential for future development [10]. However, the present utilization of a diverse array of low-carbon fuels presents certain drawbacks that impede their immediate substitution for conventional fossil fuels. Addressing this challenge entails the establishment of multi-period energy planning, which not only facilitates adaptation to demand fluctuations and evolving emission restrictions at distinct stages, but also aids in energy projection and future investment strategizing [20]. What is more, from an economic standpoint, the dual fuel propulsion system stands out as the most viable and cost-efficient alternative for container vessels in the present. By ingeniously alternating between conventional fossil fuels and renewable energy sources, this system optimally adheres to forthcoming regulatory requirements and standards [21].

1.1.2. The Optimal Decisions in Shipping

In the realm of maritime transportation, the pursuit of attaining the optimal trajectory, sailing speed, fuel consumption, and other pertinent optimal decisions constitutes the focus in cost minimization. Within this context, the exploration of optimal decisions has been undertaken to curtail operational expenses [22–25].

Several studies have been conducted to optimize sailing speed and fuel consumption for the purpose of reducing shipping costs. For the relationship between fuel consumption and sailing speed, the prevailing consensus posits that there exists a cubic relationship between fuel consumption and sailing speed [26]. Laporte et al. [27] conducted a comprehensive study on speed optimization problems within the liner transportation network, taking into account time window constraints. Arijit et al. [28] and other researchers extensively investigated the implementation of a slow streaming policy as a means to minimize fuel consumption and associated costs. The adoption of slower speeds has proven effective in reducing fuel consumption; however, it can also result in delivery delays. Moreover, reducing the sailing speed necessitates deploying more vessels for the service, thereby augmenting the operational expenses of the ships [29]. In addressing this trade-off, He et al. [30] proposed a speed optimization problem where a set of speeds for each segment of a given route is determined to minimize costs while considering time windows and speed limits for each segment. Aydin et al. [31] conducted an assessment of speed optimization in liner shipping, incorporating stochastic port times into their model. Their objective was to minimize total fuel consumption while upholding schedule reliability. Li et al. [32] proposed an innovative approach that combines the optimization of sailing routes and speeds, taking into account the interrelation between them as well as environmental factors.

In the context of model formulation involving multiple optimal decision variables, Lu et al. [33] introduce a dual-objective optimization model for ship speed, with a focus on the influence of the ECA control area, berthing costs, and AMP systems. By employing the multi-objective PSO (Particle Swarm Optimization) algorithm, the study aimed to identify the Pareto solution set that simultaneously minimizes ship operating costs and carbon emissions. Subsequently, the TOPSIS (Technique for Order Preference by Similarity to Ideal Solution) algorithm was applied to evaluate and extract the optimal compromise solution from the Pareto set. Mahsa et al. [34] formulated a bi-objective programming model that addresses the joint optimization of ship scheduling and sailing speeds for a specific shipping service. This model takes into account factors such as non-identical stream flow speeds. Iris et al. [35] focused their research on the well-known berth allocation problem (BAP). They proposed a novel mathematical formulation that extends the traditional BAP to encompass multiple ports within a shipping network. Furthermore, their model implementation demonstrates that precise speed discretization can lead to significantly improved economic and environmental outcomes. Dulebenets et al. [36] introduced an MINLP model aimed at minimizing the overall cost of liner shipping routes. Meng et al. [37] evaluated an optimization model for fuel consumption and ship speed in liner transport, accounting for deviations from the scheduled speed. Lee et al. [38] investigated a dynamic planning model to determine fuel consumption under uncertain fuel costs. The trade-off between vessel quantity and speed was further elucidated in [39–41], emphasizing the significance of selecting the optimal number of ships for liner shipping. Sheng [40] investigated optimal vessel speed and fleet size considerations for services operating within emission control areas (ECAs).

In general, prior research efforts have predominantly focused on optimizing decisions related to sailing speed, fuel consumption, deployed vessels across different routes, and fleet size. These studies aim to provide more effective decision support for shipping companies by jointly considering multiple optimal decision variables. However, limited attention has been paid to the fuel selection and corresponding sailing speeds for different legs of the journey. Our research endeavors to address this gap by integrating the latest renewable fuel policies implemented by the EU with existing research findings. In this paper, we propose an MILP model, which aims to determine the optimal fuel selection, speed decisions, and the corresponding deployment quantity of vessels, simultaneously achieving cost minimization and promoting environmental conservation and sustainable development.

1.2. Research Contributions

1. Theoretical contributions. The present research addresses a significant research gap by focusing on the optimal selection of fuel, sailing speed, and the number of ships under the newly proposed EU policy. Notably, the existing literature has overlooked this specific aspect. To the best of our knowledge, this study represents the pioneering effort in establishing mathematical models aimed at minimizing the overall costs incurred by shipping companies while considering the implications of the new EU policy. The proposed approach employs an MILP model to determine the optimal decisions for shipping companies. By conducting rigorous experiments and sensitivity analyses, this study yields specific solutions while evaluating the impacts of various parameters.
2. Practical contributions. This research contributes valuable practical insights into the development of optimal strategies for shipping companies to effectively minimize costs and ensure compliance with the new EU emissions policy. The obtained results possess practical implications for fostering sustainable growth within the shipping industry, facilitating its alignment with environmental regulations. Notably, the proposed mathematical model serves as a decision-making tool, providing shipping companies with a framework to navigate the challenges presented by the new EU emissions policy effectively.

The rest of the paper is organized as follows. Section 2 describes the research problem in detail and develops the mathematical model. Section 3 proposes solution methods for addressing the initial proposed model. Section 4 conducts experiments and sensitivity analysis. Finally, conclusions are drawn in Section 5.

The main notations used in this study are summarized in Table 1.

Table 1. Notations.

Sets	
M	Set of shipping routes, $m \in M$
I^0	Set of the legs within the non-EU areas, $i \in I^0$
I^1	Set of the legs within linking the EU and non-EU areas, $i \in I^1$
I^2	Set of the legs within the EU areas, $i \in I^2$
Parameters	
c^m	The operating cost of a vessel on each liner shipping route, m
c_{out}	The revenue of chartering out a vessel
c_{in}	The cost of chartering in a vessel
K	The total fleet size of vessels
T	The service frequency of each shipping route
μ_1	The traditional fuel price per tonne
μ_2	The renewable fuel price per tonne
b^m	The total berthing time at all ports on liner shipping route m, $m \in M$
L_i^m	The total length of the shipping route m within the i type of areas, $i = 0, 1, 2$, $m \in M$
v_0^{tm}	The sailing speed on liner shipping route m within non-EU areas with traditional fuel, $m \in M$
v_1^{tm}	The sailing speed on liner shipping route m within linking the EU and non-EU areas with traditional fuel, $m \in M$
v_2^{tm}	The sailing speed on liner shipping route m within EU areas with traditional fuel, $m \in M$
v_0^{rm}	The sailing speed on liner shipping route m within non-EU areas with renewable fuel, $m \in M$
v_1^{rm}	The sailing speed on liner shipping route m within linking the EU and non-EU areas with renewable fuel, $m \in M$
v_2^{rm}	The sailing speed on liner shipping route m within EU areas with renewable fuel, $m \in M$
v_{min}	The minimum sailing speed
v_{max}	The maximum sailing speed
J	The integer used to discretize sailing speed, $j = 0, 1, ..., J$
Function	
$f(v^3)$	Fuel consumption rate at the sailing speed of v
Decision variables	
x^m	The number of vessels to be deployed on each liner shipping route m, $m \in M$
x_{out}	The number of chartering out vessels
x_{in}	The number of chartering in vessels
l_i^{tm}	The sailing length in the i type of areas on each liner shipping route m with traditional fuel, $i = 0, 1, 2$, $m \in M$
l_i^{rm}	The sailing length in the i type of areas on each liner shipping route m with renewable fuel, $i = 0, 1, 2$, $m \in M$
z_i^{tmj}	Binary decision variable that equals 1 if vessels sail using traditional fuel on liner shipping route m in the i type of areas with speed v_j and 0 otherwise, $i = 0, 1, 2$, $m \in M$, $j = 0, 1, ..., J$
z_i^{rmj}	Binary decision variable that equals 1 if vessels sail using renewable fuel on liner shipping route m in the i type of areas with speed v_j and 0 otherwise, $i = 0, 1, 2$, $m \in M$, $j = 0, 1, ..., J$

2. Problem Description and Model Development

In this study, we consider vessel company decisions on the sailing speeds within different areas and the choice of fuel, i.e., the choice of traditional fuel and renewable fuel. Our primary objective is to assist vessel companies in achieving cost savings while simultaneously meeting the EU's 2% renewable target requirement.

We consider a shipping network with multiple routes; M denotes the set of shipping routes, $m \in M$. The liner shipping route $m \in M$ calls at several ports in Europe and Asia. The vessel company serves these routes by employing vessels equipped with dual-fuel engines, which allows them the operation on either traditional fuel or renewable fuel (though the latter option incurs higher costs). There are three distinct types of voyages in the considered container liner service network: (1) voyages within non-EU areas; (2) voyages linking the EU and non-EU areas; (3) voyages within the EU areas. We use v_0^{tm}, v_1^{tm}, and v_2^{tm} to denote the sailing speed using traditional fuel in voyages within non-EU areas, linking the EU and non-EU areas, and within the EU areas, respectively. And v_0^{rm}, v_1^{rm}, and v_2^{rm} denote the sailing speed using renewable fuel in voyages within non-EU areas, linking the EU and non-EU areas, and within the EU areas, respectively. Referring to [42], the fuel consumption and sailing speed have a cubic relationship. Therefore, we use av^3 to denote vessel fuel consumption (note that v could be v_0^{tm}, v_1^{tm}, v_2^{tm}, v_0^{rm}, v_1^{rm}, and v_2^{rm}). We use l_0^{tm}, l_1^{tm}, and l_2^{tm} to denote the total length of voyages using traditional fuel within non-EU areas, linking the EU and non-EU areas, and within the EU areas, respectively. And l_0^{rm}, l_1^{rm}, and l_2^{rm} denote the total length of voyages using renewable fuel within non-EU areas, linking the EU and non-EU areas, and within the EU areas, respectively.

The vessel company possesses a total fleet size of K vessels. Additionally, the company has the option to charter out any surplus vessels, generating additional income represented by c_{out}. On the other hand, if the existing fleet of K vessels is insufficient to serve the shipping network, the company can charter in additional vessels at a fee denoted by c_{in}. These chartering options provide the company with flexibility in managing its vessel resources and optimizing its operations based on the specific demands and requirements of the shipping network. We have $c_{in} > c_{out}$. The operating cost of vessels on each liner shipping route m is c^m.

In order to minimize the overall total costs, the vessel company must make strategic decisions regarding the following key factors:

1. The number of vessels to be deployed on each liner shipping route m, denoted by x^m;
2. The number of chartering in or chartering out vessels, denoted by x_{in} and x_{out}, respectively;
3. The sailing speed on each liner shipping route m with different types of fuel, i.e., v_0^{tm}, v_1^{tm}, v_2^{tm}, v_0^{rm}, v_1^{rm}, and v_2^{rm};
4. The sailing length on each liner shipping route m with different types of fuel, i.e., l_i^{tm} and l_i^{rm}, $i = 0, 1, 2$.

Additionally, the vessel company needs to meet the following constraints:

1. The EU's 2% renewable target requirement;
2. The service frequency of each shipping route m, denoted by T^m. That is, the time for finishing a single trip should be within T^m days;
3. The restriction of the number of vessels;
4. The restriction of vessel sailing speed.

The optimization model is formulated as follows:

[M1]

$$\min \sum_{m \in M} c^m x^m + c_{in} x_{in} - c_{out} x_{out} + \sum_{m \in M} \sum_{i=0}^{2} \left(\mu_1 \frac{l_i^{tm}}{v_i^{tm}} a (v_i^{tm})^3 + \mu_2 \frac{l_i^{rm}}{v_i^{rm}} a (v_i^{rm})^3 \right) \tag{1}$$

subject to

$$\frac{0.5a(v_1^{rm})^2 l_1^{rm} + a(v_2^{rm})^2 l_2^{rm}}{0.5a(v_1^{tm})^2 l_1^{tm} + a(v_2^{rm})^2 l_2^{tm} + 0.5a(v_1^{rm})^2 l_1^{rm} + a(v_2^{rm})^2 l_2^{rm}} \geq 0.02, \ m \in M, \tag{2}$$

$$\sum_{i=0}^{2} \left(\frac{l_i^{tm}}{v_i^{tm}} + \frac{l_i^{rm}}{v_i^{rm}} + b^m \right) \leq Tx^m, \quad m \in M, \tag{3}$$

$$\sum_{m \in M} x_m = K + x_{in} - x_{out}, \tag{4}$$

$$l_i^{tm} + l_i^{rm} = L_i^m, \quad i = 0, 1, 2, \ m \in M, \tag{5}$$

$$v_{\min} \leq v_i^{tm} \leq v_{\max}, \quad i = 0, 1, 2, \ m \in M, \tag{6}$$

$$v_{\min} \leq v_i^{rm} \leq v_{\max}, \quad i = 0, 1, 2, \ m \in M, \tag{7}$$

$$x^m \in Z_+, \quad m \in M, \tag{8}$$

$$x_{in} \in Z_+, \tag{9}$$

$$x_{out} \in Z_+, \tag{10}$$

$$l_i^{tm} \geq 0, \quad m \in M, \tag{11}$$

$$l_i^{rm} \geq 0, \quad m \in M. \tag{12}$$

Objective function (1) consists of four parts. Firstly, $\sum_{m \in M} c^m x^m$ calculates the total operation cost of vessels on m different liner shipping routes. Secondly, $c_{in} x_{in}$ represents the cost of chartering in additional vessels when the existing fleet of K vessels is insufficient to serve the shipping network; on the contrary, the third part, $c_{out} x_{out}$, is the additional revenue of chartering out surplus vessels. As Objective function (1) calculates the minimum overall total costs, we subtract this additional income. Fourthly, $\sum_{m \in M} \sum_{i=0}^{2} \left(\mu_1 \frac{l_i^{tm}}{v_i^{tm}} av_i^{tm3} + \mu_2 \frac{l_i^{rm}}{v_i^{rm}} av_i^{rm3} \right)$ represents the total fuel costs, including traditional fuel cost and renewable fuel cost, where μ_1 and μ_2 denote the traditional fuel and renewable fuel price, respectively. Constraint (2) meets the requirement of at least 2% of the renewable fuel utilized by vessels during voyages within the EU area starting from 2034. Constraints (3) restrict the service frequency of each shipping route m. Constraint (4) restricts the total number of vessels on m shipping routes, only involving the existing vessels and additional vessels that charter in or charter out, which is subtracted. Constraints (5) regulate the total voyage in every area on each shipping route m, consisting of the length of voyage using renewable fuel and traditional fuel. Constraints (6) and Constraints (7) give the domain of v_i^{tm} and v_i^{rm}, respectively, indicating the maximum and minimum of v_i^{rm} and v_i^{rm}. Constraint (8), Constraint (9) and Constraint (10) regulate the number of deployed vessels in route m, and the number of additional vessels should be positive integers. Constraints (11) and Constraints (12) restrict the length of each leg in different types of areas on m routes, and they should be positive. For parameters, $i = 0$ indicates areas within the EU; $i = 1$ represents the areas linking the EU areas and non-areas; $i = 2$ denotes areas within non-EU. μ_1 (USD/tonne) and μ_2 (USD/tonne) denote the fuel price of traditional fuel and renewable fuel. According to practice, $\mu_1 < \mu_2$. b^m represents the total berthing time at all ports. L_i^m denotes the total length of the shipping route m of different areas. The decision variables include v_i^{tm}, v_i^{rm}, l_i^{tm}, and l_i^{rm}.

3. Solution Methods

Model (M1) is hard to solve due to the operation of multiplying and dividing the decision variables. We next develop methods to address the nonlinear terms and transform (M1) into an MILP model, which improves computational efficiency.

We discretize sailing speed v by 0.01 knot. We define

$$J = \lfloor \frac{v_{\max} - v_{\min}}{0.01} \rfloor + 1, \tag{13}$$

and we set $j = 0, 1, ..., J$. Therefore, the sailing speed v can be discretized to v_j: $v_0 = v_{\min}$, $v_1 = v_{\min} + 0.01 \times 1$, $v_2 = v_{\min} + 0.01 \times 2$, ..., $v_J = \max\{v_{\max}, v_{\min} + 0.01 \times J\}$. We further adopt binary decision variables to indicate which discretized sailing speed is chosen on each shipping route m with different types of fuel. To be more specific, z_i^{tmj}, $i = 0, 1, 2$, $m \in M$, and $j = 0, ..., J$ denotes which sailing speed v_j is chosen on shipping route m on area i using traditional fuel; z_i^{rmj}, $i = 0, 1, 2$, $m \in M$, and $j = 0, ..., J$ denotes which sailing speed v_j is chosen on shipping route m on area i using renewable fuel. Therefore, we use the new binary decision variables z_i^{tmj} and z_i^{rmj} to replace v_i^{tm} and v_i^{rm}, and (M1) can be transformed into the following (M2).

[M2]

$$\min \sum_{m \in M} c^m x^m + c_{in} x_{in} - c_{out} x_{out} + a \sum_{m \in M} \sum_{i=0}^{2} \sum_{j=0}^{J} (\mu_1 l_i^{tm} z_i^{tmj} v_j^2 + \mu_2 l_i^{rm} z_i^{rmj} v_j^2) \tag{14}$$

subject to

$$\frac{0.5 \sum_{j=0}^{J} z_1^{rmj} v_j^2 l_1^{rm} + \sum_{j=0}^{J} z_2^{rmj} v_j^2 l_2^{rm}}{0.5 \sum_{j=0}^{J} z_1^{tmj} v_j^2 l_1^{tm} + \sum_{j=0}^{J} z_2^{tmj} v_j^2 l_2^{tm} + 0.5 \sum_{j=0}^{J} z_1^{rmj} v_j^2 l_1^{rm} + \sum_{j=0}^{J} z_2^{rmj} v_j^2 l_2^{rm}} \geq 0.02, \; m \in M, \tag{15}$$

$$\sum_{i=0}^{2} \sum_{j=0}^{J} \left(\frac{z_i^{tmj} l_i^{tm}}{v_j} + \frac{z_i^{rmj} l_i^{rm}}{v_j} + b^m \right) \leq T x^m, \; m \in M, \tag{16}$$

$$\sum_{m \in M} x_m = K + x_{in} - x_{out}, \tag{17}$$

$$l_i^{tm} + l_i^{rm} = L_i^m, \; i = 0, 1, 2, \; m \in M, \tag{18}$$

$$\sum_{j=0}^{J} z_i^{rmj} \leq 1, i = 0, 1, 2, \; m \in M, \tag{19}$$

$$\sum_{j=0}^{J} z_i^{tmj} \leq 1, i = 0, 1, 2, \; m \in M, \tag{20}$$

$$l_i^{tm} \leq W_i^m \sum_{j=0}^{J} z_i^{tmj}, i = 0, 1, 2 \; m \in M, \tag{21}$$

$$l_i^{rm} \leq W_i^m \sum_{j=0}^{J} z_i^{rmj}, i = 0, 1, 2 \; m \in M, \tag{22}$$

$$z_i^{tmj} \in \{0, 1\}, \; i = 0, 1, 2, \; m \in M, \; j = 0, \ldots, J, \tag{23}$$

$$z_i^{rmj} \in \{0, 1\}, \; i = 0, 1, 2, \; m \in M, \; j = 0, \ldots, J, \tag{24}$$

$$x^m \in Z_+, \ m \in M, \tag{25}$$

$$x_{in} \in Z_+, \tag{26}$$

$$x_{out} \in Z_+, \tag{27}$$

$$l_i^{tm} \geq 0, \ m \in M, \tag{28}$$

$$l_i^{rm} \geq 0, \ m \in M. \tag{29}$$

In Model (M2), we discretize sailing speed v to facilitate our research, improving computational efficiency. Specifically, we utilize v_i^{tmj} and v_i^{rmj} to represent the sailing speeds for different types of voyages using traditional fuel and renewable fuel, respectively. Furthermore, we set the value of W to L_i^m.

Broadly speaking, Model (M2) encompasses three categories of decision variables. The first category involves integer decision variables x_m, x_{in}, and x_{out}, which denote the number of vessels to be deployed on each liner shipping route m, as well as the quantities of vessels to be chartered in or out. The second category features binary decision variables z_i^{tmj} and z_i^{rmj}, comprising a total of $6mJ$ binary decision variables. When $z_i^{tmj} = 1$, the corresponding sailing speed v_i^{tmj} is selected for voyages utilizing traditional fuel, resulting in the adoption of traditional fuel consumption. On the other hand, when $z_i^{rmj} = 1$, the corresponding sailing speed v_i^{rmj} is chosen for voyages using renewable fuel. The renewable fuel and traditional fuel consumption are determined by expression av^3. Lastly, the third category consists of continuous variables l_i^{tm} and l_i^{rm}, totaling $6m$ continuous variables. By transforming Model (M1) into Model (M2), we can enhance computational efficiency.

Model (M2) is still hard to solve because of terms $l_i^{tm}z_i^{tmj}$ and $l_i^{rm}z_i^{rmj}$ in Objective function (14) and Constraints (16). As mentioned earlier, l_i^{tm}, l_i^{rm}, z_i^{tmj} and z_i^{rmj} are decision variables, so the multiplications of l_i^{tm} and z_i^{tmj}, l_i^{rm} and z_i^{rmj} lead to the formation of nonlinear terms within the objective function and constraint conditions, thereby transforming the model into an MINLP problem, which is inherently challenging to solve. As a nonlinear problem can encompass multiple feasible regions or sets of similar values for the decision variables that satisfy all constraints, each feasible region may contain multiple "peaks" (in maximization problems) or "valleys" (in minimization problems). Determining which peak is the tallest or which valley is the deepest lacks a general approach. Additionally, there can exist spurious peaks or valleys referred to as "saddle points." Due to these possibilities, nonlinear optimization methods offer limited guarantees in terms of identifying the "true" optimal solution. As a nonlinear problem can encompass multiple feasible regions or sets of similar values for the decision variables that satisfy all constraints, each feasible region may contain multiple "peaks" (in maximization problems) or "valleys" (in minimization problems). Determining which peak is the tallest or which valley is the deepest lacks a general approach. Additionally, there can exist spurious peaks or valleys referred to as "saddle points". Due to these possibilities, nonlinear optimization methods offer limited guarantees in terms of identifying the "true" optimal solution. We define $\eta_i^{tmj} = l_i^{tm}z_i^{tmj}$ and $\eta_i^{rmj} = l_i^{rm}z_i^{rmj}$. The following inequalities hold:

$$\eta_i^{tmj} \leq l_i^{tm}, \ m \in M, \ j = 0, \ldots, J, \ i = 0, 1, 2, \tag{30}$$

$$\eta_i^{tmj} \leq Q_i^m z_i^{tmj}, \ m \in M, \ j = 0, \ldots, J, \ i = 0, 1, 2, \tag{31}$$

$$\eta_i^{tmj} \geq l_i^{tm} - Q_i^m(1 - z_i^{tmj}), \ m \in M, \ j = 0,\dots,J, \ i = 0,1,2, \tag{32}$$

$$\eta_i^{tmj} \geq 0, \ m \in M, \ j = 0,\dots,J, \ i = 0,1,2, \tag{33}$$

$$\eta_i^{rmj} \leq l_i^{rm}, \ m \in M, \ j = 0,\dots,J, \ i = 0,1,2, \tag{34}$$

$$\eta_i^{rmj} \leq Q_i^m z_i^{rmj}, \ m \in M, \ j = 0,\dots,J, \ i = 0,1,2, \tag{35}$$

$$\eta_i^{rmj} \geq l_i^{rm} - Q_i^m(1 - z_i^{rmj}), \ m \in M, \ j = 0,\dots,J, \ i = 0,1,2, \tag{36}$$

$$\eta_i^{rmj} \geq 0, \ m \in M, \ j = 0,\dots,J, \ i = 0,1,2, \tag{37}$$

where Q_i^m represents an exceedingly large value that bounds the upper limit of l_i^{tm} and l_i^{rm}. We set Q_i^m to be equal to L_i^m. Typically, a smaller value of Q_i^m is preferred (while still ensuring the correctness of the model), as it usually leads to shorter computational time compared to a model with larger Q_i^m.

Regarding Constraints (30), since the binary variable z_i^{tmj} can take values of either 0 or 1, when z_i^{tmj} equals 1, Constraints (30) can be expressed as $l_i^{tmj} \leq L_i^m$, $m \in M$, $j = 0,\dots,J$, $i = 0,1,2$. This condition is evidently valid since l_i^{tm} is bounded by the total voyage length L_i^m. Conversely, when z_i^{tmj} equals 0, Constraints (30) become $0 \leq l_i^{tm}$, $m \in M$, $j = 0,\dots,J$, $i = 0,1,2$, which is also true.

Concerning Constraints (31), given the aforementioned definition, $\eta_i^{tmj} = l_i^{tm} z_i^{tmj}$, thus transforming Constraints (31) to $l_i^{tm} z_i^{tmj} \leq Q_i^m z_i^{tmj}$, $m \in M$, $j = 0,\dots,J$, $i = 0,1,2$. Consequently, Q_i^m becomes the upper bound for l_i^{tm}, equating to the total leg length L_i^m.

Regarding Constraints (32), we can discuss the case where z_i^{tmj} is equal to 1 or 0 separately. When z_i^{tmj} is equal to 1, Constraints (32) can be transformed into $l_i^{tmj} \geq l_i^{tm}$, $m \in M$, $j = 0,\dots,J$, $i = 0,1,2$. It is evident that this inequality holds. On the other hand, when z_i^{tmj} is equal to 0, the inequality becomes $0 \geq l_i^{tm} - Q_i^m$, $m \in M$, $j = 0,\dots,J$, $i = 0,1,2$, implying that the minimum value of Q_i^m is l_i^{tmj}, consistent with Constraints (31).

The validity of Constraints (33) is unquestionable, as both l_i^{tmj} and z_i^{tmj} are greater than or equal to 0. Similarly, Constraints (34)–(37) hold for the same reasons.

By Constraints (30)–(37), we can transform (M2) to the MILP model (M3).

[M3]

$$\min \sum_{m \in M} c^m x^m + c_{in} x_{in} - c_{out} x_{out} + a \sum_{m \in M} \sum_{i=0}^{2} \sum_{j=0}^{J} (\mu_1 \eta_i^{tmj} v_j^2 + \mu_2 \eta_i^{rmj} v_j^2), \tag{38}$$

subject to

$$0.5 \sum_{j=0}^{J} \eta_1^{rmj} v_j^2 + \sum_{j=0}^{J} \eta_2^{rmj} v_j^2 - 0.02 \left(0.5 \sum_{j=0}^{J} \eta_1^{tmj} v_j^2 + \sum_{j=0}^{J} \eta_2^{tmj} v_j^2 + 0.5 \sum_{j=0}^{J} \eta_1^{rmj} v_j^2 + \sum_{j=0}^{J} \eta_2^{rmj} v_j^2\right) \geq 0, \ m \in M, \tag{39}$$

$$\sum_{i=0}^{2} \sum_{j=0}^{J} \left(\frac{\eta_i^{tmj}}{v_j} + \frac{\eta_i^{rmj}}{v_j} + b^m\right) \leq T x^m, \ m \in M, \tag{40}$$

$$\sum_{m \in M} x_m = K + x_{in} - x_{out}, \tag{41}$$

$$l_i^{tm} + l_i^{rm} = L_i^m, \ i = 0,1,2, \ m \in M, \tag{42}$$

$$\sum_{j=0}^{J} z_i^{rmj} \leq 1, i = 0,1,2,\ m \in M, \tag{43}$$

$$\sum_{j=0}^{J} z_i^{tmj} \leq 1, i = 0,1,2,\ m \in M, \tag{44}$$

$$l_i^{tm} \leq W_i^m \sum_{j=0}^{J} z_i^{tmj}, i = 0,1,2,\ m \in M, \tag{45}$$

$$l_i^{rm} \leq W_i^m \sum_{j=0}^{J} z_i^{rmj}, i = 0,1,2,\ m \in M, \tag{46}$$

$$z_i^{tmj} \in \{0,1\},\ i = 0,1,2,\ m \in M,\ j = 0,...,J, \tag{47}$$

$$z_i^{rmj} \in \{0,1\},\ i = 0,1,2,\ m \in M,\ j = 0,...,J, \tag{48}$$

$$x^m \in Z_+,\ m \in M, \tag{49}$$

$$x_{in} \in Z_+, \tag{50}$$

$$x_{out} \in Z_+, \tag{51}$$

$$l_i^{tm} \geq 0,\ m \in M, \tag{52}$$

$$l_i^{rm} \geq 0,\ m \in M, \tag{53}$$

$$\eta_i^{tmj} \leq l_i^{tm},\ m \in M,\ j = 0,\ldots,J,\ i = 0,1,2, \tag{54}$$

$$\eta_i^{tmj} \leq Q_i^m z_i^{tmj},\ m \in M,\ j = 0,\ldots,J,\ i = 0,1,2, \tag{55}$$

$$\eta_i^{tmj} \geq l_i^{tm} - Q_i^m(1 - z_i^{tmj}),\ m \in M,\ j = 0,\ldots,J,\ i = 0,1,2, \tag{56}$$

$$\eta_i^{tmj} \geq 0,\ m \in M,\ j = 0,\ldots,J,\ i = 0,1,2, \tag{57}$$

$$\eta_i^{rmj} \leq l_i^{rm},\ m \in M,\ j = 0,\ldots,J,\ i = 0,1,2, \tag{58}$$

$$\eta_i^{rmj} \leq Q_i^m z_i^{rmj},\ m \in M,\ j = 0,\ldots,J,\ i = 0,1,2, \tag{59}$$

$$\eta_i^{rmj} \geq l_i^{rm} - Q_i^m(1 - z_i^{rmj}),\ m \in M,\ j = 0,\ldots,J,\ i = 0,1,2, \tag{60}$$

$$\eta_i^{rmj} \geq 0,\ m \in M,\ j = 0,\ldots,J,\ i = 0,1,2. \tag{61}$$

With the help of Constraints (30)–(37) and discretization, the original optimization model is transformed into an MILP programming model, which can be solved by the off-the-shelf optimization solvers, such as CPLEX and Gurobi. To validate our models, we design a shipping network involving four shipping routes for the experiment, setting the parameters, e.g., c^m, c_{in}, c_{out}, μ_1, and μ_2 according to practice.

4. Experiments

4.1. Experiment Settings

The experiments were run on a laptop computer equipped with 2.60 GHz of Intel Core i7 CPU and 16 GB of RAM, and Model (M3) was solved by Gurobi Optimizer 10.0.2 via Python API.

4.1.1. Selected Shipping Routes

We select four routes from Asia to northern Europe[1] to test the performance of Model (M3). Details are shown in Table 2.

Table 2. Summary of shipping routes.

Route ID	Port Rotation (City)
1	Tianjin → Dalian → Qingdao → Shanghai → Ningbo → Singapore→ Piraeus → Rotterdam → Hamburg → Antwerp → Shanghai → Tianjin
2	Busan → Ningbo → Shanghai → Yantian → Singapore → Algeciras → Dunkerque → Le Havre → Hamburg → Wilhelmshaven → Rotterdam → Port Klang → Busan
3	Shanghai → Ningbo → Xiamen → Yantian → Singapore → Felixstowe→ Zeebrugge → Gdansk → Wilhelmshaven → Singapore → Yantian → Shanghai
4	Qingdao → Shanghai → Ningbo → Yantian → Vung Tau → Singapore→ Rotterdam → Southampton → Antwerp → Le Harve → Jeddah → Singapore → Qingdao

4.1.2. Parameter Settings

We first set the values of parameters for drawing the basic results, and we conduct sensitivity analysis to examine the impacts of these parameters.

1. The operation cost c^m. Referring to [43], we first set c^m = USD 180,000 per week for a 5000-TEU (Twenty-foot Equivalent Unit) container ship.
2. The fee of chartering in a vessel c_{in}. Referring to [44], we set c_{in} = USD 120,000 per week.
3. The fee of chartering out a vessel c_{out}. Referring to [44], we set c_{out} = USD 100,000 per week.
4. The traditional fuel price μ_1. Referring to [45], we set μ_1 to be an average value of 600 (USD /tonne).
5. The renewable fuel price μ_2. Referring to [45], we set μ_2 to be an average value of 1000 (USD /tonne).
6. A company's total fleet size K. Referring to [46], we set K to be an average value of 60.
7. Referring to [43], We set $v_{\max}$ = 18 knots and $v_{\min} = 13$ knots.
8. Referring to [42], we set $f(v^3) = 0.00043 \times v^3$, $a = 0.00043$.

4.2. Basic Results

Based on the routes presented in Table 2 and the parameter settings, we conducted numerical experiments and obtained the results in Table 3. As outlined in Section 2, the decisions regarding vessel sailing speeds and fuel choices play a significant role. Therefore, we analyzed the sailing speeds with traditional and renewable fuels, as well as voyages in different types of areas. Specifically, during the course of the experiments, as we mentioned earlier, Model (M1) entails nonlinear terms and divisions between variables, rendering it a nonlinear model that poses significant challenges for solving. However, by discretizing the sailing speed and introducing variables z_i^{tmj} and z_i^{rmj}, η_i^{tmj} and η_i^{rmj}, we transformed the MINLP model (M1) into the MILP model (M3), facilitating the determination of optimal values during the experiments process. Hence, we employ Model (M3) to ascertain the optimal decisions for sailing speed, travel distance, fuel selection, and other relevant factors.

The optimal value of the objective function for Model (M3) is denoted as "OBJ," and distances are measured in nautical miles (nm). From Table 3, we observe that the number of deployed vessels is comparable across the four routes. However, Route 1 has slightly more ships (13) compared to the other routes, primarily due to its longer total voyage distance among the four routes. Additionally, the existing fleet of vessels is sufficient, leading to the chartering out of 11 ships. In general, vessels use traditional fuel in all types of areas, including those within the EU, those connecting the EU and non-EU regions, and non-EU areas. However, renewable fuel consumption is concentrated within the EU areas. As mentioned earlier, vessel companies must consider the EU's 2% renewable fuel target requirement. For voyages linking the EU and non-EU areas, half of the fuel consumption is attributed to the EU, while for voyages within the EU areas, all fuel consumption is attributed to the EU. To meet the 2% renewable fuel requirement at a lower cost, vessel companies prioritize using renewable fuel predominantly within the EU areas. This approach minimizes fuel consumption while still meeting the renewable target requirement, considering that renewable fuel prices are higher than traditional fuel prices. Additionally, a vessel exhibits a slower speed when using sustainable energy compared to its speed when utilizing traditional fuel.

Table 3. Basic results.

Route ID	Set of Legs	Total Distance (nm)	l_i^{tm} (nm)	l_i^{rm} (nm)	v_i^{tm} (knot)	v_i^{rm} (knot)	Number of Ships	OBJ ($)	x_{in}	x_{out}
1	I^0	3876.00	3876.00	0	14.78	NA	12			
	I^1	16,137.00	16,137.00	0	14.44	NA				
	I^2	3552.00	3294.65	257.35	14.12	13.62				
2	I^0	5089.00	5089.00	0	15.02	NA	11			
	I^1	15,020.00	15,020.00	0	14.99	NA				
	I^2	2269.00	2068.16	200.84	14.95	14.78		11,917,802.38	0	15
3	I^0	4885.00	4885.00	0	15.21	NA	11			
	I^1	16,704.00	16,704.00	0	15.18	NA				
	I^2	1736.00	1532.40	203.60	15.18	15.11				
4	I^0	5047.00	5047.00	0	15.27	NA	11			
	I^1	12,609.00	12,609.00	0	15.25	NA				
	I^2	4553.00	4332.35	220.65	15.21	15.11				

4.3. Sensitivity Analysis

The unit traditional fuel price and renewable fuel price may change as more policies are issued to promote the use of renewable fuel. In addition, in the basic analysis, some important parameters, e.g., the total fleet size of vessels, the operating cost per ship and the revenue of chartering out a vessel, are set to be deterministic. However, these parameters often fluctuate in real life. Therefore, sensitivity analyses on these parameters are conducted to investigate the influences of these parameters on the operation decisions. In the sensitivity analysis, the parameters are divided into three sorts. The first one is the fuel price, including traditional fuel price and renewable fuel price; the second one is the relevant cost of the vessels, involving the operating cost of a vessel, the revenue of chartering out a vessel as well as the cost of chartering in a vessel; the third one is the total fleet size of vessels.

4.3.1. Impact of the Fuel Price

Given the implementation of the EU's 2% renewable fuel policy, the selection and pricing of different fuel types have significantly impacted optimal operational decisions of vessel companies. Consequently, in this section, we specifically explore the effects of traditional and renewable fuels on such decision-making processes. In our initial analysis, the unit price of traditional fuel (μ_1) is established at 600 dollars per tonne. In sensitivity analysis, we set μ_1 varying between 500 and 800 dollars per tonne. The computational results are reported in Table 4.

According to the findings presented in Table 4, the objective value exhibits an upward trend as the price of traditional fuel increases. Furthermore, an increased allocation of vessels is observed across the four shipping routes, accompanied by a decrease in sailing speeds as traditional fuel prices soar. Given the cubic relationship between fuel consumption and sailing speed, it follows that higher vessel speeds result in greater fuel consumption. Consequently, in response to the rising unit price of traditional fuel, shipping companies are inclined to reduce sailing speeds to mitigate traditional fuel consumption and thereby achieve cost savings in their operations. Additionally, if the unit price of traditional fuel (μ_1) reaches excessively high levels, shipping companies may opt to deploy additional vessels to maintain the desired weekly service frequency.

In basic analysis, the unit price of renewable fuel (μ_2) is set at 1000 dollars per tonne. However, as stated in [21], the growing emphasis on green and sustainable development has led to the formulation of favorable policies aimed at promoting the utilization of renewable fuel. This suggests the potential future decline in renewable fuel prices. Thus, we set the range of μ_2 to span from 800 to 1000 dollars per tonne. Computational findings are presented in Table 5.

Regarding renewable fuel prices, similar patterns emerge as with traditional fuel prices, albeit to a lesser extent. Specifically, the total cost increases as the price of renewable fuel rises. However, due to the minimal proportion of renewable fuel in the overall fuel consumption, the impact of changing renewable fuel prices is relatively minor compared to the effects observed with traditional fuel.

Table 4. Impact of the unit price of traditional fuel.

μ_1 (USDton)	Route ID	Set of Legs	Total Distance (nm)	l_i^{tm} (nm)	l_i^{rm} (nm)	v_i^{tm} (knot)	v_i^{rm} (knot)	Number of Ships	OBJ (USD)	x_{in}	x_{out}
500	1	I^0	3876.00	3876.00	0	16.39	NA	11	11,024,279.65	0	16
		I^1	16,137.00	16,137.00	0	16.14	NA				
		I^2	3552.00	3308.22	243.78	15.60	15.60				
	2	I^0	5089.00	5089.00	0	15.12	NA	11			
		I^1	15,020.00	10,520.00	0	14.97	NA				
		I^2	2269.00	2046.60	222.40	14.94	14.02				
	3	I^0	4885.00	4885.00	0	15.24	NA	11			
		I^1	16,704.00	16,704.00	0	15.21	NA				
		I^2	1736.00	1516.89	219.11	14.85	14.53				
	4	I^0	5047.00	5047.00	0	15.30	NA	11			
		I^1	12,609.00	12,609.00	0	15.24	NA				
		I^2	4553.00	4334.62	218.38	15.20	15.18				

Table 4. *Cont.*

μ_1 (USD/ton)	Route ID	Set of Legs	Total Distance (nm)	l_i^{tm} (nm)	l_i^{rm} (nm)	v_i^{tm} (knot)	v_i^{rm} (knot)	Number of Ships	OBJ (USD)	x_{in}	x_{out}
600	1	I^0	3876.00	3876.00	0	14.78	NA	12	11,917,802.38	0	15
		I^1	16,137.00	16,137.00	0	14.44	NA				
		I^2	3552.00	3294.65	257.35	14.12	13.62				
	2	I^0	5089.00	5089.00	0	15.02	NA	11			
		I^1	15,020.00	10,520.00	0	14.99	NA				
		I^2	2269.00	2068.16	200.84	14.95	14.78				
	3	I^0	4885.00	4885.00	0	15.21	NA	11			
		I^1	16,704.00	16,704.00	0	15.18	NA				
		I^2	1736.00	1532.40	203.60	15.18	15.11				
	4	I^0	5047.00	5047.00	0	15.27	NA	11			
		I^1	12,609.00	12,609.00	0	15.25	NA				
		I^2	4553.00	4332.35	220.65	15.21	15.11				
700	1	I^0	3876.00	3876.00	0	14.46	NA	12	12,727,432.18	0	12
		I^1	16,137.00	16,137.00	0	14.44	NA				
		I^2	3552.00	3316.27	235.73	14.42	14.33				
	2	I^0	5089.00	5089.00	0	13.53	NA	12			
		I^1	15,020.00	10,520.00	0	13.49	NA				
		I^2	2269.00	2067.93	201.07	13.25	13.25				
	3	I^0	4885.00	4885.00	0	13.73	NA	12			
		I^1	16,704.00	16,704.00	0	13.68	NA				
		I^2	1736.00	1528.63	207.37	13.68	13.49				
	4	I^0	5047.00	5047.00	0	13.68	NA	12			
		I^1	12,609.00	12,609.00	0	13.67	NA				
		I^2	4553.00	4325.54	227.46	13.67	13.35				
800	1	I^0	3876.00	3876.00	0	13.11	NA	13	13,451,977.30	0	11
		I^1	16,137.00	16,137.00	0	13.09	NA				
		I^2	3552.00	3316.43	235.57	13.09	13.00				
	2	I^0	5089.00	5089.00	0	13.49	NA	12			
		I^1	15,020.00	15,020.00	0	13.47	NA				
		I^2	2269.00	2072.56	196.44	13.47	13.44				
	3	I^0	4885.00	4885.00	0	13.69	NA	12			
		I^1	16,704.00	16,704.00	0	13.69	NA				
		I^2	1736.00	1533.46	202.54	13.67	13.66				
	4	I^0	5047.00	5047.00	0	13.76	NA	12			
		I^1	12,609.00	12,609.00	0	13.74	NA				
		I^2	4553.00	4326.19	226.81	13.39	13.30				

Table 5. Impact of the unit price of renewable fuel.

μ_2 (USD/ton)	Route ID	Set of Legs	Total Distance (nm)	l_i^{tm} (nm)	l_i^{rm} (nm)	v_i^{tm} (knot)	v_i^{rm} (knot)	Number of Ships	OBJ (USD)	x_{in}	x_{out}
800	1	I^0	3876.00	3876.00	0	14.46	NA	12	11,901,015.30	0	15
		I^1	16,137.00	16,137.00	0	14.44	NA				
		I^2	3552.00	3316.27	235.73	14.42	14.33				
	2	I^0	5089.00	5089.00	0	15.02	NA	11			
		I^1	15,020.00	15,020.00	0	14.99	NA				
		I^2	2269.00	2061.50	207.50	14.97	14.54				
	3	I^0	4885.00	4885.00	0	15.29	NA	11			
		I^1	16,704.00	16,704.00	0	15.16	NA				
		I^2	1736.00	1531.07	204.93	15.16	15.04				
	4	I^0	5047.00	5047.00	0	15.25	NA	11			
		I^1	12,609.00	12,609.00	0	15.25	NA				
		I^2	4553.00	4335.24	217.76	15.23	15.22				
900	1	I^0	3876.00	3876.00	0	14.65	NA	12	11,909,871.43	0	15
		I^1	16,137.00	16,137.00	0	14.62	NA				
		I^2	3552.00	3276.59	275.41	14.41	13.22				
	2	I^0	5089.00	5089.00	0	15.03	NA	11			
		I^1	15,020.00	15,020.00	0	14.99	NA				
		I^2	2269.00	2071.01	197.99	14.91	14.89				
	3	I^0	4885.00	4885.00	0	15.21	NA	11			
		I^1	16,704.00	16,704.00	0	15.18	NA				
		I^2	1736.00	1533.72	202.28	15.18	15.16				
	4	I^0	5047.00	5047.00	0	15.27	NA	11			
		I^1	12,609.00	12,609.00	0	15.25	NA				
		I^2	4553.00	4335.19	217.81	15.21	15.21				
1000	1	I^0	3876.00	3876.00	0	14.78	NA	12	11,917,802.38	0	15
		I^1	16,137.00	16,137.00	0	14.44	NA				
		I^2	3552.00	3294.65	257.35	14.12	13.62				
	2	I^0	5089.00	5089.00	0	15.02	NA	11			
		I^1	15,020.00	10,520.00	0	14.99	NA				
		I^2	2269.00	2068.16	200.84	14.95	14.78				
	3	I^0	4885.00	4885.00	0	15.21	NA	11			
		I^1	16,704.00	16,704.00	0	15.18	NA				
		I^2	1736.00	1532.40	203.60	15.18	15.11				
	4	I^0	5047.00	5047.00	0	15.27	NA	11			
		I^1	12,609.00	12,609.00	0	15.25	NA				
		I^2	4553.00	4332.35	220.65	15.21	15.11				

Table 5. *Cont.*

μ_2 (USD/ton)	Route ID	Set of Legs	Total Distance (nm)	l_i^{tm} (nm)	l_i^{rm} (nm)	v_i^{tm} (knot)	v_i^{rm} (knot)	Number of Ships	OBJ (USD)	x_{in}	x_{out}
1100	1	I^0	3876.00	3876.00	0	14.75	NA	12	11,926,146.11	0	15
		I^1	16,137.00	16,137.00	0	14.41	NA				
		I^2	3552.00	3269.03	282.97	14.37	13.02				
	2	I^0	5089.00	5089.00	0	15.03	NA	11			
		I^1	15,020.00	15,020.00	0	14.98	NA				
		I^2	2269.00	2071.36	197.64	14.98	14.90				
	3	I^0	4885.00	4885.00	0	15.23	NA	11			
		I^1	16,704.00	16,704.00	0	15.18	NA				
		I^2	1736.00	1528.55	207.45	15.14	14.96				
	4	I^0	5047.00	5047.00	0	15.29	NA	11			
		I^1	12,609.00	12,609.00	0	15.25	NA				
		I^2	4553.00	4325.14	227.86	15.20	14.86				

4.3.2. Impact of Relevant Cost of Vessels

Due to the minimal impacts of fluctuations in cost of chartering in a vessel (c_{in}) and the revenue of chartering out a vessel (c_{out}) (where c_{in} resemble ship operation cost, resulting in an increase in total cost as c_{in} rises, while the total cost decreases with an upturn in c_{out}), our analysis focuses solely on ship operation cost.

In the aforementioned analysis, the c^m predetermined weekly operating cost for a vessel stands at 180,000 dollars. Nevertheless, the value of c^m is subject to fluctuation due to the impact of various unpredictable factors and risks [47], such as the global outbreak of COVID-19 in 2020, which reportedly sparked an escalation in ship operating costs [48], or the gradual impact of technological advancements, which may potentially reduce these costs. Consequently, the range of c^m is defined as 160,000 dollars to 240,000 dollars, with corresponding results meticulously documented in Table 6.

Analysis of Table 6 reveals a direct correlation between the increase in the operating cost of a vessel (c^m) and the corresponding rise in the objective value. This signifies that as the total operating costs for vessels surge, the overall objective value exhibits an upward trajectory. Furthermore, as the value of c^m intensifies, a prudent approach to saving on operating costs is witnessed through a reduction in the number of vessels deployed across the four routes.

4.3.3. Impact of the Total Fleet Size of Vessels

In the initial analysis, the total fleet size of vessels, denoted as K, is assumed to have an average value of 60. However, it is important to note that the total fleet size varies among different companies and at different stages of development. Consequently, the value of K varies according to the scale of each company. Therefore, it becomes imperative to conduct a sensitivity analysis of the total fleet size of vessels. In this experiment, we consider a range of fleet sizes, from 40 to 70 ships. It is worth mentioning that the vessel type under investigation is a 5000-TEU (Twenty-foot Equivalent Unit) container ship, which typically does not exceed 70 in number.

Table 6. Impact of the cost of operating a vessel.

c^m (USD/week)	Route ID	Set of Legs	Total Distance (nm)	l_i^{tm} (nm)	l_i^{rm} (nm)	v_i^{tm} (knot)	v_i^{rm} (knot)	Number of Ships	OBJ (USD)	x_{in}	x_{out}
160,000	1	l^0	3876.00	3876.00	0	14.49	NA	12	11,012,790.23	0	13
		l^1	16,137.00	16,137.00	0	14.49	NA				
		l^2	3552.00	3318.64	233.36	14.43	13.00				
	2	l^0	5089.00	5089.00	0	15.00	NA	11			
		l^1	15,020.00	10,520.00	0	14.99	NA				
		l^2	2269.00	2072.71	196.29	14.98	13.22				
	3	l^0	4885.00	4885.00	0	13.73	NA	12			
		l^1	16,704.00	16,704.00	0	13.71	NA				
		l^2	1736.00	1524.35	211.65	13.63	13.38				
	4	l^0	5047.00	5047.00	0	13.74	NA	12			
		l^1	12,609.00	12,609.00	0	13.66	NA				
		l^2	4553.00	4331.94	221.06	13.61	13.62				
180,000	1	l^0	3876.00	3876.00	0	14.78	NA	12	11,917,802.38	0	15
		l^1	16,137.00	16,137.00	0	14.44	NA				
		l^2	3552.00	3294.65	257.35	14.12	13.62				
	2	l^0	5089.00	5089.00	0	15.02	NA	11			
		l^1	15,020.00	10,520.00	0	14.99	NA				
		l^2	2269.00	2068.16	200.84	14.95	14.78				
	3	l^0	4885.00	4885.00	0	15.21	NA	11			
		l^1	16,704.00	16,704.00	0	15.18	NA				
		l^2	1736.00	1532.40	203.60	15.18	15.11				
	4	l^0	5047.00	5047.00	0	15.27	NA	11			
		l^1	12,609.00	12,609.00	0	15.25	NA				
		l^2	4553.00	4332.35	220.65	15.21	15.11				
200,000	1	l^0	3876.00	3876.00	0	14.47	NA	12	12,817,094.15	0	15
		l^1	16,137.00	16,137.00	0	14.45	NA				
		l^2	3552.00	3307.81	244.19	14.38	14.07				
	2	l^0	5089.00	5089.00	0	15.03	NA	11			
		l^1	15,020.00	10,520.00	0	14.98	NA				
		l^2	2269.00	2071.36	197.64	14.98	14.90				
	3	l^0	4885.00	4885.00	0	15.21	NA	11			
		l^1	16,704.00	16,704.00	0	15.18	NA				
		l^2	1736.00	1532.93	203.07	15.18	15.13				
	4	l^0	5047.00	5047.00	0	15.28	NA	11			
		l^1	12,609.00	12,609.00	0	15.25	NA				
		l^2	4553.00	4331.03	221.97	15.20	15.06				

Table 6. *Cont.*

c^m (USD/week)	Route ID	Set of Legs	Total Distance (nm)	l_i^{tm} (nm)	l_i^{rm} (nm)	v_i^{tm} (knot)	v_i^{rm} (knot)	Number of Ships	OBJ (USD)	x_{in}	x_{out}
220,000	1	l^0	3876.00	3876.00	0	16.13	NA	11	15,448,312.66	0	16
		l^1	16,137.00	16,137.00	0	16.10	NA				
		l^2	3552.00	3316.00	236.00	16.05	15.96				
	2	l^0	5089.00	5089.00	0	15.00	NA	11			
		l^1	15,020.00	10,520.00	0	15.00	NA				
		l^2	2269.00	2071.08	197.92	14.91	14.89				
	3	l^0	4885.00	4885.00	0	15.21	NA	11			
		l^1	16,704.00	16,704.00	0	15.19	NA				
		l^2	1736.00	1530.73	205.27	15.08	15.04				
	4	l^0	5047.00	5047.00	0	15.28	NA	11			
		l^1	12,609.00	12,609.00	0	15.24	NA				
		l^2	4553.00	4332.58	220.42	15.23	15.12				
240,000	1	l^0	3876.00	3876.00	0	14.49	NA	11	13,706,839.57	0	16
		l^1	16,137.00	16,137.00	0	14.44	NA				
		l^2	3552.00	3316.00	236.00	14.39	14.28				
	2	l^0	5089.00	5089.00	0	15.05	NA	11			
		l^1	15,020.00	10,520.00	0	14.98	NA				
		l^2	2269.00	2071.08	197.92	14.93	14.92				
	3	l^0	4885.00	4885.00	0	13.73	NA	11			
		l^1	16,704.00	16,704.00	0	13.68	NA				
		l^2	1736.00	1530.73	205.27	13.67	13.54				
	4	l^0	5047.00	5047.00	0	13.71	NA	11			
		l^1	12,609.00	12,609.00	0	13.67	NA				
		l^2	4553.00	4332.58	220.42	13.63	13.48				

Analyzing Table 7, we observe that as the total fleet size of vessels increases, the objective value decreases. This is attributed to a shift in the vessel situation of companies, transforming it from having inadequate vessels to having surplus ships. Thus, additional revenue is generated through chartering out surplus vessels. Moreover, it is notable that the allocation of ships to the four shipping routes remains unchanged regardless of variations in the total number of ships. The shipping companies only need to determine the number of ships to be chartered out or chartered in based on the balance between the sum of ships allocated to the four routes and the company's existing total number of ships.

Table 7. Impact of the total fleet size of vessels.

K	Route ID	Set of Legs	Total Distance (nm)	l_i^{tm} (nm)	l_i^{rm} (nm)	v_i^{tm} (knot)	v_i^{rm} (knot)	Number of Ships	OBJ (USD)	x_{in}	x_{out}
40	1	I^0	3876.00	3876.00	0	14.47	NA	12	14,017,651.72	5	0
		I^1	16,137.00	16,137.00	0	14.44	NA				
		I^2	3552.00	3316.36	235.64	14.41	14.33				
	2	I^0	5089.00	5089.00	0	15.13	NA	11			
		I^1	15,020.00	10,520.00	0	14.98	NA				
		I^2	2269.00	2057.03	211.97	14.80	14.34				
	3	I^0	4885.00	4885.00	0	15.22	NA	11			
		I^1	16,704.00	16,704.00	0	15.18	NA				
		I^2	1736.00	1532.53	203.47	15.15	15.11				
	4	I^0	5047.00	5047.00	0	15.28	NA	11			
		I^1	12,609.00	12,609.00	0	15.25	NA				
		I^2	4553.00	4331.78	221.22	15.21	15.09				
50	1	I^0	3876.00	3876.00	0	14.47	NA	12	12,917,358.26	0	5
		I^1	16,137.00	16,137.00	0	14.44	NA				
		I^2	3552.00	3315.71	236.29	14.41	14.31				
	2	I^0	5089.00	5089.00	0	15.03	NA	11			
		I^1	15,020.00	10,520.00	0	15.02	NA				
		I^2	2269.00	2066.97	202.03	14.72	14.71				
	3	I^0	4885.00	4885.00	0	15.22	NA	11			
		I^1	16,704.00	16,704.00	0	15.18	NA				
		I^2	1736.00	1532.53	203.47	15.15	15.11				
	4	I^0	5047.00	5047.00	0	15.27	NA	11			
		I^1	12,609.00	12,609.00	0	15.25	NA				
		I^2	4553.00	4327.58	225.42	15.22	14.19				
60	1	I^0	3876.00	3876.00	0	14.78	NA	12	11,917,802.38	0	15
		I^1	16,137.00	16,137.00	0	14.44	NA				
		I^2	3552.00	3294.65	257.35	14.12	13.62				
	2	I^0	5089.00	5089.00	0	15.02	NA	11			
		I^1	15,020.00	10,520.00	0	14.99	NA				
		I^2	2269.00	2068.16	200.84	14.95	14.78				
	3	I^0	4885.00	4885.00	0	15.21	NA	11			
		I^1	16,704.00	16,704.00	0	15.18	NA				
		I^2	1736.00	1532.40	203.60	15.18	15.11				
	4	I^0	5047.00	5047.00	0	15.27	NA	11			
		I^1	12,609.00	12,609.00	0	15.25	NA				
		I^2	4553.00	4332.35	220.65	15.21	15.11				

Table 7. *Cont.*

K	Route ID	Set of Legs	Total Distance (nm)	l_i^{tm} (nm)	l_i^{rm} (nm)	v_i^{tm} (knot)	v_i^{rm} (knot)	Number of Ships	OBJ (USD)	x_{in}	x_{out}
70	1	l^0	3876.00	3876.00	0	14.47	NA	12	10,918,026.70	0	25
		l^1	16,137.00	16,137.00	0	14.44	NA				
		l^2	3552.00	3319.10	232.90	14.41	14.32				
	2	l^0	5089.00	5089.00	0	15.02	NA	11			
		l^1	15,020.00	15,020.00	0	14.99	NA				
		l^2	2269.00	2072.07	196.93	14.95	14.75				
	3	l^0	4885.00	4885.00	0	15.25	NA	11			
		l^1	16,704.00	16,704.00	0	15.25	NA				
		l^2	1736.00	1533.01	202.99	14.46	14.22				
	4	l^0	5047.00	5047.00	0	15.29	NA	11			
		l^1	12,609.00	12,609.00	0	15.25	NA				
		l^2	4553.00	4334.11	218.89	15.22	15.08				

5. Conclusions

This research investigates the shipping company's optimal strategy regarding fuel selection, sailing speed, ship deployment, and the number of ships chartered in or chartered out, considering the EU's proposed new policy. Initially, we present an innovative MINLP model, subsequently converting it into an MILP model through the application of advanced mathematical techniques. Through rigorous experiments, our proposed model demonstrates its efficacy, leading to the following conclusions: (i) Due to the lower prices compared to those of renewable fuels, traditional fuels are employed in all three types of regions, while renewable fuels are selectively used only in specific segments within the EU area. (ii) In terms of sailing speed, given the cubic relationship between speed and fuel consumption, as well as the price disparities between the two fuel types, ships tend to operate at higher speeds when using traditional fuels, thus reducing operational costs compared to utilizing sustainable fuels. (iii) The number of vessels employed and the mileage of routes are interconnected, with decisions regarding vessel chartered in or chartered out contingent upon striking a balance between the total number of vessels employed to each route and the overall fleet size of vessels. Furthermore, we analyze the influence of variations in fuel prices, relevant costs of vessels, and the total fleet size of vessels on optimal decisions. In general, increases in fuel prices, vessel-related costs, and fleet size result in rising total costs. Specifically, when the prices of both fuel types increase, shipping companies, seeking cost reduction, tend to lower sailing speeds to minimize fuel consumption, while also employing a higher number of vessels to satisfy weekly service frequencies. Moreover, when the operational costs of vessels increase, the number of vessels employed decreases. It is worth noting that changes in the total fleet size of vessels do not affect the number of vessels employed to individual routes but rather impact decisions regarding vessel leasing and chartering.

This investigation offers valuable insights into shipping company strategies under the new policy. Overall, our study contributes to the understanding of how shipping companies can make informed decisions in response to the EU's new policy. By providing efficient solution methods and examining the sensitivity of optimal decisions to various factors, our research offers practical guidance for navigating the challenges and opportunities presented by the policy.

Despite the significant contributions of this study, there exist certain limitations that require further investigation.

Firstly, the proposed model assumes static conditions and does not account for dynamic factors such as changing market trends, weather conditions, or evolving regulations. Incorporating these dynamic elements would be valuable for a more comprehensive analysis.

Secondly, our research focuses primarily on the shipping company's perspective and optimal decision-making. However, future studies could consider the broader impact of the EU's policy on the maritime industry as a whole, including the implications for sustainability, environmental protection, and the overall supply chain efficiency.

Additionally, our study assumes perfect information availability and precise parameter estimations. In reality, uncertain data and imperfect information are common challenges. Future research endeavors could explore robust optimization techniques or apply stochastic programming to address these uncertainties and enhance the reliability of decision-making processes.

In conclusion, this study provides valuable insights into the optimal strategies for shipping companies under the new EU policy. However, addressing the aforementioned limitations and pursuing further research in the suggested directions will yield a more comprehensive understanding of the complex interplay between fuel selection, sailing speed, and fleet management, contributing to the advancement of sustainable and efficient maritime operations.

Author Contributions: Conceptualization, H.W., Y.L. and S.W.; methodology, H.W., Y.L., S.W. and L.Z.; software, H.W. and Y.L.; validation, H.W. and Y.L.; formal analysis, H.W. and Y.L.; writing—original draft preparation, H.W. and Y.L.; writing—review and editing, L.Z. and S.W.; visualization, Y.L.; supervision, L.Z. and S.W. All authors have read and agreed to the published version of the manuscript.

Funding: This research received no external funding.

Institutional Review Board Statement: Not applicable.

Informed Consent Statement: Not applicable.

Data Availability Statement: Data are contained within the article.

Conflicts of Interest: The authors declare no conflict of interest.

Note

[1] https://www.cma-cgm.com/products-services/flyers, accessed on 1 August 2023.

References

1. Cao, M.; Liu, Y.; Gai, T.; Zhou, M.; Fujita, H.; Wu, J. A comprehensive star rating approach for cruise ships based on interactive group decision making with personalized individual semantics. *J. Mar. Sci. Eng.* **2022**, *10*, 638. [CrossRef]
2. Du, Y.; Chen, Y.; Li, X.; Schönborn, A.; Sun, Z. Data fusion and machine learning for ship fuel efficiency modeling: Part II–Voyage report data, AIS data and meteorological data. *Commun. Transp. Res.* **2022**, *2*, 100073. [CrossRef]
3. Du, Y.; Chen, Y.; Li, X.; Schönborn, A.; Sun, Z. Data fusion and machine learning for ship fuel efficiency modeling: Part III–Sensor data and meteorological data. *Commun. Transp. Res.* **2022**, *2*, 100072. [CrossRef]
4. Li, X.; Du, Y.; Chen, Y.; Nguyen, S.; Zhang, W.; Schönborn, A.; Sun, Z. Data fusion and machine learning for ship fuel efficiency modeling: Part I–Voyage report data and meteorological data. *Commun. Transp. Res.* **2022**, *2*, 100074. [CrossRef]
5. Riall, ; A.I.; Bouman, E.A.; Lindstad, E.; Strømman, A.H. State-of-the-art technologies, measures, and potential for reducing GHG emissions from shipping—A review. *Transp. Transp. Res. Part D Transp. Environ.* **2017**, *52*, 408–421.
6. Zweers, B.G.; Bhulai, S.; van der Mei, R.D. Planning hinterland container transportation in congested deep-sea terminals. *Flex. Serv. Manuf. J.* **2021**, *33*, 583–622. [CrossRef]
7. Kizilay, D.; Eliiyi, D.T. A comprehensive review of quay crane scheduling, yard operations and integrations thereof in container terminals. *Flex. Serv. Manuf. J.* **2021**, *33*, 1–42. [CrossRef]
8. He, J.; Yan, N.; Zhang, J.; Yu, Y.; Wang, T. Battery electric buses charging schedule optimization considering time-of-use electricity price. *J. Intell. Connect. Veh.* **2022**, *5*, 138–145. [CrossRef]
9. IMO. Marine Environment Protection Committee (MEPC 80), 3–7 July 2023. 2023. Available online: https://www.imo.org/en/MediaCentre/MeetingSummaries/Pages/MEPC-80.aspx (accessed on 11 September 2023).
10. Harahap, F.; Nurdiawati, A.; Conti, D.; Leduc, S.; Urban, F. Renewable marine fuel production for decarbonised maritime shipping: Pathways, policy measures and transition dynamics. *J. Clean. Prod.* **2023**, *415*, 137906. [CrossRef]

11. Lloyd's List. EU to Set 2% Renewable Marine Fuel Target to Cut Carbon Emissions. 2023. Available online: https://lloydslist-maritimeintelligence-informa-com.ezproxy.lb.polyu.edu.hk/LL1144425/EU-to-set-2-renewable-marine-fuel-target-to-cut-carbon-emissions (accessed on 1 August 2023).

12. Wang, H.; Ouyang, M.; Li, J.; Yang, F. Hydrogen fuel cell vehicle technology roadmap and progress in China. *J. Automot. Saf. Energy* **2022**, *13*, 211.

13. Psaraftis, H.N.; Zis, T.; Lagouvardou, S. A comparative evaluation of market based measures for shipping decarbonization. *Marit. Transp. Res.* **2021**, *2*, 100019. [CrossRef]

14. Malmborg, F.V. Advocacy coalitions and policy change for decarbonisation of international maritime transport: The case of FuelEU maritime. *Marit. Transp. Res.* **2023**, *4*, 100091. [CrossRef]

15. European Parliament. Revision of the Energy Taxation Directive. 2021. Available online: https://taxation-customs.ec.europa.eu/green-taxation-0/revision-energy-taxation-$directive_en$ (accessed on 21 August 2023).

16. European Parliament. DIRECTIVE (EU) 2018/2001 of the European Parliament and of the Council of 11 December 2018 on the Promotion of the Use of Energy from Renewable Sources. 2018. Available online: https://eur-lex.europa.eu/legal-content/EN/TXT/PDF/?uri=CELEX:32018L2001 (accessed on 21 August 2023).

17. Wang, Y.; Cao, Q.; Liu, L.; Wu, Y.; Liu, H.; Gu, Z.; Zhu, C. A review of low and zero carbon fuel technologies: Achieving ship carbon reduction targets. *Sustain. Energy Technol. Assess.* **2022**, *54*, 102762. [CrossRef]

18. Benti, N.E.; Aneseyee, A.B.; Geffe, C.A.; Woldegiyorgis, T.A.; Gurmesa, G.S.; Bibiso, M.; Asfaw, A.A.; Milki, A.W.; Mekonnen, Y.S. Biodiesel production in Ethiopia: Current status and future prospects. *Sci. Afr.* **2023**, *19*, e01531. [CrossRef]

19. Staffell, I.; Balcombe, P.; Kerdan, I.G.; Speirs, J.F.; Brandon, N.P.; Hawkes, A.D. How can LNG-fuelled ships meet decarbonisation targets? An environmental and economic analysis. *Energy* **2021**, *227*, 120462.

20. Jia, X.; Xu, T.; Zhang, Y.; Li, Z.; Tan, R.; Aviso, K.B.; Wang, F. An improved multi-period algebraic targeting approach to low carbon energy planning. *Energy* **2023**, *268*, 126627. [CrossRef]

21. Ammar, N.R.; Seddiek, I.S. Hybrid/dual fuel propulsion systems towards decarbonization: Case study container ship. *Ocean. Eng.* **2023**, *281*, 114962. [CrossRef]

22. Wu, T.; Xu, M. Modeling and optimization for carsharing services: A literature review. *Multimodal Transp.* **2022**, *1*, 100028. [CrossRef]

23. Huang, D.; Wang, S. A two-stage stochastic programming model of coordinated electric bus charging scheduling for a hybrid charging scheme. *Multimodal Transp.* **2022**, *1*, 100006. [CrossRef]

24. Zhao, H.; Meng, Q.; Wang, Y. Robust container slot allocation with uncertain demand for liner shipping services. *Flex. Serv. Manuf. J.* **2022**, *34*, 551–579. [CrossRef]

25. Zhang, J.; Long, D.Z.; Wang, R.; Xie, C. Impact of Penalty Cost on Customers' Booking Decisions. *Prod. Oper. Manag.* **2021**, *30*, 1603–1614. [CrossRef]

26. Adl, ; R.; Cariou, P.; Wolff, F. Optimal ship speed and the cubic law revisited: Empirical evidence from an oil tanker fleet. *Transp. Res. Part E Logist. Transp. Rev.* **2020**, *140*, 101972.

27. Laporte, G.; Fagerholt, K.; Norstad, I. Reducing fuel emissions by optimizing speed on shipping routes. *Transp. Logist. Environ.* **2020**, *61*, 523–529.

28. Arijit, D.; Alok, C.; Metin, T.; Manoj, K.T. Bunkering policies for a fuel bunker management problem for liner shipping networks. *Eur. J. Oper. Res.* **2021**, *289*, 927–937.

29. Wang, X.; Teo, C.C. Integrated hedging and network planning for container shipping's bunker fuel management. *Marit. Econ. Logist.* **2013**, *15*, 172–196. [CrossRef]

30. He, Q.; Zhang, X.; Nip, K. Speed optimization over a path with heterogeneous arc costs. *Transp. Res. Part B Methodol.* **2017**, *104*, 198–214. [CrossRef]

31. Aydin, N.; Lee, H.; Mansouri, S.A. Speed optimization and bunkering in liner shipping in the presence of uncertain service times and time windows at ports. *Eur. J. Oper. Res.* **2017**, *259*, 143–154. [CrossRef]

32. Wang, K.; Li, J.; Huang, L.; Ma, R.; Jiang, X.; Yuan, Y.; Mwero, N.A.; Negenborn, R.R.; Sun, P.; Yan, X. A novel method for joint optimization of the sailing route and speed considering multiple environmental factors for more energy efficient shipping. *Ocean. Eng.* **2020**, *216*, 107591. [CrossRef]

33. Lu, J.; Wu, X.; Wu, Y. The construction and application of dual-objective optimal speed model of liners in a changing climate: Taking Yang Ming route as an example. *J. Mar. Sci. Eng.* **2023**, *11*, 157. [CrossRef]

34. Mahsa, L.; Ali, A.A.; Saba, N. A new model for simultaneously optimizing ship route, sailing speed, and fuel consumption in a shipping problem under different price scenarios. *Appl. Ocean. Res.* **2021**, *113*, 102725.

35. Venturini, G.; Iris, C.; Kontovas, C.A.; Larsen, A. The multi-port berth allocation problem with speed optimization and emission considerations. *Transp. Res. Part D Transp. Environ.* **2017**, *54*, 142–159. [CrossRef]

36. Dulebenets, M.A. Minimizing the Total Liner Shipping Route Service Costs via Application of an Efficient Collaborative Agreement. *IEEE Trans. Intell. Transp. Syst.* **2019**, *20*, 123–136. [CrossRef]

37. Wang, Y.; Meng, Q.; Du, Y. Liner container seasonal shipping revenue management. *Transp. Res. Part E Methodol.* **2015**, *82*, 141–161. [CrossRef]

38. Ghosh, S.; Lee, L.H.; Ng, S.H. Bunkering decisions for a shipping liner in an uncertain environment with service contract. *Eur. J. Oper. Res.* **2015**, *224*, 792–802. [CrossRef]

39. Ng, M.W. Vessel speed optimisation in container shipping: A new look. *J. Oper. Res. Soc.* **2019**, *70*, 541–547. [CrossRef]
40. Sheng, D.; Meng, Q.; Li, Z. Optimal vessel speed and fleet size for industrial shipping services under the emission control area regulation. *Transp. Res. Part C Emerg. Technol.* **2019**, *105*, 37–53. [CrossRef]
41. Ng, N.W. Bounds on ship deployment in container shipping with time windows. *J. Oper. Res. Soc.* **2019**, *72*, 1252–1258. [CrossRef]
42. Wang, S.; Meng, Q. Sailing speed optimization for container ships in a liner shipping network. *Transp. Res. Part E Logist. Transp. Rev.* **2012**, *48*, 701–714. [CrossRef]
43. Wang, Y.; Wang, S. Deploying, scheduling, and sequencing heterogeneous vessels in a liner container shipping route. *Transp. Res. Part E* **2021**, *151*, 102365. [CrossRef]
44. Lai, X.; Wu, L.; Wang, S.; Wang, K.; Wang, F. Robust ship fleet deployment with shipping revenue management. *Transp. Res. Part B* **2022**, *161*, 169–196. [CrossRef]
45. ShipBunker. 2023. Available online: https://shipandbunker.com/prices (accessed on 31 August 2023).
46. Meng, Q.; Wang, T. A scenario-based dynamic programming model for multi-period liner ship fleet planning. *Transp. Res. Part E Logist. Transp. Rev.* **2011**, *47*, 401–413. [CrossRef]
47. Wu, J.; Chen, Y.; Gai, T.; Liu, Y.; Li, Y.; Cao, M. A new leader–follower public-opinion evolution model for maritime transport incidents: A case from Suez Canal Blockage. *J. Mar. Sci. Eng.* **2022**, *10*, 2006. [CrossRef]
48. Drewry. COVID-19 Drives up Ship Operating Costs. 2020. Available online: https://www.drewry.co.uk/news/news/covid-19-drives-up-ship-operating-costs (accessed on 30 July 2023).

Journal of
*Marine Science
and Engineering*

MDPI

Article

Feasibility Assessment of Alternative Clean Power Systems onboard Passenger Short-Distance Ferry

Ahmed G. Elkafas [1,2,*], Massimo Rivarolo [1], Stefano Barberis [1] and Aristide F. Massardo [1]

[1] Thermochemical Power Group (TPG), DIME, University of Genova, 16145 Genova, Italy;
    massimo.rivarolo@unige.it (M.R.); stefano.barberis@unige.it (S.B.); massardo@unige.it (A.F.M.)
[2] Department of Naval Architecture and Marine Engineering, Faculty of Engineering, Alexandria University,
    Alexandria 21544, Egypt
*   Correspondence: ahmed.elkafas@edu.unige.it

Abstract: In order to promote low-carbon fuels such as hydrogen to decarbonize the maritime sector, it is crucial to promote clean fuels and zero-emission propulsion systems in demonstrative projects and to showcase innovative technologies such as fuel cells in vessels operating in local public transport that could increase general audience acceptability thanks to their showcase potential. In this study, a short sea journey ferry used in the port of Genova as a public transport vehicle is analyzed to evaluate a "zero emission propulsion" retrofitting process. In the paper, different types of solutions (batteries, proton exchange membrane fuel cell (PEMFC), solid oxide fuel cell (SOFC)) and fuels (hydrogen, ammonia, natural gas, and methanol) are investigated to identify the most feasible technology to be implemented onboard according to different aspects: ferry daily journey and scheduling, available volumes and spaces, propulsion power needs, energy storage/fuel tank capacity needed, economics, etc. The paper presents a multi-aspect analysis that resulted in the identification of the hydrogen-powered PEMFC as the best clean power system to guarantee, for this specific case study, a suitable retrofitting of the vessel that could guarantee a zero-emission journey.

Keywords: fuel cells; decarbonization; total cost; short-sea navigation; battery; hydrogen

Citation: Elkafas, A.G.; Rivarolo, M.;
Barberis, S.; Massardo, A.F.
Feasibility Assessment of Alternative
Clean Power Systems onboard
Passenger Short-Distance Ferry. *J.
Mar. Sci. Eng.* **2023**, *11*, 1735.
https://doi.org/10.3390/
jmse11091735

Academic Editors: Jian Wu and
Xianhua Wu

Received: 3 August 2023
Revised: 28 August 2023
Accepted: 31 August 2023
Published: 2 September 2023

1. Introduction

In recent decades, the escalating environmental challenges posed by traditional fossil fuel-powered maritime transportation have stimulated a global pursuit of eco-friendly alternatives. Also, they were stimulated by the International Maritime Organization's (IMO) ambitious targets and challenges in 2018 [1] which aim to reduce the total annual greenhouse gas (GHG) emissions by at least 50% by 2050 compared to 2008. These targets are boosted in the revised 2023 IMO strategy [2] to achieve net-zero emissions from ships by or close to 2050 with suggested milestones for lowering GHG emissions by 20–30% in 2030, and 70–80% in 2040, both in contrast to levels in 2008. Regarding other ship emissions, Annex VI of MARPOL [3] poses limitations on nitrogen oxides (NOx), sulfur oxide (SOx), and particulate matter (PM), obliging maritime operators (ship owner, ship manager, ship craft, etc.) to engage in deep thought about current/future fuel choice and propulsion/power generation system technology [4]. The only approach to ensure a cleaner future for the maritime industry appears to be to look at alternative fuels and clean power systems rather than simply acting on an exhaust after-treatment system [5].

Particularly looking at vessels operating in coastal areas and maritime urban environments, emissions reductions are becoming more and more important [6], where nearly 70% of pollutant emissions are estimated to occur within 400 km of coastlines and where 45% of the world's population resides [7].

In order to reduce vessels' emissions and decarbonize the shipping sector [8,9], different measures could be put in place [10,11], acting at different levels of the vessels. The ship emission reduction measures include optimizing the efficiency of the ship engine by using waste heat recovery [12,13], working on alternative propulsion technologies such as wing sails [14] or electric hybrid propulsion [15], reducing ship resistance via trim optimization [16] or an enhanced vessel design [17], and using voyage optimization measures such as onboard energy management [18] or slow steaming concept [19,20].

Nevertheless, in order to achieve relevant emissions reduction as targeted by the IMO, it is crucial to promote alternative fuels for vessels such as natural gas [21,22], biofuels [23], and hydrogen-based e-fuels [24].

The identification of the most suitable alternative fuels in shipping depends on many aspects, technical (fuels' own thermo-physical properties that pose a limitation on fuel storage onboard and need for a specific on-board/on-shore bunkering/refueling infrastructure) and non-technical (safety, regulatory, classification, etc.) with impacts on economic (fuel price, investment, and operational costs), environmental (emissions, well-to-tank life cycle performance) and social (availability, politics, public opinion, etc.) aspects [25–28].

At the same time, the choice of energy and power propulsion systems depends on the type of vessel, its journey profile, and its own shipbuilding/environment features (space and volumes onboard, buoyancy needs, etc.) [8,29]. In order to solve this multi-aspect problem, different tools and approaches have been promoted to guide shipping operators in the identification of the most relevant and sustainable options when looking at a retrofitting/newly built vessel construction project targeting low emissions [30]. The goal of these approaches is to identify the most suitable/worthy-of-investigation technological solution, looking at both the vessel and targeted clean power system/fuel peculiarities [31].

Particularly looking at inland waterways and short sea shipping segments [32], and thanks to demonstrations driven by different EU-funded research projects [33–35], fuel cell (FC) systems are gaining more and more interest as a promising solution for maritime applications, as they are characterized by a high efficiency and low level of emissions, noise, and vibrations [35]. For all of these reasons and in accordance with the FC main peculiarities presented in [36,37], the authors already reviewed the recent research development/commercial products of FC systems [35] and investigated the use of FCs onboard different types of vessels, from cruise ships [38] to research vessels [39–41].

Research Novelty

Starting from these previous research works, as well as inspired by other FCs equipped vessel R&D (research and development) works targeting different types of FC [7,42,43] and different fuels [44,45], in the current paper the authors will investigate the possibility of retrofitting an existing small-scale ferry operating in the port of Genova as a public transportation vehicle for citizens.

While different R&D studies investigated the possibility of applying different fuels [27] towards zero emission vessels also looking at batteries [46], different types of FCs [47], and hydrogen-based energy systems onboard vessels [48], this paper targets a small-scale vessel used for urban transport. As well as this paper highlights the uniqueness of the application by studying the possibility of installing different energy systems (also looking at different types of FC) and fuel types both from an economic, energy, and onboard integration point of view, proposing a multi-aspect retrofitting methodology and step-by-step approach.

The choice of targeting this type of vessel, as already highlighted in previous research work [49], for the proposed retrofitting project has been driven by three reasons: (1) the fact that small ferries/vessels and short sea journey vessels looking at their journey profile would require a limited power capacity of the propulsion system and fuel volumes to be stored onboard, thus overcoming main limitations related to a large hydrogen tank needing to be integrated onboard; (2) the fact that this type of vessel, looking at their journey profiles, have very frequent and precise scheduling, thus enabling potential recurrent and easy-to-plan refueling (thus further reducing the amount of fuel to be stored onboard); (3) the fact that this type of vessel operates in the urban environment where emission limitation is more urgent and where the showcase of the effectiveness of FC technologies onboard ferries could have a higher social impact in terms of public awareness.

For this purpose, this research paper delves into the application of alternative zero-emission power systems onboard small vessels used for public transport working on short-sea navigation. In this paper, different solutions will be investigated such as full battery systems and FCs fed by different fuels like pure hydrogen, ammonia, liquified natural gas (LNG), and methanol (MeOH). The paper aims to present a retrofitting case study that will be analyzed by studying multi-aspect reasons that could favor one technology instead of another. The technical feasibility aims to look at the vessel's journey, energy needs, available volume/weights onboard, and potential daily refueling opportunities. Moreover, the paper will study the impact of applying clean power systems on the overall design of the case study from the system's weight and size perspectives.

The multi-aspect analysis involves the economic feasibility of using clean power systems to achieve the decarbonization of ships working on short-sea navigation. This study will identify the most economically viable clean power system through cost assessment indicators such as net present value (NPV), levelized cost of energy (LCOE), return on investment (ROI), and marginal abatement cost (MAC).

2. Case Study Description

The proposed case study is one of the passenger ferries working in short-sea navigation through Genoa city in Italy as part of the public-transport offering of the municipality. This type of vessel can be seen as an entry point/showcase for clean propulsion solutions as they do not require large storage onboard, and they have a large audience impact thus potentially increasing public awareness of clean maritime technologies. The ship is called "Rodi Jet—NaveBus" which navigates between the west side of Genoa (Pegli) and the ancient port in the city centre (the old port—Porto Antico) [50]. The maximum capacity of the ship is 362 passengers and access to the ship is guaranteed through a ramp 2.2 m long and 85 cm wide. The main specifications of the NaveBus are summarized in Table 1 [51].

Table 1. Characteristics of the case study (NaveBus).

Parameter	Unit	Value
Maximum number of passengers	(-)	362
Length overall	(m)	28.6
Breadth	(m)	6.92
Depth	(m)	2.34
Draught	(m)	1.14
Maximum displacement	(tons)	84.3
Maximum design speed	(knots)	20
Service speed	(knots)	10.2
Main engine type	(-)	2 × Caterpillar 3412
Main engine power	(kW)	2 × 895
Fuel tank capacity	(tons)	7.4

The NaveBus is propelled by using two fixed-pitch propellers (FPP) powered by two diesel engines from Caterpillar 3412, each one has a maximum rated output power, footprint volume, and weight equal to 895 kW, 2.45 m^3, and 1.9 tons, respectively. Moreover, there is an engine room with dimensions of 6.1 m (L) × 6.5 m (W) × 2 m (H) where the following components are located: two Caterpillar main engines, two gearboxes, a control panel, and two auxiliary generators (rated 18 kW and 6 kW) for hoteling and service generation.

Next to the engine room, there is a room containing two diesel storage tanks to be filled with 7.2 tons of marine diesel oil (MDO) at a maximum, each tank has the following dimensions (2.9 m (L) × 1.24 m (W) × 1.21 m (H)) with a footprint volume equal to 4.4 m^3.

The ship is designed to operate at a maximum speed of 20 knots, but the actual service speed (as understood thanks to an interview with the local crew) for the investigated journey and route is approximately 10.2 knots in sailing mode as a maximum. This service speed at normal weather conditions can be achieved by using approximately 40% of the installed engines' rated power (approximately 715 kW).

The NaveBus is characterized by having a specific navigational route for its trips which start in Porto Antico and sail along the Ligurian coast to Pegli, in a sea area protected from heavy winds and waves by the port of Genova Coastal Dam. The navigational route is described in Figure 1. The distance between the two ports/terminals area is 6 nautical miles (nm).

Figure 1. Navigational route of the NaveBus between Porto Antico and Pegli. (A-B) maneuvering at Porto Antico terminal, (B-C) crossing Porto Antico channel, (C-D) Sailing mode, (D-E) entrance of Pegli terminal, (E-F) maneuvering at Pegli terminal.

As previously mentioned, this ferry is part of the public transport offering of the city of Genova and it works integrated with other modes of public transportation such as buses and metro as it sails eight times every day during rush hours as follows: three consecutive morning trips, two trips in the afternoon, and three consecutive trips in the evening.

The operational profile of the NaveBus is characterized by different modes: passenger loading at Porto Antico port, maneuvering the ferry away from the Porto Antico terminal area (A-B), crossing the Porto Antico Channel to the west side (B-C), sailing through the Ligurian Sea (C-D), entrance of the Pegli terminal (D-E), maneuvering at the Pegli terminal (E-F), and passenger unloading. The operational profile for one day is described in Figure 2a by plotting the relation between the ferry speed and actual time in a day (daily scheduling of the NaveBus), while the operational modes of one round trip are described in Figure 2b in terms of ferry speed versus the time spent at each mode.

As shown in Figure 2, the trip that takes off from Pegli and ends in Porto Antico is quite like the other trip from Porto Antico to Pegli in terms of spending time and ship speed. Moreover, the recorded energy requirement for the two trips is the same (thanks to an interview with the local crew).

Based on the limitation of the available weight and volume onboard the ship, the current study will design a clean power system for the implementation of three consecutive trips without refueling (or recharging in the case of a full battery electric system). This is acceptable according to the daily journey profile as shown in Figure 2a; therefore, three one-way trips (OWTs) can be considered as a functional unit of the current study, to evaluate the power capacity needs and fuel needs. Looking at the daily scheduling, there is indeed the possibility to foresee two refueling/recharging periods in Porto Antico at midday and another one in the evening according to the ship's operational schedule depicted in Figure 2a.

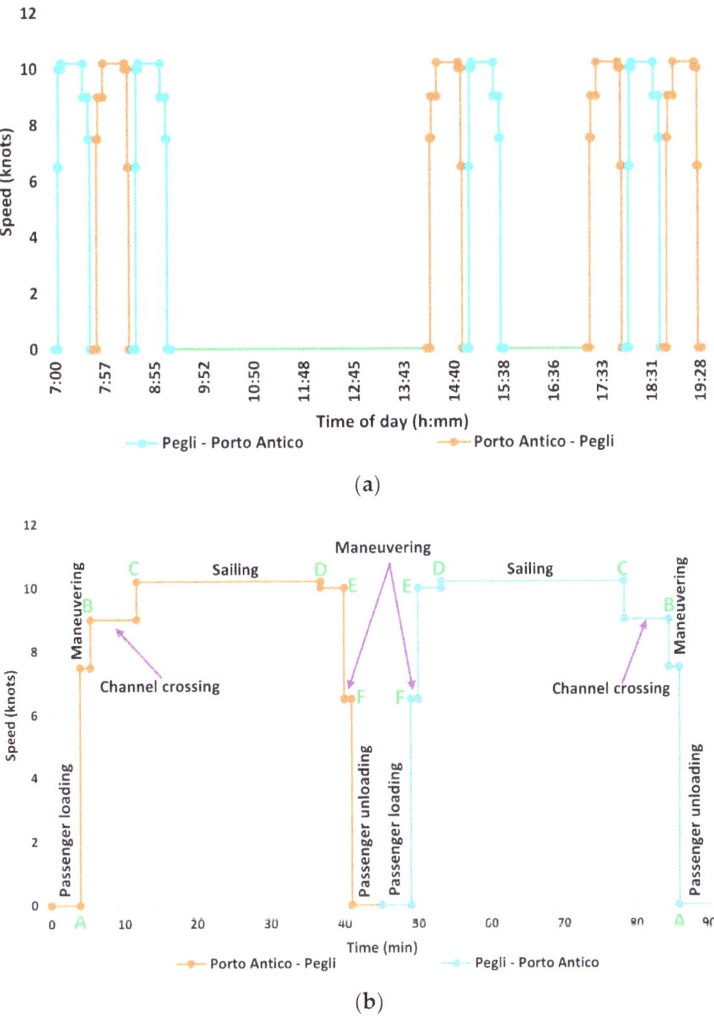

Figure 2. The operational profile of the NaveBus for (**a**) one day, (**b**) one round trip.

The paper aims to investigate the feasibility of replacing conventional diesel engines with alternative clean power systems based on FC and battery technologies. Figure 3 provides a schematic illustration of the different clean power systems (and different potential options in terms of fuel per each power system considered) taken into consideration for this study and the potential component combinations.

As shown in Figure 3, there are five categories: fuel storage system, fuel processing equipment, power generation system, power conditioning equipment, and propulsion/auxiliary system. For the power generation system, three different technologies are considered (proton exchange membrane fuel cell (PEMFC), solid oxide fuel cell (SOFC), and battery system), while looking at the fuel storage system onboard, the following fuels are considered: hydrogen, ammonia, LNG, and MeOH, with related fuel processing equipment. The power conditioning equipment is composed of a DC/DC converter, and a DC/AC inverter, while the fuel-processing equipment is based on the fuel type and consists of an ammonia cracker, natural gas reformer, and MeOH reformer. Regarding the propulsion system, the electric motor is selected to exist in all the proposed cases to deliver the required power to the FPP.

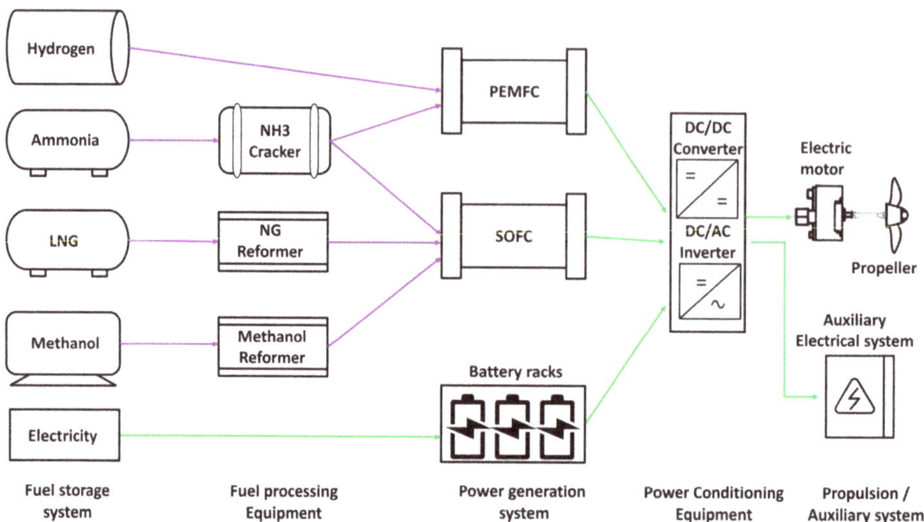

Figure 3. Conceptual diagram for clean power systems design considered in the current study.

3. Feasibility Assessment Method

The paper presents a multi-aspect analysis to determine the most feasible clean power system to be implemented onboard the case study according to different aspects: ferry daily journey and scheduling, available volumes and spaces, propulsion power needs, energy storage/fuel tank capacity needed, economics, etc. Different types of clean power systems (batteries, PEMFC, and SOFC) are proposed to be investigated to assess and evaluate their effectiveness onboard from an economic and design perspective. Therefore, this target can be investigated and accomplished by using the following methodology as shown in Figure 4.

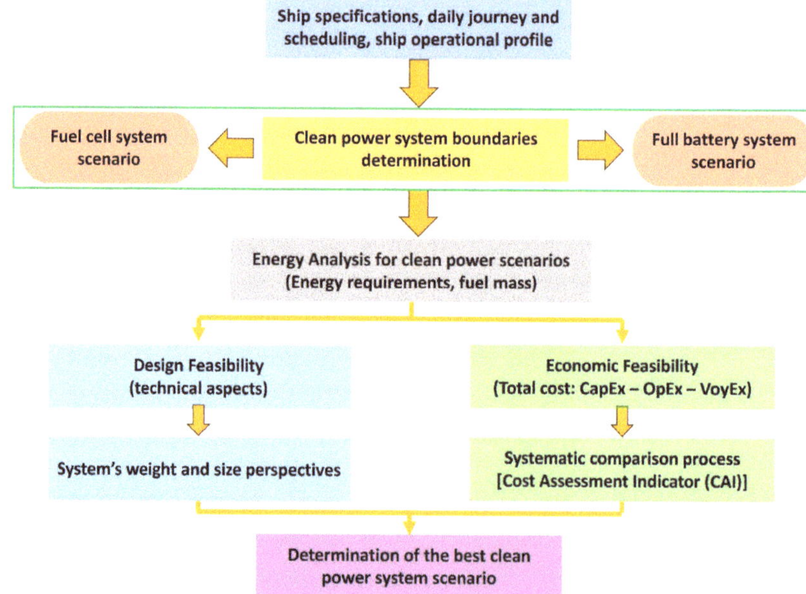

Figure 4. Overview of the steps applied in the assessment methodology.

The first step is to consider the input data at the start of the assessment procedure including the ship design specifications (available volumes, surface, weights), the ship operational profile based on voyage details, daily journey/scheduling, and the power requirements. Second, the power system boundaries in terms of its power capacity needs and fuel needs must be determined and it must be identified whether it is a FC-based system or a full battery electric power system. This step includes the identification of the methods to evaluate the energy requirement and fuel mass for the identified power system. The energy requirements for each potential clean power system scenario are then determined using an energy analysis. After that, the feasibility assessment is divided into economic and design aspects, the latter one intends to look at volume/spaces onboard and assess the clean power system's weight and volume, while the economic feasibility aspect is applied to the case study considering the total costs that contain capital expenses (CapEx), operational expenses (OpEx), and voyage expenses (VoyEx). Followed by a systematic comparison process that will be studied by using cost assessment indicators (CAI) as shown in Figure 5.

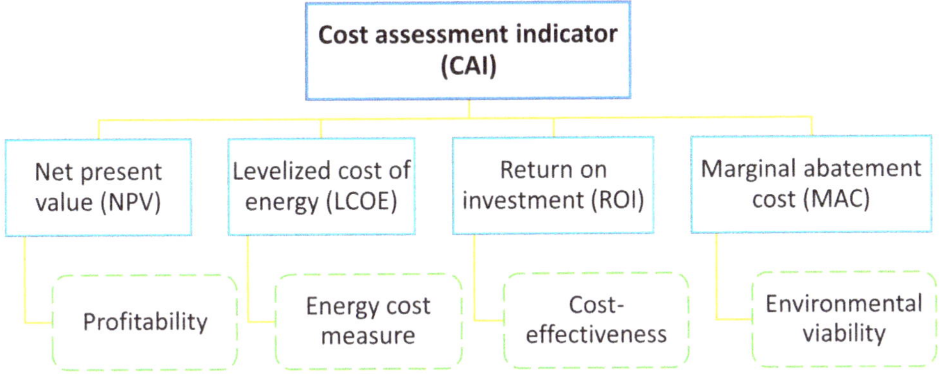

Figure 5. Identification of cost assessment indicators and their impact categories.

These indicators are used to quantify and reflect the performance of the power system from an economical perspective such as NPV, LCOE, ROI, and MAC. The last step of the assessment methodology is to present the results of the multi-aspect analysis and identify the best clean power system to guarantee a suitable retrofitting of the vessel and pledge a zero-emission journey to the ferry.

3.1. System Boundary Determination

System boundary determination is considered the crucial step in the methodology as it includes the methods that must be followed to calculate energy capacity and fuel consumption. This procedure is divided into two different subsections based on the applied power system (FC system and full battery system).

3.1.1. Fuel Cell System Scenario

The first step in system boundary determination is to calculate the required power capacity for the propulsion system and auxiliary system covered by using the FC system that will enable the definition of the number of FC modules/systems to be installed and foreseen onboard the ship while also looking at the typical commercial FC module/system power capacity. It is assumed that the current vessel's power requirement for propulsion and auxiliary systems (diesel-powered ferry) can be considered constant for the new clean energy system under investigation, except for the additional load related to auxiliaries when using the FC system by adding an extra factor.

Using the same approach, fuel consumption can be calculated by considering the FC efficiency, the type of fuel, and additional equipment for fuel processing. The FC efficiency

(η_{FC}) depends on the FC technology type and the required load factor. Also, the quantity of fuel depends on the fuel type which differs in the lower heating value (LHV) measured in kWh/kg. The required fuel mass by using the FC system (FM_{FC-f}) to perform the identified functional unit (three OWTs) can be calculated as shown in Equation (1).

$$FM_{FC-f} = \sum_{J_{owt}=1}^{J_{owt}=3} \sum_{OM=s}^{OM} \frac{(1+f_{se,f})*P_{OM,owt}*T_{OM,owt}}{\eta_{FC}*LHV_f} \tag{1}$$

where FM_{FC-f} is measured in kg, $f_{se,f}$ is the extra factor for the supplementary equipment (se), subscript (f) refers to the fuel type, $P_{OM,owt}$ is the average required power for each operational mode (OM) in the particular trip measured in (kW), J_{owt} is the number of OWT considered in the calculation, and $T_{OM,owt}$ is the duration of the operational mode at each trip measured in hours. The operational modes can be classified into several modes (s) such as maneuvering, sailing, etc., as presented in Section 2.

If ammonia is used as a hydrogen storage media onboard, more equipment for fuel processing such as a cracker and purifier have to be considered for ammonia's catalytic decomposition and the purification of residual ammonia to deliver pure hydrogen into the PEMFC [52]. Thus, the efficiency of the auxiliaries has to be considered once calculating the ammonia consumption as shown in Equation (2).

$$FM_{PEMFC-NH3} = \sum_{J_{owt}=1}^{J_{owt}=3} \sum_{OM=s}^{OM} \frac{(1+f_{se,f})*P_{OM,owt}*T_{OM,owt}}{\eta_{FC}*\eta_{cr}*\eta_{pu}*LHV_{H_2}*X_H} \tag{2}$$

where η_{cr} and η_{pu} are the efficiencies of the cracker and purifier that are assumed to be 80% and 90%, respectively, while (X_H) refers to the hydrogen content in ammonia, i.e., 17.8% [43,53].

Once investigating FC integration onboard, the start-up period has to be considered too, particularly for SOFC [54,55]. For this purpose, it is proposed to install a battery rack onboard the ship to cover the heating-up energy required for the FC during the start-up period. The function of the battery rack is to heat up the system to reach its operating temperature; after that, the FC generates the required electricity to propel the ship. For PEMFC, the heating-up energy depends on the fuel used, since the utilization of ammonia as a hydrogen carrier requires more heating energy than using pure hydrogen due to the presence of a cracker and purifier. Therefore, the formula in Equation (3) can be used to determine the battery energy capacity required during the starting-up period for covering the heating-up energy of the FC system.

$$HEC_{BT,f} = 1.5*P_{FC}*HEF_{FC,f} \tag{3}$$

where $HEC_{BT,f}$ is the required battery energy capacity in kWh, subscript BT refers to the battery, P_{FC} is the installed power of the FC, and $HEF_{FC,f}$ is the heating-up energy factor of the FC system measured in kWh/kW; its value varies with the FC type as shown in [43,56]. The capacity is proposed to be increased by 50% for considering the safety and battery's state of charge issues.

3.1.2. Full Battery System Scenario

The battery rack is the key component of the power system under investigation, and its capacity must be adequate to ensure that the ship can travel a specific path. The lithium-ion battery type is proposed to be investigated in the current study as it has unmatched qualities compared to other types such as a high-energy capacity, lowered self-discharging rate, quick charging capability, and high number of battery cycles [57,58]. The installed battery rack's energy capacity must be raised by 20% due to slow battery deterioration, which causes a capacity drop of up to 20% of its original capacity [59]. Additionally, the installed battery rack's energy capacity has to be raised by an additional 30% (10% for safety and 20% to keep the minimal level of capacity) [59]. Consequently, the battery rack's energy capacity (EC_{BT}) has to coincide with the energy requirement of the ship to perform the functional unit and has to be raised by an overall percentage equal to 50% for the reasons listed before. Thus, it can be calculated as shown in Equation (4).

$$EC_{BT} = \sum_{J_{owt}=1}^{J_{owt}=3} \sum_{OM=s}^{OM} 1.5*P_{OM,owt}*T_{OM,owt} \tag{4}$$

Since the battery lifetime can be given and expressed as the number of battery cycles, the batteries are replaced several times based on the ship's lifetime and the number of trips per year. The number of replacements ($N_{RE,BT}$) can be calculated as shown in Equation (5).

$$N_{RE,BT} = \frac{LT_{ship}*J_{owt,ann}}{3*Y_{Bc}} - 1 \tag{5}$$

where LT_{ship} is the lifetime of ship in years, $J_{owt,ann}$ is the number of OWT annually (ann), and Y_{Bc} is the number of battery cycles. The first part of Equation (5) is divided by three as each battery cycle is assumed to cover three consecutive trips as a functional unit for the case study. Moreover, a subtraction of one exists in Equation (5) that indicates the initial installation of batteries in the investment phase.

3.2. Total Cost Assessment Method

The economic evaluation of different power systems can be performed by using the total cost assessment which considers the total costs of a power system configuration during the ship's lifetime. In this study, the total costs have been divided into three terms CapEx, OpEx, and VoyEx.

Firstly, the CapEx represents the investment and installation costs of the power system. The OpEx includes the maintenance/operating and replacement costs. Moreover, VoyEx denotes the costs of fuel consumption/electricity onboard the ship annually. Therefore, the total cost of the clean power system can be calculated as shown in Equation (6).

$$TC_{cps} = CapEx_{cps} + \sum_{n=1}^{LT} OpEx_{cps,n} + \sum_{n=1}^{LT} VoyEx_{cps,n} \tag{6}$$

where TC_{cps} is the total cost of a clean power system (cps) over its lifetime, and (n) is the number of years in the ship's lifetime (LT).

3.2.1. Capital Expenses (CapEx)

The CapEx is the total investment cost of the clean power system. The proposed clean power system is composed of five categories which are the power generation system, fuel storage system, power conditioning equipment, fuel processing equipment, and electric motors. The power generation system can be PEMFC, SOFC, or battery racks. The fuel storage system cost is an important component of a clean energy system's CapEx because the fuel is different from conventional marine fuels, especially in the case of a power system's replacement like the current study. The cost of power conditioning equipment includes the cost of a DC/DC converter and DC/AC inverter. The formula that is used to calculate the CapEx of the proposed clean power system is shown in Equation (7).

$$CapEx_{cps} = CF_{ps} \times P_{ps} + CF_{fss} \times FC_{FC-f} + \left(\sum_{pce} CF_{pce} \times P_{ps} \right) + CF_{em} \times P_{ps} + \left(\sum_{oc} CF_{oc} \times P_{ps} \right) \tag{7}$$

where P_{ps} is the rated power of the proposed power system measured in kW in the case of PEMFC or SOFC and measured in kWh in the case of full battery electric system. In Equation (7), there are cost factors (CF) that vary with the components as follows: CF_{ps} is the cost factor of the power system measured in EUR/kW or EUR/kWh, CF_{fss} is the cost factor of the fuel storage system measured in EUR/kg-fuel, CF_{pce} is the cost factor of the power conditioning equipment measured in EUR/kW, CF_{em} is the cost factor of the electric motor measured in EUR/kW, and CF_{oc} is the cost factor of other components such as the reformer and cracker measured in EUR/kW. The cost factors of the power system components and their major technical parameters are shown in Table 2.

Table 2. Investment cost factors and technical parameters for power system components.

Component	Cost Factor (CF)	Technical Parameter	Reference
PEMFC	1500 EUR/kW	η_{peak} = 55%, Z_{FC} = 20,000 h	[60]
SOFC	5000 EUR/kW	η_{peak} = 60%, Z_{FC} = 20,000 h	[60]
Battery	210 EUR/kWh	Y_{BC} = 5000	[58,61]
DC/DC converter	120 EUR/kW	η = 98%, LT = 25 years	[62]
Electric motor	250 EUR/kW	η = 96%, LT = 25 years	[48,63]
Hydrogen tank	480 EUR/kg$_{H2}$	LT = 25 years	[64,65]
LNG reformer	370 EUR/kW	LT = 25 years	[66]
MeOH reformer	475 EUR/kW	LT = 25 years	[62]
Ammonia cracker	250 EUR/kW	LT = 25 years	[67]

The cost factors for storage tanks of ammonia, methanol, and LNG can be calculated by using the mathematical equations available in [30] that correlate the required storage capacity with the cost of the storage tank.

3.2.2. Operational Expenses (OpEx)

The second term in the total cost is the OpEx which includes the maintenance cost of the power system annually and the replacement cost of some parts of the power system over its lifetime. For all power systems, it is assumed that maintenance costs are associated with a growth rate of 2% annually. The maintenance cost of an electric battery power system is taken as 1% of its CapEx per year [62]. While the batteries must be replaced a certain number of times during the ship's lifetime as calculated before in Equation (5), the cost of this can be calculated based on the forecasted average price in [61].

Similarly, the operational expenses of the FC system include maintenance costs and replacement costs. The annual maintenance cost of PEMFC and SOFC is assumed to be 2% of their CapEx [62]. The crucial parameter in the FC power system's OpEx is the replacement cost as it is dependent on the FC lifetime and the forecasted number of replacements over the ship's lifetime. Based on the literature review [56,68], the lifetime of FCs is approximately 20,000 h and their stacks must be replaced after implementing these operational hours, while the Balance-of-Plant (BoP) system of FC racks demonstrates a prolonged lifespan compared to its stack; therefore, only the replacement of FC stacks will be considered. As a result of the forecasted development in the market size of FCs in the transportation sector, the FC prices may be reduced to about half of today's price [64]. Hence, the replacement cost of FC stacks is set to be 50% of its CapEx [64]. The number of replacements of the FC power system ($N_{RE,FC}$) can be calculated as shown in Equation (8).

$$N_{RE,FC} = \frac{LT_{ship} * J_{owt,ann} * T_{owt}}{Z_{FC}} - 1 \tag{8}$$

where LT_{ship} is the lifetime of the ship in years, $J_{owt,ann}$ is number of OWTs per year, T_{owt} is the duration of OWT in (h), and Z_{FC} is the lifetime of FC in hours. In the formula, there is a subtraction of one indicating the initial installation of FC in the investment step.

For the power conditioning equipment and the electric motor, there is no replacement required due to the high expected lifetime, but the annual operation and maintenance cost could be taken as 1% of its CapEx [62,69].

3.2.3. Voyage Expenses (VoyEx)

In the current paper, the VoyEx is based on the annual consumption of the energy carrier and its type. To estimate the annual VoyEx, it must be calculated by multiplying the energy carrier (fuel or electricity) consumption per trip by its price (EUR/kWh), then making a summation over all the trips per year as shown in Equation (9).

$$VoyEx_{cps} = \sum_{J_{owt}=1}^{J_{owt,ann}} ECC_{cps,owt} * CF_{ec} \tag{9}$$

where $VoyEx_f$ is the annual VoyEx in EUR/year, ECC is the energy carrier/fuel consumption measured in kWh, and CF_{ec} is the cost of the energy carrier measured in EUR/kWh. The cost of hydrogen, ammonia, LNG, and methanol based on the recent prices are 100, 81, 58, and 61 EUR/MWh, respectively [56,70,71], while the electricity cost is assumed to be 226 EUR/MWh based on the average price in the last five years in Italy [72]. Due to the lack of information available regarding the bunkering operation fees of alternative fuels, these fees have been neglected in the current study.

3.3. Cost Assessment Indicators

By bringing the entire expenses of various power systems down to the Net Present Value (NPV), it is possible to compare their total costs with each other. According to Equation (10), the NPV of the proposed clean power system (cps) is evaluated.

$$NPV_{cps} = CapEx_{cps} + \sum_{n=1}^{LT} \frac{OpEx_{cps,n}}{(1+d)^n} + \sum_{n=1}^{LT} \frac{VoyEx_{cps,n}}{(1+d)^n} \tag{10}$$

where d implies a discount rate that is set at 5%, and n is the number of years. Furthermore, LCOE can be used to compare different alternative power systems in terms of the energy cost measure on a consistent basis. LCOE depends mainly on the NPV of the clean power system and the total energy generated from it. LCOE can be calculated as shown in Equation (11) [73], in which $E_{owt,n}$ is the total energy generated in MWh during OWT through the year (n).

$$LCOE_{cps} = \frac{NPV_{cps}}{\sum_{n=1}^{LT} \frac{\sum_{J_{owt}=1}^{J_{owt,ann}} E_{owt,n}}{(1+d)^n}} \tag{11}$$

Additionally, the return on investment (ROI) may be calculated in order to gain a sense of the cost-effectiveness of the clean power system, as presented in Equation (12).

$$ROI_{cps} = \frac{\sum_{n=1}^{n=20} \left(OpEx_{DP,n} - OpEx_{cps,n} \right) + \sum_{n=1}^{n=20} \left(VoyEx_{DP,n} - VoyEx_{cps,n} \right) - CapEx_{cps}}{CapEx_{cps}} \tag{12}$$

As shown in Equation (12), the ROI is based on the difference between the operational and voyage expenses of each clean power system and the diesel-powered system (DP). The annual operational costs of diesel power systems are assumed to be 5 EUR/kW (2% of CapEx per year) as reported in [62], while its VoyEx is based on the recent price of diesel fuel (1.85 EUR/Liter) in Genoa refueling stations [74].

Moreover, MAC is a crucial indicator in the economic and environmental assessment, specifically in the case of evaluating the financial rationale for pursuing and investing in a clean power system [75,76]. The MAC is defined as the ratio between the costs or savings that would be incurred from the retrofitting process of the vessel's power system and the abated emissions over the lifetime of the ship that could guarantee a zero-emission journey to the ferry [77]. This indicator can be calculated for each power system by considering the total capital costs and the annual costs/savings discounted over the lifetime of the clean power system. The total capital costs and the annual costs discounted to the present value are expressed as NPV which can be calculated by using Equation (10). For the case study of retrofitting process, the savings result from the removed OpEx and VoyEx of the diesel-powered system discounted to the present value. In this study, the formula shown in Equation (13) can be used to calculate the MAC [78].

$$\text{MAC}_{\text{cps}} = \frac{\text{NPV}_{\text{cps}} - \left(\sum_{n=1}^{LT} \frac{\text{OpEx}_{\text{DP},n} + \text{VoyEx}_{\text{DP},n}}{(1+d)^n}\right)}{\sum_{n=1}^{LT} \left(E_{CO_2,\text{DP}} - E_{CO_2,\text{cps}}\right)} \tag{13}$$

where $E_{CO_2,\text{DP}}$ is the annual carbon dioxide (CO_2) emissions displaced from the diesel-powered system that is proposed to be replaced and $E_{CO_2,\text{cps}}$ is the annual CO_2 emissions resulting from the clean power system, if available. The displaced annual CO_2 emissions from the diesel-powered system can be calculated by multiplying the annual diesel fuel consumption by the CO_2 emission factor (3.206 kg-CO_2/kg-fuel) [79]. On the other hand, there are no CO_2 emissions from all power systems that are proposed in the paper except the SOFC system powered by natural gas: its emission rate is equal to 308 kgCO_2/MWh$_e$ (per each electricity unit produced) as reported in the datasheet of Bloom Energy [80]. Moreover, the SOFC system powered by methanol has a significant CO_2 emission resulting from the methanol reforming process that can be calculated as shown in [81,82].

Since cost-assessment indicators depend on various assumptions such as the costs of fuel energy and electricity, sensitivity analysis is proposed to be applied for discussion regarding the credibility of the results. In the sensitivity analysis, the fuel and electricity costs are proposed to be varied by $\pm 30\%$, with an increment of 10%.

3.4. Design Feasibility Assessment Method

The feasibility assessment based on the design perspective intends to evaluate the weight and volume of the clean power system to assess its viability to be installed onboard. Similar to the economic feasibility, the weight and volume of the proposed clean power system is based on the weight/volume of each component as shown in Equations (14) and (15).

$$W_{\text{cps}} = \frac{P_{\text{ps}}}{GD_{\text{ps}}} + W_{\text{fss}} + \left(\sum_{\text{pce}} \frac{P_{\text{ps}}}{GD_{\text{pce}}}\right) + \frac{P_{\text{ps}}}{GD_{\text{em}}} + W_{\text{oc}} \tag{14}$$

$$V_{\text{cps}} = \frac{P_{\text{ps}}}{VD_{\text{ps}}} + V_{\text{fss}} + \left(\sum_{\text{pce}} \frac{P_{\text{ps}}}{VD_{\text{pce}}}\right) + \frac{P_{\text{ps}}}{VD_{\text{em}}} + V_{\text{oc}} \tag{15}$$

where W_{cps} and V_{cps} are weight in (kg) and volume in (m^3) of the clean power system. The weight and volume of each component is expressed in terms of its gravimetric power density (GD) and volumetric power density (VD) as (GD) is measured in kW/kg, while (VD) is measured in kW/m^3. As shown in Equations (14) and (15), the weight and volume are based on the rated power of the proposed power system (P_{ps}).

Regarding the fuel storage system, its weight and volume (W_{fss} and V_{fss}) can be calculated based on the mathematical functions in [30] that correlate the required storage capacity with the weight and volume of the storage tank. These functions were created using extensive market research, literature studies, and confidential discussions with the authors' research group's industry partners. There are mathematical functions for storage tanks of hydrogen, ammonia, LNG, and MeOH; moreover, there are other functions for fuel-processing equipment to calculate its weight (W_{oc}) and volume (V_{oc}).

Based on the technical specifications gathered by the authors in [35] for the commercial products of PEMFC and SOFC, there is a suitable PEMFC commercial product from Ballard [83] called Fcwave, which is designed for maritime applications and certified by DNV to be employed in marine environments. For the SOFC, there is a commercial system available from Bloom Energy called Energy Server 5 [80] which is utilized for stationary applications and not certified yet to be employed in marine environments. Regarding the full battery scenario, there is a commercial product available in the market from Corvus that is called Corvus Dolphin Energy [84], and it can be selected for the design feasibility assessment of the case study. The technical specification of the selected commercial products is shown in Table 3 with a focus on the volumetric density and gravimetric density.

Table 3. Technical specifications of commercial products for PEMFC, SOFC, and battery.

Parameter	PEMFC [83]	SOFC [80]	Battery Pack [84]
Supplier	Ballard	Bloom Energy	Corvus
Rated power (kW)	200	330	132.5 kWh
Voltage range (V)	350–720	480	576–797
Physical dimensions L $\times$ W $\times$ H (m)	$1.21 \times 0.74 \times 2.2$	$5.5 \times 2.6 \times 2.1$	$1.85 \times 0.5 \times 0.67$
Weight (kg)	1000	15,800	782
Volumetric density (kW/m^3)	101.4	11.12	214 kWh/m^3
Gravimetric density (kW/kg)	0.2	0.021	0.169 kWh/kg

The weight and volume of the electric motor can be calculated based on the commercial products available from ABB [85] that are fulfilled with classification societies' requirements. Furthermore, the weight and volume of the power conditioning equipment such as DC/DC converters can be calculated based on a commercial product available in [86]. It is considered to use the same power condition equipment for all clean power systems.

4. Results and Discussion

This section investigates the feasibility results of replacing the conventional diesel power system onboard the NaveBus by using three alternative power system scenarios. The section is divided into three subsections, the first one is related to the energy analysis results based on the alternative power system scenarios, while the second and third subsections represent the feasibility assessment results from the economic and design points of view, respectively.

4.1. Energy Analysis Results

The energy analysis is a critical investigation, especially in the replacement of a conventional system with a clean one such as that proposed in this paper, because the number of installed FC/battery modules depends on the required power/energy to accomplish the identified functional unit. Moreover, energy evaluation is crucial for designing a suitable fuel storage system, especially by using alternative fuels characterized by a different volumetric density compared to conventional diesel fuel. Based on the collected data onboard the ferry and the operational profile described in Figure 2, the electric power and energy requirements for the case study are shown in Table 4.

Table 4. Electric power and energy requirements to perform a one-way trip.

Operational Mode	Time (min)	Rated Power (kW)	Energy Consumption (kWh)
A-B	1.46	258	6.3
B-C	6.28	458	47.9
C-D	25.16	750	314.4
D-E	3.10	681	35.2
E-F	1.02	169	2.9
Total	37		407

In fact, the FC efficiency varies with the rated power and the load factor for each operational mode. For the current study, this variation can be between 46.5% and 53.5% for PEMFC, while the efficiency varies between 51.5% and 58.5% for the SOFC scenario. On the other hand, the energy capacity of the full battery scenario must be increased by 50% more than the estimated energy at each operational mode as presented in Section 3.1.2. The results of the electric energy requirements based on the FC system and full battery scenarios are shown in Figure 6.

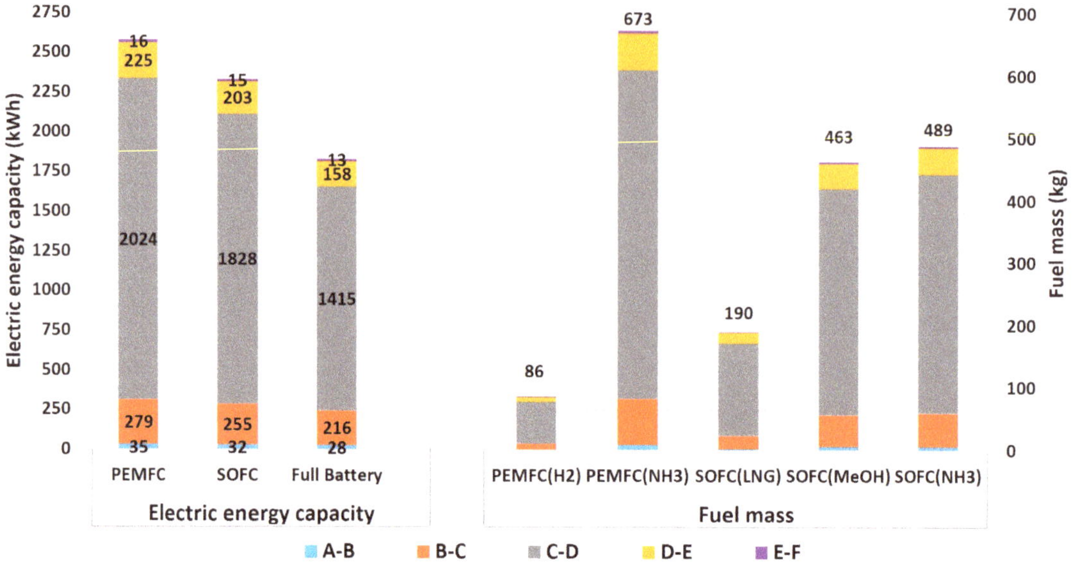

Figure 6. The required electric energy capacity (**left side**) and fuel mass (**right side**) to perform three OWTs based on different scenarios for each operational mode.

As shown in Figure 6 (left side), the total electric energy capacity to accomplish the functional unit of the case study (three OWTs) by using PEMFC and SOFC must be 2579 kWh and 2333 kWh, respectively, while the total battery capacity must be 1830 kWh. The sailing operational mode (C-D) contributes about 78% of the total required energy capacity for all scenarios.

The required fuel mass can be calculated for the PEMFC and SOFC scenarios by considering the LHV of each fuel. The results of fuel mass are shown in Figure 6 (right side) after adding a design margin of 10% for all cases.

As shown in Figure 6 (right side), the total required fuel mass to accomplish three trips in the case of the PEMFC powered by hydrogen and ammonia is 86 kg and 673 kg, respectively. Furthermore, in the case of SOFC scenario, the required mass of LNG, MeOH and ammonia is 190 kg, 463 kg, and 489 kg, respectively.

4.2. Economic Feasibility Assessment Results

In this subsection, the results of the economic feasibility for the alternative power systems are presented based on the methodology discussed in Section 3. The results of the total cost of alternative clean power systems are presented in Figure 7, in which different options are assessed based on different cost categories that include the capital cost of the power system. Moreover, it includes the total expected OpEx and VoyEx over the lifetime of the ship (that is assumed to be 20 years).

As shown in Figure 7, the VoyEx expenses have the highest contribution to the total cost over other expenses for all PEMFC scenarios and the full battery scenario; this is because of the high expenses of the fuel/electricity. The ammonia-powered SOFC system demands a significant quantity of ammonia; hence, its VoyEx costs are greater than the SOFC systems operated by LNG or methanol. Although SOFC power systems have a higher efficiency than PEMFC and require less fuel to generate electricity onboard the ship, the CapEx of SOFC systems powered by different fuels is higher than the CapEx of PEMFC systems and full battery systems. On the other hand, the OpEx of the full battery system is quite similar to the PEMFC systems, while it is lower than the OpEx of SOFC systems. The OpEx of SOFC systems is constant when using different fuels, as it depends on the replacement cost and maintenance cost that is independent of the type of fuel.

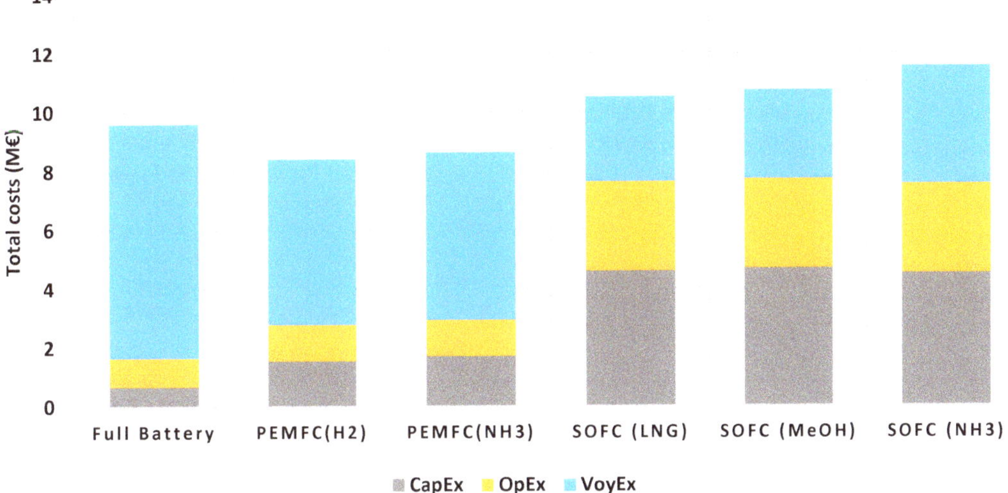

Figure 7. The cost assessment results of different alternative power systems.

To figure out the profitability of the clean power system, the total costs are converted to the NPV to compare the alternative options with each other as presented in Figure 8. The results show that the hydrogen-powered PEMFC system is the best option in terms of NPV for the replacement of the existing power system onboard the NaveBus, as its NPV is equal to EUR 5.8 million which is lower than other clean power systems. Moreover, the NPV of ammonia powered PEMFC system and full battery system is EUR 6 million and 6.2 million, respectively. Due to the high VoyEx of the full battery system, the NPV of a full battery system is greater than that of a PEMFC system.

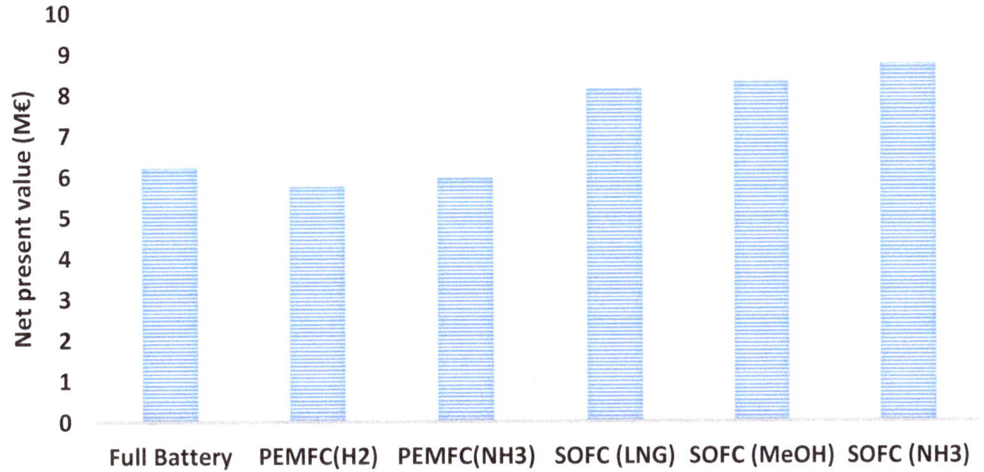

Figure 8. Total cost comparison between clean power systems in terms of the net present value.

The NPV of SOFC systems depends on the fuel type and ranges from around EUR 8.1 million to 8.8 million. When a hydrogen PEMFC system is compared to SOFC power systems in terms of NPV, it is lower than them by about EUR 2.4–3 million depending on the type of fuel used inside the SOFC. This is mainly because of the high capital expenditure of the SOFC power system and the low OpEx of the PEMFC power system.

Furthermore, it is important to assess the energy cost level of the different clean power systems that can be determined by calculating the LCOE as presented in Section 3.3. The results reveal that the LCOE of the full battery system is equal to 427 EUR/MWh, while the LCOE of the PEMFC powered by hydrogen and ammonia is 396 EUR/MWh and 410 EUR/MWh, respectively. On the other hand, the LCOE of the SOFC power system varies between 558 EUR/MWh and 600 EUR/MWh based on the fuel type. Therefore, the hydrogen-powered PEMFC system is the most economically feasible option for the retrofitting process onboard the NaveBus as it has the lowest cost of energy, followed by the full battery system scenario.

The cost-effectiveness of retrofitting the conventional diesel power system onboard the NaveBus with a clean power system is evaluated by using the concept of ROI as discussed in Section 3.3. The ROI is calculated based on the operational and voyage expenses difference between the diesel power system and the proposed clean power system over the lifetime of the ship. The results are shown in Figure 9.

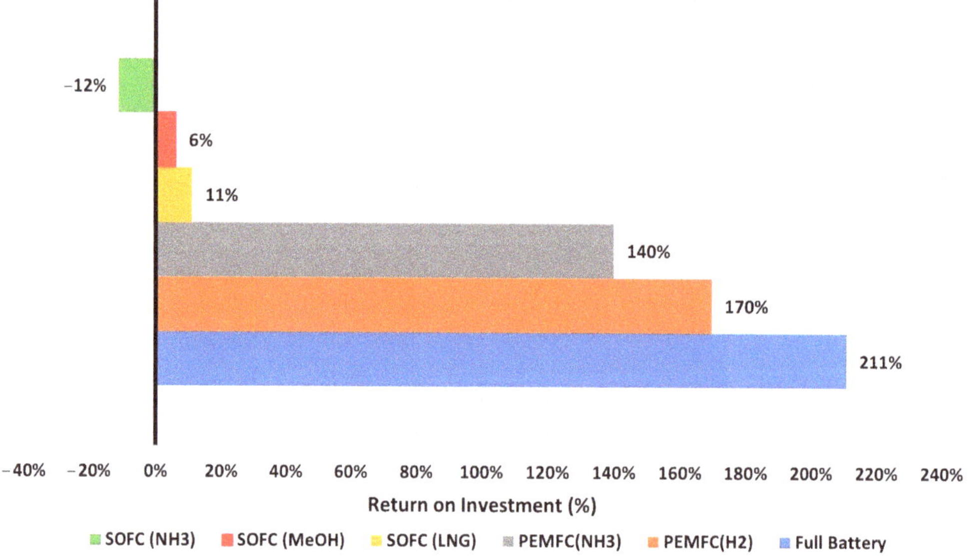

Figure 9. Return on investment of different clean power systems.

As shown in Figure 9, the ROI varies with the power system whether it is a PEMFC, SOFC, or full battery system. The results show that the full battery system scenario has the greatest profitability trend over other alternatives, as its ROI equal to 211% with an annualized ROI equal to 5.8%, while the ROI of a PEMFC operated by hydrogen and ammonia is 170% and 140%, with an annualized ROI equal to 5.1% and 4.5%, respectively. The PEMFC system and full battery system accomplish a high profitability because of their low CapEx at the initial investment and the low OpEx compared to the high diesel fuel costs that were eliminated by the retrofitting process. On the other hand, the SOFC system operated by ammonia has a negative ROI that proves that the system will not accomplish a profit over the entire lifetime of the ship due to its high CapEx and VoyEx.

The environmental viability of an alternative power system must be assessed when considering the retrofitting of a ship's power system and investing in a clean power system as an emission reduction option. Therefore, the MAC of each scenario is calculated to figure out its environmental viability and ease the investment decision. The MAC is based on the NPV results and the expected displaced CO_2 emissions when applying the alternative power system. Figure 10 shows the MAC results of different power systems on the *y*-axis, while the total CO_2 emissions abated over the lifetime are shown on the *x*-axis. The negative values of MAC indicate the amount of cost-saving that was achieved to abate 1 ton of CO_2, while the positive values refer to the cost required to abate 1 ton of CO_2.

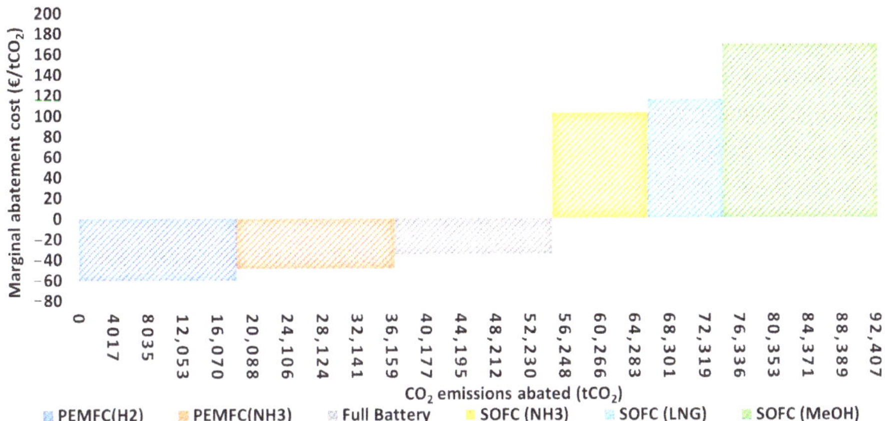

Figure 10. Marginal abatement cost for reducing CO_2 emissions by using different alternative clean power systems.

A total of 18,255 tons of CO_2 are expected to be displaced over the lifetime by using a full battery system, a PEMFC operated by hydrogen or ammonia, and an SOFC system operated by ammonia. As shown in Figure 10, the hydrogen-powered PEMFC system has the highest environmental viability, as it abates a high amount of emissions at the lowest price. Moreover, the hydrogen PEMFC will guarantee a saving equal to 60 EUR per ton of CO_2 abated; therefore, this scenario should be prioritized for the retrofitting process. This scenario is followed by PEMFC operated by ammonia and the full battery system, as they also guarantee a saving equal to 49 and 35 EUR/ton-CO_2, respectively.

Although the capital cost of the SOFC system operated by different fuels is almost the same and their MAC has a positive value, the MAC when using ammonia (103 EUR/ton-CO_2) is better than for SOFC operated by LNG and methanol (115 and 169 EUR/ton-CO_2). The SOFC system operated by LNG and methanol displaces an amount of CO_2 emissions equal to 11,041 and 8516 tons, respectively, over the entire lifetime which is equal to 60% and 47% of the abated CO_2 emissions by other systems.

The variation in clean power system energy carriers (fuels and electricity) costs has impacts on VoyEx, of course. Such variations may arise due to uncertainty in their market prices (both quite volatile in recent years); therefore, it is crucial to discuss the credibility of the results by applying a sensitivity analysis. The effect of changing fuel and electricity costs on LCOE for each clean power system is shown in Figure 11.

As shown in Figure 11, by increasing the prices of energy carriers/fuels for different systems, the LCOE value is increased by a significant amount, especially the full battery and PEMFC systems as their LCOE increases by about 24% and 18%, respectively, when their energy carrier/fuel costs increase by 30%. This significant increasing trend resulted from the higher contribution of the VoyEx parameter in the total costs of the full battery and PEMFC systems. On the other hand, the LCOE of SOFC systems is increased by about 7–9% when fuel costs rise by 30%, as their VoyEx parameter has a lower contribution to their total costs when compared to their CapEx.

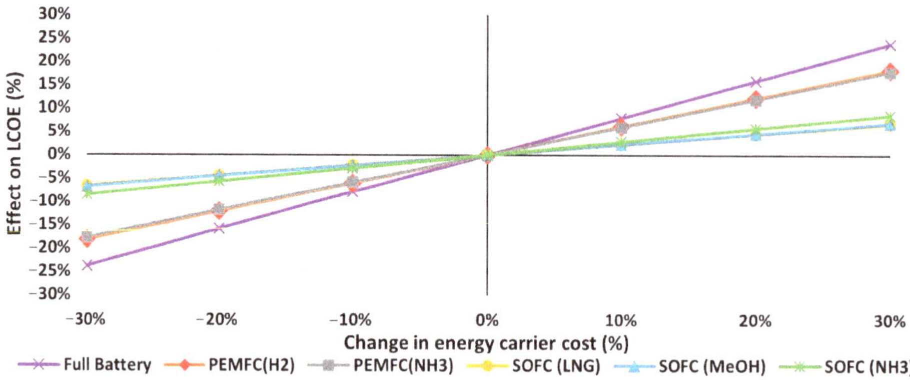

Figure 11. Sensitivity analysis results focusing on the effect of energy carrier/fuel costs on the LCOE of the considered system.

Moreover, the sensitivity analysis proves the credibility of the baseline case results, which show that with increasing or decreasing fuel costs as shown in Figure 11, the hydrogen-powered PEMFC has the lowest LCOE (468 EUR/MWh with a fuel cost change of +30%), followed by ammonia-powered PEMFC (483 EUR/MWh with a fuel cost change of +30%), and the full battery system (529 EUR/MWh with an electricity cost change of +30%), while the LCOE of the SOFC system powered by LNG, methanol, and ammonia (with a fuel cost change of +30%) is 595, 608, and 651 EUR/MWh, respectively.

4.3. Design Feasibility Assessment Results

This subsection investigates the feasibility results of replacing the conventional diesel power system onboard the NaveBus by using three alternative power system scenarios from the design point of view. For the retrofitting process of the NaveBus, there are some components that will be removed such as two diesel main engines, two gearboxes, two auxiliary generators, and two diesel fuel tanks. The estimated weight and volume of these removed parts is 16 tons and 26 m^3.

By looking at the power and energy analysis results, the required power for the NaveBus at the maximum service speed reaches 750 kW, while the required battery energy capacity to perform three OWTs is 1830 kWh based on the full battery system scenario. Therefore, the weight and volume of the different power generation systems can be evaluated as discussed in Section 3.4. Moreover, the fuel mass has been calculated as shown in Figure 6 and by applying the mathematical formulas in [30], the weight and volume of the storage tanks can be calculated. The weight and volume of the different clean power systems are described in Figure 12 after applying the formula in Equations (14) and (15).

As shown in Figure 12, the most feasible scenario to guarantee a suitable retrofitting of the vessel in terms of weight and volume is the hydrogen-powered PEMFC system followed by the full battery system and the PEMFC system powered by ammonia. The ferry's power system can be retrofitted by a hydrogen-powered PEMFC system without issues, as its weight and volume (11 tons and 17 m^3) are lower than the removed parts' weight and volume by a considerable percentage (31.3% and 34.6%). Moreover, the weight of the full battery system and PEMFC system powered by ammonia is 15 tons, while their volume is 11 m^3 and 25 m^3, respectively.

Although the SOFC has the advantage of fuel flexibility and higher electrical efficiencies than the PEMFC systems, the results showed that the power system based on SOFC is not feasible from the design point of view to be fitted inside the NaveBus as a propulsion system because of the limitation in volume and weight. The total weight of the SOFC system fueling by different fuels varies between 49 tons and 51 tons, while its volume is

88 m^3, where the power generation unit contributes about 75–80% of the total weight and 83% of the total volume.

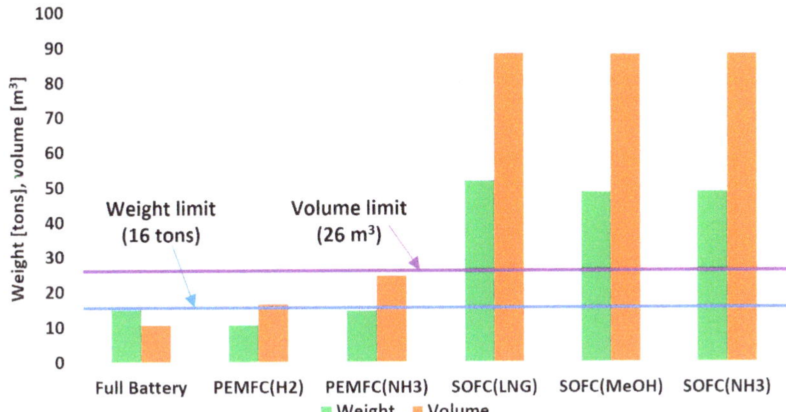

Figure 12. Weight and volume results of different power system scenarios.

For the hydrogen-powered PEMFC scenario (the most feasible scenario), the required power for the NaveBus can be covered using four modules of Ballard Fcwave [83] that can deliver a rated power of up to 800 kW. Therefore, Figure 13 shows the block diagram of the propulsion system including the main components such as the FC modules, battery rack, DC/DC converters, DC/AC inverter, electric motor, and FPP. The four PEMFC modules are distributed through the engine room. Each PEMFC rack supplies power to the DC bus main switchboard (SWBD) through a DC/DC converter. The propulsion power is delivered to the electric motor through a DC/AC inverter.

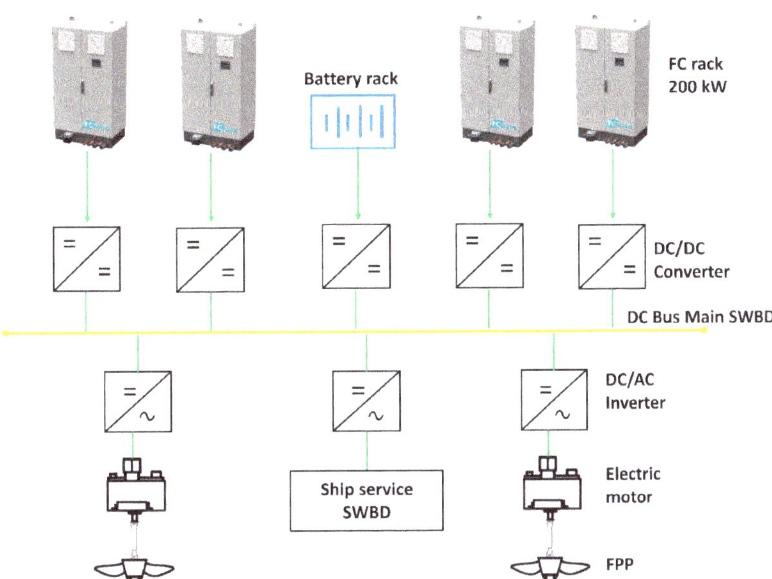

Figure 13. Design of propulsion system electrical architecture block diagram based on PEMFC.

Moreover, the ship service switchboard is supplied with electric power through a DC/AC inverter to deliver the required power for hoteling services such as lighting, fans, pump, etc. There is a small lithium-ion battery rack to support the PEMFC system during the transient loads and to cover the heating up energy required for the FC during the start-up period. The battery rack can be charged when the FC racks provide the load to propel the ship.

Furthermore, for the full battery system, based on the technical specifications of the battery rack available from Corvus [84] and that mentioned in Table 3, the required energy capacity of 1830 kWh can be covered by using 14 packs (each pack includes 16 modules, while the capacity of each module is 8.3 kWh).

Figure 14 shows the block diagram of the full electrical battery propulsion system including the main components such as the battery racks, DC/DC converters, DC/AC inverter, DC bus main switchboard, electric motor, and FPP. There are seven racks installed on each side of the ship to keep its stability and connected by the main SWBD through a DC/DC converter.

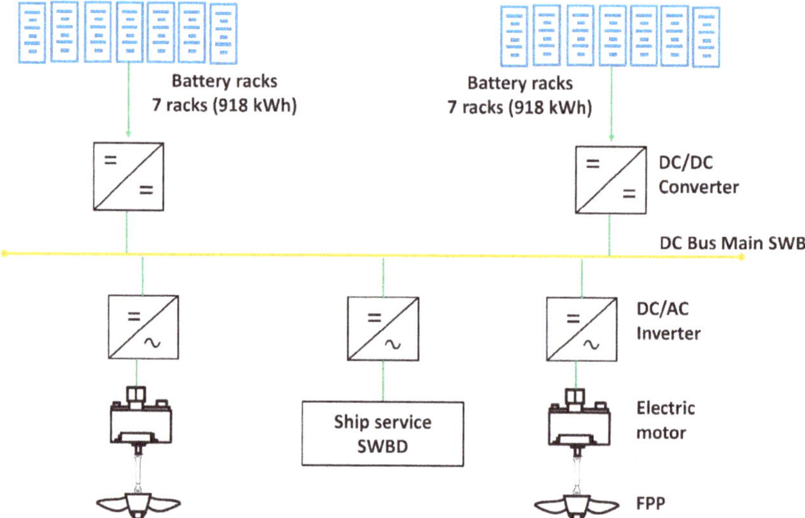

Figure 14. Full battery power system electrical architecture block diagram.

5. Conclusions

The current paper investigates the feasibility of replacing the conventional power system by using alternative clean power systems onboard one of the passenger ferries belonging to the short-sea navigation fleet. The case study is a ferry implementing a short sea journey and used in the port of Genova as a public transport vehicle. The ferry is analyzed to evaluate a "zero emission propulsion" retrofitting process. This type of ship is considered a suitable case for examining the viability of innovative technologies in the maritime industry because of their low energy consumption, their large audience impact, and their navigational routes which are close to the ports and the shore. The investigated clean power systems include fuel cell technologies and a full battery electric system. PEMFC, a low-temperature FC technology, and SOFC, a high-temperature FC, are both examined in this paper. The PEMFC is proposed to be operated by using alternative clean fuels such as hydrogen and ammonia, while LNG, MeOH, and ammonia are evaluated to power the SOFC.

The paper assesses the feasibility of installing the alternative clean power system instead of the diesel-powered system from the energy, design, and economic perspectives. The design assessment approach includes the assessment of the system's weight and size with an emphasis on defining the system components, while cost-assessment indicators were employed to assess the viability of the power system from an economical perspective. The following is a summary of the study's major results:

- The total fuel energy capacity to accomplish the target three OWTs by using PEMFC and SOFC must be 2579 kWh and 2333 kWh, respectively, while the required battery energy capacity is 1830 kWh for the full battery system scenario.
- Among the options taken into consideration, the PEMFC system fueled by hydrogen and ammonia has the lowest total costs at EUR 8.4 million and 8.6 million, respectively. However, due to its high voyage expenses (electricity cost), the full battery system scenario has a total cost of roughly EUR 9.6 million.
- Despite the fact that the SOFC is more fuel-efficient and takes less fuel to produce electricity, the total cost assessment showed that the power system based on SOFC has higher total costs than other solutions.
- The results showed that the LCOE of the PEMFC system powered by hydrogen and ammonia is 396 EUR/MWh and 410 EUR/MWh, while the full battery system's LCOE is 427 EUR/MWh. On the other hand, the LCOE of the SOFC power system varies depending on the fuel type and its value is between 558 and 600 EUR/MWh.
- The results indicated that the cost-effectiveness of retrofitting the conventional diesel power system onboard the NaveBus by the full battery system scenario is viable and achieves an ROI equal to 211%. Moreover, the PEMFC system operated by hydrogen and ammonia has a high profitability trend, as the ROI is 170% and 140%, respectively.
- The PEMFC system powered by hydrogen has the best environmental viability over other options, since it achieves a high reduction in CO_2 over the lifetime of the ship with a saving of 60 EUR/ton-CO_2; hence, this scenario should be given priority during the retrofitting process.
- From a weight and volume perspective, the hydrogen-powered PEMFC system is considered the best clean power system to guarantee, for this specific case study, a suitable retrofitting of the diesel power system This is because its weight is lower by 31.3% compared to the removed parts' weight, and its volume is lower by 34.6% compared to the removed parts' volume.
- The performed design feasibility study indicated that the power system based on SOFC technology could not be fitted inside the case study because of the limitation in volume and weight that is available onboard. The total weight of the SOFC system fueling by different fuels varies between 49 tons and 52 tons, while its volume is 88 m^3.

Even though the paper primarily focuses on the NaveBus as a case study, the technical and economic feasibility assessment methodology that has been developed is generally applicable to other short-distance passenger ferries to achieve the decarbonization of ships working on short-sea navigation. For larger vessels, particularly looking at economic methodology, it would be important to consider savings in terms of CO_2 taxes (as the vessel could be subject to ETS) and externalities (e.g., savings form sanitary systems expenses due to lower NOx, SOx, and PM emissions in urban environments).

Author Contributions: Conceptualization, A.G.E. and S.B.; methodology, A.G.E.; software, A.G.E.; validation, M.R. and A.F.M.; formal analysis, A.G.E.; investigation, A.G.E. and S.B.; resources, S.B. and M.R.; data curation, A.G.E.; writing—original draft preparation, A.G.E. and S.B.; writing—review and editing, S.B. and M.R.; visualization, A.G.E. and M.R.; supervision, A.F.M. All authors have read and agreed to the published version of the manuscript.

Funding: This research received no external funding.

Institutional Review Board Statement: Not applicable.

Informed Consent Statement: Not applicable.

Data Availability Statement: Not applicable.

Acknowledgments: The authors would like to acknowledge the technical support from the Liguria Via Mare Consortium and NaveBus crew for providing us with the required data for the case study.

Conflicts of Interest: The authors declare no conflict of interest.

Nomenclature

Abbreviations

BoP	Balance of plant
CAI	Cost-assessment indicators
CO_2	Carbon dioxide
DC	Direct current
DNV	Det Norske Veritas
ETS	Emission trading system
EU	European Union
FC	Fuel cell
FPP	Fixed-pitch propellers
GHG	Greenhouse gases
IMO	International Maritime Organization
LCOE	Levelized cost of energy
LNG	Liquefied natural gas
MAC	Marginal abatement cost
MARPOL	International Convention for the Prevention of Pollution from Ships
MDO	Marine diesel oil
MeOH	Methanol
NOx	Nitrogen oxide
NPV	Net present value
OWT	One-way trip
PEMFC	Proton exchange membrane fuel cell
PM	Particulate matter
ROI	Return on investment
R&D	Research and development
SOFC	Solid oxide fuel cell
SOx	Sulfur oxide

Variables

CapEx	Capital expenses (EUR)
CF	Cost factor (EUR/kW or EUR/kWh or EUR/kg-fuel)
d	Discount rate (%)
E	Electricity energy generated (kWh)
EC	Energy capacity (kWh)
ECC	Energy carrier consumption (kWh)
f	Extra factor for the supplementary equipment (-)
FM	Fuel mass (kg)
GD	Gravimetric density (kW/kg)
HEC	Heating up energy capacity (kWh)
HEF	Heating up energy factor (kWh/kW)
J	Number of one-way trips (-)
LHV	Lower heating value (kWh/kg)
LT	Lifetime (years)
n	Number of years (year)
N	Number of replacements (-)
P	Rated power of fuel cell (kW)
OpEx	Operating expenses (EUR)
T	Operational time [hour]
V	Volume (m^3)
VD	Volumetric density (kW/m^3)
VoyEx	Voyage expenses (EUR)
W	Weight (kg)
X_H	Hydrogen content in ammonia (%)
Y	Number of battery cycles (-)
Z	Lifetime of fuel cell (hours)
η	Efficiency (%)

Subscripts

ann	Annual
BC	Battery cycles
BT	Battery
cps	Clean power system
cr	Cracker
DP	Diesel-powered system
ec	Energy carrier
em	Electric motor
f	Fuel type
FC	Fuel cell
fss	Fuel storage system
oc	Other components
OM	Operational mode
pce	Power conditioning equipment
ps	Power system
pu	Purifier
se	Supplementary equipment
RE	Replacement

References

1. IMO. *Initial IMO Strategy on Reduction of GHG Emissions from Ships*; Resolution MEPC.304(72); International Maritime Organization (IMO): London, UK, 2018. Available online: https://wwwcdn.imo.org/localresources/en/KnowledgeCentre/IndexofIMOResolutions/MEPCDocuments/MEPC.304(72).pdf (accessed on 20 July 2023).
2. IMO. *2023 IMO Strategy on Reduction of GHG Emissions from Ships*; Resolution MEPC.377(80); International Maritime Organization (IMO): London, UK, 2023; pp. 1–18. Available online: https://www.imo.org/en/OurWork/Environment/Pages/2023-IMO-Strategy-on-Reduction-of-GHG-Emissions-from-Ships.aspx (accessed on 27 July 2023).
3. IMO. *Amendments to the Annex of the Protocol of 1997 to Amend the International Convention for the Prevention of Pollution from Ships, 1973, as Modified by the Protocol of 1978*; Resolution MEPC.203(62); International Maritime Organization (IMO): London, UK, 2013. Available online: https://wwwcdn.imo.org/localresources/en/KnowledgeCentre/IndexofIMOResolutions/MEPCDocuments/MEPC.203(62).pdf (accessed on 20 July 2023).
4. Schinas, O.; Stefanakos, C.N. Selecting Technologies towards Compliance with MARPOL Annex VI: The Perspective of Operators. *Transp. Res. D Transp. Environ.* **2014**, *28*, 28–40. [CrossRef]
5. Brynolf, S.; Magnusson, M.; Fridell, E.; Andersson, K. Compliance Possibilities for the Future ECA Regulations through the Use of Abatement Technologies or Change of Fuels. *Transp. Res. D Transp. Environ.* **2014**, *28*, 6–18. [CrossRef]
6. Viana, M.; Hammingh, P.; Colette, A.; Querol, X.; Degraeuwe, B.; de Vlieger, I.; van Aardenne, J. Impact of Maritime Transport Emissions on Coastal Air Quality in Europe. *Atmos. Environ.* **2014**, *90*, 96–105. [CrossRef]
7. Wu, P.; Bucknall, R. Hybrid Fuel Cell and Battery Propulsion System Modelling and Multi-Objective Optimisation for a Coastal Ferry. *Int. J. Hydrogen Energy* **2020**, *45*, 3193–3208. [CrossRef]
8. Reusser, C.A.; Pérez Osses, J.R. Challenges for Zero-Emissions Ship. *J. Mar. Sci. Eng.* **2021**, *9*, 1042. [CrossRef]
9. Mallouppas, G.; Yfantis, E.A. Decarbonization in Shipping Industry: A Review of Research, Technology Development, and Innovation Proposals. *J. Mar. Sci. Eng.* **2021**, *9*, 415. [CrossRef]
10. RICARDO. Technological, Operational and Energy Pathways for Maritime Transport to Reduce Emissions Towards 2050. Available online: https://www.concawe.eu/publication/technological-operational-and-energy-pathways-for-maritime-transport-to-reduce-emissions-towards-2050-concawe-review-31-1/ (accessed on 1 August 2023).
11. Bouman, E.A.; Lindstad, E.; Rialland, A.I.; Strømman, A.H. State-of-the-Art Technologies, Measures, and Potential for Reducing GHG Emissions from Shipping—A Review. *Transp. Res. D Transp. Environ.* **2017**, *52*, 408–421. [CrossRef]
12. Theotokatos, G.; Rentizelas, A.; Guan, C.; Ancic, I. Waste Heat Recovery Steam Systems Techno-Economic and Environmental Investigation for Ocean-Going Vessels Considering Actual Operating Profiles. *J. Clean. Prod.* **2020**, *267*, 121837. [CrossRef]
13. Olaniyi, E.O.; Prause, G. Investment Analysis of Waste Heat Recovery System Installations on Ships' Engines. *J. Mar. Sci. Eng.* **2020**, *8*, 811. [CrossRef]
14. Li, C.; Wang, H.; Sun, P. Numerical Investigation of a Two-Element wingsail for Ship Auxiliary Propulsion. *J. Mar. Sci. Eng.* **2020**, *8*, 333. [CrossRef]
15. Elkafas, A.G.; Shouman, M.R. A Study of the Performance of Ship Diesel-Electric Propulsion Systems from an Environmental, Energy Efficiency, and Economic Perspective. *Mar. Technol. Soc. J.* **2022**, *56*, 52–58. [CrossRef]
16. Elkafas, A.G. Advanced Operational Measure for Reducing Fuel Consumption Onboard Ships. *Environ. Sci. Pollut. Res.* **2022**, *29*, 90509–90519. [CrossRef] [PubMed]
17. Spinelli, F.; Mancini, S.; Vitiello, L.; Bilandi, R.N.; De Carlini, M. Shipping Decarbonization: An Overview of the Different Stern Hydrodynamic Energy Saving Devices. *J. Mar. Sci. Eng.* **2022**, *10*, 574. [CrossRef]

18. Ren, Y.; Zhang, L.; Shi, P.; Zhang, Z. Research on Multi-Energy Integrated Ship Energy Management System Based on Hierarchical Control Collaborative Optimization Strategy. *J. Mar. Sci. Eng.* **2022**, *10*, 1556. [CrossRef]
19. Elkafas, A.G.; Shouman, M.R. Assessment of Energy Efficiency and Ship Emissions from Speed Reduction Measures on a Medium Sized Container Ship. *Trans. R. Inst. Nav. Archit. Int. J. Marit. Eng.* **2021**, *163*, 121–132. [CrossRef]
20. Elkafas, A.G.; Rivarolo, M.; Massardo, A.F. Environmental Economic Analysis of Speed Reduction Measure Onboard Container Ships. *Environ. Sci. Pollut. Res.* **2023**, *30*, 59645–59659. [CrossRef] [PubMed]
21. Ruiz Zardoya, A.; Oregui Bengoetxea, I.; Lopez Martinez, A.; Loroño Lucena, I.; Orosa, J.A. Methodological Design Optimization of a Marine LNG Internal Combustion Gas Engine to Burn Alternative Fuels. *J. Mar. Sci. Eng.* **2023**, *11*, 1194. [CrossRef]
22. Elkafas, A.G.; Khalil, M.; Shouman, M.R.; Elgohary, M.M. Environmental Protection and Energy Efficiency Improvement by Using Natural Gas Fuel in Maritime Transportation. *Environ. Sci. Pollut. Res.* **2021**, *28*, 60585–60596. [CrossRef]
23. Sagin, S.; Karianskyi, S.; Madey, V.; Sagin, A.; Stoliaryk, T.; Tkachenko, I. Impact of Biofuel on the Environmental and Economic Performance of Marine Diesel Engines. *J. Mar. Sci. Eng.* **2023**, *11*, 120. [CrossRef]
24. Nerheim, A.R.; Æsøy, V.; Holmeset, F.T. Hydrogen as a Maritime Fuel–Can Experiences with LNG Be Transferred to Hydrogen Systems? *J. Mar. Sci. Eng.* **2021**, *9*, 743. [CrossRef]
25. Brynolf, S.; Fridell, E.; Andersson, K. Environmental Assessment of Marine Fuels: Liquefied Natural Gas, Liquefied Biogas, Methanol and Bio-Methanol. *J. Clean. Prod.* **2014**, *74*, 86–95. [CrossRef]
26. Elkafas, A.G.; Elgohary, M.M.; Shouman, M.R. Implementation of Liquefied Natural Gas as a Marine Fuel to Reduce the Maritime Industries Climate Impact. *Nav. Eng. J.* **2022**, *134*, 125–135.
27. Elkafas, A.G.; Rivarolo, M.; Massardo, A.F. Assessment of Alternative Marine Fuels from Environmental, Technical, and Economic Perspectives Onboard Ultra Large Container Ship. *Trans. R. Inst. Nav. Archit. Int. J. Marit. Eng.* **2022**, *164*, 125–134. [CrossRef]
28. Shim, H.; Kim, Y.H.; Hong, J.P.; Hwang, D.; Kang, H.J. Marine Demonstration of Alternative Fuels on the Basis of Propulsion Load Sharing for Sustainable Ship Design. *J. Mar. Sci. Eng.* **2023**, *11*, 567. [CrossRef]
29. Vladimir, N.; Bakica, A.; Perčić, M.; Jovanović, I. Modular Approach in the Design of Small Passenger Vessels for Mediterranean. *J. Mar. Sci. Eng.* **2022**, *10*, 117. [CrossRef]
30. Rivarolo, M.; Rattazzi, D.; Magistri, L.; Massardo, A.F. Multi-Criteria Comparison of Power Generation and Fuel Storage Solutions for Maritime Application. *Energy Convers. Manag.* **2021**, *244*, 114506. [CrossRef]
31. Kim, K.; Park, K.; Roh, G.; Choung, C.; Kwag, K.; Kim, W. Proposal of Zero-Emission Tug in South Korea Using Fuel Cell/Energy Storage System: Economic and Environmental Long-Term Impacts. *J. Mar. Sci. Eng.* **2023**, *11*, 540. [CrossRef]
32. van Biert, L.; Visser, K. Chapter 3—Fuel Cells Systems for Sustainable Ships. In *Sustainable Energy Systems on Ships*; Baldi, F., Coraddu, A., Mondejar, M.E., Eds.; Elsevier: Amsterdam, The Netherlands, 2022; pp. 81–121, ISBN 978-0-12-824471-5. [CrossRef]
33. FLAGSHIPS Projects. Available online: https://flagships.eu/about/ (accessed on 24 November 2022).
34. Wang, X.; Zhu, J.; Han, M. Industrial Development Status and Prospects of the Marine Fuel Cell: A Review. *J. Mar. Sci. Eng.* **2023**, *11*, 238. [CrossRef]
35. Elkafas, A.G.; Rivarolo, M.; Gadducci, E.; Magistri, L.; Massardo, A.F. Fuel Cell Systems for Maritime: A Review of Research Development, Commercial Products, Applications, and Perspectives. *Processes* **2023**, *11*, 97. [CrossRef]
36. van Biert, L.; Godjevac, M.; Visser, K.; Aravind, P.V. A Review of Fuel Cell Systems for Maritime Applications. *J. Power Sources* **2016**, *327*, 345–364. [CrossRef]
37. Xing, H.; Stuart, C.; Spence, S.; Chen, H. Fuel Cell Power Systems for Maritime Applications: Progress and Perspectives. *Sustainability* **2021**, *13*, 1213. [CrossRef]
38. Rivarolo, M.; Rattazzi, D.; Lamberti, T.; Magistri, L. Clean Energy Production by PEM Fuel Cells on Tourist Ships: A Time-Dependent Analysis. *Int. J. Hydrogen Energy* **2020**, *45*, 25747–25757. [CrossRef]
39. Cavo, M.; Rivarolo, M.; Gini, L.; Magistri, L. An Advanced Control Method for Fuel Cells—Metal Hydrides Thermal Management on the First Italian Hydrogen Propulsion Ship. *Int. J. Hydrogen Energy* **2023**, *48*, 20923–20934. [CrossRef]
40. Elkafas, A.G.; Barberis, S.; Rivarolo, M. Thermodynamic Analysis for SOFC/ICE Integration in Hybrid Systems for Maritime Application. *E3S Web Conf.* **2023**, *414*, 02002. [CrossRef]
41. Cavo, M.; Gadducci, E.; Rattazzi, D.; Rivarolo, M.; Magistri, L. Dynamic Analysis of PEM Fuel Cells and Metal Hydrides on a Zero-Emission Ship: A Model-Based Approach. *Int. J. Hydrogen Energy* **2021**, *46*, 32630–32644. [CrossRef]
42. Bassam, A.M.; Phillips, A.B.; Turnock, S.R.; Wilson, P.A. An Improved Energy Management Strategy for a Hybrid Fuel Cell/Battery Passenger Vessel. *Int. J. Hydrogen Energy* **2016**, *41*, 22453–22464. [CrossRef]
43. Kim, K.; Roh, G.; Kim, W.; Chun, K. A Preliminary Study on an Alternative Ship Propulsion System Fueled by Ammonia: Environmental and Economic Assessments. *J. Mar. Sci. Eng.* **2020**, *8*, 183. [CrossRef]
44. Qu, J.; Feng, Y.; Zhu, Y.; Wu, B.; Wu, Y.; Xiao, Z.; Zheng, S. Assessment of a Methanol-Fueled Integrated Hybrid Power System of Solid Oxide Fuel Cell and Low-Speed Two-Stroke Engine for Maritime Application. *Appl. Therm. Eng.* **2023**, *230*, 120735. [CrossRef]
45. Lee, G.N.; Kim, J.M.; Jung, K.H.; Park, H.; Jang, H.S.; Lee, C.S.; Lee, J.W. Environmental Life-Cycle Assessment of Eco-Friendly Alternative Ship Fuels (MGO, LNG, and Hydrogen) for 170 GT Nearshore Ferry. *J. Mar. Sci. Eng.* **2022**, *10*, 755. [CrossRef]
46. Wang, H.; Boulougouris, E.; Theotokatos, G.; Zhou, P.; Priftis, A.; Shi, G. Life Cycle Analysis and Cost Assessment of a Battery Powered Ferry. *Ocean. Eng.* **2021**, *241*, 110029. [CrossRef]

47. de-Troya, J.J.; Álvarez, C.; Fernández-Garrido, C.; Carral, L. Analysing the Possibilities of Using Fuel Cells in Ships. *Int. J. Hydrogen Energy* **2016**, *41*, 2853–2866. [CrossRef]
48. Korberg, A.D.; Brynolf, S.; Grahn, M.; Skov, I.R. Techno-Economic Assessment of Advanced Fuels and Propulsion Systems in Future Fossil-Free Ships. *Renew. Sustain. Energy Rev.* **2021**, *142*, 110861. [CrossRef]
49. Dall'Armi, C.; Micheli, D.; Taccani, R. Comparison of Different Plant Layouts and Fuel Storage Solutions for Fuel Cells Utilization on a Small Ferry. *Int. J. Hydrogen Energy* **2021**, *46*, 13878–13897. [CrossRef]
50. MarineTraffic. Rodi Jet. Available online: https://www.marinetraffic.com/en/ais/details/ships/shipid:5792556/mmsi:247209800/imo:8012140/vessel:RODI_JET (accessed on 24 June 2023).
51. LiguriaViaMare. Rodi Jet Motor Vessel. Available online: https://www.whalewatchliguria.it/en/news-31/whale-watching-rodi-jet-motor-vessel.html (accessed on 24 June 2023).
52. Zhao, J.F.; Liang, Q.C.; Liang, Y.F. Simulation and Study of PEMFC System Directly Fueled by Ammonia Decomposition Gas. *Front. Energy Res.* **2022**, *10*, 819939. [CrossRef]
53. Giddey, S.; Badwal, S.P.S.; Munnings, C.; Dolan, M. Ammonia as a Renewable Energy Transportation Media. *ACS Sustain. Chem. Eng.* **2017**, *5*, 10231–10239. [CrossRef]
54. Wang, Y.; Ruiz Diaz, D.F.; Chen, K.S.; Wang, Z.; Adroher, X.C. Materials, Technological Status, and Fundamentals of PEM Fuel Cells—A Review. *Mater. Today* **2020**, *32*, 178–203. [CrossRef]
55. Bicer, Y.; Dincer, I. Clean Fuel Options with Hydrogen for Sea Transportation: A Life Cycle Approach. *Int. J. Hydrogen Energy* **2018**, *43*, 1179–1193. [CrossRef]
56. Perčić, M.; Vladimir, N.; Jovanović, I.; Koričan, M. Application of Fuel Cells with Zero-Carbon Fuels in Short-Sea Shipping. *Appl. Energy* **2022**, *309*, 118463. [CrossRef]
57. Nuchturee, C.; Li, T.; Xia, H. Energy Efficiency of Integrated Electric Propulsion for Ships—A Review. *Renew. Sustain. Energy Rev.* **2020**, *134*, 110145. [CrossRef]
58. EMSA. Study on Electrical Energy Storage for Ships, Battery Systems for Maritime Applications—Technology, Sustainability and Safety, European Maritime Safety Agency. Available online: http://www.emsa.europa.eu/sustainable-shipping/new-technologies/download/6186/4507/23.html (accessed on 23 July 2023).
59. Perčić, M.; Frković, L.; Pukšec, T.; Ćosić, B.; Li, O.L.; Vladimir, N. Life-Cycle Assessment and Life-Cycle Cost Assessment of Power Batteries for All-Electric Vessels for Short-Sea Navigation. *Energy* **2022**, *251*, 123895. [CrossRef]
60. Clean Hydrogen JU. Strategic Research and Innovation Agenda (SRIA) 2021–2027. *Agenda* **2022**, *2021*, 2027. Available online: https://www.clean-hydrogen.europa.eu/about-us/key-documents/strategic-research-and-innovation-agenda_en (accessed on 23 July 2023).
61. Ioannis, T.; Dalius, T.; Natalia, L. *Li-Ion Batteries for Mobility and Stationary Storage Applications—Scenarios for Costs and Market Growth*; Publications Office of the European Union: Luxembourg, 2018; ISBN 9789279972546. Available online: https://publications.jrc.ec.europa.eu/repository/handle/JRC113360 (accessed on 22 July 2023).
62. Kanchiralla, F.M.; Brynolf, S.; Malmgren, E.; Hansson, J.; Grahn, M. Life-Cycle Assessment and Costing of Fuels and Propulsion Systems in Future Fossil-Free Shipping. *Environ. Sci. Technol.* **2022**, *56*, 12517–12531. [CrossRef]
63. de Vries, N. Safe and Effective Application of Ammonia as a Marine Fuel, Delft of Technology. Master's Thesis, Delft University of Technology, Delft, The Netherlands, 2019.
64. Bellotti, D.; Rivarolo, M.; Magistri, L. A Comparative Techno-Economic and Sensitivity Analysis of Power-to-X Processes from Different Energy Sources. *Energy Convers. Manag.* **2022**, *260*, 115565. [CrossRef]
65. Ikäheimo, J.; Kiviluoma, J.; Weiss, R.; Holttinen, H. Power-to-Ammonia in Future North European 100% Renewable Power and Heat System. *Int. J. Hydrogen Energy* **2018**, *43*, 17295–17308. [CrossRef]
66. Becker, W.L.; Braun, R.J.; Penev, M.; Melaina, M. Design and Technoeconomic Performance Analysis of a 1MW Solid Oxide Fuel Cell Polygeneration System for Combined Production of Heat, Hydrogen, and Power. *J. Power Sources* **2012**, *200*, 34–44. [CrossRef]
67. Baldi, F.; Brynolf, S.; Maréchal, F. The Cost of Innovative and Sustainable Future Ship Energy Systems. In Proceedings of the 32nd International Conference on Efficiency, Cost, Optimization, Simulation and Environmental Impact of Energy Systems, Wroclaw, Poland, 23 June 2019.
68. Ma, S.; Lin, M.; Lin, T.E.; Lan, T.; Liao, X.; Maréchal, F.; Van herle, J.; Yang, Y.; Dong, C.; Wang, L. Fuel Cell-Battery Hybrid Systems for Mobility and off-Grid Applications: A Review. *Renew. Sustain. Energy Rev.* **2021**, *135*, 110119. [CrossRef]
69. Grzesiak, S. Alternative Propulsion Plants for Modern LNG Carriers. *New Trends Prod. Eng.* **2018**, *1*, 399–407. [CrossRef]
70. Argus. Hydrogen and Future Fuels, Market News, Analysis and Prices. Available online: https://www.argusmedia.com/en/power/argus-hydrogen-and-future-fuels (accessed on 24 July 2023).
71. Argus. Sample Report: Argus Marine Fuels. Available online: https://www.argusmedia.com/en/oil-products/argus-marine-fuels (accessed on 24 July 2023).
72. EuroStat. Electricity Prices for Non-Household Consumers. Available online: https://ec.europa.eu/eurostat/web/energy/database (accessed on 28 July 2023).
73. Aldersey-Williams, J.; Rubert, T. Levelised Cost of Energy—A Theoretical Justification and Critical Assessment. *Energy Policy* **2019**, *124*, 169–179. [CrossRef]
74. GlobalPetrolPrices Genoa Diesel Prices. Available online: https://www.globalpetrolprices.com/Italy/Genoa/diesel_prices/ (accessed on 15 July 2023).

75. Nepomuceno de Oliveira, M.A.; Szklo, A.; Castelo Branco, D.A. Implementation of Maritime Transport Mitigation Measures According to Their Marginal Abatement Costs and Their Mitigation Potentials. *Energy Policy* **2022**, *160*, 112699. [CrossRef]

76. Wu, S.; Miao, B.; Chan, S.H. Feasibility Assessment of a Container Ship Applying Ammonia Cracker-Integrated Solid Oxide Fuel Cell Technology. *Int. J. Hydrogen Energy* **2022**, *47*, 27166–27176. [CrossRef]

77. Ejder, E.; Arslanoğlu, Y. Evaluation of Ammonia Fueled Engine for a Bulk Carrier in Marine Decarbonization Pathways. *J. Clean. Prod.* **2022**, *379*, 134688. [CrossRef]

78. Eide, M.S.; Longva, T.; Hoffmann, P.; Endresen, Ø.; Dalsøren, S.B. Future Cost Scenarios for Reduction of Ship CO_2 Emissions. *Marit. Policy Manag.* **2011**, *38*, 11–37. [CrossRef]

79. IMO. *Fourth IMO Greenhouse Gas Study*; International Maritime Organization: London, UK, 2020; pp. 197–212. Available online: https://www.imo.org/en/ourwork/Environment/Pages/Fourth-IMO-Greenhouse-Gas-Study-2020.aspx (accessed on 10 July 2023).

80. BloomEnergy. The Bloom Energy Server 5. Available online: https://www.bloomenergy.com/resource/bloom-energy-server/ (accessed on 4 June 2023).

81. Lee, H.; Jung, I.; Roh, G.; Na, Y.; Kang, H. Comparative Analysis of On-Board Methane and Methanol Reforming Systems Combined with HT-PEM Fuel Cell and CO2 Capture/Liquefaction System for Hydrogen Fueled Ship Application. *Energies* **2020**, *13*, 224. [CrossRef]

82. Byun, M.; Lee, B.; Lee, H.; Jung, S.; Ji, H.; Lim, H. Techno-Economic and Environmental Assessment of Methanol Steam Reforming for H2 Production at Various Scales. *Int. J. Hydrogen Energy* **2020**, *45*, 24146–24158. [CrossRef]

83. Ballard. Marine Modules. Available online: https://www.ballard.com/fuel-cell-solutions/fuel-cell-power-products/marine-modules (accessed on 3 September 2022).

84. Corvus. Energy storage solutions - Corvus Dolphin Energy. Available online: https://corvusenergy.com/products/energy-storage-solutions/corvus-dolphin-energy/ (accessed on 20 May 2023).

85. ABB. Low Voltage Water Cooled Motors. Available online: https://new.abb.com/motors-generators/iec-low-voltage-motors/process-performance-motors/water-cooled-motors (accessed on 15 July 2023).

86. Foripower. DC/DC Converter. Available online: https://www.foripower.com/index_84.aspx?lcoid=84 (accessed on 15 July 2023).

Journal of
Marine Science and Engineering

Article

Sustainable Maritime Transportation Operations with Emission Trading

Haoqing Wang [1], Yuan Liu [2], Fei Li [3,*] and Shuaian Wang [1]

1 Department of Logistics and Maritime Studies, Faculty of Business, The Hong Kong Polytechnic University, Hong Kong; haoqing.wang@connect.polyu.hk (H.W.); shanswang@polyu.edu.hk (S.W.)
2 School of Economics and Management, Wuhan University, Wuhan 430072, China; 2020301052156@whu.edu.cn
3 Department of Mechanical Engineering, Tsinghua University, Beijing 100190, China
* Correspondence: lf19@mails.tsinghua.edu.cn

Abstract: The European Union (EU) has recently approved the inclusion of shipping in its Emissions Trading System, aiming to foster sustainable development within the shipping industry. While this new policy represents a significant step towards reducing carbon emissions, it also poses challenges for shipping companies, particularly in terms of operation costs. To assist shipping companies in devising optimal strategies under the new policy, this study proposes new techniques to determine the optimal solutions for sailing speed and the number of ships on the route, covering both EU and non-EU areas. Additionally, we demonstrate how to adjust these optimal decisions in response to changes in charged fees, fuel prices, and weekly operational costs of ships. This research offers innovative insights into the optimal decision-making process for shipping companies under the new EU policy and serves as a valuable decision-making tool to minimize total costs.

Keywords: maritime transport; green shipping; sailing speed; Emissions Trading System

Citation: Wang, H.; Liu, Y.; Li, F.; Wang, S. Sustainable Maritime Transportation Operations with Emission Trading. *J. Mar. Sci. Eng.* **2023**, *11*, 1647. https://doi.org/10.3390/jmse11091647

Academic Editor: Mihalis Golias

Received: 7 August 2023
Revised: 20 August 2023
Accepted: 21 August 2023
Published: 23 August 2023

1. Introduction

Maritime transportation has progressively gained recognition as the principal mode of conveyance for international trade [1,2]. Nevertheless, the shipping industry has faced mounting criticism in recent years, primarily due to the heightened focus on carbon emissions. This criticism stems from the industry's heavy reliance on the combustion of fossil fuels, leading to substantial emissions of carbon dioxide (CO_2) and contributing to the exacerbation of global warming and climate change. The Fourth International Maritime Organization (IMO) greenhouse gas (GHG) study provides notable insights into this issue [3]. It reveals that the cumulative CO_2 emissions originating from maritime shipping increased from 962 million tonnes in 2012 to 1056 million tonnes in 2018, signifying approximately 3% of the total global anthropogenic CO_2 emissions during the period spanning from 2012 to 2018. Under this challenging circumstance, many countries and international organizations have put forward carbon-emission-reduction policies to promote the sustainable development of the shipping industry, such as double carbon goals in China (carbon peaking and carbon neutrality) [4], and the "Fit for 55" plan launched by the European Union (EU) [5]. These policies have been strategically formulated to effectively attain substantial emissions reductions, with the primary objective of achieving a minimum 40% reduction in emissions by the year 2030 and a subsequent aim to curtail total annual GHG emissions by no less than 50% by 2050, as cited in IMO [6]. Moreover, the overarching goal of these policies is to actively encourage the widespread implementation and adoption of cutting-edge technologies, fuels, and alternative energy sources that possess zero or near-zero GHG emissions profiles, as stated in the same source [6].

On 18 April 2023, the EU Parliament approved the inclusion of shipping in its Emissions Trading System (ETS) [7]. This decision represents a significant step toward addressing carbon emissions from ships and promoting sustainability within the shipping industry. Specifically, shipping companies need to buy 100% of emissions for voyages within the EU and 50% of emissions for voyages into or out of the EU starting in 2026. This decision represents a significant step towards addressing carbon emissions from ships and promoting sustainability within the shipping industry. Suppose that there is a container shipping route starting from Shanghai (SHA) port, passing through Singapore (SIN) port, Rotterdam (RTM) port, Hamburg (HH) port, and SIN port, and finally returning to SHA port (see Figure 1 for an illustration):

$$\text{SHA} \rightarrow \text{SIN} \rightarrow \text{RTM} \rightarrow \text{HH} \rightarrow \text{SIN} \rightarrow \text{SHA}. \tag{1}$$

The reported emissions are 100% charged between the RTM port and the HH port, 50% charged between SIN port and RTM port and between HH port and SIN port, and 0% charged in the remaining legs. Obviously, this policy aims to make shipping companies consider methods to reduce emissions within the EU area because more emissions generate more costs.

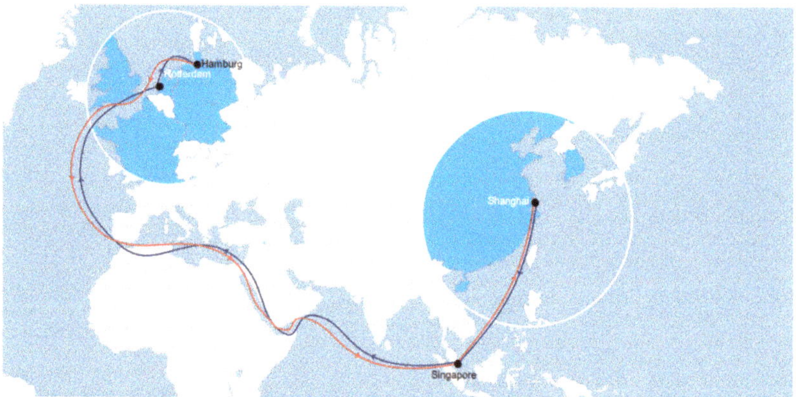

Figure 1. An example of a shipping route.

In this study, we propose mathematical models to facilitate optimal decision-making that seeks to minimize shipping companies' costs while adhering to the newly introduced EU policy on emissions. To the best of our knowledge, our research stands as the pioneering study that takes into account this recently promulgated policy by the EU. Specifically, our study answers the following research questions:

1. What are the optimal sailing speeds within the EU and non-EU areas that minimize the shipping company's total costs under the new EU policy on emissions?
2. What is the optimal number of ships to be equipped in the shipping route that leads to the lowest total costs while adhering to the emissions reduction requirements set by the EU's policy?
3. How do the optimal sailing speeds within EU and non-EU areas, as well as the optimal number of ships equipped in the shipping route, vary with changes in the charged fee for emissions, fuel price, and weekly operational costs of ships?

To address the three aforementioned research questions, we first propose a nonlinear optimization model, which presents challenges in terms of its complexity and solving difficulty. Leveraging the structural characteristics of the model, we establish two propositions that allow us to reduce its scale significantly. Then, we transform the nonlinear optimization model into an integer programming (IP) model by discretizing decision variables. This IP model can be solved using off-the-shelf optimization solvers. Finally, we conduct experiments and sensitivity analysis to examine the model performance regarding the changes in parameters.

1.1. Literature Review

1.1.1. Carbon Emission Reduction Policies in Shipping

From the perspective of policies' content, carbon emission reduction policies mainly focus on carbon emission allowance and tax. The implementation of the cap and trade (C&T) system within the shipping industry has garnered extensive attention from various stakeholders [8]. This system establishes a fixed emission target that is coupled with market flexibility. It allocates a specific number of carbon emission allowances (CEAs) to each participant within the system. In order to effectively curb carbon emissions in maritime transport, Zhu et al. [9] conducted an investigation into the strategies and performance of CEAs among shipping companies under the C&T mechanism. Their research findings serve as valuable guidance for multiple stakeholders, aiding them in formulating their own carbon-emission-reduction strategies, including determining the optimal carbon price and the overall carbon emissions targets. In addition, carbon allowance allocation in the shipping industry under the Energy Efficiency Design Index (EEDI) and the non-EEDI is explored. Chang and Huang [10] compared the carbon allowance and cost difference for shipping vessels that follow or ignore the guidelines of the EEDI. It has been proven that carbon tax exerts an important influence on carbon emission reduction. Based on previous scholarly investigations, the efficacy of a carbon tax primarily hinges on its extensive scope, inherent simplicity, and reduced uncertainty pertaining to future carbon pricing [11]. Moreover, its regulatory alignment with the existing governance framework and adherence to the "polluter pays principle" bolster its standing as a more suitable instrument for curtailing carbon emissions [12]. In the study conducted by Heine and Gäde [13], a novel hybrid mechanism was introduced, encompassing a cargo-based tax levied on international shipping emissions alongside a bunker levy targeting domestic shipping emissions. Notably, this approach incorporates the establishment of a default ship efficiency benchmark while incentivizing ship owners to operate energy-efficient vessels through subsidies. Consequently, such a hybrid tax regime facilitates the attainment of a global consensus. Additionally, various incentive-based carbon tax policies, such as the comparison between tonnage tax and conventional profit tax regimes [14], serve as catalysts for investment promotion and the advancement of green transformation within the maritime industry. For instance, existing tonnage tax regimes heavily subsidize international shipping activities [15]. Moreover, decarbonization continues to dominate the medium-term agenda of the maritime sector, exerting a profound influence on pivotal investments and strategic deliberations in the shipping industry [16]. So far, there have been plenty of options for decarbonizing the shipping sector, such as "slow steaming" [17] and measures that enhance energy efficiency [18].

At the regional level, the European Union (EU) is at the forefront of implementing effective measures to address carbon emissions in maritime shipping. On 18 April 2023, the legislative bodies of the EU reached a significant consensus by incorporating shipping into the Emission Trading System (ETS) [19]. Pending the EU's final approval, vessels above 5000 gross tonnages (GTs) that are engaged in the commercial transportation of cargo or passengers within the EU will be obligated to obtain and surrender emission allowances for their CO_2 emissions starting from 2024. Furthermore, by 2034, these ships will need to ensure that at least 2% of their fuel mix consists of specific renewable fuels [19]. This development will inevitably impose increasing compliance costs on the shipping industry. Moreover, in line with EU members' commitment to promoting renewable energy, the European Commission has introduced the Inducement Prize, aimed at encouraging the adoption of renewable fuels in retrofitted container ships [20].

The trend of carbon emissions reduction in the shipping industry is gaining momentum as green and sustainable development practices are widely recognized and promoted. Countries, regions, and international organizations are formulating policies to regulate shipping companies' development and encourage them to take measures to reduce carbon emissions. This growing emphasis on environmental responsibility reflects the global commitment to combat climate change and foster a more sustainable future for the shipping industry.

1.1.2. Optimal Decisions in Shipping

Against the backdrop of carbon emission reduction, a range of policies profoundly impact optimal decisions in shipping. The management of ship operations has been a focal point of prior research, primarily focusing on ship routing, ship deployment, and ship sailing speed. With regard to shipping routing, Lin and Tsai [21] delved into the intricate problem of ship routing and freight assignment in daily liner shipping operations, introducing a Lagrangian relaxation technique. Moreover, Lin and Chang [22] employed a decomposition algorithm that incorporates the Lagrangian factor to optimize route selection and freight-allocation decisions. Exploring the optimization of the shipping network with respect to carbon dioxide emissions charges on the Asia–Europe route, Dai et al. [23] conducted an in-depth analysis. In terms of ship deployment, Zhen et al. [24] devised a nonlinear mixed-integer programming model to facilitate strategic ship deployment, taking into account the stochastic nature of the ships' weight distribution. Additionally, Gu et al. [12] explored the maritime carbon-trading mechanism within the context of conventional green ship deployment, discovering its limited effectiveness in reducing short-term carbon dioxide emissions. Notably, there exist studies that consider shipping demand uncertainty, with researchers proposing stochastic optimization models and robust optimization models [25] to address uncertainties in decision-making. Furthermore, the optimization of ship sailing speed has garnered attention. For instance, Wang and Xu [26] tackled the optimization problem of sailing speed during voyages, accounting for distinct carbon-emission-taxation regimes. A mixed-integer programming model was established by Sheng et al. [27] to explore optimal ship speed and size when traversing emission control areas such as the EU region. Moreover, several research efforts have pursued holistic decision-making approaches, considering the interplay of multiple factors. Wang et al. [28] advanced a sophisticated mathematical programming model, aiming to concurrently optimize ship routes and the interconnected cargo-allocation schemes. The model itself is effectively solved by transforming it into an equivalent mixed-integer linear program, allowing for efficient computational analysis. Furthermore, researchers such as Koza [29] and Ozcan and Eliiyi [30] developed distinct algorithms to streamline the optimization of service scheduling, encompassing crucial elements like transit time and container volume, alongside cargo allocation strategies within the realm of liner shipping. In addition, Giovannini and Psaraftis [31] undertook the optimization of various factors, including shipping speed, the number of ships deployed, and service frequency, to maximize the average daily profit of liner shipping companies, illustrating a holistic approach to decision-making in the industry.

Substantially, the optimal decisions mainly focus on the deployed ships, sailing speeds, and sailing routes, as well as the connections between them. To obtain optimal decisions, the IP model and other derived models are proposed to provide decision information for stakeholders, especially for shipping companies' operators. In general, carbon-emission-reduction policies are stricter and diversified all over the world as the goal of green transformation and sustainable development in maritime shipping, especially in the EU area. These policies have a significant influence on ship deployment and sailing speed, which means the optimal decisions will change with newly introduced policies. Our study takes into account a recently published policy by the EU, proposing an IP model to obtain the optimal sailing speed and deployed ships in different routes.

1.2. Research Contributions

The theoretical and practical contributions of our research are summarized as follows.

1. Theoretical contributions. This study addresses a research gap, as existing literature has not focused on the optimal decisions of sailing speed and the number of ships under the newly proposed EU policy. To the best of our knowledge, this is the

first study to establish mathematical models aimed at minimizing the total costs of shipping companies while considering the implications of the new EU policy. The proposed approach involves a nonlinear optimization model to determine the shipping company's optimal decisions. By leveraging the unique structure of the optimization problem under the new EU policy, two propositions are proven. We further transform the nonlinear model into a solvable IP model. Through experiments and sensitivity analyses, specific solutions are obtained, and the impacts of different parameters are tested.

2.   Practical contributions. This study contributes valuable insights into optimal strategies for shipping companies to minimize costs and comply with the new EU emissions policy. The results have practical implications for the sustainable development of the shipping industry and its adherence to environmental regulations. The proposed mathematical model can serve as a decision tool for shipping companies facing the new EU emissions policy.

The rest of the paper is organized as follows. Section 2 describes the research problem in detail and develops the mathematical model. Section 3 proposes solution methods for addressing the initial proposed model. Section 4 presents the experiments and sensitivity analysis that were conducted. Finally, conclusions are drawn in Section 5.

The main notations used in this study are summarized in Table 1.

Table 1. Notations.

Sets	
I	Set of ports of call in a shipping route, $i \in I$
I^{EU}	Set of ports in the EU area, $i \in I^{EU}$
I^{NEU}	Set of ports outside the EU area, $i \in I^{NEU}$
I^0	Set of the legs on which emissions are 0 charged, $i \in I^0$
I^1	Set of the legs on which emissions are 50% charged, $i \in I^1$
I^2	Set of the legs on which emissions are 100% charged, $i \in I^2$
Parameters	
c	The fixed cost for each ship per week
μ	The fuel price per tonne
β	The charged fee of emissions per tonne
γ	The conversion rate of fuel consumption and emissions
Q	The emission per hour during berthing
L_i	The distance of leg i
t_i	The berthing time at port i
$\hat{v}_i$	The sailing speed on leg i
$v_{\min}$	The minimum sailing speed
$v_{\max}$	The maximum sailing speed
X	The integer used to discretize sailing speed, $x = 0, 1, ..., X$
v_k^x	The discretized sailing speed, $k = 0, 1, 2$
Function	
$f(\hat{v}_i^3)$	Fuel consumption at the sailing speed of $\hat{v}_i$
Decision variables	
z	The number of deployed ships in a route
y_k^x	Binary decision variable that equals 1 if ships sail with speed v_k^x and 0 otherwise

2. Problem Description and Model Development

We use set I to denote the set of ports of call in one shipping route covering both EU and non-EU areas, and $i \in I$ denotes port i and also denotes leg i. Since ships need to return to the original port, there is no need to denote the final port. For example, $I = \{1, 2, 3, 4, 5\}$ in (1). $i = 1$ indicates the SH port; $i = 2$ and $i = 5$ all represent the SIN port because the ship calls at the SIN port twice in this route. $i = 1$ also indicates the leg between the SHA port and the SIN port, and $i = 5$ represents the leg between the SIN port and the SHA port.

For the shipping companies operating container ships on this route, their decision involves the number of ships used on this route (denoted by z) and the sailing speed of these ships in each leg i (denoted by $\hat{v}_i$). The distance of leg i is denoted by L_i. The berthing time at each port i is denoted by t_i. Certainly, the emissions generated during the berthing at EU ports are subject to a full charge, as mandated by the new policy. According to widely recognized domain knowledge, speed and fuel consumption are cubically related [32]. That is, fuel consumption equals $f(\hat{v}_i^3)$, where function $f(\hat{v}_i^3)$ maps sailing speed to fuel consumption and $f(\hat{v}_i^3) = a\hat{v}_i^3$. And we assume the emissions during the berthing are Qt_i, where Q is a constant representing the emission per hour during the berthing. To facilitate model construction, we define set I^{EU} and set I^{NEU}. The set I^{EU} comprises ports within the EU area, while the set I^{NEU} includes ports outside the EU area. Furthermore, we define sets I^0, I^1, and I^2, where I^0 denotes the legs on which emissions are 0% charged, I^1 denotes the legs on which emissions are 50% charged, and I^2 denotes the legs on which emissions are 100% charged. Taking the route in Figure 1 as an example, the set I^{EU} includes RTM and HH, the set I^{NEU} consists of SHA and SIN, the set I^0 comprises the leg from SHA to SIN and the leg from SIN to SHA, the set I^1 includes the leg from SIN to RTM and the leg form HH to SIN, and the set I^2 comprises the leg from RTM to HH. Obviously, we have $I = I^{EU} \cup I^{NEU} = I^0 \cup I^1 \cup I^2$. To minimize the total costs, the shipping company's optimal decision problem can be formulated as follows:

[M1]

$$\min cz + \mu\left(\sum_{i\in I} Qt_i + \sum_{i\in I} \frac{L_i}{\hat{v}_i}f(\hat{v}_i^3)\right) + \beta\gamma\left(\sum_{i\in I^{EU}} Qt_i + 0.5\sum_{i\in I^1} \frac{L_i}{\hat{v}_i}f(\hat{v}_i^3) + \sum_{i\in I^2} \frac{L_i}{\hat{v}_i}f(\hat{v}_i^3)\right) \quad (2)$$

subject to

$$\sum_{i\in I}\left(\frac{L_i}{\hat{v}_i} + t_i\right) \le 168z \quad (3)$$

$$v_{\min} \le \hat{v}_i \le v_{\max}, i \in I \quad (4)$$

$$z \in Z_+. \quad (5)$$

The objective function (2) involves three parts. Firstly, cz calculates the total fixed costs of ships. Secondly, $\mu(\sum_{i\in I} Qt_i + \sum_{i\in I} \frac{L_i}{\hat{v}_i}f(\hat{v}_i^3))$ represents the total fuel costs, where μ is the fuel price. Thirdly, $\beta\gamma(\sum_{i\in I^{EU}} Qt_i + 0.5\sum_{i\in I^1} \frac{L_i}{\hat{v}_i}f(\hat{v}_i^3) + \sum_{i\in I^2} \frac{L_i}{\hat{v}_i}f(\hat{v}_i^3))$ calculates the emission tax, where γ represents the conversion rate of fuel consumption and emissions and β represents the charged fee. Constraint (3) restricts the weekly service frequency. Constraint (4) gives the domain of $\hat{v}_i$, indicating the maximum and minimum of $\hat{v}_i$. Constraint (5) requires that the number of deployed ships in a route should be a positive integer.

3. Solution Methods

Model [M1] is complex to solve due to different $\hat{v}_i$ values in different legs, which means a significant number of decision variables and sophisticated algorithms. In addition, the term $\frac{L_i}{\hat{v}_i}$ is nonlinear, making it harder to obtain the optimal solution in this model. Jointly considering the model characteristics and sailing speeds in different areas, we put forward Proposition 1 to reduce the number of decision variables and further discretize sailing speed to linearize the proposed [M1].

Proposition 1. *In the same type of leg, the sailing speed remains consistent. Specifically, $\hat{v}_i$ for $i \in I^0$ is the same, $\hat{v}_i$ for $i \in I^1$ is the same, and $\hat{v}_i$ for $i \in I^2$ is the same.*

Proof. To simplify the notation, we define $\hat{v} = (\hat{v}_1, \hat{v}_2, ..., \hat{v}_{|I|})$, where $|I|$ denotes the total number of legs in set I. We further use $F(z, \hat{v})$ to denote the objective function (2). The objective function (2) is actually a monotonically increasing function of z and $\hat{v}_i$ because

$$\frac{\partial F(z, \hat{v})}{\partial z} = c > 0 \tag{6}$$

$$\frac{\partial F(z, \hat{v})}{\partial \hat{v}_i} = 2\mu L_i \hat{v}_i > 0, \ i \in I^0 \tag{7}$$

$$\frac{\partial F(z, \hat{v})}{\partial \hat{v}_i} = 2\mu L_i \hat{v}_i + \beta\gamma L_i > 0, \ i \in I^1 \tag{8}$$

$$\frac{\partial F(z, \hat{v})}{\partial \hat{v}_i} = 2\mu L_i \hat{v}_i + 2\beta\gamma L_i > 0, \ i \in I^2. \tag{9}$$

Suppose that there are two decision variables $\hat{v}_{i^\#}$ and $\hat{v}_{i'}$, $i^\# \in I^0$, $i' \in I^0$, and $\hat{v}_{i^\#} \neq \hat{v}_{i'}$ satisfies Constraints (3) and (4). We use $L_{i^\#}$ and $L_{i'}$ to denote the corresponding lengths of legs of $\hat{v}_{i^\#}$ and $\hat{v}_{i'}$, respectively. We can always find optimal solutions $\bar{v}_{i^\#}$ and $\bar{v}_{i'}$ equaling $\frac{L_{i^\#}+L_{i'}}{\frac{L_{i^\#}}{\hat{v}_{i^\#}}+\frac{L_{i'}}{\hat{v}_{i'}}}$ that satisfy Constraints (3) and (4) and generate a smaller value of $F(z, \hat{v}_i)$.

To facilitate the proof process, we define $t = \frac{L_{i^\#}}{\hat{v}_{i^\#}} + \frac{L_{i'}}{\hat{v}_{i'}}$. Therefore, we have $\hat{v}_{i'} = \frac{L_{i'}}{t - \frac{L_{i^\#}}{\hat{v}_{i^\#}}}$.

The objective function aims to minimize the following formula because $i^\# \in I^0$ and $i' \in I^0$:

$$L_{i^\#}\hat{v}_{i^\#}^2 + L_{i'}\hat{v}_{i'}^2, \tag{10}$$

which is

$$L_{i^\#}\hat{v}_{i^\#}^2 + L_{i'}\left(\frac{L_{i'}}{t - \frac{L_{i^\#}}{\hat{v}_{i^\#}}}\right)^2. \tag{11}$$

We take the derivative with respect to $\hat{v}_{i^\#}$:

$$\frac{d\left[L_{i^\#}\hat{v}_{i^\#}^2 + L_{i'}\left(\frac{L_{i'}}{t - \frac{L_{i^\#}}{\hat{v}_{i^\#}}}\right)^2\right]}{d\hat{v}_{i^\#}} = L_{i^\#}\hat{v}_{i^\#} - L_{i'}^3\left[(t - \frac{L_{i^\#}}{\hat{v}_{i^\#}})^{-3}\frac{L_{i^\#}}{\hat{v}_{i^\#}}\right] = 0. \tag{12}$$

The $\bar{v}_{i^\#}$ satisfying the above equation should be

$$\bar{v}_{i^\#} = \frac{L_{i^\#} + L_{i'}}{t} = \frac{L_{i^\#} + L_{i'}}{\frac{L_{i^\#}}{\hat{v}_{i^\#}} + \frac{L_{i'}}{\hat{v}_{i'}}} \tag{13}$$

and thus

$$\bar{v}_{i'} = \frac{L_{i^\#} + L_{i'}}{\frac{L_{i^\#}}{\hat{v}_{i^\#}} + \frac{L_{i'}}{\hat{v}_{i'}}}. \tag{14}$$

Therefore, $\bar{v}_{i^\#} = \bar{v}_{i'} = \frac{L_{i^\#}+L_{i'}}{\frac{L_{i^\#}}{\hat{v}_{i^\#}}+\frac{L_{i'}}{\hat{v}_{i'}}}$ should be the optimal solutions. Next, we analyze how $\bar{v}_{i^\#}$ and $\bar{v}_{i'}$ satisfy Constraints (3) and (4).

First, we have

$$\frac{L_{i^\#}}{\hat{v}_{i^\#}} + t_{i^\#} + \frac{L_{i'}}{\hat{v}_{i'}} + t_{i'} + \sum_{i \in I \setminus \{i^\#, i'\}} (\frac{L_i}{\hat{v}_i} + t_i) \leq 168z. \tag{15}$$

Because $\overline{v}_{i^\#} = \overline{v}_{i'} = \frac{L_{i^\#}+L_{i'}}{\frac{L_{i^\#}}{\overline{v}_{i^\#}} + \frac{L_{i'}}{\overline{v}_{i'}}}$, we have the following relationship:

$$
\begin{aligned}
\frac{L_{i^\#}}{\hat{v}_{i^\#}} + \frac{L_{i'}}{\hat{v}_{i'}} - \left(\frac{L_{i^\#}}{\overline{v}_{i^\#}} + \frac{L_{i'}}{\overline{v}_{i'}} \right) &= \frac{L_{i^\#}}{\hat{v}_{i^\#}} + \frac{L_{i'}}{\hat{v}_{i'}} - \left(\frac{L_{i^\#}}{\frac{L_{i^\#}+L_{i'}}{\frac{L_{i^\#}}{\hat{v}_{i^\#}}+\frac{L_{i'}}{\hat{v}_{i'}}}} + \frac{L_{i'}}{\frac{L_{i^\#}+L_{i'}}{\frac{L_{i^\#}}{\hat{v}_{i^\#}}+\frac{L_{i'}}{\hat{v}_{i'}}}} \right) \\
&= \frac{L_{i'}\hat{v}_{i^\#} + L_{i^\#}\hat{v}_{i'}}{\hat{v}_{i^\#}\hat{v}_{i'}} - \frac{L_{i'}\hat{v}_{i^\#} + L_{i^\#}\hat{v}_{i'}}{\hat{v}_{i^\#}\hat{v}_{i'}} \\
&= 0,
\end{aligned}
\tag{16}
$$

which indicates that $\overline{v}_{i^\#}$ and $\overline{v}_{i'}$ satisfy Constraint (3).

In terms of Constraints (4), we have:

$$
v_{\min} \leq \hat{v}_{i^\#} \leq v_{\max}
\tag{17}
$$

$$
v_{\min} \leq \hat{v}_{i'} \leq v_{\max}.
\tag{18}
$$

With loss of generality, we suppose that $\hat{v}_{i^\#} > \hat{v}_{i'}$. Therefore, we have the following relationship:

$$
\overline{v}_{i^\#} = \overline{v}_{i'} = \frac{L_{i^\#}+L_{i'}}{\frac{L_{i^\#}}{\hat{v}_{i^\#}}+\frac{L_{i'}}{\hat{v}_{i'}}} = \frac{\hat{v}_{i'}\hat{v}_{i^\#}(L_{i^\#}+L_{i'})}{L_{i'}\hat{v}_{i^\#}+L_{i^\#}\hat{v}_{i'}} < \frac{\hat{v}_{i'}\hat{v}_{i^\#}(L_{i^\#}+L_{i'})}{(L_{i'}+L_{i^\#})\hat{v}_{i'}} \leq v_{\max},
\tag{19}
$$

and

$$
\overline{v}_{i^\#} = \overline{v}_{i'} = \frac{L_{i^\#}+L_{i'}}{\frac{L_{i^\#}}{\hat{v}_{i^\#}}+\frac{L_{i'}}{\hat{v}_{i'}}} = \frac{\hat{v}_{i'}\hat{v}_{i^\#}(L_{i^\#}+L_{i'})}{L_{i'}\hat{v}_{i^\#}+L_{i^\#}\hat{v}_{i'}} > \frac{\hat{v}_{i'}\hat{v}_{i^\#}(L_{i^\#}+L_{i'})}{(L_{i'}+L_{i^\#})\hat{v}_{i^\#}} \geq v_{\min}.
\tag{20}
$$

Therefore, we prove that $\overline{v}_{i^\#} = \overline{v}_{i'} = \frac{L_{i^\#}+L_{i'}}{\frac{L_{i^\#}}{\hat{v}_{i^\#}}+\frac{L_{i'}}{\hat{v}_{i'}}}$ should be the optimal solutions when $i \in I^0$. By the same logic, the optimal values of $\hat{v}_i$, $i \in I^1$ must be the same, and the optimal values of $\hat{v}_i$, $i \in I^1$ must be the same. $\square$

Taking advantage of Proposition 1, we can reduce the number of decision variables in Model [M1]. That is, for each type of leg, we only need to decide on one optimal sailing speed. We use v_0, v_1, and v_2 to denote the sailing speed on each type of leg. Model [M1] can be converted to the following model.

[M2]

$$
\min cz + \mu \left(\sum_{i \in I^0} \frac{L_i}{v_0} f(v_0^3) + \sum_{i \in I^1} \frac{L_i}{v_1} f(v_1^3) + \sum_{i \in I^2} \frac{L_i}{v_2} f(v_2^3) \right) + \beta\gamma \left(0.5 \sum_{i \in I^1} \frac{L_i}{v_1} f(v_1^3) + \sum_{i \in I^2} \frac{L_i}{v_2} f(v_2^3) \right) + C
\tag{21}
$$

subject to

$$
\sum_{i \in I^0} \left(\frac{L_i}{v_0} + t_i \right) + \sum_{i \in I^1} \left(\frac{L_i}{v_1} + t_i \right) + \sum_{i \in I^2} \left(\frac{L_i}{v_2} + t_i \right) \leq 168z
\tag{22}
$$

$$
v_{\min} \leq v_0 \leq v_{\max}
\tag{23}
$$

$$
v_{\min} \leq v_1 \leq v_{\max}
\tag{24}
$$

$$
v_{\min} \leq v_2 \leq v_{\max}
\tag{25}
$$

$$
z \in Z_+,
\tag{26}
$$

where $C = \mu \sum_{i \in I} Qt_i + \beta\gamma \sum_{i \in I^{EU}} Qt_i$, which is a constant and does not affect the optimal solutions. However, Model [M2] is still difficult to solve because of the nonlinear terms $\frac{L_i}{v_0}$, $\frac{L_i}{v_1}$, and $\frac{L_i}{v_2}$. Referring to [33], we discretize sailing speed with 0.1 knots. We define

$$X = \lfloor \frac{v_{\max} - v_{\min}}{0.1} \rfloor + 1. \tag{27}$$

We set $x = 0, 1, ..., X$. Therefore, the sailing speed $v_k, k = 0, 1, 2$ can be discretized to $v_k^0 = v_{\min}, v_k^1 = v_{\min} + 0.1 \times 1, v_k^2 = v_{\min} + 0.1 \times 2, ..., v_k^X = \max\{v_{\max}, v_{\min} + 0.1 \times X\}$. We introduce binary decision variables $y_k^x, k = 0, 1, 2$, and $x = 0, ..., X$ to indicate which discretized sailing speed is chosen. Specifically, $y_k^x = 1$ means the corresponding sailing speed v_k^x is selected and 0 otherwise. With this newly introduced binary decision variable, we can transform Model [M2] to the following IP model:

[M3]

$$\min cz + \mu \left(\sum_{i \in I^0} \sum_{x=0}^{X} y_0^x \frac{L_i}{v_0^x} f(v_0^{x3}) + \sum_{i \in I^1} \sum_{x=0}^{X} y_1^x \frac{L_i}{v_1^x} f(v_1^{x3}) + \sum_{i \in I^2} \sum_{x=0}^{X} y_2^x \frac{L_i}{v_2^x} f(v_2^{x3}) \right)$$
$$+ \beta\gamma \left(0.5 \sum_{i \in I^1} \sum_{x=0}^{X} y_1^x \frac{L_i}{v_1^x} f(v_1^{x3}) + \sum_{i \in I^2} \sum_{x=0}^{X} y_2^x \frac{L_i}{v_2^x} f(v_2^{x3}) \right) + C \tag{28}$$

subject to

$$\sum_{i \in I^0} (t_i + \sum_{x=0}^{X} y_0^x \frac{L_i}{v_0^x}) + \sum_{i \in I^1} (t_i + \sum_{x=1}^{X} y_1^x \frac{L_i}{v_1^x}) + \sum_{i \in I^2} (t_i + \sum_{x=2}^{X} y_2^x \frac{L_i}{v_2^x}) \leq 168z \tag{29}$$

$$\sum_{x=0}^{X} y_k^x = 1, k = 0, 1, 2 \tag{30}$$

$$y_k^x \in \{0, 1\}, k = 0, 1, 2, \ x = 0, ..., X \tag{31}$$

$$z \in Z_+. \tag{32}$$

Model [M3] has two types of decision variables: the first one is the integer decision variable z, which means how many ships should be used in a route; the second one is the binary decision variable y_k^x and we have a total of $3X$ binary decision variables. If $y_k^x = 1$, the corresponding sailing speed v_k^x is selected, and the fuel consumption is decided. Therefore, the objective function and the constraints are linear and the decision variables are all integers, which means that we transform Model [M2] to IP model [M3].

We can prove that the optimal solutions of Model [M3] satisfy Proposition 2.

Proposition 2. *We use v_0^*, v_1^*, and v_2^* to denote the optimal values of sailing speed within non-EU areas, linking non-EU and EU areas, and within EU areas, respectively. We must have $v_0^* \geq v_1^* \geq v_2^*$.*

Proof. We use z^* to denote the optimal value of the decision variable z. Based on [M3], the value of the objective function (28) is

$$cz^* + \mu a \left(\sum_{i \in I^0} L_i v_0^{*2} + \sum_{i \in I^1} L_i v_1^{*2} + \sum_{i \in I^2} L_i v_2^{*2} \right) + \beta\gamma a \left(0.5 \sum_{i \in I^1} L_i v_1^{*2} + \sum_{i \in I^2} L_i v_2^{*2} \right) + C. \tag{33}$$

And we also have

$$\sum_{i \in I^0} (t_i + \frac{L_i}{v_0^*}) + \sum_{i \in I^1} (t_i + \frac{L_i}{v_1^*}) + \sum_{i \in I^2} (t_i + \frac{L_i}{v_2^*}) \leq 168z^* \tag{34}$$

$$v_{\min} \leq v_0^* \leq v_{\max} \tag{35}$$

$$v_{\min} \leq v_1^* \leq v_{\max} \tag{36}$$

$$v_{\min} \leq v_2^* \leq v_{\max}. \tag{37}$$

There is a trade-off between the objective function (33) and Constraint (34). To be more specific, Constraint (34) tends to generate greater values of v_0^*, v_1^*, and v_2^*. However, minimizing the objective function (33) tends to generate smaller values of v_0^*, v_1^*, and v_2^*. Therefore, there are two cases in the optimal solutions. The first case is that Constraint (34) is binding. That is,

$$\sum_{i \in I^0} (t_i + \frac{L_i}{v_0^*}) + \sum_{i \in I^1} (t_i + \frac{L_i}{v_1^*}) + \sum_{i \in I^2} (t_i + \frac{L_i}{v_2^*}) = 168z^*. \tag{38}$$

And the second case is that the optimal solutions equal the minimum sailing speed, i.e.,

$$v_0^* = v_1^* = v_2^* = v_{\min}. \tag{39}$$

This is because the coefficients of v_0^*, v_1^*, and v_2^* are μa, $\mu a + 0.5\beta\gamma a$, and $\mu a + \beta\gamma a$, respectively. Under the condition of satisfying Equation (38), the optimal solutions that satisfy $v_0^* \geq v_1^* \geq v_2^*$ can achieve the minimum value of the objective function (33) in the first case. Moreover, the optimal solutions in the second case also satisfy $v_0^* \geq v_1^* \geq v_2^*$. $\square$

4. Experiments

4.1. Experiment Settings

With the help of Proposition 1 and discretization, we transform the original optimization model into an IP programming model with the minimum number of decision variables, which can be solved using the off-of-shelf optimization solvers, such as CPLEX and Gurobi. We here introduce the selected container shipping routes for the experiment and how to set the parameters, e.g., c, μ, β, γ, and Q according to practice.

The experiments were run on a laptop computer equipped with 2.60 GHz of Intel Core i7 CPU and 16 GB of RAM, and Model [M3] was solved using the Gurobi Optimizer 10.0.2 via Python API.

4.1.1. Selected Shipping Routes

We select two routes from Asia to northern Europe[1] to test the performance of Model [M3]. These two routes play a pivotal role in fostering communication between Asia and Europe. Remarkably, certain ports within these routes occupy a paramount position within the realm of international transportation of goods, including notable ones like Singapore and Rotterdam. Moreover, for comprehensive insights into the distances of various segments within the routes, we relied on some authoritative websites[2]. Details are shown in Table 2, and the names of EU ports are bolded. The travel distances (in nautical miles) of these two routes, i.e., L_i, are shown in Figure 2 and Figure 3, respectively.

Table 2. Summary of shipping routes.

Route ID	Port Rotation (City)
1	Busan → Ningbo → Shanghai → Yantian → Singapore → **Algeciras** → **Dunkerque** → **Le Havre** → **Hamburg** → **Wilhelmshaven** →**Rotterdam** → Port Klang → Busan
2	Tianjin → Dalian → Qingdao → Shanghai → Ningbo → Singapore→ **Piraeus** → **Rotterdam** → **Hamburg** → **Antwerp** → Shanghai → Tianjin

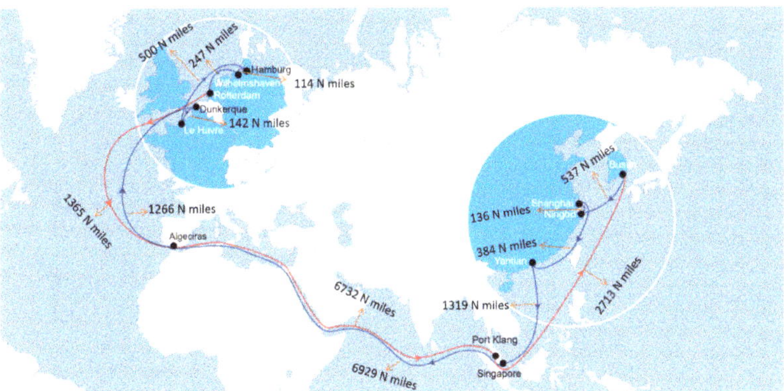

Figure 2. Shipping route 1.

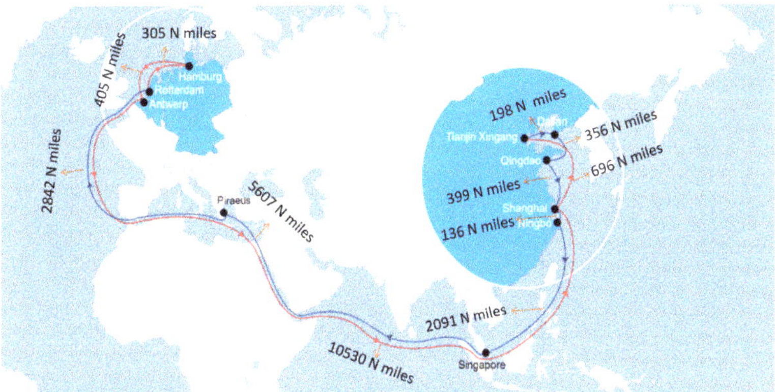

Figure 3. Shipping route 2.

4.1. Parameter Settings

We first set the values of parameters for drawing the basic results, and we conduct sensitivity analysis to examine the impacts of these parameters.

1. The fixed cost c. Referring to [34], we first set $c = 180,000$ per week for a 5000-TEU (twenty-foot equivalent unit) container ship.
2. The fuel price μ. Referring to [35], we set μ to be an average value of 600 (USD/tonne).
3. The charged fee of emissions β. EU ETS allowance prices closed at USD 102 per tonne on April 17, according to Ice Exchange data [7].
4. The conversion rate of fuel consumption and emissions γ is set to 3.15 [36].
5. Referring to [32], we set $f(v^3) = 0.00043 \times v^3$.
6. The berthing time at port i t_i^3: Busan—1.1 days; Ningbo—1.5 days; Shanghai—1.0 day; Yantian—0.6 day; Singapore—1.0 day; Algeciras—0.7 day; Dunkerque—1.6 days; Le Havre—0.8 day; Hamburg—1.4 days; Wilhelmshaven—1.1 days; Rotterdam—1.3 days; Tianji—1.2 days; Dalian—1.5 days; Qingdao—1.5 days; Antwerp—1.3 days.
7. We set the emissions per hour during the berthing to be 2 tonnes; i.e., $Q = 2$.
8. We set $v_{\max} = 18$ knots and $v_{\min} = 10$ knots.

4.2. Basic Results

Using the routes in Table 2, we conducted numerical experiments and report the results in Table 3. As defined in Section 2, the legs within the EU area are represented by I^2, the legs into or out of the EU areas are represented by I^1, and the remaining legs are represented by I^0. The optimal value of the objective function of [M3] is represented by "OBJ". From Table 2,

Route 1 is equipped with more ships compared to Route 2, which indicates that more ships are needed for a longer route to maintain the weekly service frequency (the total distance of Route 1 is 22,258 nautical miles, and the total distance of Route 2 is 23,565 nautical miles). The ship's sailing speed is the highest during the non-EU legs, followed by the legs linking the EU and non-EU areas, and finally, it is the lowest within the EU area, which validates Proposition 2. For instance, the sailing speed in EU legs on Route 1 is 11.6 knots, while in the legs linking the EU and non-EU areas, it reaches 12.1 knots, and in non-EU legs, it reaches its peak at 13.0 knots. Indeed, the rationale behind varying sailing speeds is to manage emissions effectively with the aim of minimizing costs. As ships increase their sailing speed, they also generate higher emissions, which subsequently results in a higher charged fee. The policy of charging fees based on emissions aims to encourage shipping companies to adopt more environmentally friendly practices and optimize their sailing speeds to minimize their carbon footprint. By aligning charging with emissions, the policy incentivizes the adoption of sustainable measures.

We next use Route 2 as a computational instance for the following sensitivity analysis.

Table 3. Basic results.

Route ID	Set of Legs	Legs	Total Distance (Nautical Mile)	Sailing Speed (knot)	Number of Ships	OBJ (USD)
1	I^0	Busan → Ningbo → Shanghai → Yantian → Singapore; Port Klang → Busan	3901	13.0	13	3,924,499.1
	I^1	Singapore → Algeciras; Rotterdam → Port Klang	15,020	12.1		
	I^2	Algeciras → Dunkerque → Le Havre → Hamburg → Wilhelmshaven → Rotterdam	2269	11.6		
2	I^0	Tianjin → Dalian → Qingdao → Shanghai → Ningbo → Singapore	3876	12.8	14	4,166,763.3
	I^1	Singapore→ Piraeus; Antwerp → Shanghai	16,137	12.0		
	I^2	Piraeus→Rotterdam → Hamburg → Antwerp	3552	11.1		

4.3. Sensitivity Analysis

As concerns about carbon emissions intensify, the levied fees associated with emissions are subject to potential changes as more countries prioritize addressing the issue of carbon emissions. Moreover, in the fundamental analysis, certain critical parameters, such as the unit fuel price and the weekly fixed cost per ship, are assumed to be deterministic. Nevertheless, these parameters actually often experience fluctuations. Consequently, sensitivity analyses are undertaken to examine the impacts of these parameters on operational decisions, considering their dynamic nature in real-world scenarios.

4.3.1. Impact of the Charged Fee of Emissions

This study first investigates the impact of the charged fee of emissions β on the operation decisions. With more emphasis on the carbon-emission problems and sustainable development recently, the charged fee of emissions has become an important tool to control carbon emissions and exerts a significant influence on shipping companies' decisions. This means the value of β may change in reality. So the sensitivity analysis of β is necessary. In this experiment, we set the fee of the emissions range from 80 to 170 USD per tonne. Given current trends, it is anticipated that the EU area will indeed implement stricter policies for carbon emissions control in the future. As a result, a higher emissions fee is expected to be imposed on shipping companies operating within the EU region.

Computational results are summarized in Table 4, where we can find that the incremental surge in the charged emissions fee engenders an upward trajectory in the objective value, primarily driven by the simultaneous amplification in the cumulative fee levied. In addition, when the fee charged for emissions increases, more ships are needed and the sailing speed and fuel consumption decreases, because the faster the ships sail, the more fuel they will consume, leading to more emissions, which means more costs. So shipping companies will decrease sailing speed to reduce emissions.

Table 4. Impact of the charged fee of emissions on the operation decision.

β (USD/ton)	Set of Legs	Sailing Speed (knot)	Fuel Consumption (ton)	Number of Ships	OBJ (USD)
80	I^0	12.8	0.90	14	4,101,154.6
	I^1	11.9	0.72		
	I^2	11.5	0.65		
90	I^0	12.8	0.90	14	4,131,128.5
	I^1	12.0	0.74		
	I^2	11.1	0.59		
100	I^0	12.8	0.90	14	4,160,824.1
	I^1	12.0	0.74		
	I^2	11.1	0.59		
110	I^0	12.8	0.90	14	4,190,519.8
	I^1	12.0	0.74		
	I^2	11.1	0.59		
120	I^0	13.3	1.01	14	4,220,180.5
	I^1	11.9	0.72		
	I^2	11.1	0.59		
130	I^0	13.3	1.01	14	4,249,612.9
	I^1	11.9	0.72		
	I^2	11.1	0.59		
140	I^0	13.3	1.01	14	4,279,045.4
	I^1	11.9	0.72		
	I^2	11.1	0.59		
150	I^0	12.1	0.76	15	4,305,595.6
	I^1	11.1	0.59		
	I^2	10.2	0.46		
160	I^0	12.1	0.76	15	4,331,828.7
	I^1	11.0	0.57		
	I^2	10.2	0.46		
170	I^0	12.1	0.76	15	4,358,061.8
	I^1	11.0	0.57		
	I^2	10.2	0.46		
180	I^0	12.1	0.76	15	4,384,294.9
	I^1	11.0	0.57		
	I^2	10.2	0.46		

4.3.2. Impact of the Fuel Price

Within the framework of the fundamental analysis, the deterministic assumption sets the unit price of fuel (μ) at 600 USD/ton. However, to account for the inherent volatility observed in real-life scenarios, this sensitivity analysis considers a range of values for μ, spanning from 570 to 700 USD/ton. This range is determined based on the minimum and maximum fuel prices recorded between September 2020 and July 2023, amounting to 579.00 and 690.50 USD/ton, respectively [35]. The findings of this analysis are succinctly presented in Table 5.

Examining Table 5, it becomes evident that a direct relationship exists between the fuel price and the objective value, whereby an increase in fuel price leads to a corresponding increase in the objective value due to amplified total fuel costs. Additionally, as the fuel price increases, a higher number of ships are necessitated and, consequently, the sailing speed diminishes. This phenomenon arises from the cubic relationship between fuel consumption and sailing speed. Heightened fuel prices prompt shipping companies to explore methods of curbing fuel consumption, thereby resulting in a decrease in sailing speed. Furthermore, in the event that the unit price of fuel becomes excessively exorbitant, shipping companies may opt to employ more ships in order to fulfill the weekly service frequency requirement while operating the vessels at reduced speeds.

Table 5. Impact of the fuel price on the operation decisions.

μ (USD/ton)	Set of Legs	Sailing Speed (knot)	Number of Ships	OBJ (USD)
570	I^0	12.8	14	4,099,569.9
	I^1	12.0		
	I^2	11.1		
580	I^0	12.8	14	4,121,967.7
	I^1	12.0		
	I^2	11.1		
590	I^0	12.8	14	4,144,365.5
	I^1	12.0		
	I^2	11.1		
600	I^0	12.8	14	4,166,763.3
	I^1	12.0		
	I^2	11.1		
610	I^0	12.8	14	4,189,161.1
	I^1	12.0		
	I^2	11.1		
620	I^0	12.8	14	4,211,558.85
	I^1	12.0		
	I^2	11.1		
630	I^0	12.8	14	4,233,956.7
	I^1	12.0		
	I^2	11.1		
640	I^0	12.8	14	4,256,354.4
	I^1	12.0		
	I^2	11.1		
650	I^0	12.8	14	4,278,752.2
	I^1	12.0		
	I^2	11.1		
660	I^0	12.1	15	4,300,885.9
	I^1	10.9		
	I^2	10.6		
670	I^0	12.1	15	4,321,062.5
	I^1	10.9		
	I^2	10.6		
680	I^0	12.1	15	4,341,239.1
	I^1	10.9		
	I^2	10.6		

Table 5. *Cont.*

μ (USD/ton)	Set of Legs	Sailing Speed (knot)	Number of Ships	OBJ (USD)
690	I^0	12.1	15	4,361,415.6
	I^1	10.9		
	I^2	10.6		
700	I^0	12.1	15	4,381,592.2
	I^1	10.9		
	I^2	10.6		

4.3.3. Impact of the Weekly Fixed Cost per Ship

In this study, the investigation focuses on exploring the influence of the weekly fixed cost per ship on operational decisions. The predetermined value for the weekly fixed cost per ship is set at 180,000 USD, which aligns with the configuration employed in [34]. However, it is worth noting that the value of c can significantly vary due to various factors such as the impact of epidemics or other unforeseen circumstances [37], or it can even experience substantial reductions as a result of technological advancements. Consequently, the value range for c is established between 60,000 and 300,000 USD, and the corresponding outcomes are documented in Table 6.

Based on the findings presented in Table 6, it becomes apparent that the objective value exhibits an upward trend as the weekly fixed cost per ship increases. This outcome can be attributed to the fact that with a higher weekly fixed operating cost, fewer ships are deployed within the route networks to mitigate the overall operational expenses. Notably, due to the fixed service frequency, liner ships are compelled to navigate toward their destinations at swifter speeds.

Table 6. Impact of the weekly fixed cost per ship.

c (USD/week)	Set of Legs	Sailing Speed (knot)	Number of Ships	OBJ (USD)
60,000	I^0	10.6	16	2,309,423.4
	I^1	10.2		
	I^2	10.0		
80,000	I^0	10.6	16	2,629,423.4
	I^1	10.2		
	I^2	10.0		
100,000	I^0	10.6	16	2,949,423.4
	I^1	10.2		
	I^2	10.0		
120,000	I^0	10.6	16	3,269,423.4
	I^1	10.2		
	I^2	10.0		
140,000	I^0	12.1	15	3,579,676.8
	I^1	11.0		
	I^2	10.2		
160,000	I^0	12.1	15	3,879,676.8
	I^1	11.0		
	I^2	10.2		
180,000	I^0	12.8	14	4,166,763.3
	I^1	12.0		
	I^2	11.1		

Table 6. *Cont.*

c (USD/week)	Set of Legs	Sailing Speed (knot)	Number of Ships	OBJ(USD)
200,000	l^0	12.8	14	4,446,763.3
	l^1	12.0		
	l^2	11.1		
220,000	l^0	14.0	13	4,722,375.4
	l^1	13.1		
	l^2	12.2		
240,000	l^0	14.0	13	4,982,375.3
	l^1	13.1		
	l^2	12.2		
260,000	l^0	14.0	13	5,242,375.4
	l^1	13.1		
	l^2	12.2		
280,000	l^0	14.0	13	5,502,375.4
	l^1	13.1		
	l^2	12.2		
300,000	l^0	15.5	12	5,748,343.4
	l^1	14.4		
	l^2	13.6		

5. Conclusions

This study delves into the optimal strategies of shipping companies regarding sailing speeds and the number of ships on shipping routes, taking into account the policy recently proposed by the European Union—mandating the purchase of 100% of emissions for voyages within the EU and 50% of emissions for voyages to and from the EU commencing in 2026. Initially, we develop a non-linear optimization model, which we subsequently transform into an efficient IP model by introducing two propositions and discretizing decision variables. This transformation facilitates the utilization of effective solution methods. Through comprehensive experimentation, we demonstrate the efficacy of our proposed model and solution techniques. Moreover, we conduct an in-depth analysis of the effects ensuing from changes in the charged fee of emissions, fuel price, and weekly operational costs of ships on optimal decision making. In general, the total cost experiences an increase when the charged fee of emissions, fuel price, and weekly operational costs arise. Specifically, as the charged fee of emissions or fuel price escalates, shipping companies tend to reduce sailing speeds to curtail fuel consumption, while simultaneously augmenting the deployment of ships on routes to ensure compliance with weekly service frequency requirements. Conversely, in the event of an increase in weekly operational costs, fewer vessels are deployed to mitigate overall expenses.

Overall, our study significantly adds to the comprehension of how shipping enterprises can strategically formulate well-informed decisions as a proactive response to the EU's innovative policy landscape. By offering novel solution methodologies and conducting meticulous analyses on the susceptibility of optimal choices to diverse influencing factors, our research offers practical guidance for navigating the challenges and opportunities presented by the policy.

Author Contributions: Conceptualization, H.W., Y.L., and S.W.; methodology, H.W., S.W., Y.L. and F.L.; software, H.W. and Y.L.; validation, Y.L. and F.L.; formal analysis, Y.L. and F.L.; writing—original draft preparation, H.W. and Y.L.; writing—review and editing, F.L. and S.W.; visualization, Y.L.; supervision, F.L. and S.W. All authors have read and agreed to the published version of the manuscript.

Funding: This research received no external funding.

Institutional Review Board Statement: Not applicable.

Informed Consent Statement: Not applicable.

Data Availability Statement: Data are contained within the article.

Conflicts of Interest: The authors declare no conflict of interest.

Notes

1. https://www.cma-cgm.com/products-services/flyers, accessed on 4 August 2023
2. http://port.sol.com.cn/licheng.asp, accessed on 4 August 2023
3. https://www.econdb.com/maritime/ports/, accessed on 5 August 2023

References

1. Wang, S.; Yin, J.; Khan, R.U. Dynamic Safety Assessment and Enhancement of Port Operational Infrastructure Systems during the COVID-19 Era. *J. Mar. Sci. Eng.* **2023**, *11*, 1008. [CrossRef]
2. Wu, Y.; Huang, Y.; Wang, H.; Zhen, L. Joint planning of fleet deployment, ship refueling, and speed optimization for dual-fuel ships considering methane slip. *J. Mar. Sci. Eng.* **2022**, *10*, 1690. [CrossRef]
3. IMO. Fourth IMO Greenhouse Gas Study. 2020. Available online: https://www.imo.org/en/OurWork/Environment/Pages/Fourth-IMO-Greenhouse-GasStudy2020.aspx (accessed on 5 August 2023).
4. Central People's Government of the People's Republic of China. Suggestions on Complete, Accurate and Comprehensive Implementation of the New Development Concept to Do a Good Job in the Work of Carbon Peak Carbon Neutrality. 2021. Available online: https://www.gov.cn/zhengce/2021-10/24/content_5644613.htm (accessed on 4 August 2023).
5. The European Parliament. Revision of the Renewable Energy Directive: Fit for 55 Package. 2023. Available online: https://www.europarl.europa.eu/thinktank/en/document/EPRS_BRI(2021)698781#:~:text=The%20%27fit%20for%2055%27%20package%20is%20part%20of,EU%20deliver%20the%20new%2055%20%25%20GHG%20target (accessed on 4 August 2023).
6. IMO. 2023 IMO Strategy on Reduction of GHG Emissions from Ships. 2023. Available online: https://www.imo.org/en/OurWork/Environment/Pages/2023-IMO-Strategy-on-Reduction-of-GHG-Emissions-from-Ships.aspx (accessed on 5 August 2023).
7. Lloyd's List. EU Parliament Approves Shipping's Inclusion in Regional Carbon Market. 2023. Available online: https://lloydslist-maritimeintelligence-informa-com.ezproxy.lb.polyu.edu.hk/LL1144780/EU-Parliament-approves-shippings-inclusion-in-regional-carbon-market (accessed on 10 July 2023).
8. Wu, M.; Li, K.X.; Xiao, Y.; Yuen, K.F. Carbon emission trading scheme in the shipping sector: Drivers, challenges, and impacts. *Mar. Policy* **2022**, *138*, 104989. [CrossRef]
9. Zhu, M.; Shen, S.; Shi, W. Carbon emission allowance allocation based on a bi-level multi-objective model in maritime shipping. *Ocean. Coast. Manag.* **2023**, *241*, 106665. [CrossRef]
10. Chang, C.C.; Huang, P.C. Carbon allowance allocation in the shipping industry under EEDI and non-EEDI. *Sci. Total Environ.* **2019**, *678*, 341–350. [CrossRef]
11. Chen, S., Zheng, S.; Syo, C. Policies focusing on market-based measures towards shipping decarbonization: Designs, impacts and avenues for future research. *Transp. Policy* **2023**, *137*, 109–124. [CrossRef]
12. Gu, Y.; Wallace, S.W.; Wang, X. Can an Emission Trading Scheme really reduce CO_2 emissions in the short term? Evidence from a maritime fleet composition and deployment model. *Transp. Res. Part D Transp. Environ.* **2019**, *74*, 318–338. [CrossRef]
13. Heine, D.; Gäde, S. Unilaterally removing implicit subsidies for maritime fuels: A mechanism to unilaterally tax maritime emissions while satisfying extraterritoriality, tax competition and political constraints. *Int. Econ. Econ. Policy* **2018**, *15*, 523–545. [CrossRef]
14. Tvedt, J.; Wergeland, T. Tax regimes, investment subsidies and the green transformation of the maritime industry. *Transp. Res. Part D Transp. Environ.* **2023**, *119*, 103741. [CrossRef]
15. Merk, O.M. Quantifying tax subsidies to shipping. *Marit. Econ. Logist.* **2020**, *22*, 517–535. [CrossRef]
16. Bouman, E.A.; Lindstad, E.; Rialland, A.I.; Strømman, A.H. State-of-the-art technologies, measures, and potential for reducing GHG emissions from shipping—A review. *Transp. Res. Part D Transp. Environ.* **2017**, *52*, 408–421. [CrossRef]
17. Lin, D.Y.; Juan, C.J.; Ng, M. Evaluation of green strategies in maritime liner shipping using evolutionary game theory. *J. Clean. Prod.* **2021**, *279*, 123268. [CrossRef]
18. Adland, R.; Cariou, P.; Jia, H.; Wolff, F.C. The energy efficiency effects of periodic ship hull cleaning. *J. Clean. Prod.* **2018**, *178*, 1–13. [CrossRef]
19. European Parliament. Cutting Emissions from Planes and Ships: EU Actions Explained. 2023. Available online: https://www.europarl.europa.eu/news/en/headlines/society/20220610STO32720/cutting-emissions-from-planes-and-ships-eu-actions-explained (accessed on 5 August 2023).
20. Panoutsou, C.; Germer, S.; Karka, P.; Papadokostantakis, S.; Kroyan, Y.; Wojcieszyk, M.; Maniatis, K.; Marchand, P.; Landalv, I. Advanced biofuels to decarbonise European transport by 2030: Markets, challenges, and policies that impact their successful market uptake. *Energy Strategy Rev.* **2021**, *34*, 100633. [CrossRef]
21. Lin, D.-Y.; Tsai, Y.-Y. The ship routing and freight assignment problem for daily frequency operation of maritime liner shipping. *Transp. Res. Part E Logist. Transp. Rev.* **2014**, *67*, 52–70. [CrossRef]

22. Lin, D.Y.; Chang, Y.T. Ship routing and freight assignment problem for liner shipping: Application to the Northern Sea Route planning problem. *Transp. Res. Part E Logist. Transp. Rev.* **2018**, *110*, 47–70. [CrossRef]
23. Dai, W.L.; Fu, X.; Yip, T.L.; Hu, H.; Wang, K. Emission charge and liner shipping network configuration—An economic investigation of the Asia-Europe route. *Transp. Res. Part A Policy Pract.* **2018**, *110*, 291–305. [CrossRef]
24. Zhen, L.; Hu, Y.; Wang, S.; Laporte, G.; Wu, Y. Fleet deployment and demand fulfillment for container shipping liners. *Transp. Res. Part Methodol.* **2019**, *120*, 15–32. [CrossRef]
25. Zhao, Y.; Chen, Z.; Lim, A.; Zhang, Z. Vessel deployment with limited information: Distributionally robust chance constrained models. *Transp. Res. Part B Methodol.* **2022**, *161*, 197–217. [CrossRef]
26. Wang, C.; Xu, C. Sailing speed optimization in voyage chartering ship considering different carbon emissions taxation. *Comput. Ind. Eng.* **2015**, *89*, 108–115. [CrossRef]
27. Sheng, D.; Meng, Q.; Li, Z.C. Optimal vessel speed and fleet size for industrial shipping services under the emission control area regulation. *Transp. Res. Part C Emerg. Technol.* **2019**, *105*, 37–53. [CrossRef]
28. Wang, H.; Wang, S.; Meng, Q. Simultaneous optimization of schedule coordination and cargo allocation for liner container shipping networks. *Transp. Res. Part E Logist. Transp. Rev.* **2014**, *70*, 261–273. [CrossRef]
29. Koza, D.F. Liner shipping service scheduling and cargo allocation. *Eur. J. Oper. Res.* **2019**, *275*, 897–915. [CrossRef]
30. Ozcan, S.; Eliiyi, D.T.; Reinhardt, L.B. Cargo allocation and vessel scheduling on liner shipping with synchronization of transshipments. *Appl. Math. Model.* **2020**, *77*, 235–252. [CrossRef]
31. Giovannini, M.; Psaraftis, H.N. The profit maximizing liner shipping problem with flexible frequencies: Logistical and environmental considerations. *Flex. Serv. Manuf. J.* **2019**, *31*, 567–597. [CrossRef]
32. Wang, S.; Meng, Q. Sailing speed optimization for container ships in a liner shipping network. *Transp. Res. Part E Logist. Transp. Rev.* **2012**, *48*, 701–714. [CrossRef]
33. Wang, H.; Yan, R.; Au, M.H.; Wang, S.; Jin, Y.J. Federated learning for green shipping optimization and management. *Adv. Eng. Inform.* **2023**, *56*, 101994. [CrossRef]
34. Wu, Y.; Huang, Y.; Wang, H.; Zhen, L.; Shao, W. Green Technology Adoption and Fleet Deployment for New and Aged Ships Considering Maritime Decarbonization. *J. Mar. Sci. Eng.* **2022**, *11*, 36. [CrossRef]
35. ShipBunker. 2023. Available online: https://shipandbunker.com/prices (accessed on 28 July 2023).
36. Offshore Energy. 2023. Available online: https://www.offshore-energy.biz/marine-benchmark-2020-global-shipping-co2-emissions-down-1/ (accessed on 28 July 2023).
37. Drewry. COVID-19 Drives up Ship Operating Costs. 2020. Available online: https://www.drewry.co.uk/news/news/covid-19-drives-up-ship-operating-costs (accessed on 30 July 2023).

Journal of
*Marine Science
and Engineering*

MDPI

Review

A Systematic Literature Review of Maritime Transportation Safety Management

Minqiang Xu [1], Xiaoxue Ma [2], Yulan Zhao [2] and Weiliang Qiao [3,*]

[1] School of Marxism, Dalian Maritime University, Dalian 116026, China; xuminqiang@dlmu.edu.cn
[2] School of Maritime Economics and Management, Dalian Maritime University, Dalian 116026, China; maxx1020@dlmu.edu.cn (X.M.)
[3] Marine Engineering College, Dalian Maritime University, Dalian 116026, China
* Correspondence: xiaoqiao_fang@dlmu.edu.cn

Abstract: Maritime transportation plays a critical role in global trade, and studies on maritime transportation safety management are of great significance to the sustainable development of the maritime industry. Consequently, there has been an increasing trend recently in studies on maritime transportation safety management, especially in terms of safety risk analysis and emergency management. Therefore, the general idea of this article is to provide a detailed literature review of maritime transportation safety management based on 186 articles in the Web of Science (WOS) database published from 2011 to 2022. The purposes of this article are as follows: (1) to provide a statistics-based description and conduct a network-based bibliometric analysis on the basis of the collected articles; (2) to summarize the methodologies/technologies employed in maritime transportation safety management spatiotemporally; and (3) to propose four potential research perspectives in terms of maritime transportation safety management. Based on the findings and insights obtained from the bibliometric and systematic review, the development of a resilient maritime transportation system could be facilitated by means of data- or intelligence-driven technologies, such as scenario representation, digital twinning, and data simulation. In addition, the issues facing intelligent maritime shipping greatly challenge the current maritime safety management system due to the co-existence of intelligent and non-intelligent maritime operation.

Keywords: maritime transportation; emergency management; safety risk analysis; bibliometrics; research perspective

Citation: Xu, M.; Ma, X.; Zhao, Y.; Qiao, W. A Systematic Literature Review of Maritime Transportation Safety Management. *J. Mar. Sci. Eng.* **2023**, *11*, 2311. https://doi.org/10.3390/jmse11122311

Academic Editor: Claudio Ferrari

Received: 19 October 2023
Revised: 21 November 2023
Accepted: 28 November 2023
Published: 6 December 2023

1. Introduction

Maritime transportation is critical for global trade, and over 80% of global goods are delivered by ocean shipping [1]. According to the *"Review of Maritime Transport 2022"* issued by the United Nations Conference on Trade and Development, the global commercial fleet has increased sharply in the last three decades, which reflects an increase in global maritime transportation activities. The safety-related issues associated with maritime transportation are highly concerned with minimizing maritime accidents and their impacts on human life and the ocean environment. For this purpose, various risks involved in maritime transportation must be controlled to an acceptable/tolerable level [2]. In addition, in the case of heavy casualties or large-scale oil spill pollution, effective emergency management is critical to reduce the damage caused by these events. Meanwhile, search and rescue (SAR) requirements at sea also require effective emergency responses to reduce the loss of human life. To maintain maritime transportation safety at a satisfactory level, the International Maritime Organization (IMO) has taken proactive measures to promote safety in the maritime industry [3–5], such as the International Convention for the Safety of Life at Sea (SOLAS), the International Convention on Standards of Training, Certification and Watchkeeping for Seafarers (STCW), and the International Safety Management Code

(ISM). The International Convention on Maritime Search and Rescue has also come into use to enhance emergency collaboration globally. At the national level, various regulations and measures have also been adopted. For instance, the UK government supports the promotion of autonomous ships and invests actively in the research and development of new space technologies. The Danish government is improving the maritime information and communications technology (ICT) infrastructure under the regulatory framework of the European Union (EU) and the IMO. The Korean government has strengthened maritime safety through digital technology and seeks to create a big data platform in the maritime and fisheries sectors. However, catastrophic consequences in terms of human life losses, damage to commodities, and environmental pollution are still frequently reported. For instance, according to a report issued by the Ministry of Transport of the P. R. of China, there were a total of 237 human lives lost or missing and 83 vessels sunk in 2018 [6]. The European Maritime Safety Agency (EMSA) also reported a total of 230 vessel losses during 2011–2018 [7].

The container ship grounding accident that happened in the Suez Canal on March 23, 2021 prompted the public to rethink safety and emergency issues in the maritime industry [8]. Human errors, technical failures, and mechanical breakdowns are highlighted as the main root causes of maritime accidents [9]. In addition, the absence of effective emergency management also contributes to unexpected loss of life and damage, all of which impede the sustainable development of the maritime industry. Therefore, the present study sought to systemically review the work conducted to clear these obstacles impeding the sustainable development of the maritime industry. Several similar reviews have been conducted by scholars [10–19], mainly focused on the theme of risk analysis, methodologies, or factors analysis, and these review articles are summarized in Table 1. Unlike previous literature reviews concentrating on the analysis of accident causes and risk assessment to prevent the reoccurrence of maritime accidents, in the present study, we focused on the thematic coverage of safety risk analyses and emergency issues in maritime transportation aspects, such as shipping lanes, maritime supply chains, and maritime operations. Solutions for these issues are essential for maritime industry continuity and handling unforeseen disruptions, such as natural disasters, heavy maritime casualties, and terrorist attacks. The primary significance of this study was to assess the state of existing knowledge on maritime transportation safety management and to suggest future research perspectives, which might facilitate scholars and practitioners in promoting the sustainable development of the maritime industry.

Table 1. Review articles related to maritime safety in recent years (2012–2023).

Publication	Research Theme
[20]	Risk assessment models
[16]	Risk analysis
[11]	Risk analysis
[19]	Expert elicitation and BN modeling
[21]	Maritime transport policy
[10]	Fire and explosion accidents
[14]	Models and computational algorithms
[22]	Marine fuels
[23]	Navigation data visualization
[24]	Cyber risk perception
[18]	Human and organizational factors analysis
[15]	Human factors and safe performance
[25]	Resilience
[13]	Human reliability analysis
[12]	Risk assessment methods

The remainder of this study is organized as follows: Section 2 presents the data source and methodology. Section 3 shows the results of the descriptive statistical analysis. The network-based bibliometric analysis is described in Section 4. Section 5 summarizes the methodologies/technologies employed for maritime safety and emergency studies. A discussion on potential research perspectives is provided in Section 6, and finally, conclusions are drawn in Section 7.

2. Data Source and Bibliometric Methods

2.1. Data Source

The scientific publication data used in this study were collected from the Web of Science (WOS) core collection database, one of the most comprehensive multi-disciplinary content search platforms for academic research. Only journal articles were included in this study. The determination process for this data source is shown in Figure 1.

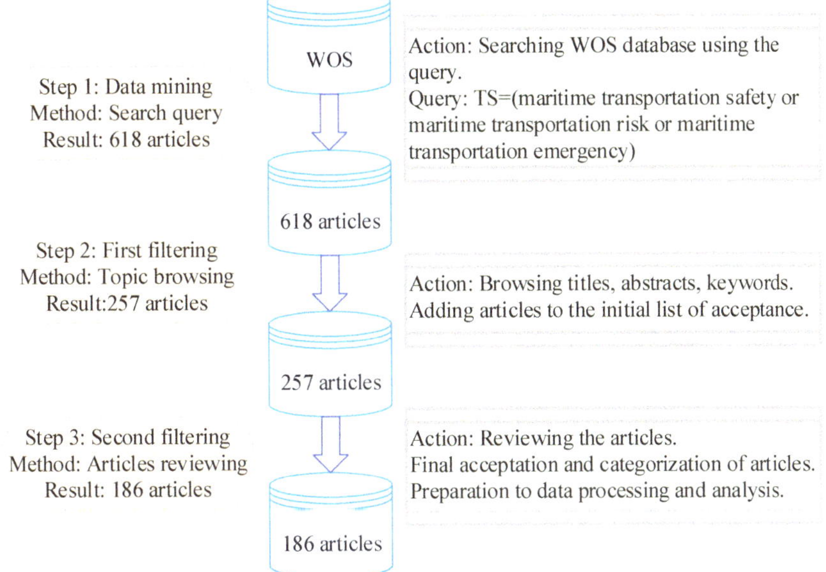

Figure 1. Literature retrieval process.

Step 1—Obtaining the original dataset by using data mining. According to the discussion in Section 1, the safety issues involved in maritime transportation can be interpreted from the aspects of risk analysis and emergency management; therefore, "maritime transportation safety" or "maritime transportation risk" or "maritime transportation emergency" was selected as the search query to identify the records from the database. In the present study, the two WOS core collections of *Science Citation Index Expanded (SCIE)* and *Social Sciences Citation Index (SSCI)* were selected as the database. The time span was set from 1 January 2011 to 31 December 2022, and the precise search was set as disabled. Finally, a total of 618 journal articles were identified by implementing the search query.

Step 2—Preparing the data sample by the first filtering. All of the articles retrieved from WOS in Step 1 were investigated by focusing on the title, abstract, and keywords, the results of which led to the development of a new dataset for further filtering. According to the research purpose mentioned in Section 1, an article was considered relevant for further analysis if any aspects of maritime transportation safety management were investigated, such as maritime transportation risk analysis/assessment, maritime transportation safety management/strategies, and maritime transportation emergency management/response. As a result, 257 articles were identified.

Step 3—Determining the final data sample to be analyzed by the second filtering. All of the articles obtained in Step 2 were reviewed one by one to verify their conformance within the scope of this study. Finally, 186 articles were selected for the final literature review. Additionally, all of the collected articles were categorized on the basis of their research perspective, as presented in Table 2. In some cases, an article was assigned to more than one category. For instance, those articles concentrating on emergency decision analysis by means of maritime network modeling were assigned to the categories of maritime network/system and decision analysis.

Table 2. Categories of the studies in terms of maritime transportation safety management.

Category	Description
Maritime network/system	Assessment of the resilience of maritime transportation networks; exploring the vulnerability of transportation networks; maritime transportation system risk assessment and safety analysis; designing maritime safety management systems.
Polar navigation	Decision making on process risk of Arctic route; optimizing the management for Arctic mass rescue events; interfering ship navigation process safety in Arctic waters.
Intelligent/unmanned navigation	Assessment of the potential impact of unmanned vessels; risk assessment of the operations of maritime autonomous ships.
Marine environment	Analyzing oil spill risk assessments; predicting the oil spill's trajectory; studying optimal scheduling of emergency resources for maritime oil spills; causes of oil spill pollution; providing a scientific basis for targeted strategic oil spill emergency planning.
Accident causation analysis	Exploring the causal factors of marine accidents; evaluation of the prediction of marine accident consequences; human error assessment.
Port/supply chain	Improving the resilience strategies of ports/supply chain; managing port operational efficiency.
Decision analysis	Optimization of maritime emergency material allocation; studies on emergency evacuation management; improving emergency management operations.
Other	All other studies not specified above.

2.2. Bibliometric Methods

Early discussion of bibliometrics began in the 1950s [26], during which bibliometrics were used to study or measure academic research through the scientific publications stored or indexed in large bibliographic databases [27]. Total scientific output, number of citations, keywords, authors, and institutions are typical indicators. The results of such analysis can be visualized in various forms, such as maps or networks, to describe datasets in a clear way. Such mapping analysis of academic research is becoming a popular method to gain insight into the field of scientific activity through the representation of bibliometric indicators [28]. Its popularity mainly lies in the advancement, availability, and accessibility of bibliometric software, such as CiteSpace 5.5.R2 and VOSviewer 1.6.13 [29]. Bibliometric software can be used to analyze data samples in a very pragmatic way, which has thus increased the academic interest in bibliometric analysis. Furthermore, with the help of software analysis, the bibliometric methodology has been widely applied in various fields, such as medicine [30], agriculture [31], business strategy [32], and marine development [33,34].

Generally, bibliometric analysis is widely used to characterize the internal structural relationship of the collected articles by means of: (1) obtaining changes in the number of articles in the field over the years; (2) showing the cooperative relationship of countries, institutions, or authors and visualizing the research team throughout the circle of research; (3) locating high-impact journals, institutions, and authors and finding the most influential authors, research institutions, or journals; (4) grasping popular research topics, gathering statistics on high-frequency keywords over the years, and finding research hotspots and development trend; (5) analyzing the citation relationship network, sorting the research development context, and assisting literature reviews. In the present study, bibliometric analysis was used to comprehensively understand the research hotspots and development trends of maritime transportation safety and emergency management from a global perspective.

3. Descriptive Statistical Analysis

3.1. Publication and Citation Distribution Analysis

Citation and publication metrics can be used to demonstrate the importance of maritime safety and emergency research. Figure 2 illustrates information about the development of the number of total publications and citations. The peak number of citations can be observed in 2021, with 654 citations. New publications need time to catch up regarding the number of citations, so it is natural that the number of citations should decrease after a period of time. Meanwhile, the evolution of annual publications shows a gradual increasing trend, which indicates that the sustainable development of maritime transportation has received increasing academic attention. International organizations, such as IMO and the International Association of Lighthouse Authorities (IALA), have put forward suggestions on the use of specific risk analysis and management tools [11], which may be an explanation for the significant increase in production.

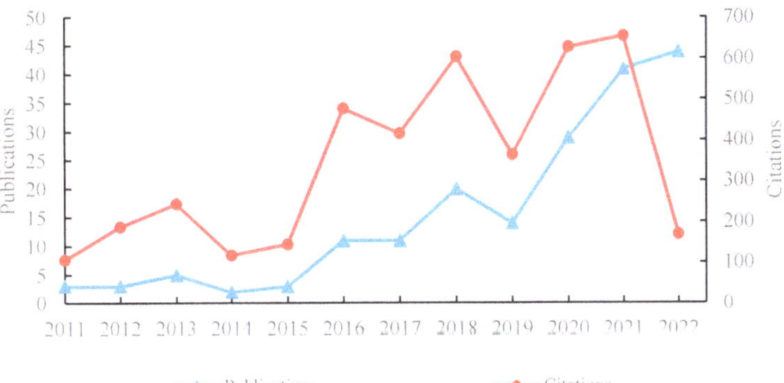

Figure 2. Number of publications and citations.

Detailed information of each considered category is presented in Figure 3 in terms of citations, publications, publication years, and affiliations. Each radar plot represents a different aspect of the conducted analysis. According to Figure 3, the maximum citation frequency was observed in the "Other" category. In terms of publications, the "Maritime network/system" category had the highest number of articles published, with 41 articles. With regard to the year of publication, the maritime transportation network and decision analysis topics were analyzed relatively early. Additionally, in this study, more than 80% of all of the articles were contributed by research universities.

3.2. Country and Institution Distribution Analysis

A total of 49 countries contributed to maritime transportation safety and emergency management over the time span of the study, and the top 10 countries with the highest scientific productions from 2011 to 2022 are presented in Figure 4, including 5 European countries, 2 Asian countries, 2 North American countries, and 1 Oceanian country. Furthermore, as seen in Figure 4, the most highly published articles were written by scholars from China, England, and Finland. Of the articles, 44% were published by Chinese scholars. China has been paying attention to research in the field of maritime transportation safety and emergency management since 2007 [35]. Within China, the top three productive universities were Dalian Maritime University (19%), Wuhan University Technology (14%), and Shanghai Maritime University (10%). Meanwhile, 14% of the articles were from England, ranking second, with Liverpool John Moores University (12%) and University Oxford (2%) being the main research institutions regarding safety and emergency management for maritime transportation. Lastly, 12% of articles came from Finland, ranking third, with Aalto University (9%) dominating in research intensity. Additionally, Turkey, Canada, the USA, Singapore, Poland, Australia, and Portugal contributed to the sustainable development of maritime transportation in the aspects of safety and emergency management.

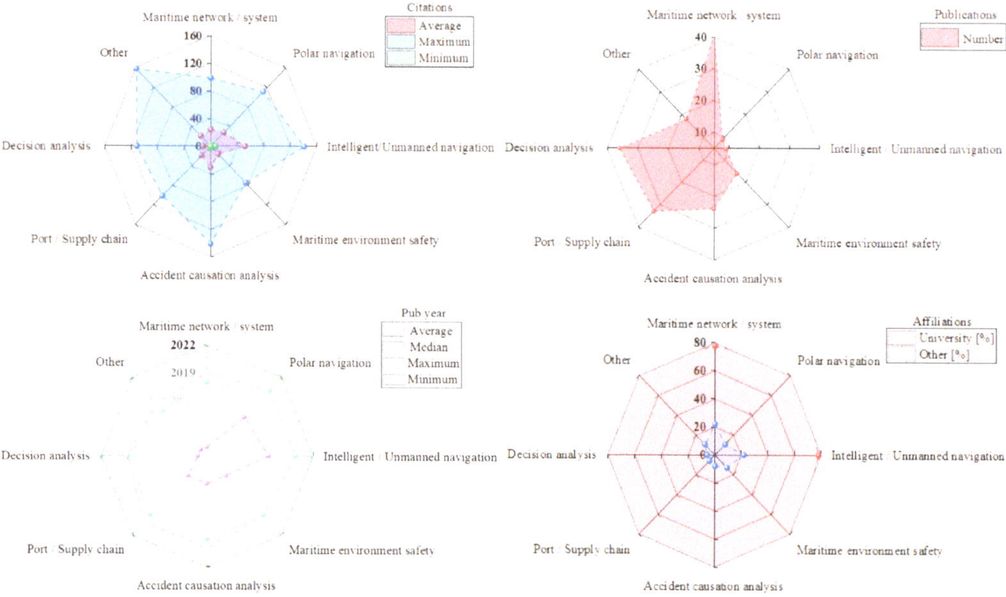

Figure 3. Radar plots for the citations, publications, publication years, and affiliations for each category.

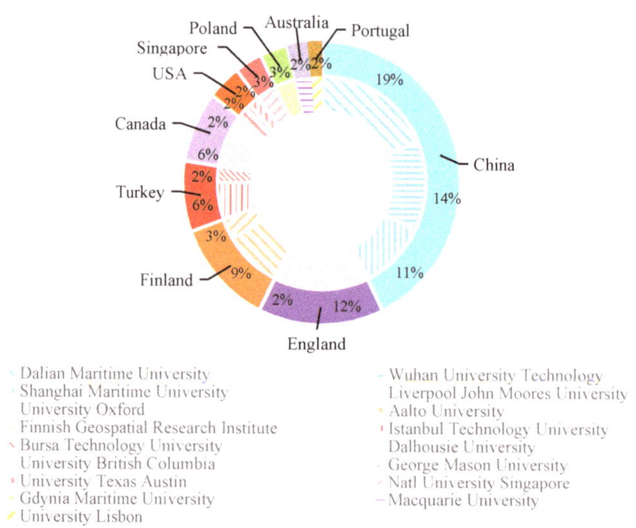

Figure 4. Counties and affiliations on WOS by number of articles published.

A co-operation diagram of the affiliated institutions distributed by countries is shown in Figure 5. According to the co-authorship among the different institutions, it is worth noting that Dalian Maritime University cooperated with scholars from Liverpool John Moores University and Shanghai Maritime University. Wuhan University of Technology engaged in academic co-operation with Liverpool John Moores University, Alto University, and Dalhousie University. Overall, this highlights that researchers actively collaborate with cross-regional institutions in maritime transportation safety and emergency management.

Figure 5. Co-operation between research institutions.

3.3. Influential Journal and Study Analysis

3.3.1. Influential Journal Analysis

In the field of maritime transportation safety and emergency management, a total of 94 journals published relevant articles, of which the journal with the highest citations was *Reliability Engineering & System Safety*, with 697 citations, as shown in Figure 6. Focusing on shipping logistics and policy, *Ocean Engineering* was the second most-cited journal, with 582 citations, indicating its high influence for maritime transportation safety and emergency management. Subsequently, we noticed that *Transportation Research Part E-logistics and Transportation Review, Safety Science*, and *Risk Analysis* had higher citation frequencies. These journals have made important academic contributions to the development of maritime transportation safety and emergency management and are widely recognized as high-quality journals related to maritime transportation safety and emergency issues. In addition, journals, such as *Maritime Policy & Management, Ocean & Coastal Management, Ocean Engineering,* and *Reliability Engineering & System Safety*, mainly focus on addressing maritime accidents (e.g., oil spills and collisions), resilience assessment [36], and vulnerability analysis [37] with quantitative methods, such as DBN and Markov chain [38–41].

The top 10 journals with the highest number of publications are presented in Figure 7. According to Figure 7, *Ocean Engineering* ranks first, with 21 publications. In second position is *Reliability Engineering & System Safety*, with 16 publications, followed by *Maritime Policy & Management*. The number of articles in the top three journals accounts for 26.34% of the total number of articles. The majority of these journals are related to transportation, operations research, and management science, highlighting the theme of maritime transportation safety and emergency management.

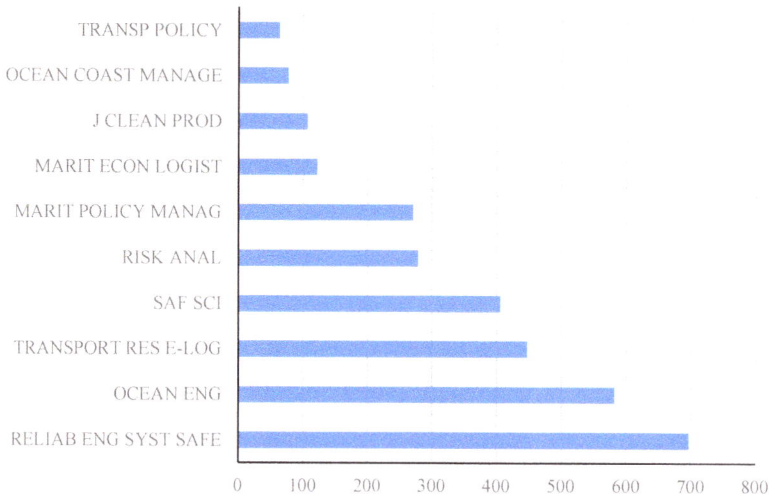

Figure 6. Top 10 journals in terms of global citations.

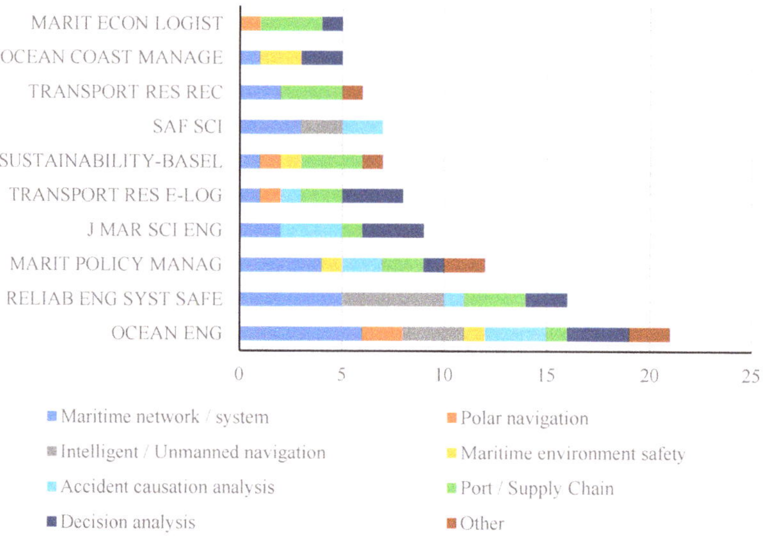

Figure 7. Number of publications by journal.

3.3.2. Influential Scholar Analysis

A total of 674 authors were included in 186 articles collected in this study. Core authors can conduct continuous research and have certain influence in their respective research field, and the number of publications is the most intuitive indicator. The price formula is considered a quantitative standard for selecting core authors, as shown in Equation (1):

$$M = 0.749\sqrt{N_{max}} \tag{1}$$

where N_{max} represents the number of articles published by the most productive author during a study period of time; M denotes the minimum number of articles published by the core author. According to Equation (1), the core author published no less than three articles. The authors' production over time is depicted in Figure 8. The color code used denotes an average number of citations aggregated for articles published in a given year.

The core authors in the field of maritime safety and emergency management published 67 articles in total, accounting for 36.02% of the 186 publications. It can also be seen that Yang ZaiLi conducted maritime safety and emergency management research in early 2013 [42]. The increase in the number of articles published by core authors remained consistent with the overall trend in the number of published articles in the analyzed topic. It is noted that some authors interrupted their research on maritime safety and emergency management; however, they returned to this study field after a few years. According to the query used in the WOS database and collected dataset, Lv Jing and Fu Shanshan had a four-year gap between their articles related to maritime safety and emergency management. Meanwhile, we found that these core authors frequently used the BN and CN methods to study maritime transportation safety issues [43–46] and mainly focused on oil spill problems [47]. In addition, a few authors were instrumental in studying maritime resilience and vulnerability assessment [48,49].

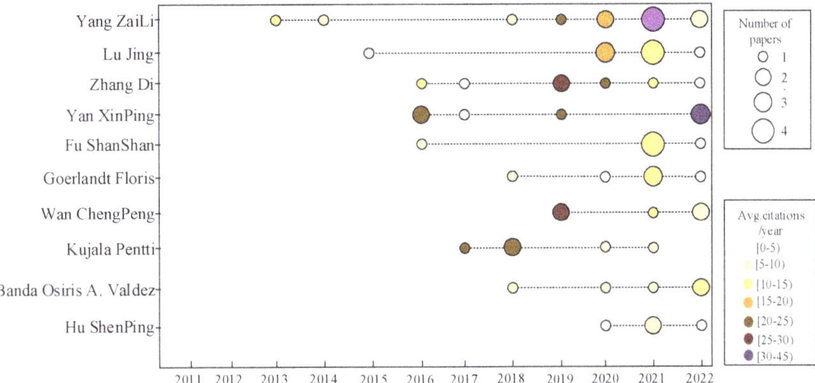

Figure 8. Top authors' production over time with the number of citations in a given year.

3.3.3. Highly Cited Articles

Citation frequency is the most commonly used indicator to measure the quality of literature, and Table 3 shows the top 10 citation frequency rankings. The highly cited articles related to maritime transportation safety and emergency management were mainly concentrated in journals, such as *Transportation Research Part E-logistics And Transportation Review, Risk Analysis*, and *Safety Science*. A total of 10 highly cited articles were cited 1150 times. Specifically, "An Overview of Maritime Waterway Quantitative Risk Assessment Models" published in *Risk Analysis* [20] had the highest number of citations (i.e., 158), and an average citation rate of 15 times per year. In addition, "Towards the assessment of potential impact of unmanned vessels on maritime transportation safety" was ranked as the second citation, in which the Human Factors Analysis and Classification System (HFACS) was adopted to study maritime accident causations (e.g., collision and grounding) [50]. Furthermore, "Marine transportation risk assessment using Bayesian Network: Application to Arctic waters" attracted much attention in terms of understanding the risk of maritime accidents [51]. These highly cited articles have laid the foundation for subsequent scholarly research in the maritime transportation safety and emergency management.

Table 3. Top 10 articles in terms of global citations.

Article	Source	Total Citations	CY
An Overview of Maritime Waterway Quantitative Risk Assessment Models [20]	*Risk Analysis*	158	14
Towards the assessment of potential impact of unmanned vessels on maritime transportation safety [50]	*Reliability Engineering & System Safety*	141	21
A Human and Organisational Factors (HOFs) analysis method for marine casualties using HFACS-Maritime Accidents (HFACS-MA) [42]	*Safety Science*	141	13
Marine transportation risk assessment using Bayesian Network: Application to Arctic waters [51]	*Ocean Engineering*	112	20
Multi-objective decision support to enhance environmental sustainability in maritime shipping: A review and future directions [52]	*Transportation Research Part E-logistics And Transportation Review*	111	13
An advanced fuzzy Bayesian-based FMEA approach for assessing maritime supply chain risks [53]	*Transportation Research Part E-logistics And Transportation Review*	105	23
Disruptions and resilience in global container shipping and ports: the COVID-19 pandemic versus the 2008–2009 financial crisis [54]	*Maritime Economics & Logistics*	102	39
A quality function deployment approach to improve maritime supply chain resilience [55]	*Transportation Research Part E-logistics And Transportation Review*	100	13
Maritime Transportation Risk Assessment of Tianjin Port with Bayesian Belief Networks [56]	*Risk Analysis*	99	13
A marine accident analysing model to evaluate potential operational causes in cargo ships [57]	*Safety Science*	81	12

CY: Citations per article per year on average.

4. Network-Based Bibliometric Analysis

4.1. Keyword Analysis

The keywords that appear in the titles and abstracts of scientific articles are important descriptions of the key contents. In this section, keywords were extracted from the titles and abstracts of the scientific publications from the WOS dataset using an automatic keyword recognition method [58], and a keyword map was visualized by using VOSviewer [59]. In Figure 9, a keyword co-occurrence diagram depicts only those keywords that appeared in at least three different articles, as a threshold was adopted for its visualization. A simple descriptive statistical analysis showed that the frequency distribution of keywords was very uneven, with only a few keywords appearing with high frequency and many keywords appearing with relatively low frequency. For example, there were only nine keywords that occurred at least 10 times in the considered dataset. These keywords were "resilience" (38), "model" (32), "framework" (27), "vulnerability" (24), "Bayesian network" (16), "optimization" (16), "impact" (14), "risk" (12), and "safety" (12).

Keyword burst citation analysis can be used to detect whether a specific research topic is hot or not. Generally, notable increases in a research field are characterized by citation bursts in publications. Keyword citation bursts can show the emerging topics in the maritime safety and emergency management field. In our study, in order to better understand the research trends of maritime transportation safety and emergency management, the evolution of keyword hotspots was analyzed from the perspective of keyword emergence. In the past 12 years, there were 17 different bursting keywords in maritime safety and emergency management publications. Table 4 lists these 17 bursting keywords with their strength and time span, as keyword emergence is typically divided into five phases by time.

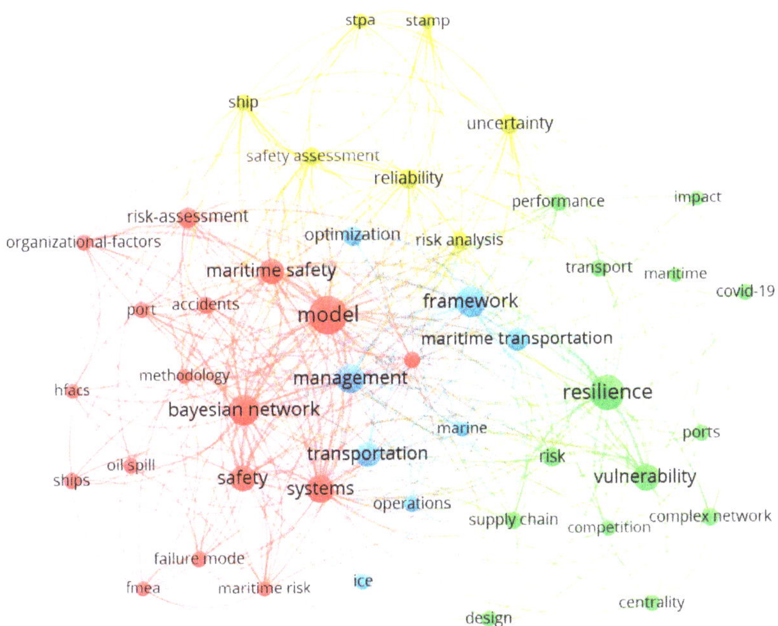

Figure 9. Co-operation between research topics.

Table 4. Top 17 keywords with the strongest citation bursts.

Keywords	Strength	Start	End	2011 to 2022
Accident	3.0334	2016	2017	
Maritime transportation	1.8375	2017	2018	
Transportation	1.3173	2018	2019	
Optimization	0.5586	2018	2019	
Disaster	2.2001	2018	2019	
Network	1.4665	2018	2019	
Vulnerability	0.3527	2018	2019	
Emergency evacuation	2.2001	2018	2019	
Port	0.731	2019	2020	
Management	1.3672	2019	2020	
Operation	1.3958	2020	2021	
Maritime accident	1.1139	2020	2021	
Risk	1.4033	2020	2021	
BN	3.1174	2020	2021	
Recovery	1.1139	2020	2021	
Sustainability	0.8203	2021	2022	
Maritime safety	0.8203	2021	2022	

(1) The first stage (2016–2018): Mainly focused on maritime accident causation analysis. The keywords that emerged were accident and maritime transportation.

(2) The second stage (2018–2019): The study of maritime transportation safety and emergency management boomed. To prevent any disasters caused by activities (e.g., grounding and oil spills), emergency evacuation under different scenarios was studied by using mathematical modeling and optimization methods. Maritime transportation network resilience and maritime transportation vulnerability identification were also an-

alyzed. The main keywords that emerged were transportation, optimization, network, vulnerability, and emergency evacuation.

(3) The third stage (2019–2020): Mainly focused on the maritime port network resilience, the risks of port disruptions, and maritime transportation management strategies. The main keywords that emerged were port and management.

(4) The fourth stage (2020–2021): Mostly focused on the risk evaluation and recovery strategies of maritime transport systems. The main keywords that emerged were operation, maritime accident, risk, BN, and recovery.

(5) The fifth stage (2021–2022): Mainly focused on maritime transportation safety and sustainability. In this stage, researchers analyzed the safety of maritime transportation from different perspectives, such as the development of port vulnerability assessment (PVA) frameworks and the proposal of optimal resilience models in maritime transportation systems. The main keywords that emerged were sustainability and maritime safety.

4.2. Co-Cited Analysis

4.2.1. Journal Co-Citations

The co-citation frequency of journals is the main indicator for measuring the attractiveness of academic journals. Therefore, we further analyzed journal sources by drawing a network map of co-citations with the WOSviewer application, where the relationship between two different publications could be assessed based on the number of articles citing both journals. This helped to visualize the inter-relationships between the sources of the journal, as shown in Figure 10, in which the link strength is the frequency with which two journals appear in one publication simultaneously. The most relevant sources were *Reliability Engineering & System Safety*, *Safety Science*, and *Maritime policy & Management*, which were closely linked based on the most citations. We noticed that the cluster centered by *Reliability Engineering & System Safety* was the largest, and its number of published articles ranked second, with 16 articles. A similar situation was also applicable to *Risk analysis* and *Accident Analysis & Prevention*, both of which had common citation characteristics. In addition, there was a large number of co-citations in other journals, such as *Expert System with Applications* and *Computers & Industrial Engineering*, making the field of maritime transportation safety and emergency management comprehensive and interdisciplinary.

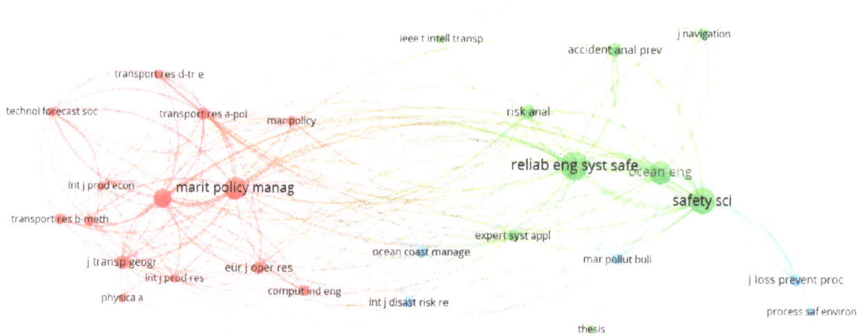

Figure 10. Co-citation network of journals cited by publications.

4.2.2. Article Co-Citations

Two articles establish a co-citation relationship when they appear in the references of another article simultaneously. As a result, a co-citation map could be established by using WOSviewer 1.6.13, which is illustrated in Figure 11. This figure shows the minimum number of a cited reference to be at least 10 times; the size of the node represents the citation frequency, and the line between two nodes indicates that they have been cited at least once. In general, if two articles establish a co-citation relationship, they are more or less similar.

It can be seen that the published articles in the field of maritime transportation safety and emergency management can be grouped into three categories.

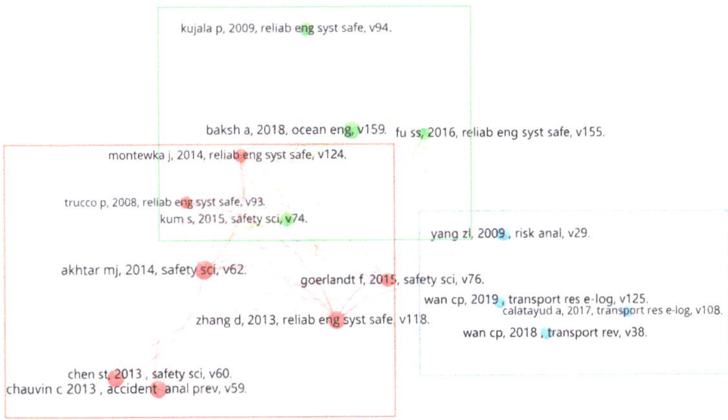

Figure 11. Co-citation network of references cited by publications.

Green: Highly co-cited articles mainly analyzing the possibility, risk, and cause of maritime accidents, such as collision, foundering, and grounding [51,60–62]. It is noteworthy that such studies were mainly focused on Arctic waters.

Red: Highly co-cited articles mainly studying the risk quantification [63–65] and human and organizational factor (HOF) analysis [42,66–68] for maritime transportation systems.

Blue: Highly co-citated articles mostly developing fuzzy logic methods to analyze maritime security assessments [53,69]. In addition, maritime transportation vulnerability was described by Calatayud et al. [70] from a multiplex network perspective.

4.2.3. Author Co-Citations

Author co-citation analysis is another interesting topic. We set the threshold as 15, and 36 authors were selected. The results were visualized by using WOSviewer 1.6.13, where a node represents an author and a line is established when two authors are cited in one article. The distance between two nodes reflects the degree of similarity to the authors' field of study. The most influential authors can be observed in Figure 12. It should be noted that there were some articles that cited the data provided by institutes like the IMO and the UNCTAD; in these cases, the institutes that provided the data were regarded as authors. According to the results, Goerlandt, Montewka, Yang, Akyuz, IMO, Wu, Hollnagel, Zhang, Banda, and Ducruet were the most-co-cited authors.

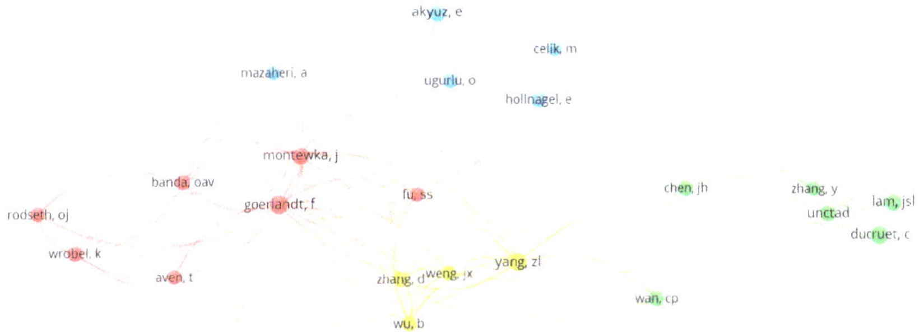

Figure 12. Mapping of the author co-citations.

The first category represents the authors in the field of human factors analysis, including Akyuz, Hollnagel, and Celik. Specifically, to achieve maritime safety, Akyuz and Celik proposed the HFACS combined with a cognitive map (CM) in maritime accident analysis [71], as well as human error assessment and reduction technique (HEART) [72] methods, which were developed as a marine-specific approach to quantify human error. Hollnagel focused on evaluating the human factors based on the FRAM approach [73].

The second category represents the authors in the field of risk influencing factors analysis of Arctic waters, including Zhang, Fu, and Khan. Specifically, Zhang and Fu mainly used BN [74] and the analytical hierarchy process (AHP) [75] to analyze the potential risk factors in Arctic shipping. As a co-author, Khan also adopted DBN [76] and BN [77] to identify and classify contributing risk factors for Arctic waters. Additionally, an updated Nagel–Schrekenberg (NaSch) model of Arctic marine convoy traffic integrated with a BN-based probabilistic approach was used to predict the maximum waterway density for the safe flow of traffic and the collision probability during a convoy [78], highlighting the implementation of advanced technology as being crucial in enhancing safe navigation at sea.

The third category represents the authors in the field of safety management, including Goerlandt, Montewka, and Banda. Specifically, Goerlandt was the most-co-cited author in the field of maritime transportation and mainly engaged in research on maritime risk analysis and management, involving oil spill preparedness, planning, and development of tools for maritime accidents. As the co-authors, Goerlandt and Montewka presented a review and analysis of risk definitions, perspectives, and scientific approaches to risk analysis [11], as well as applied BN modeling for probabilistic risk quantification [63]. Banda developed a formal safety assessment (FSA) to assess and manage the risk of winter navigation operations [79].

The fourth category represents the authors in the field of risk assessment and accident prevention, including Wan, Yang, and Wu. Specifically, Wan and Yang applied an advanced fuzzy Bayesian-based FMEA approach to assess maritime supply chain risks [53]. Wu developed a modified cognitive reliability and error analysis method (CREAM) for estimating the human error probability in maritime accidents [80].

The fifth category represents the authors in the field of maritime network and vulnerability analysis, including Ducruet, Zhang, and Berle. Specifically, Ducruet focused on the maritime network characteristics [81–83]. Zhang frequently used the geo-spatial techniques of kernel density estimation (KDE) to identify accident-prone sea areas [84,85]. To assess the vulnerability of maritime transportation, formal vulnerability assessment (FVA) methodology was employed by Berle et al. [86] and Berle et al. [87].

5. Methodology for Maritime Safety and Emergency Management

5.1. Overview of the Research Methods

In this study, the main research methods used in the 186 articles were identified by means of a manual review. Figure 13 illustrates an overview of the research methods used in maritime transportation safety and emergency management.

There are still traditional risk assessment methods that remain widely used, despite the emergence of new methods in recent years. The early studies in maritime accident research usually adopted very basic methods, such as interviews and surveys analysis, while recent studies often used multi-disciplinary approaches and comprehensive analyses. Many different approaches have been developed to address maritime transportation safety and emergency management problems. Recently, new methods that have appeared in maritime safety and emergency management research include STPA, cognitive reliability error analysis method (CREAM), DBN, emergency assessment-based simulation [88], probabilistic risk assessment-based simulation [89], resilience assessment-based simulation [90], and mathematical modeling and optimization methods, such as non-linear optimization and enhanced particle swarm optimization (EPSO) models [91], multi-objective particle swarm algorithm [47], and dynamic multi-objective optimization model [92]. At present, machine learning is introduced to improve maritime safety and management.

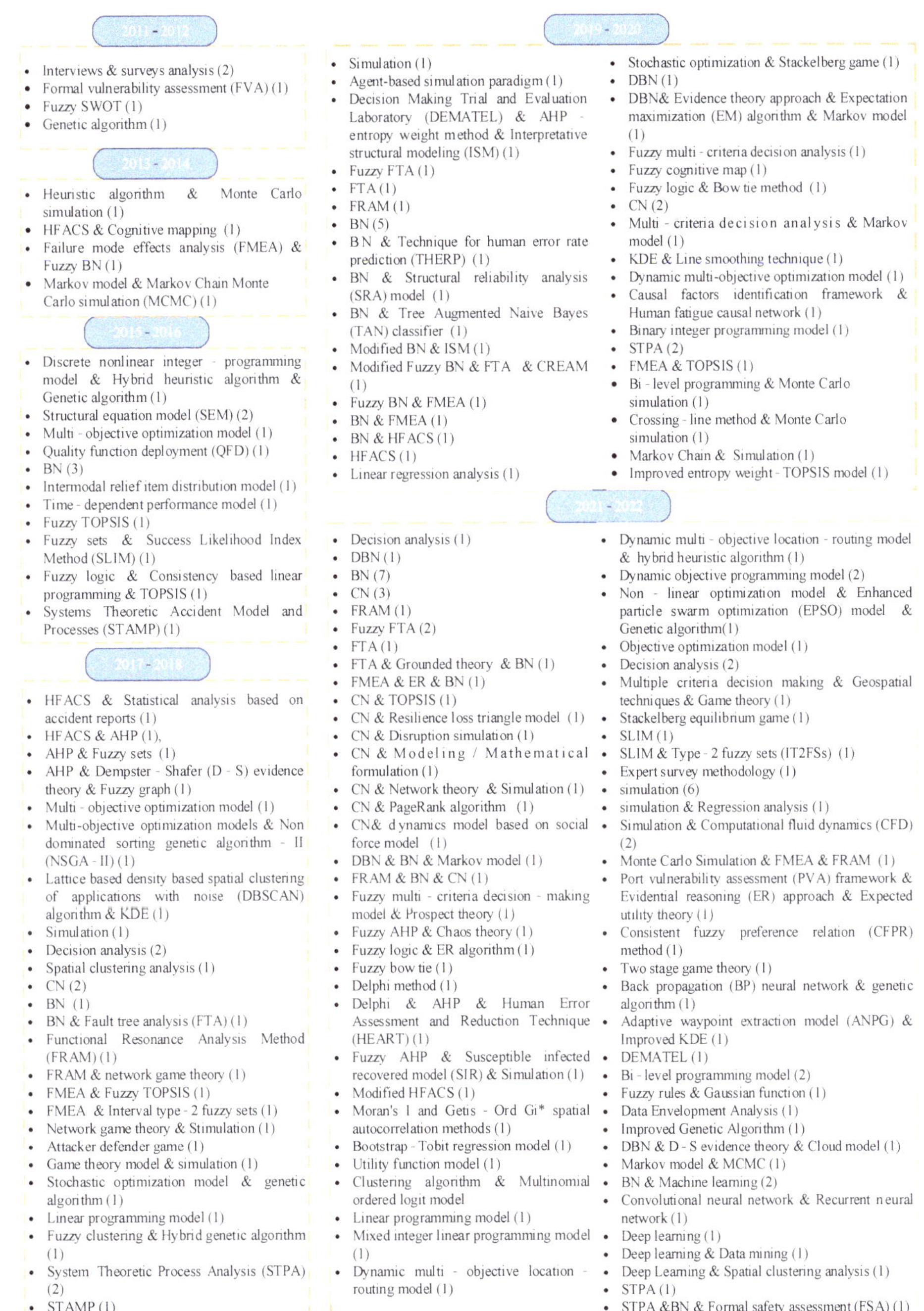

Figure 13. Overview of the methods employed in maritime safety and emergency management.

Model extension has occurred alongside the introduction of new models to this research area. Ung [93] extended the CREAM approach by incorporating BN and FTA in a fuzzy environment.

Chen et al. [42] proposed the HFACS, which has been used to identify human errors in maritime accidents. Akyuz combined the HFACS approach with the analytic hierarchy process (AHP) to evaluate potential operational causes in maritime accidents [57]. Uğurlu et al. integrated the HFACS and BN to analyze maritime collision, grounding, and sinking accidents [94]. To improve port safety, BN and FMEA were combined to assess the criticality of the hazardous events by [95]. Yuan et al. [96] combined BN and FTA to study the causal factors in emergency processes in response to fire accidents for oil gas storage. Likewise, Wang et al. [97] used BN and FTA to assess the critical risk factors in ship fire accidents. Then, Abaei et al. [98] linked BN and machine learning to analyze the resilience of unattended machinery plants in autonomous ships.

Figure 14 shows the statistics of the main research methods used in the literature. More than half of the articles used quantitative analysis to study maritime transportation safety and emergency management problems. Meanwhile, it can be seen that BN, fuzzy logic, simulation, CN, FMEA, FTA, game theory, machine learning, FRAM, HFACS, STPA, Markov model, and DBN are the most commonly used measurement methods in the field of maritime transportation safety and emergency management.

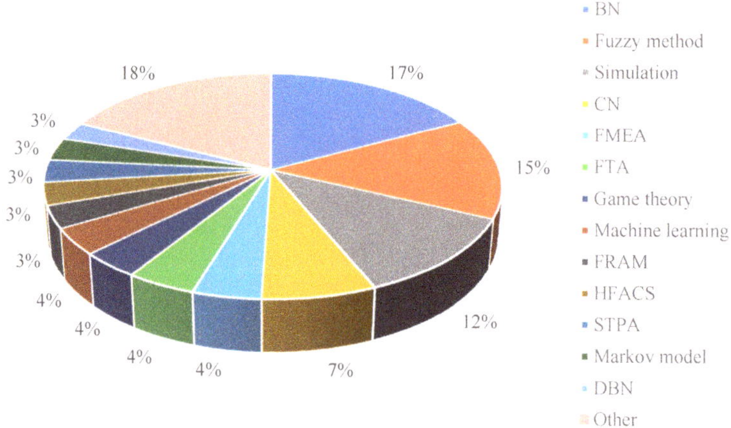

Figure 14. Statistics of the main research methods used in the literature.

According to Figure 14, the most frequently used method was the BN method. The application of the BN method mainly focuses on the following two aspects: (1) study of the causal correlation degree of factors from the accident causation theory perspective [96]; (2) risk prediction carried out using the BN model [44,53]. Maritime transportation has great uncertainty, which is affected by system complexity, environmental factors, human factors, and organizational factors [68]. BN is a suitable method for risk assessment and decision making. Furthermore, BN can replace FTA as a classification method and can take into account the joint effect of several events. This is the reason why the BN model is popular in the field of maritime transportation safety and emergency management. However, data availability is one of the biggest problems in calculating the failure rate in the maritime industry. In order to solve this limitation, fuzzy methods are widely introduced to deal with the uncertain data. Simulation is the third most frequently used tool; the risk of maritime accidents has a probabilistic attribute, and simple statistical data are not sufficient to explain and predict the risk of accidents over time. The simulation method can be used to analyze the influence of many uncertain factors. Faghih-Roohi et al. [99] combined Monte Carlo simulation and the Markov model to estimate the probability of maritime transport accidents for the first time. Huang et al. [100] adopted the Monte Carlo method to calculate the probability of a ship crossing the channel boundary. Zou and Chen [101] used Monte Carlo simulation to assess the resilience of the maritime supply chain and analyzed the impact of interruption scenarios for maritime transportation systems.

Table 5 provides a comparison of the main methodological features used in this study: (1) Quantitative—used to distinguish whether the method is quantitative or qualitative. (2) Interactivity—referring to the mechanism of interaction between the factors. (3) Interpretability—providing a specific path for the propagation of risk factors. (4) Decoupling—considering the contribution of each factor function in the case of multiple factors completing the task in a cooperative manner. (5) Memorable—considering the impact of task completion results on other subsequent tasks. Not all risk behaviors or states will necessarily contribute to serious consequences, and only when certain conditions are met will the next stage of risk events or accidents be triggered. (6) Sequential—considering the sequence of events. (7) Scalability—the ability to combine with other methods. (8) Extensibility—the ability to handle large-scale parameters.

Table 5. Comparison of the main methodological features used in the literature.

Methods	Quantitative	Interactivity	Interpretability	Decoupling	Memorable	Sequential	Scalability	Extendibility
BN [97]	✓	✓	✓			✓	✓	
Fuzzy method [93]	✓						✓	
Simulation [102]	✓	✓	✓	✓	✓		✓	✓
CN [49]	✓	✓	✓	✓			✓	✓
FMEA [99]	✓		✓				✓	
FTA [93]	✓		✓			✓	✓	
Game theory [103]	✓	✓	✓				✓	
Machine learning [98]	✓	✓			✓		✓	✓
FRAM [89]		✓	✓	✓			✓	
HFACS [94]							✓	
STPA [104]		✓	✓				✓	
Markov model [44]	✓	✓	✓				✓	
DBN [44]	✓	✓	✓			✓	✓	

5.2. The Progressive Trend of Research Methods

Figure 15 illustrates the progressive trend of the primary methods and models utilized in maritime transportation safety and emergency management between 2011 and 2022. The advancement of technology has expanded the application scope and enhanced their accuracy in various scenarios. Between 2011 and 2013, statistical analysis and framework-based analysis were the predominant methods employed in maritime transportation safety and emergency management. Traditional risk analysis techniques, such as FVA, SWOT, and HFACS, were widely applied during this period. Heuristic algorithms were used to solve complex problems by iteratively exploring and evaluating a large search space. From 2014 to 2016, fuzzy BN, Markov model, MCMC, mathematical modeling, and STAMP were employed to identify the maritime transportation system risk. From 2017 to 2019, CN, cluster analysis, multi-objective optimization models, simulation, game theory, STPA, and ISM were introduced to study maritime transportation safety and emergency management issues. Human reliability analysis methods, such as CREAM and THERP, were employed to identify human errors in maritime risk. More specific and detailed research methods were used to evaluate maritime safety and emergency management. From 2020 to 2022, methods, such as DBN and dynamic programming models, were utilized to assess risks and optimized paths in the maritime safety and emergency management field. Additionally, the development of research methods has facilitated the application of machine learning algorithms, such as BP neural networks, convolutional neural networks, and recurrent neural networks in maritime safety and emergency management.

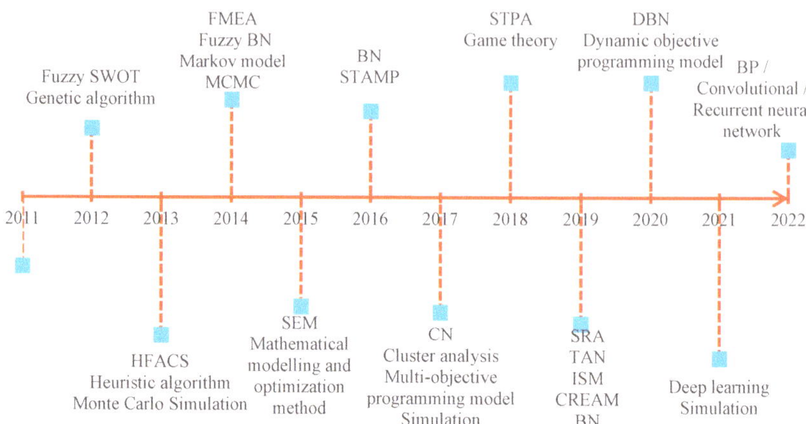

Figure 15. The progressive trend of research methods.

5.3. Spatial Interaction Visualization of Research Methods

Figure 16 shows the spatial interaction visualization of research methods. Let graph $G(V, E, W)$ be the representation of the main research method network. In this graph, V is the set of nodes representing methods, E is the set of edges representing the linking between methods, and W is the weight of edge. Graph G is represented as a weighted adjacency matrix **A**, whose elements are $a_{ij} = w_{ij}$. If link $(i, j) \in E, i, j \in V, a_{ij} = w_{ij}$, where w_{ij} indicates the number of methods connecting method i and method j; otherwise, $a_{ij} = 0$. Finally, the spatial interaction visualization of main research methods is connected in a CN, as given Figure 16.

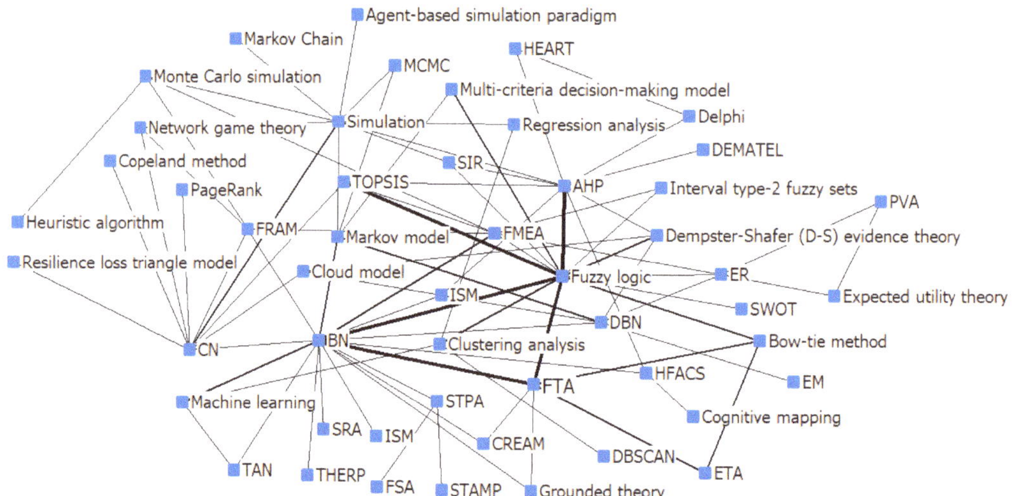

Figure 16. Spatial interaction visualization of the methods and models used in the literature.

Fuzzy logic can deal with uncertainty and vagueness and can be integrated with other methods to handle imprecise inputs. Fuzzy set analysis has been widely used together with methods, such as FTA, SWOT, FMEA, ER, bow-tie method, BN, and AHP. For example, Zaib et al. [105] analyzed human error using a fuzzy FTA. Jiang et al. [48] used the fuzzy evidential reasoning (ER) algorithm to estimate the vulnerability of straits or canals in maritime transportation. Furthermore, Fuzzy TOPSIS was combined with FMEA to analyze port risks [106], while interval type-2 fuzzy sets were integrated with FMEA to conduct oil spill risk assessments [38].

By combining BN with other methods, a more comprehensive analysis method can be adopted, such as BN-FMEA [95], BN-FTA [97], and BN machine learning [98]. The use of BN in risk assessments has made significant advancements; scholars have begun to explore the integration of time-sliced temporal data into BN models, known as DBN [12], which is used to model the evolution of a system over time. For instance, Jiang and Lu [44] presented a DBN model for assessing the dynamic risk of maritime accidents.

In recent years, scholars have attempted to combine CN with other methods to study the maritime transportation network from the perspective of resilience. Yang and Liu [90] constructed the Maritime Silk Road shipping network using the CN method and then used disruption simulations to analyze the resilience of the Maritime Silk Road transportation network, identifying dominant and weak port nodes. Wan et al. [49] used the resilience loss triangle model to analyze the performance of liner shipping networks (LSNs) during recovery, and the rationality and feasibility of the developed indicators in LSN-aided decision making were tested from the recovery strategies based on the degree of centrality, closeness of the degree of centrality, and betweenness centrality. Poo and Yang [107] assessed the global shipping network focusing on climate resilience by using a methodology that combined CN and a ship routing optimization model.

Simulation methods were used to study the operation of real-world or theoretical processes or systems in various pre-defined environments for different purposes (e.g., numerical testing and exploring new states) [108]. Simulations provide a more comprehensive understanding and accurate prediction of the impacts resulting from different game strategies. Game theory and simulation have been combined to discuss the impact of investment behavior on maritime transportation. In order to provide insights into resilience improvements for maritime transportation, Chen et al. [104] used the network game theory to investigate the impact of participants' investment decisions on maritime logistics network resilience and simulated participants' investment strategies in the face of catastrophic accidental explosions, labor strikes, and terrorist attacks. Liu et al. [109] applied the game theory model to study the pre-disaster investment strategies of two neighboring seaports and conducted a numerical simulation to evaluate the stability of a co-operation mechanism.

This study revealed that a new trend in recent years is the use of combined methods and coupled analysis. The application of combined methods and coupled analysis can enhance maritime transportation safety and management; furthermore, BN, fuzzy logic, CN, and simulation are generally combined with various methods.

6. Discussion and Future Research

6.1. Quantitative and Systematic Assessment of Maritime Transportation System Resilience

Due to the frequent occurrence of natural and man-made disasters, the concept of resilience is gradually emerging. Resilience was first proposed by Holling in the field of ecosystem research [110]; then, it began being widely used in other fields, including economics [111], psychology [112], and system engineering [113]. The core concept of resilience is the ability of a system to resist and recover performance from unexpected disruption events. The hypothetical system performance of the curves under normal conditions and in the face of destructive events can be referred to [108], which attempts to incorporate as many resilience features as possible. For maritime transport systems, it is critical to mix methods and tools to compare the results of maritime resilience under disruptive events. To date, some studies have assessed maritime resilience from the network topology viewpoint [90], but few have included performance indicators via quantitative evaluation. In the future, the metrics used to evaluate maritime resilience are worthy of our consideration. Furthermore, it is necessary to develop new indicator assessment frameworks and incorporate the features of maritime resilience. Although the TOSE (technical, organizational, social, and economic)–R4 (robustness, redundancy, resourcefulness, and rapidity) framework for assessing community earthquake resilience has been studied for many years, relevant research framework on maritime resilience is still in its infancy. The TOSE–4R framework provides a reference for maritime resilience analysis. More importantly, how to analyze the dynamic interaction process between maritime resilience and external perturbations based on actual scenarios is worth studying. However, it should be noted that the Russia–Ukraine war has greatly disturbed the shipping market; thus, a potential research direction is to discuss maritime resilience in the context of the Russia–Ukraine war on the future research agenda.

6.2. Data- or Intelligence-Driven Technologies for Maritime Safety and Emergency Management

In terms of maritime safety modeling, in the early stage, expert knowledge remains an essential data source when essential data are unavailable or incomplete from relevant investigations; however, expert knowledge is argued to be subjective and uncertain [114]. The data-driven approach as an emerging method that reduces the subjectivity of research and is more consistent with the actual situation; nevertheless, in the face of diverse data types and extensive data sources, obtaining and extracting effective information is a challenge. Industry 4.0, which is the deep integration of information and intelligence, can be considered the current trend of data exchange in manufacturing processes and automation. Industry 4.0. After Industry 4.0, intelligent technology has penetrated into various industries, including maritime transportation areas, and the Internet of Things, big data analysis, artificial intelligence, and cloud computing are the key focus directions. Intelligent technology is used to process massive data, extract valuable information, and enhance the intelligent capacity of maritime transportation safety and emergency management. With the aid of advanced intelligent technology and concepts, combining intelligent data and historical data with machine learning and other model algorithms to realize intelligent safety and emergency management is worth exploring. First, knowledge related to safety and emergency management, including event information, warning rules, and processing procedures, can be represented. Second, the semantics related to safety and emergency management can be parsed and reasoned to identify the risks and hidden dangers. Finally, an intelligent monitoring, assessment, and early warning system can be established to realize the real-time monitoring and early warning of maritime transport safety and emergency management.

6.3. Scenario Representation and Digital Twins Are Becoming Critical and Practical for Maritime Emergency Research

Maritime emergencies occur frequently, causing great damage to the environment and society. When a maritime emergency occurs, an immediate response is important for minimizing the damage. There needs to be a balance between the focus on preventative safety efforts and the extent of emergency preparedness provisions. Currently, existing research focuses on the location of the emergency supplies reserve base, rescue resource allocation, and configuration optimization of salvage vessels [91,115]. Moreover, most research has taken oil spills as a case to analyze emergency resource dispatching [92,116]. Notably, in the future, comparative analyses of the application effects under different emergency scenario levels will be important for improving the efficiency of maritime emergency salvage. Moreover, climate change and the development of sea routes have increased maritime activity in the Arctic, which increases risks of maritime accidents, such as oil spills, collisions, and explosions. To cope with the complexity and uncertainty of Arctic maritime emergency operations caused by humans or nature in the Arctic, future studies can extend the digital twin technology into maritime emergency management, which can reduce uncertainties in the emergency operation process and optimize the integration of maritime emergency management. To achieve this objective, actual data are mapped to a digital twin; after connecting with the digital twin, the response entities from different stages dynamically adjust based on changes in the unexpected events, thereby continuously optimizing the effectiveness of maritime emergency response effects.

6.4. Issues Associated with Intelligent Maritime Shipping Require Urgent Attention for a Sustainable Maritime Industry

Despite continually improving safety records, shipping is considered a dangerous industry with a high rate of fatal injuries and high consequences of maritime disasters, with investigations of the underlying causes of marine accidents tending to point to human error as the single greatest contributor, estimated to be involved in 75–96% of all accidents [117]. Furthermore, due to the high impact of human errors, autonomous ship development has become an important issue in the shipping industry. At present, multiple studies are focusing on the risk identification of autonomous ships. For example, Wróbel et al. proposed a model for the safety assessment of autonomous merchant ships [118], Fan et al. identified the factors influencing navigational risk for autonomous ships [119], and a framework for risk modeling was outlined for autonomous ships by Utne et al. [120]. However, thus far, few research works have quantified the risk management techniques for autonomous ships, and the percentage of accidents that can be prevented by risk management techniques has not been quantified. In this context, safety management frameworks are the direction of autonomous ships moving forward. To achieve autonomous ships that are safe at sea, the first step is to establish a mapping relationship between safety management framework functions and risk evolution characteristics. Embedding a multi-agent model for human–autonomous ship interaction follows this, with a subsequent consequence assessment of the application of safety management technology, which is crucial based on autonomous shipping scenarios.

7. Conclusions

An updated literature review can identify the hotspots and trends of a thematic discussion. In this study, we used systematic and bibliometric reviews to examine 186 articles published on maritime transportation safety management between 2011 and 2022, which allowed to provide a comprehensive summary and mapping of this topic. The results show that most authors examined maritime safety and emergency management from risk assessment, emergency resource optimization, vulnerability analysis, and resilience measurement perspectives. This study also provided a specific elaboration on the main research methods. Maritime transportation safety management assessment methods have undergone a qualitative analysis and quantitative evaluation. Most journals contribute articles using BN, fuzzy theory, and simulation methods. Recently, studies have focused on mixed methods and advanced tools (e.g., deep learning and dynamic programming models) to analyze maritime safety and emergency management issues. Analyzing maritime safety management publications can be helpful, especially in assisting researchers in effectively finding suitable directions for upcoming research. This study provides an initial, comprehensive, and systematic evaluation of the literature on maritime safety management. The findings serve as the foundation for further investigation of maritime safety management concerns. This review article provides guidance for essential future research topics, such as maritime resilience, data-driven analysis, emergency evacuation, and autonomous ship navigation safety development.

However, there are some limitations to our research. First, the bibliometric approach can only provide a high-level overview of the maritime transportation safety and emergency management field. Therefore, a systematic literature review is recommended to gain more detailed insights. Second, using other databases, such as the Springer or Scopus database, may produce different results. Third, the high citation rate indicates that these articles have made significant contributions to the development of maritime transportation safety management; however, high citation rates do not mean correctness of the academic research results. Focusing on highly cited articles may lead to the neglect of important emerging research topics.

Author Contributions: M.X., Methodology and formal analysis; X.M., data curation and formal analysis; Y.Z., data curation, investigation, and writing—original draft; W.Q., conceptualization, writing—editing and review, and investigation. All authors have read and agreed to the published version of the manuscript.

Funding: This research was funded by the Postdoctoral Funding of China (2022M720626), Science Foundation of Liaoning (Grant No. 2022-BS-095), Youth Talent Support Foundation of Dalian (Grant No. 2021RQ043), and the Fundamental Research Funds for the Central Universities (3132023530; 3132023625).

Institutional Review Board Statement: Not applicable.

Informed Consent Statement: Not applicable.

Data Availability Statement: Data are contained within the article.

Conflicts of Interest: The authors declare no conflict of interest.

References

1. UNCTAD. Review of Maritime Transport 2022. 2022. Available online: https://unctad.org/system/files/official-document/rmt2022_en.pdf (accessed on 18 October 2023).
2. Arici, S.S.; Akyuz, E.; Arslan, O. S Application of fuzzy bow-tie risk analysis to maritime transportation: The case of ship collision during the STS operation. *Ocean Eng.* **2020**, *217*, 107960. [CrossRef]
3. Størkersen, K.V.; Antonsen, S.; Kongsvik, T. One size fits all? Safety management regulation of ship accidents and personal injuries. *J. Risk Res.* **2017**, *20*, 1154–1172. [CrossRef]
4. Yoo, Y.; Park, H.S. Qualitative risk assessment of cybersecurity and development of vulnerability enhancement plans in consideration of digitalized ship. *J. Mar. Sci. Eng.* **2021**, *9*, 565. [CrossRef]
5. Zhang, L.; Wang, H.; Meng, Q.; Xie, H. Ship accident consequences and contributing factors analyses using ship accident investigation reports. *Proc. Inst. Mech. Eng. Part O-J. Risk Reliab.* **2019**, *233*, 35–47. [CrossRef]
6. MOT. *Statistical Bulletin on the Development of Transportation Industry*; Ministry of Transport of China: Beijing, China, 2019.
7. EMSA. *Annual Overview of Marine Casualties and Incidents*; European Maritime Safety Agency: Lisbon, Portugal, 2019.
8. Ramos, K.G.; Rocha, I.C.N.; Cedeno, T.D.D.; Dos Santos Costa, A.C.; Ahmad, S.; Essar, M.Y.; Tsagkaris, C. Suez Canal blockage and its global impact on healthcare amidst the COVID-19 pandemic. *Int. Marit. Health* **2021**, *72*, 145–146. [CrossRef] [PubMed]
9. Celik, M.; Cebi, S. Analytical HFACS for investigating human errors in shipping accidents. *Accid. Anal. Prev.* **2009**, *41*, 66–75. [CrossRef] [PubMed]
10. Baalisampang, T.; Abbassi, R.; Garaniya, V.; Khan, F.; Dadashzadeh, M. Review and analysis of fire and explosion accidents in maritime transportation. *Ocean Eng.* **2018**, *158*, 350–366. [CrossRef]
11. Goerlandt, F.; Montewka, J. Maritime transportation risk analysis: Review and analysis in light of some foundational issues. *Reliab. Eng. Syst. Saf.* **2015**, *138*, 115–134. [CrossRef]
12. Huang, X.; Wen, Y.; Zhang, F.; Han, H.; Huang, Y.; Sui, Z. A review on risk assessment methods for maritime transport. *Ocean Eng.* **2023**, *279*, 114577. [CrossRef]
13. Kandemir, C.; Celik, M. A systematic literature review and future insights on maritime and offshore human reliability analysis. *Proc. Inst. Mech. Eng. Part M J. Eng. Marit. Environ.* **2023**, *237*, 3–19. [CrossRef]
14. Lim, G.J.; Cho, J.; Bora, S.; Biobaku, T.; Parsaei, H. Models and computational algorithms for maritime risk analysis: A review. *Ann. Oper. Res.* **2018**, *271*, 765–786. [CrossRef]
15. Moreno, F.C.; Gonzalez, J.R.; Muro, J.S.; Maza, J.G. Relationship between human factors and a safe performance of vessel traffic service operators: A systematic qualitative-based review in maritime safety. *Saf. Sci.* **2022**, *155*, 105892. [CrossRef]
16. Ozbas, B. Safety Risk Analysis of Maritime Transportation Review of the Literature. *Transp. Res. Rec.* **2013**, *2326*, 32–38. [CrossRef]
17. Talley, W.K. Maritime transportation research: Topics and methodologies. *Marit. Policy Manag.* **2013**, *40*, 709–725. [CrossRef]
18. Wu, B.; Yip, T.L.; Yan, X.; Guedes Soares, C. Review of techniques and challenges of human and organizational factors analysis in maritime transportation. *Reliab. Eng. Syst. Saf.* **2022**, *219*, 108249. [CrossRef]
19. Zhang, G.; Thai, V.V. Expert elicitation and Bayesian Network modeling for shipping accidents: A literature review. *Saf. Sci.* **2016**, *87*, 53–62. [CrossRef]
20. Li, S.; Meng, Q.; Qu, X. An overview of maritime waterway quantitative risk assessment models. *Risk Anal.* **2012**, *32*, 496–512. [CrossRef] [PubMed]
21. Suárez-Alemán, A. Short sea shipping in today's Europe: A critical review of maritime transport policy. *Marit. Econ. Logist.* **2016**, *18*, 331–351. [CrossRef]
22. Al-Enazi, A.; Okonkwo, E.C.; Bicer, Y.; Al-Ansari, T. A review of cleaner alternative fuels for maritime transportation. *Energy Rep.* **2021**, *7*, 1962–1985. [CrossRef]
23. Laera, F.; Fiorentino, M.; Evangelista, A.; Boccaccio, A.; Manghisi, V.M.; Gabbard, J.; Gattullo, M.; Uva, A.E.; Foglia, M.M. Augmented reality for maritime navigation data visualisation: A systematic review, issues and perspectives. *J. Navig.* **2021**, *74*, 1073–1090. [CrossRef]

24. Larsen, M.H.; Lund, M.S. Cyber risk perception in the maritime domain: A systematic literature review. *IEEE Access* **2021**, *9*, 144895–144905. [CrossRef]
25. Gu, B.; Liu, J. A systematic review of resilience in the maritime transport. *Int. J. Logist. Res. Appl.* **2023**, 1–22. [CrossRef]
26. Wallin, J.A. Bibliometric methods: Pitfalls and possibilities. *Basic Clin. Pharmacol. Toxicol.* **2005**, *97*, 261–275. [CrossRef] [PubMed]
27. Gutierrez-Salcedo, M.; Angeles Martinez, M.; Moral-Munoz, J.A.; Herrera-Viedma, E.; Cobo, M.J. Some bibliometric procedures for analyzing and evaluating research fields. *Appl. Intell.* **2018**, *48*, 1275–1287. [CrossRef]
28. Gil, M.; Wróbel, K.; Montewka, J.; Goerlandt, F. A bibliometric analysis and systematic review of shipboard Decision Support Systems for accident prevention. *Saf. Sci.* **2020**, *128*, 104717. [CrossRef]
29. Donthu, N.; Kumar, S.; Mukherjee, D.; Pandey, N.; Lim, W.M. How to conduct a bibliometric analysis: An overview and guidelines. *Bus. Res.* **2021**, *133*, 285–296. [CrossRef]
30. Liang, Y.; Li, Y.; Zhao, J.; Wang, X.; Zhu, H.; Chen, X. Study of acupuncture for low back pain in recent 20 years: A bibliometric analysis via CiteSpace. *J. Pain Res.* **2017**, *2017*, 951–964. [CrossRef]
31. Wei, J.; Liang, G.; Alex, J.; Zhang, T.; Ma, C. Research progress of energy utilization of agricultural waste in China: Bibliometric analysis by citespace. *Sustainability* **2020**, *12*, 812. [CrossRef]
32. Kumar, S.; Sureka, R.; Lim, W.M.; Kumar Mangla, S.; Goyal, N. What do we know about business strategy and environmental research? Insights from Business Strategy and the Environment. *Bus. Strategy Environ.* **2021**, *30*, 3454–3469. [CrossRef]
33. Hu, H.; Xue, W.; Jiang, P.; Li, Y. Bibliometric analysis for ocean renewable energy: An comprehensive review for hotspots, frontiers, and emerging trends. *Renew. Sustain. Energy Rev.* **2022**, *167*, 112739. [CrossRef]
34. Wu, M.; Jiang, Y.; Kwong, R.W.; Brar, S.K.; Zhong, H.; Ji, R. How do humans recognize and face challenges of microplastic pollution in marine environments? A bibliometric analysis. *Environ. Pollut.* **2021**, *280*, 116959. [CrossRef] [PubMed]
35. Chang, C.C.; Chen, S.Y.; Yu, S.C. Security management in maritime supply chain—The case of liner shipping companies. In Proceedings of the 1st International Symposium on Technology Innovation, Risk Management and Supply Chain Management, Beijing, China, 1–3 November 2007; pp. 76–85.
36. Zhou, X. A comprehensive framework for assessing navigation risk and deploying maritime emergency resources in the South China Sea. *Ocean Eng.* **2022**, *248*, 110797. [CrossRef]
37. Jiang, M.; Lu, J.; Qu, Z.; Yang, Z. Port vulnerability assessment from a supply Chain perspective. *Ocean Coast. Manag.* **2021**, *213*, 105851. [CrossRef]
38. Akyuz, E.; Celik, E. A quantitative risk analysis by using interval type-2 fuzzy FMEA approach: The case of oil spill. *Marit. Policy Manag.* **2018**, *45*, 979–994. [CrossRef]
39. Li, X.; Tang, W. Structural risk analysis model of damaged membrane LNG carriers after grounding based on Bayesian belief networks. *Ocean Eng.* **2019**, *171*, 332–344. [CrossRef]
40. Wen, T.; Gao, Q.; Chen, Y.; Cheong, K.H. Exploring the vulnerability of transportation networks by entropy: A case study of Asia-Europe maritime transportation network. *Reliab. Eng. Syst. Saf.* **2022**, *226*, 108578. [CrossRef]
41. Wu, B.; Zhao, C.; Yip, T.L.; Jiang, D. A novel emergency decision-making model for collision accidents in the Yangtze River. *Ocean Eng.* **2021**, *223*, 108622. [CrossRef]
42. Chen, S.T.; Wall, A.; Davies, P.; Yang, Z.; Wang, J.; Chou, Y.-H. A Human and Organisational Factors (HOFs) analysis method for marine casualties using HFACS-Maritime Accidents (HFACS-MA). *Saf. Sci.* **2013**, *60*, 105–114. [CrossRef]
43. Fan, S.; Yang, Z.; Blanco-Davis, E.; Zhang, J.; Yan, X. Analysis of maritime transport accidents using Bayesian networks. *Proc. Inst. Mech. Eng. Part O-J. Risk Reliab.* **2020**, *234*, 439–454. [CrossRef]
44. Jiang, M.; Lu, J. Maritime accident risk estimation for sea lanes based on a dynamic Bayesian network. *Marit. Policy Manag.* **2020**, *47*, 649–664. [CrossRef]
45. Liu, H.; Tian, Z.; Huang, A.; Yang, Z. Analysis of vulnerabilities in maritime supply chains. *Reliab. Eng. Syst. Saf.* **2018**, *169*, 475–484. [CrossRef]
46. Zhang, M.; Zhang, D.; Yao, H.; Zhang, K. A probabilistic model of human error assessment for autonomous cargo ships focusing on human-autonomy collaboration. *Saf. Sci.* **2020**, *130*, 104838. [CrossRef]
47. Zhang, L.; Lu, J.; Yang, Z. Optimal scheduling of emergency resources for major maritime oil spills considering time-varying demand and transportation networks. *Eur. J. Oper. Res.* **2021**, *293*, 529–546. [CrossRef]
48. Jiang, M.; Wang, B.; Hao, Y.; Chen, S.; Lu, J. Vulnerability assessment of strait/canals in maritime transportation using fuzzy evidential reasoning approach. *Risk Anal.* **2022**, *43*, 1795–1810. [CrossRef] [PubMed]
49. Wan, C.; Tao, J.; Yang, Z.; Zhang, D. Evaluating recovery strategies for the disruptions in liner shipping networks: A resilience approach. *Int. J. Logist. Manag.* **2022**, *33*, 389–409. [CrossRef]
50. Wrobel, K.; Montewka, J.; Kujala, P. Towards the assessment of potential impact of unmanned vessels on maritime transportation safety. *Reliab. Eng. Syst. Saf.* **2017**, *165*, 155–169. [CrossRef]
51. Baksh, A.A.; Abbassi, R.; Garaniya, V.; Khan, F. Marine transportation risk assessment using Bayesian Network: Application to Arctic waters. *Ocean Eng.* **2018**, *159*, 422–436. [CrossRef]
52. Mansouri, S.A.; Lee, H.; Aluko, O. Multi-objective decision support to enhance environmental sustainability in maritime shipping: A review and future directions. *Transp. Res. Part E-Logist. Transp. Rev.* **2015**, *78*, 3–18. [CrossRef]
53. Wan, C.; Yan, X.; Zhang, D.; Qu, Z.; Yang, Z. An advanced fuzzy Bayesian-based FMEA approach for assessing maritime supply chain risks. *Transp. Res. Part E-Logist. Transp. Rev.* **2019**, *125*, 222–240. [CrossRef]

54. Notteboom, T.; Pallis, T.; Rodrigue, J.P. Disruptions and resilience in global container shipping and ports: The COVID-19 pandemic versus the 2008–2009 financial crisis. *Marit. Econ. Logist.* **2021**, *23*, 179–210. [CrossRef]
55. Lam, J.S.L.; Bai, X. A quality function deployment approach to improve maritime supply chain resilience. *Transp. Res. Part E-Logist. Transp. Rev.* **2016**, *92*, 16–27. [CrossRef]
56. Zhang, J.; Teixeira, A.P.; Guedes Soares, C.; Yan, X.; Liu, K. Maritime Transportation Risk Assessment of Tianjin Port with Bayesian Belief Networks. *Risk Anal.* **2016**, *36*, 1171–1187. [CrossRef] [PubMed]
57. Akyuz, E. A marine accident analysing model to evaluate potential operational causes in cargo ships. *Saf. Sci.* **2017**, *92*, 17–25. [CrossRef]
58. Van Eck, N.J.; Waltman, L.; Noyons, E.C.M.; Buter, R.K. Automatic term identification for bibliometric mapping. *Scientometrics* **2010**, *82*, 581–596. [CrossRef]
59. Van Eck, N.J.; Waltman, L. Software survey: VOSviewer, a computer program for bibliometric mapping. *Scientometrics* **2010**, *84*, 523–538. [CrossRef] [PubMed]
60. Fu, S.; Zhang, D.; Montewka, J.; Yan, X.; Zio, E. Towards a probabilistic model for predicting ship besetting in ice in Arctic waters. *Reliab. Eng. Syst. Saf.* **2016**, *155*, 124–136. [CrossRef]
61. Kujala, P.; Hanninen, M.; Arola, T.; Ylitalo, J. Analysis of the marine traffic safety in the Gulf of Finland. *Reliab. Eng. Syst. Saf.* **2009**, *94*, 1349–1357. [CrossRef]
62. Kum, S.; Sahin, B. A root cause analysis for Arctic Marine accidents from 1993 to 2011. *Saf. Sci.* **2015**, *74*, 206–220. [CrossRef]
63. Goerlandt, F.; Montewka, J. A framework for risk analysis of maritime transportation systems: A case study for oil spill from tankers in a ship-ship collision. *Saf. Sci.* **2015**, *76*, 42–66. [CrossRef]
64. Montewka, J.; Ehlers, S.; Goerlandt, F.; Hinz, T.; Tabri, K.; Kujala, P. A framework for risk assessment for maritime transportation systems-A case study for open sea collisions involving RoPax vessels. *Reliab. Eng. Syst. Saf.* **2014**, *124*, 142–157. [CrossRef]
65. Zhang, D.; Yan, X.P.; Yang, Z.L.; Wall, A.; Wang, J. Incorporation of formal safety assessment and Bayesian network in navigational risk estimation of the Yangtze River. *Reliab. Eng. Syst. Saf.* **2013**, *118*, 93–105. [CrossRef]
66. Akhtar, M.J.; Utne, I.B. Human fatigue's effect on the risk of maritime groundings—A Bayesian Network modeling approach. *Saf. Sci.* **2014**, *62*, 427–440. [CrossRef]
67. Chauvin, C.; Lardjane, S.; Morel, G.; Clostermann, J.P.; Langard, B. Human and organisational factors in maritime accidents: Analysis of collisions at sea using the HFACS. *Accid. Anal. Prev.* **2013**, *59*, 26–37. [CrossRef] [PubMed]
68. Trucco, P.; Cagno, E.; Ruggeri, F.; Grande, O. A Bayesian Belief Network modelling of organisational factors in risk analysis: A case study in maritime transportation. *Reliab. Eng. Syst. Saf.* **2008**, *93*, 845–856. [CrossRef]
69. Yang, Z.L.; Wang, J.; Bonsall, S.; Fang, Q.G. Use of Fuzzy Evidential Reasoning in Maritime Security Assessment. *Risk Anal.* **2009**, *29*, 95–120. [CrossRef] [PubMed]
70. Calatayud, A.; Mangan, J.; Palacin, R. Vulnerability of international freight flows to shipping network disruptions: A multiplex network perspective. *Transp. Res. Part E Logist. Transp. Rev.* **2017**, *108*, 195–208. [CrossRef]
71. Akyuz, E.; Celik, M. Utilisation of cognitive map in modelling human error in marine accident analysis and prevention. *Saf. Sci.* **2014**, *70*, 19–28. [CrossRef]
72. Akyuz, E.; Celik, M. A hybrid human error probability determination approach: The case of cargo loading operation in oil/chemical tanker ship. *J. Loss Prev. Process Ind.* **2016**, *43*, 424–431. [CrossRef]
73. Franca, J.E.M.; Hollnagel, E.; dos Santos, I.J.A.L.; Haddad, A.N. FRAM AHP approach to analyse offshore oil well drilling and construction focused on human factors. *Cogn. Technol. Work.* **2020**, *22*, 653–665. [CrossRef]
74. Fu, S.; Yu, Y.; Chen, J.; Xi, Y.; Zhang, M. A framework for quantitative analysis of the causation of grounding accidents in arctic shipping. *Reliab. Eng. Syst. Saf.* **2022**, *226*, 108706. [CrossRef]
75. Fu, S.; Yan, X.; Zhang, D.; Zhang, M. Risk influencing factors analysis of Arctic maritime transportation systems: A Chinese perspective. *Marit. Policy Manag.* **2018**, *45*, 439–455. [CrossRef]
76. Khan, B.; Khan, F.; Veitch, B. A Dynamic Bayesian Network model for ship-ice collision risk in the Arctic waters. *Saf. Sci.* **2020**, *130*, 104858. [CrossRef]
77. Khan, B.; Khan, F.; Veitch, B.; Yang, M. An operational risk analysis tool to analyze marine transportation in Arctic waters. *Reliab. Eng. Syst. Saf.* **2018**, *169*, 485–502. [CrossRef]
78. Khan, B.; Khan, F.; Veitch, B. A cellular automation model for convoy traffic in Arctic waters. *Cold Reg. Sci. Technol.* **2019**, *164*, 102783. [CrossRef]
79. Banda, O.A.V.; Goerlandt, F.; Kuzmin, V.; Kujala, P.; Montewka, J. Risk management model of winter navigation operations. *Mar. Pollut. Bull.* **2016**, *108*, 242–262. [CrossRef] [PubMed]
80. Wu, B.; Yan, X.; Wang, Y.; Soares, C.G. An Evidential Reasoning-Based CREAM to Human Reliability Analysis in Maritime Accident Process. *Risk Anal.* **2017**, *37*, 1936–1957. [CrossRef] [PubMed]
81. Ducruet, C. Network diversity and maritime flows. *Transp. Geogr.* **2013**, *30*, 77–88. [CrossRef]
82. Ducruet, C. Port specialization and connectivity in the global maritime network. *Marit. Policy Manag.* **2022**, *49*, 1–17. [CrossRef]
83. Ducruet, C.; Rozenblat, C.; Zaidi, F. Ports in multi-level maritime networks: Evidence from the Atlantic (1996–2006). *Transp. Geogr.* **2010**, *18*, 508–518. [CrossRef]
84. Zhang, Y.; Sun, X.; Chen, J.; Cheng, C. Spatial patterns and characteristics of global maritime accidents. *Reliab. Eng. Syst. Saf.* **2021**, *206*, 107310. [CrossRef]

85. Zhang, Y.; Zhai, Y.; Chen, J.; Xu, Q.; Fu, S.; Wang, H. Factors Contributing to Fatality and Injury Outcomes of Maritime Accidents: A Comparative Study of Two Accident-Prone Areas. *J. Mar. Sci. Eng.* **2022**, *10*, 1945. [CrossRef]

86. Berle, O.; Asbjornslett, B.E.; Rice, J.B. Formal Vulnerability Assessment of a maritime transportation system. *Reliab. Eng. Syst. Saf.* **2011**, *96*, 696–705. [CrossRef]

87. Berle, O.; Norstad, I.; Asbjornslett, B.E. Optimization, risk assessment and resilience in LNG transportation systems. *Supply Chain. Manag.-Int. J.* **2013**, *18*, 253–264. [CrossRef]

88. Zhu, M.; Huang, L.; Huang, Z.; Shi, F.; Xie, C. Hazard analysis by leakage and diffusion in Liquefied Natural Gas ships during emergency transfer operations on coastal waters. *Ocean Coast. Manag.* **2022**, *220*, 106100. [CrossRef]

89. Fu, S.; Yu, Y.; Chen, J.; Han, B.; Wu, Z. Towards a probabilistic approach for risk analysis of nuclear-powered icebreakers using FMEA and FRAM. *Ocean Eng.* **2022**, *260*, 112041. [CrossRef]

90. Yang, Y.; Liu, W. Resilience analysis of maritime silk road shipping network structure under disruption simulation. *Mar. Sci. Eng.* **2022**, *10*, 617. [CrossRef]

91. Sun, Y.; Ling, J.; Chen, X.; Kong, F.; Hu, Q.; Biancardo, S.A. Exploring Maritime Search and Rescue Resource Allocation via an Enhanced Particle Swarm Optimization Method. *J. Mar. Sci. Eng.* **2022**, *10*, 906. [CrossRef]

92. Huang, X.; Ren, Y.; Zhang, J.; Wang, D.; Liu, J. Dynamic Scheduling Optimization of Marine Oil Spill Emergency Resource. *J. Coast. Res.* **2020**, *107*, 437–442. [CrossRef]

93. Ung, S.T. Evaluation of human error contribution to oil tanker collision using fault tree analysis and modified fuzzy Bayesian Network based CREAM. *Ocean Eng.* **2019**, *179*, 159–172. [CrossRef]

94. Uğurlu, Ö.; Yıldız, S.; Loughney, S.; Wang, J.; Kuntchulia, S.; Sharabidze, I. Analyzing collision, grounding, and sinking accidents occurring in the Black Sea utilizing HFACS and Bayesian networks. *Risk Anal.* **2020**, *40*, 2610–2638. [CrossRef]

95. Alyami, H.; Lee, P.T.W.; Yang, Z.; Riahi, R.; Bonsall, S.; Wang, J. An advanced risk analysis approach for container port safety evaluation. *Marit. Policy Manag.* **2014**, *41*, 634–650. [CrossRef]

96. Yuan, C.; Cui, H.; Tao, B.; Ma, S. Cause factors in emergency process of fire accident for oil–gas storage and transportation based on fault tree analysis and modified Bayesian network model. *Energy Environ.* **2018**, *29*, 802–821. [CrossRef]

97. Wang, L.; Wang, J.; Shi, M.; Fu, S.; Zhu, M. Critical risk factors in ship fire accidents. *Marit. Policy Manag.* **2021**, *48*, 895–913. [CrossRef]

98. Abaei, M.M.; Hekkenberg, R.; BahooToroody, A.; Banda, O.V.; van Gelder, P. A probabilistic model to evaluate the resilience of unattended machinery plants in autonomous ships. *Reliab. Eng. Syst. Saf.* **2022**, *219*, 108176. [CrossRef]

99. Faghih-Roohi, S.; Xie, M.; Ng, K.M. Accident risk assessment in marine transportation via Markov modelling and Markov Chain Monte Carlo simulation. *Ocean Eng.* **2014**, *91*, 363–370. [CrossRef]

100. Huang, J.C.; Nieh, C.Y.; Kuo, H.C. Risk assessment of ships maneuvering in an approaching channel based on AIS data. *Ocean Eng.* **2019**, *173*, 399–414. [CrossRef]

101. Zou, Q.; Chen, S. Resilience Modeling of Interdependent Traffic-Electric Power System Subject to Hurricanes. *J. Infrastruct. Syst.* **2020**, *26*, 04019034. [CrossRef]

102. Jon, M.H.; Kim, Y.P.; Choe, U. Determination of a safety criterion via risk assessment of marine accidents based on a Markov model with five states and MCMC simulation and on three risk factors. *Ocean Eng.* **2021**, *236*, 109000. [CrossRef]

103. Chen, H.; Lam, J.S.L.; Liu, N. Strategic investment in enhancing port-hinterland container transportation network resilience: A network game theory approach. *Transp. Res. Part B-Methodol.* **2018**, *111*, 83–112. [CrossRef]

104. Chaal, M.; Bahootoroody, A.; Basnet, S.; Banda, O.A.V.; Goerlandt, F. Towards system-theoretic risk assessment for future ships: A framework for selecting Risk Control Options. *Ocean Eng.* **2022**, *259*, 111797. [CrossRef]

105. Zaib, A.; Yin, J.; Khan, R.U. Determining Role of Human Factors in Maritime Transportation Accidents by Fuzzy Fault Tree Analysis (FFTA). *J. Mar. Sci. Eng.* **2022**, *10*, 381. [CrossRef]

106. Şenel, M.; Şenel, B.; Havle, C.A. Risk analysis of ports in Maritime Industry in Turkey using FMEA based intuitionistic Fuzzy TOPSIS Approach. *ITM Web Conf.* **2018**, *22*, 01018. [CrossRef]

107. Poo, M.C.-P.; Yang, Z. Optimising the resilience of shipping networks to climate vulnerability. *Marit. Policy Manag.* **2022**, 1–20. [CrossRef]

108. Wan, C.; Yang, Z.; Zhang, D.; Yan, X.; Fan, S. Resilience in transportation systems: A systematic review and future directions. *Transp. Rev.* **2018**, *38*, 479–498. [CrossRef]

109. Liu, N.; Gong, Z.; Xiao, X. Disaster prevention and strategic investment for multiple ports in a region: Cooperation or not. *Marit. Policy* **2018**, *45*, 585–603. [CrossRef]

110. Holling, C.S. Resilience and stability of ecological systems. *Annu. Rev. Ecol. Syst.* **1973**, *4*, 1–23. [CrossRef]

111. Fiksel, J. Sustainability and resilience: Toward a systems approach. *Sustain. Sci. Pract. Policy* **2006**, *2*, 14–21. [CrossRef]

112. Luthar, S.S.; Cicchetti, D.; Becker, B. The construct of resilience: A critical evaluation and guidelines for future work. *Child Dev.* **2000**, *71*, 543–562. [CrossRef]

113. Hosseini, S.; Barker, K. Modeling infrastructure resilience using Bayesian networks: A case study of inland waterway ports. *Comput. Ind. Eng.* **2016**, *93*, 252–266. [CrossRef]

114. Li, H.; Ren, X.; Yang, Z. Data-driven Bayesian network for risk analysis of global maritime accidents. *Reliab. Eng. Syst. Saf.* **2023**, *230*, 108938. [CrossRef]

115. Ai, Y.-f.; Lu, J.; Zhang, L.-l. The optimization model for the location of maritime emergency supplies reserve bases and the configuration of salvage vessels. *Transp. Res. Part E Logist. Transp. Rev.* **2015**, *83*, 170–188. [CrossRef]
116. Zhang, L.; Lu, J.; Yang, Z. Dynamic optimization of emergency resource scheduling in a large-scale maritime oil spill accident. *Comput. Ind. Eng.* **2021**, *152*, 107028. [CrossRef]
117. Allianz. *Safety and Shipping Review 2020: An Annual Review of Trends and Developments in Shipping Losses and Safety*; Allianz Global Corporate & Specialty: Munich, Germany, 2020.
118. Wróbel, K.; Montewka, J.; Kujala, P. Towards the development of a system-theoretic model for safety assessment of autonomous merchant vessels. *Reliab. Eng. Syst. Saf.* **2018**, *178*, 209–224. [CrossRef]
119. Fan, C.; Wróbel, K.; Montewka, J.; Gil, M.; Wan, C.; Zhang, D. A framework to identify factors influencing navigational risk for Maritime Autonomous Surface Ships. *Ocean Eng.* **2020**, *202*, 107188. [CrossRef]
120. Utne, I.B.; Rokseth, B.; Sørensen, A.J.; Vinnem, J.E. Towards supervisory risk control of autonomous ships. *Reliab. Eng. Syst. Saf.* **2020**, *196*, 106757. [CrossRef]

Journal of
*Marine Science
and Engineering*

Technical Note

Port Digital Twin Development for Decarbonization: A Case Study Using the Pusan Newport International Terminal

Jeong-On Eom [1], Jeong-Hyun Yoon [1], Jeong-Hum Yeon [2] and Se-Won Kim [1,*]

[1] Department of Intelligent Mechatronics Engineering, Sejong University, Seoul 05006, Republic of Korea; jeongon@sju.ac.kr (J.-O.E.); dbswjdgus9@gmail.com (J.-H.Y.)
[2] Research Department of Port, Busan Port Authority, Busan 48943, Republic of Korea; jhyeon@busanpa.com
* Correspondence: sewonkim@sejong.ac.kr; Tel.: +82-10-3123-7176

Abstract: The maritime industry is a major carbon emission contributor. Therefore, the global maritime industry puts every effort into reducing carbon emissions in the shipping chain, which includes vessel fleets, ports, terminals, and hinterland transportation. A representative example is the carbon emission reduction standard mandated by the International Maritime Organization for international sailing ships to reduce carbon emissions this year. Among the decarbonization tools, the most immediate solution for reducing carbon emissions is to reduce vessel waiting time near ports and increase operational efficiency. The operation efficiency improvement in maritime stakeholders' port operations can be achieved using data. This data collection and operational efficiency improvement can be realized using a digital twin. This study develops a digital twin that measures and reduces carbon emissions using the collaborative operation of maritime stakeholders. In this study, the authors propose a data structure and backbone scheduling algorithm for a port digital twin. The interactive scheduling between a port and its vessels is investigated using the digital twin. The digital twin's interactive scheduling for the proposed model improved predictions of vessel arrival time and voyage carbon emissions. The result of the proposed digital twin model is compared to an actual operation case from the Busan New Port in September 2022, which shows that the proposed model saves over 75 % of the carbon emissions compared with the case.

Keywords: carbon emission; port digital twin; just-in-time arrival; vessel digital twin

Citation: Eom, J.-O.; Yoon, J.-H.; Yeon, J.-H.; Kim, S.-W. Port Digital Twin Development for Decarbonization: A Case Study Using the Pusan Newport International Terminal. *J. Mar. Sci. Eng.* **2023**, *11*, 1777. https://doi.org/10.3390/jmse11091777

Academic Editors: Jian Wu and Xianhua Wu

Received: 12 August 2023
Revised: 4 September 2023
Accepted: 8 September 2023
Published: 11 September 2023

1. Introduction

1.1. Background

In the face of mounting environmental challenges, one critical issue that has emerged as a defining concern for the maritime industry is decarbonization. The global shipping sector, which is responsible for transporting approximately 90% of the world's goods, plays a pivotal role in international trade and economic prosperity. However, this essential industry has also been a significant contributor to greenhouse gas emissions and climate change. As the consequences of climate change become increasingly evident, the urgency to address maritime decarbonization is reaching a tipping point. Transitioning toward cleaner, more sustainable practices within the maritime sector has become an imperative shared by governments, industry stakeholders, and environmental advocates alike.

Decarbonization within the maritime industry hinges on the seamless coordination of stakeholders in the shipping supply chain. To achieve this, a system capable of monitoring, sharing, and scheduling the operations of each actor becomes essential. One such system is the digital twin, a concept that holds promise as a suitable system for this purpose.

A digital twin represents a virtual counterpart of a physical object, system, or process. It is continually updated with real-time data from its physical counterpart. The term 'digital twin' was first introduced at the SME (Society of Manufacturing Engineering) conference in Troy, Michigan in October 2002 [1]. Initially conceived within the context of product lifecycle management (PLM), the concept evolved over time, transitioning from PLM to the Mirrored Spaces Model, then to the Information Mirroring Model, and finally, in 2010, it assumed the name 'Digital Twin' [2]. The strength of digital twins lies in their ability to replicate tasks performed in the physical world, commencing from the support and operational stages of the product life cycle. Unlike physical spaces with a single instance, their power resides in their capacity to manifest an infinite number of instances in the digital realm. Digital twin applications become pertinent when an object system is too complex and vast to construct in a real-scale test facility, thereby mitigating high costs. Examples range from simulating entire cities to ports, airports, and industrial plants. Real-time remote monitoring and effective decision-making are facilitated with the development of digital twin cities (DTCs), underpinned by core technologies such as surveying and mapping, building information modeling, 5G-enabled Internet of Things (IoT), blockchain, and collaborative computing [3]. The concept of DTCs holds the potential to enhance not only urban planning, disaster management, construction, and transportation but also the efficiency and sustainability of logistics, energy consumption, and communication.

Given the colossal scale of maritime shipping chain components, conducting tests for informed decision-making is exceedingly challenging. For instance, when managers at a container terminal seek to experiment with a trial scheme to reduce carbon emissions from vessels berthing at the terminal by developing a berth allocation policy, they face formidable financial barriers to conducting validation tests with real ships and berths, as real-scale operational trials incur substantial costs. In contrast, a digital twin model that can be used for simulation testing is a more cost-effective alternative, requiring only the initial investment to construct the model.

Maritime shipping plays an indispensable role in global trade. However, it also exerts a significant influence on greenhouse gas emissions, which are a primary driver of climate change. In 2018, the International Maritime Organization (IMO) reported that shipping accounted for approximately 2.89% of the global greenhouse gas emissions [4]. As the demand for maritime shipping continues to surge, emissions follow suit, presenting a considerable predicament for the shipping industry. The industry must now seek solutions to curtail emissions while upholding its essential role in global trade. Recent regulations on CO_2 emissions have been reinforced on a global scale. To counterbalance the cumulative greenhouse gases released into the atmosphere by numerous vessels, the IMO introduced new CO_2 regulations aimed at steering the maritime shipping chain toward "net-zero" emissions during the 80th MEPC (Marine Environment Protection Committee) meeting. Concurrently, the European Union (EU) imposed additional taxes on vessels emitting CO_2 while sailing in EU waters.

The significance of digital twins extends beyond shipping and encompasses port infrastructures as well. A port digital twin serves as a digital replica of a physical port in the real world, encompassing vessels, quay cranes, yard tractors, and hinterland transportation. These digital models allow port stakeholders to monitor and forecast operational efficiency in real time, thereby reducing energy consumption and promoting greener port operations. Moreover, digital twins enable the identification of optimal energy efficiency measures using simulation and data analysis, contributing to emissions reduction and the realization of sustainable port operations. Consequently, a port digital twin is recognized as an indispensable element in attaining carbon neutrality at the port level.

In contrast with the conventional berth planning method used in commercial digital port solutions, which does not incorporate real-time data from moving objects such as ships, terminal equipment, and hinterland transportation, the digital twin model proposed in this study leverages current data for real-time-based simulation and decision-making, thus eliminating time delays. This approach enables efficient operation planning using real-time-based simulation and forecasting.

This study implements a port digital twin to reduce CO_2 emissions from vessels and terminals. It showcases the simulation model, the data structure, and case studies that compare CO_2 reduction performance with and without the developed digital twin.

Furthermore, it presents real terminal case results demonstrating carbon emission reductions achieved with the application of the digital twin at the Pusan Newport International Terminal (PNIT).

1.2. Literature Review

A digital twin is a research field wherein a twin of a real-world physical entity is made in digital space. As complexity and integration become defining characteristics of various subsystems, the demand for digital twins continues to grow. Particularly, digital twins are essential for testing and simulating intricate and interconnected operations. Augustine [5] underscored the application of digital twins in diverse projects, including space initiatives and aircraft development. Taylor et al. [6] delineated domains where digital twins are useful, with a focus on the manufacturing sector.

Given the intricate dynamics and large-scale operations involving multiple stakeholders such as shipping companies, terminals, tugboats, pilot boats, hinterland trucks, and port authorities, ports represent ideal environments for implementing digital twins. Recent research efforts have aimed to develop specialized digital twins tailored to the unique requirements of port areas. Hofmann and Branding [7] advocated for the implementation of the Internet of Things (IoT) and cloud-based digital twins to support real-time decision-making in port operations. The International Maritime Organization (IMO) has suggested the adoption of port community systems to enhance communication between ships and ports during recent facilitation committee meetings [8]. The Digital Container Shipping Association (DCSA) and the Maritime & Port Authority of Singapore (MPA) have also bolstered communication infrastructures between ships and ports to reduce CO_2 emissions and improve ship arrival and departure efficiency [9]. The digital twin emerges as a pivotal infrastructure for facilitating data exchange between ships and ports, thus enhancing their interaction.

The maritime industry has embraced digital twins as a valuable tool for validation and integrated simulation. For instance, Liu, Zhou et al. [10] applied digital twin tools to analyze variations in ship voyage performance. Stoumpos et al. [11] performed research to develop high-fidelity digital twins as integrated models for modeling dual-fuel engines and ship control systems. Gao et al. [12] harnessed digital twins for automated storage scheduling in container terminals. Wang et al. [13] integrated digital twins into management infrastructure within smart port contexts. Wang, Hu, and Liu [14] asserted that digital twins are apt tools for managing shipping industry processes and outlined their potential application in the smart port concept.

Digital twin technology has also made inroads into the shipbuilding industry, primarily for performance analysis. Fonseca and Gaspar [15] proposed data modeling for digital twin ships. Coraddu et al. [16] estimated ship fouling using a data-driven digital twin model. Danielsen-Haces [17] introduced a comprehensive digital twin model for simulating electricity-driven model vessels. Vasstein [18] proposed a high-fidelity digital twin framework for testing autonomous vessels, while Raza et al. [19] applied digital twins to an application framework for autonomous ship development.

The existing digital twin research has predominantly focused on either ships [10,11,15–17,19] or terminal yards [7,12]. Even when digital twins have been proposed for an entire port infrastructure [13,14], they often omit critical interfaces between ships, ports, terminals, and dynamic objects. This study strives to establish a coupled operational digital twin platform that facilitates interaction among vessels, terminal assets, and port authorities. Real-world equipment and situational data from actual ports are harnessed to create a functional port digital twin and its associated simulation algorithm.

Research into port call optimization delves into resolving scheduling challenges between ships and ports. Initiatives such as port collaborative decision-making (PortCDM) and just-in-time arrival (JITA) have aimed to reduce ship, tug, and terminal waiting times. Unfortunately, these concepts have not been widely adopted within the industry due to technical limitations of the digital infrastructure among port members and data standardization issues. Jahn and Scheidweiler [20] sought to optimize port calls by exchanging estimated time of arrivals (ETAs), while Cho et al. [21] proposed the development of a digital infrastructure to facilitate communication and data sharing among port stakeholders.

Ports represent significant hubs for addressing carbon dioxide emissions, given the substantial emissions from various moving objects such as ships, yard tractors, container trailers, and quay cranes. Congestion among these moving objects exacerbates carbon emissions in ports. To mitigate congestion and reduce excessive carbon emissions, seamless communication between these objects becomes paramount. Sarantakos, Bowkett et al. [22] introduced digital infrastructure aimed at improving a port's carbon emissions profile. Alamoush, Ölçer, and Ballini [23] conducted a review of the existing regulations and incentives for potentially reducing CO_2 emissions in port areas. Both the European Union (EU) and the United States have established emission control areas (ECAs) to regulate gas emissions in nearshore and port areas. Additionally, the IMO's data collection system (DCS) and the EU's monitoring, reporting, and verification (MRV) framework have been instrumental in monitoring and controlling CO_2 emissions.

This study's prime contribution lies in the application of digital twins within the maritime industry, specifically addressing challenges involving terminals, ships, and tugboats. Most previous berth scheduling models primarily focus on one or two elements, often overlooking complex considerations. For instance, Lu and Le [24] focused on equipment planning in the yard to reduce costs, while Ismail et al. [25] considered the berth's surroundings in their berth planning. In order to solve an unexpected delay in a ship's arrival, which affects berth operation plans in addition to terminal conditions, Du, Xu, and Chen [26] solved a berth assignment problem by considering the ship's delay probabilistically. Their study adjusted the time buffer for each ship and supported planner decision-making with a diagram showing the frequency distribution of the time buffer. Xiang, Liu, and Miao [27] addressed uncertainty in berthing schedules using discrete scenarios and accounting for ship arrival and sailing times. Park, Cho, and Lee [28] introduced time buffers to accommodate uncertain ship arrival times in berth scheduling.

In contrast to previous studies that often treated ship arrival times as probabilistic or inserted time buffers into plans to account for ship and terminal scheduling uncertainties, this study leverages real-time ship location data for accurate ship arrival time prediction and terminal operation data to plan schedules, considering a terminal's operational status. By collecting real-time data from ships and terminals and incorporating it into schedule planning, this study enhances operational efficiency using dynamic data integration. Compared with previous port call optimization research, this study distinguishes itself by focusing on schedule optimization among ships, terminals, and tugboats, thus encompassing more than just terminal yard scheduling problems.

1.3. Physical Asset in the Real World—Pusan Newport International Container Terminal

The focus of this study was the Pusan Newport International Terminal (PNIT) located in the Busan New Port of South Korea, which is recognized as the seventh largest container port globally. In 2022, the volume of containers handled at the Busan port amounted to approximately 22 million. To construct a fully operational technical system for the digital twin, real-time operational data from PNIT was harnessed. The development of this system was underpinned by various technologies: Unity for 3D object visualization, the Oracle Database for database management, Java for web application development, and Python for simulation modeling, as illustrated in Figure 1.

Geographic and climatic data, including geometric details, weather forecasts, bathymetry, and static and dynamic elements of PNIT, were transformed into digital representations. Dynamic elements included vessels, quay cranes, yard cranes, yard tractors, and containers, with PNIT typically accommodating 400 vessel arrivals and handling an average of 8,000,000 containers per year. PNIT boasts a 1.2 km quay wall and 27-yard blocks for container stacking. Figure 2 provides an overhead view of the case study, highlighting the PNIT terminal within the broader context of the Busan New Port.

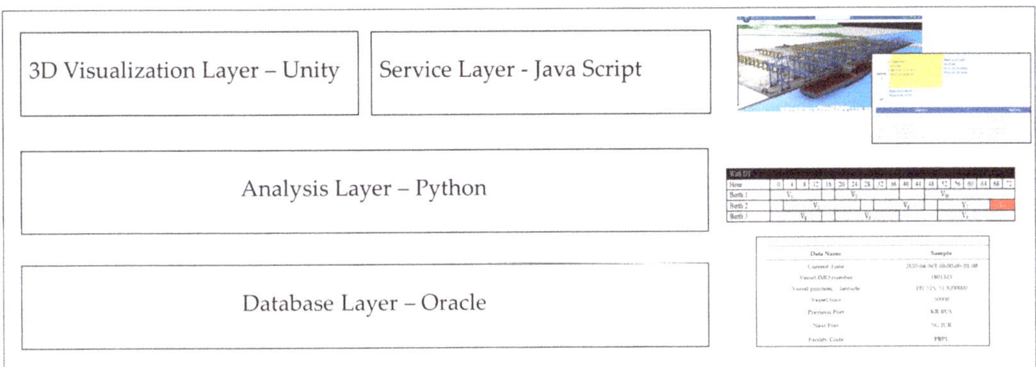

Figure 1. Port digital twin structure and implementation.

Figure 2. Layout of the Pusan Newport International Terminal (PNIT).

Figure 3 illustrates the outcome of this study, showcasing digitally replicated dynamic elements such as vessels, cranes, containers, buildings, roads, gateways, and trailers.

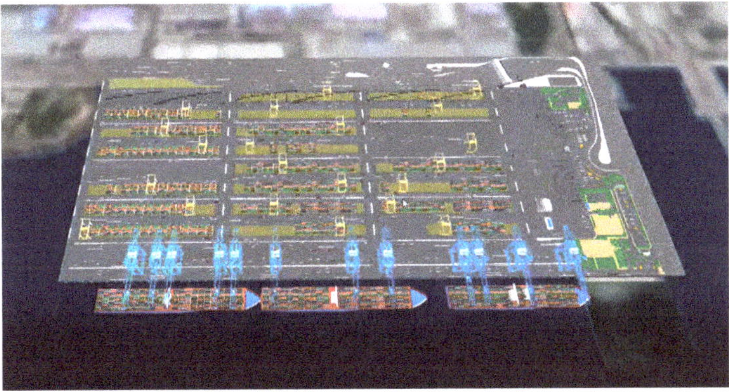

Figure 3. The digital twin of PNIT.

The initial step in this study involved digitally reproducing all real physical assets within the digital twin model. The selection of these digital assets was driven by their capacity to emulate real-world phenomena. Table 1 lists the digital assets incorporated into the target digital twin.

Table 1. Digital twin model.

Digital Assets	Number	Detail
Ship	600	200~10,000 TEU container carrier
Quay crane	12	Outreach 65~70 m
Yard crane	42	6 TEU height
Tug	8	10 m length
Yard tractor	82	220 horsepower (HP)
Container	20,000/day	20 ft, 40 ft container
Truck	100	20 ft, 40 ft trailer
Area	625 km^2	Yard, anchorage, and hinterland

Figure 4 showcases a representative digital twin model from Table 1, including ships, yard cranes, and quay cranes. Each digital asset's position, speed, and identification were meticulously recorded within the digital twin. Additionally, each object features a cost model that calculated the performance of the digital twin.

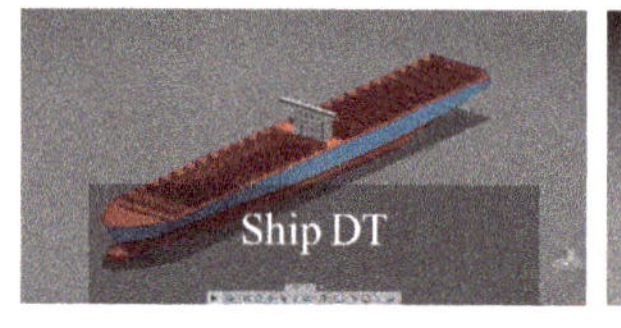

Figure 4. Digital twin subjects.

Traditional port system scheduling often operates independently in silos, where the actions of one object do not correspond with others. When delays occur, other related objects can become idle while waiting for congestion to subside, which can exacerbate CO_2 emissions. Recognizing this issue, the concept of chained scheduling has been proposed by multiple authors [13,14] to address downtime. For instance, when unexpected delays due to weather or prior port conditions occur, a ship may inform the terminal operator and shipping agent via email. However, if changing the original berth schedule to accommodate the delay is more efficient, but the terminal operator cannot make the change without confirmation from another shipper, it becomes challenging to resolve this issue within the current system. Nevertheless, digital twins (DTs) offer a potential solution. DTs can be accessible to all port members, including the port, terminal, and tug operators. This open platform allows everyone to monitor ship arrivals and departures and share their schedules with others using the DT schedule model. Such a system can serve as a valuable tool for collaborative scheduling among port stakeholders.

2. Materials and Methods

2.1. Analysis Framework

This study proposes a comprehensive five-step framework for the development of a port digital twin model, as depicted in Figure 5. The research formulation for the port's digital twin is elucidated in Section 2.2, providing insights into the problem formulation and defining the scheduling decision-making problem within the context of the port's digital twin. Section 2.3 delves into data structure modeling, while Section 2.5.1 explores interactive scheduler modeling, a pivotal development in this study. This section meticulously describes the scheduling interactions among all stakeholders, including ships, tugs, and terminal berth scheduling. In Section 2.5.2, we delve into the visualization aspect of the digital twin. Subsequently, Section 3 highlights a case study that investigates the impact of using the digital twin in port scheduling.

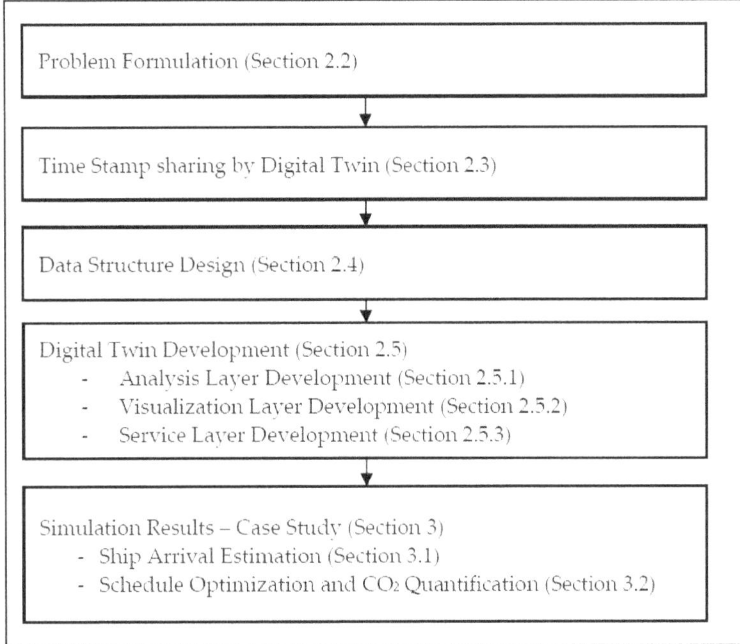

Figure 5. Schematic diagram showing the research methodology.

This study offers three significant contributions to its topic. First, it introduces the port digital twin development process, shedding light on the intricacies of its creation. Second, it puts forth a robust data structure and analysis framework essential for facilitating collaborative scheduling aimed at reducing carbon emissions using interconnected scheduling. Third, this study presents a digital twin model that addresses the conventional berth planning problem in operations research. While conventional berth planning research often assumes shipping delays as probability distributions, the developed model can instantaneously resolve berth scheduling based on real-time data concerning the port's moving objects.

2.2. Problem Formulation

Unexpected delays can lead to additional carbon emissions at a port, as ships must run their generators while waiting at anchor to supply electricity. Furthermore, the time spent waiting at anchor before berthing can result in ship biofouling, increased fuel consumption, and carbon emissions. One common scenario contributing to delays is a shift in ship arrival time, which can be caused by adverse weather conditions or delays in cargo operations at a previous port.

The port of Busan, for example, sees an average delay of at least 4 hours, contributing to additional carbon emissions. Conventional cargo loading and unloading methods exacerbate sequential delays of ships docking at the same berth, further increasing carbon emissions. These unexpected delays could be mitigated by sharing information about delays; however, the current operational method only exchanges arrival time stamps twice before a vessel's arrival per voyage. In contrast, this study proposes continuous time stamp exchanges, allowing for more frequent communication—every 5 min, leading to 288 communications between ships and the terminal per day under the port digital twin system. This seamless communication enables the early detection of delays for vessels and terminals, ultimately reducing CO_2 emissions.

Another significant contribution of this study lies in streamlining the communication process for arrival timestamps. Figures 6 and 7 illustrate the difference between the legacy method and the proposed approach. The legacy port system typically requires three steps to generate ETA (estimated time of arrival), RTA (required time of arrival), and PTA (planned time of arrival). ETA represents the expected arrival time of a vessel at its destination, as reported to the terminal based on the voyage plan. RTA is the time required for a ship to arrive at a berth, as determined by the terminal. PTA is derived from timestamp exchanges between a ship and terminal, signifying a mutually agreed-upon arrival time. In general, the required arrival time forms the basis for contracts between a ship and a terminal. If the ship arrives later than the RTA, demurrage charges may apply.

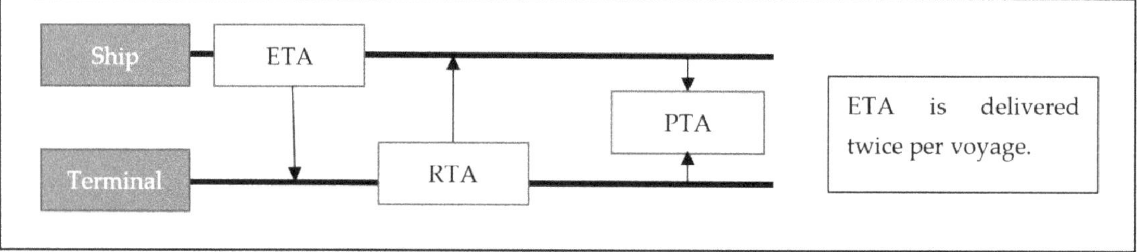

Figure 6. Arrival timestamp decision process of a legacy system.

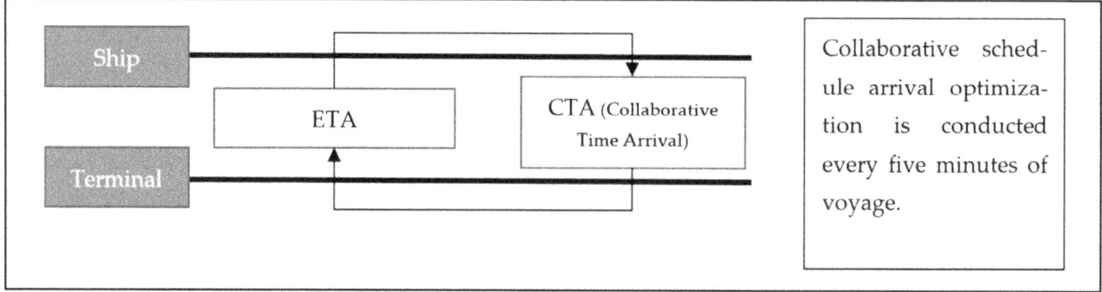

Figure 7. Collaborative scheduling of the digital twin.

In contrast, the proposed method simplifies the timestamp decision-making process by combining two optimization problems into one coupled optimization problem that simultaneously determines the timestamps for both a ship and a terminal. This method offers the advantage of considering CO_2 reduction for both vessels and terminals, thereby reducing overall CO_2 emissions in a port area.

The collaborative time of arrival (CTA) is the timestamp that minimizes overall CO_2 emissions by optimizing the arrival times of ships and a terminal's berth schedule concurrently. The following formulation presents the combined optimization problem based on the mixed integer linear programming (MILP) model for berth planning, which determines the collaborative time of arrival (CTA) that minimizes CO_2 emissions for both vessels and terminals. The model's decision variables encompass ship arrival time at a berth and berthing positions, constituting the berth plan.

- Decision variables

    BT_i (i-th vessel's berth start time) and BP_i (i-th vessel's berth place).

- Objective function

$$Minimize\ (vessel\ CO_2(BT_i, BP_i) + terminal\ CO_2(BP_i))\qquad(1)$$

- Constraints

$$BT_i + dwell\ time < BT_{i+1} \tag{2}$$

$$BT_i < RTA_i \tag{3}$$

where RTA_i is the required time of arrival.

$$0 < BP_i < 80 \tag{4}$$

$$\sum L_i < 1200\ (Berth\ Length) \tag{5}$$

where L_i is the *i*-th vessel's overall length.

$$H_i < H_{quay} \tag{6}$$

where H_i is the i-th vessel's height and H_{quay} is the height of the quay crane.

Constraint set (2) ensures that berthing times between vessels do not overlap, and constraint set (3) guarantees that the arrival time can be adjusted by setting the RTA of the ship to be later than the berthing start time. Constraint sets (4)–(6) ensure that the ship can be docked, considering the berthing bitt, the length of the berth, and the height of the quay crane.

- Objective Function Modeling

$$essel\ CO_2(BT_i, BP_i)[ton] = \sum(voyage\ CO_2(S(i)) + waiting\ CO_2(BT_i - WT_i)) \tag{7}$$

where $S(i)$ = *ship speed* = *Distance* $_i/(BT_i - Present\ Time)$ and WT_i is the waiting time of the *i*-th vessel.

Ship CO_2 modeling was derived from Kim, Son, and Yoon's works [29–31]. The ship cost modeling was based on the objective function of the autonomous vessel's route decision-making model. RTA_i is a constraint in the route decision-making algorithm.

$$Terminal\ CO_2(BP_i)\,[ton] = \sum yard\ tractor\ CO_2(BP_i, BP_{optimal,i}) \tag{8}$$

where if $BP_i == BP_{optimal,i}$ then =0;

Else, yard tractor $CO_2 = \sum cargo(j) * (BP_i - BP_{optimal,i})$;

where *cargo(j)* is the *j*-th container cargo.

The calculation for yard tractor CO_2 emissions considers the container cargo's target yard position, which aligns with the vessel's berth position BP_i. To minimize carbon emissions, it is optimal for a ship's berthing location ($BP_{optimal,i}$) and the yard location to be close. However, if a ship's berthing position changes due to a shift in the berthing order or if berthing does not occur at the optimal position, carbon emissions increase as yard tractors move the container cargo to the target yard position.

2.3. Timestamp Exchanging Using the Port Digital Twin

One of the primary benefits of the digital twin infrastructure is that all stakeholders with access to the digital twin can observe the movements and schedules of other parties in real time. Figure 8 illustrates the main distinction between the proposed digital twin model and the legacy system. The proposed model uses a continuous schedule-sharing approach between a ship and a terminal with satellite communication. In contrast, the legacy system only exchanges timestamps twice per voyage. Consequently, if there is a deviation in the arrival time, it cannot be promptly reflected in the schedule, potentially leading to additional congestion.

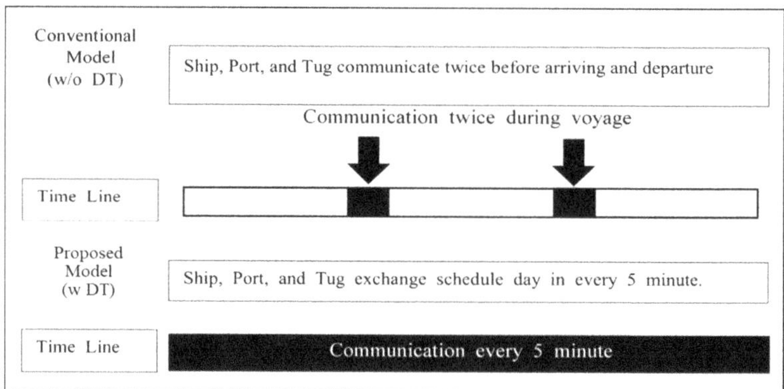

Figure 8. Communication scheme difference between the legacy system and digital twin.

A ship's arrival estimator can provide continuous predictions of the ship's arrival. The terminal can then create a CO_2-minimized schedule based on this real-time estimated time of arrival (ETA). ETA calculations utilize data from the automatic identification system (AIS), environmental forecasts for the voyage route, and historical voyage data [29–31]. This research introduces a schedule exchange structure between the ship, terminal, and tug fleet, enabling the schedule optimization process to occur every five minutes, totaling 288 times per day.

In the digital twin model, the primary aim was to establish continuous, intermediate communication among the terminal, ship, and tug fleet. Each party updates its schedule in real time and proactively reacts to changes in the schedule, ensuring that all stakeholders can monitor each other's movements and plans.

In contrast, the legacy system struggles to deliver changes in the terminal's berth schedule to the vessel in a timely manner. As a result, the ship cannot adjust its speed during the voyage, leading to additional fuel consumption and excessive CO_2 emissions at the port. This is because the ship relies on auxiliary diesel engines to generate electricity for accommodation, and fouling affects its hull. Furthermore, for berth planners, without a digital twin, manually monitoring and updating ship arrival and departure times can be challenging. Consequently, adjusting schedules in the event of delays becomes a complex task. Table 2 summarizes the differences in the schedule optimization methods used by port stakeholders in the legacy model compared to the proposed digital twin (DT) model.

Table 2. Schedule optimization method.

Entity	Without DT (Legacy Model)	With DT (Proposed Model)
Ship	ETAs updated twice during the voyage	Collaborative time of arrival generated every five minutes during the voyage
Terminal	First-come-first-serve	Mixed integer linear programming
Tug	First-come-first-serve	Mixed integer linear programming

2.4. Data Structure Design

One of the primary contributions of this paper is the proposed data structure for the port's digital twin. This data structure has the capacity to generate all the necessary schedule decision-making processes required for ship berthing using the digital twin itself. It not only monitors the current movements of all objects within the digital twin but also provides contextual information to predict the next movements of these objects. Figure 9 illustrates the data structure of the digital twin and depicts the relationships among the data. For the sake of brevity, detailed data that can be used to generate a collaborative schedule for the port's digital twin are summarized in Tables 3–5. The complete tables, including all their contents, are provided in Appendix A Tables A1–A3.

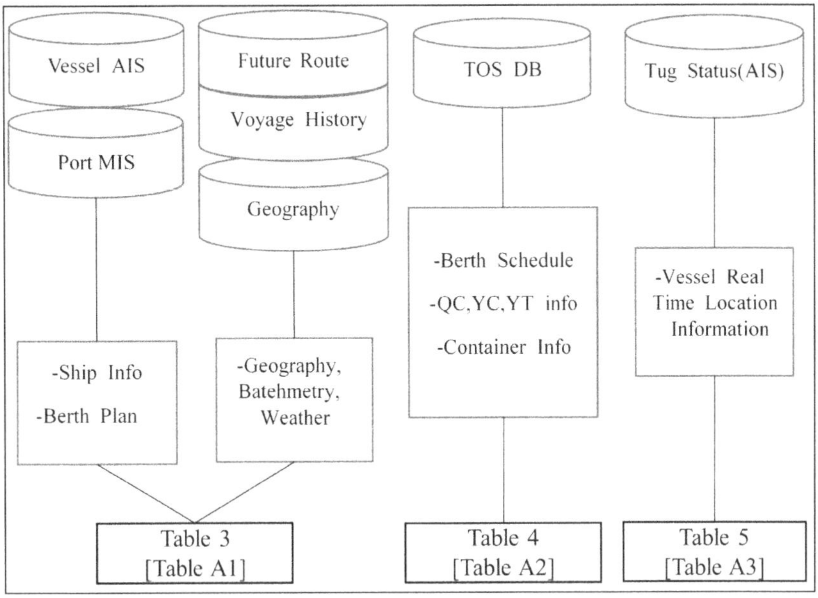

Figure 9. Database structure for a digital twin.

Table 3. Ship data for the DT collaborative scheduler.

Data Name	Sample	Standard
Current Time	2020-04-06T 08:00:00 + 02:00	ISO8601
Vessel IMO number	1801323	IMO
Vessel position, latitude	192.515, 51.9200000	ISO 6709:2008
⋮	⋮	⋮
(Continued in Appendix A)		

Table 4. Terminal data for the DT collaborative scheduler.

Data Name	Sample	Standard
Current time	2020-04-06T 08:00:00 + 02:00	ISO8601
Vessel IMO number	1801323	IMO
Vessel tons	50,000	Gross ton
⋮	⋮	⋮
(Continued in Appendix A)		

Table 5. Carbon factors by fuel type.

Fuel Type	C_F (t-CO$_2$/t-fuel)	Carbon Content
MDO	3.206	0.8744
HFO	3.114	0.8493
LNG	2.766	0.7500
Methanol	1.375	0.3750

A distinctive feature of the digital twin, particularly in the context of large-scale simulation, is real-time data synchronization. To ensure the validity of real-time data, the digital twin maintains a latency of less than one second, achieved with the use of 5G communication technology among port objects such as vessels, quay cranes, yard cranes, and trailers. Additionally, the primary analysis logic incorporates a function that evaluates the quality of communication among these objects. This seamless communication network for the digital twin, reflecting real-world phenomena, interfaces with the smart ship platform, the terminal operating system, and the truck management system. Satellite communication is used for offshore ship–port communication, while a 5G network is utilized near the shore. Equipment status data, including that from quay cranes (QCs), yard trucks (YTs), yard cranes (YCs), and hinterland trucks, are collected using 5G-RTK (real-time kinematic) global positioning system (GPS) devices. Terminal operation data are periodically retrieved using the terminal operating system (TOS) via the internal network. These datasets are stored in the digital twin's database at the terminal and inform the decision-making process. Figure 10 provides a visual representation of the data flow and communication method.

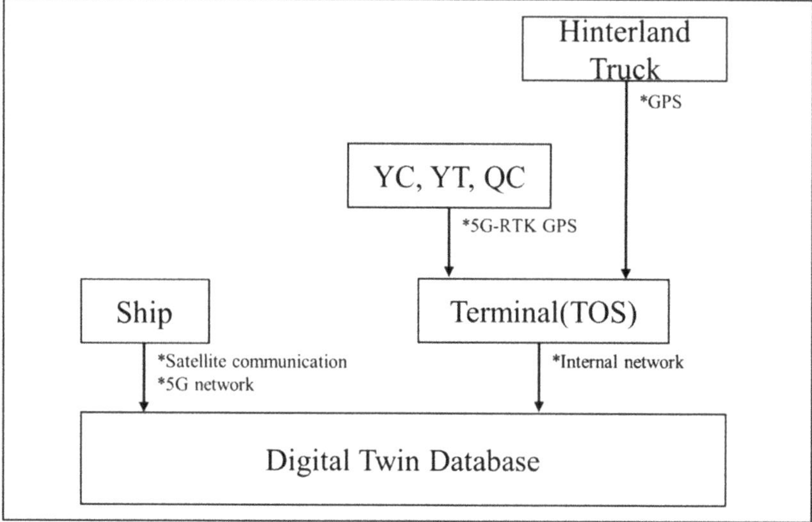

* communication method

Figure 10. Data flow method used in the digital twin.

2.4.1. Ship Data for the DT Collaborative Scheduler

Table 3 summarizes the data obtained from vessels in the port digital twin. The majority of these data pertain to the voyage. The information in Table 3 serves as the basis for creating the CO_2-optimal vessel time schedule.

2.4.2. Terminal Data for the DT Collaborative Scheduler

Table 4 presents data relevant to terminal berth allocation. These data are explained and exemplified, allowing readers to reproduce the berth allocation algorithm in the DT model.

2.5. Digital Twin Development

The port digital twin has four layers: visualization, service, analysis, and database, as presented in Figure 1. The database layer was previously explained in Section 2.4. Therefore, we will delve into the development of the analysis, visualization, and service layers in this section.

2.5.1. Analysis Layer Development—Collaborative Scheduler

For the analysis layer, we focus on the development of the scheduler module, as detailed in this section. The scheduler is the core component responsible for optimizing schedules for vessels, berths, and tugs. The problem formulation for the scheduler was introduced in Section 2.2. Traditionally, scheduling for ships, berths, and tugs has been conducted independently and discretely. In a previous study by Park and Kim [32], ship arrival was treated as an unknown value following a uniform distribution due to the inability to predict real-time vessel arrival estimates. In contrast, this study aims to predict vessel arrival in real time and feed this information to the berth and tug schedulers, enabling real-time schedule optimization.

Previous studies [32–34] often assumed ship arrivals followed a probability distribution, resulting in certain delays. Consequently, terminals needed to allocate buffer time to accommodate unexpected arrivals and departure delays. The proposed digital twin (DT) continuously monitors and shares estimated time of arrival (ETA) and estimated time of departure (ETD) data with all stakeholders, eliminating the need for buffer time. This creates opportunities for schedule optimization.

Kim et al. [29] proposed an optimal vessel routing method based on three-dimensional dynamic programming (3DDP). This method allows for the calculation of ETAs, corresponding fuel consumption, and CO_2 emissions. The output of the ship schedule optimization is the collaborative time of arrival (CTA), constrained by the required time of arrival (RTA) for berth voyage optimization. Voyage optimization can only select route candidates that meet the RTA condition. The digital twin's ship scheduling algorithm aims to calculate ship ETA and corresponding fuel consumption, considering a total of 600 vessels that visited the PNIT more than once. The berth plan comprises ETAs of expected vessel arrivals at the PNIT.

The digital twin addresses this challenge using continuous state updates between ships and the terminal. The DT provides information on ship arrival and departure status, as well as terminal berth availability. This enables ships to adjust their speed, while terminal and hinterland transportation can prepare for variable ship arrivals.

- Vessel Scheduler Optimizer

Ship ETA can be calculated using Equation (9). AIS (automatic identification system) data indicate a vessel's current position. The remaining distance to the berth location is the remaining distance for the vessel. This study uses the route estimation method developed by Kim and Yoon [29,31], which selects a similar route based on a target vessel's voyage history. The berth plan is constructed based on real-time vessel positions, eliminating the need for probabilistic assumptions about ship arrival times.

This study proposes a vessel arrival scheduler that minimizes its fuel oil consumption using the following steps:

1. Setting up voyage planning constraints including available berth time (BT_i), maximum speed (S_{max}), weather, and geography.
2. Generating a grid.
3. Assigning weather forecast data to the grid.
4. Determining bathymetry and geo-fencing data.
5. Finding the optimal route based on past voyage history.
6. Calculating the ship's fuel oil consumption based on route selection.
7. Estimating carbon emissions using fuel consumption data.

$$BT_i = D_{remain,i}/S_{optimal,i} \tag{9}$$

where $D_{remain,i}$ is the remaining distance of the i-th vessel and $S_{optimal,i}$ is the optimal speed of the i-th vessel.

$$Fuel\ Oil\ Consumption[ton] = Power_{ME}(S_{optimal,i}) \times SFOC_{ME} \times VT_i + Power_{GE}(S_{optimal,i}) \times SFOC_{GE} \times WT_i \tag{10}$$

where $SFOC$ is specific fuel oil consumption, VT_i is the voyage time of the i-th vessel, ME is the main engine, GE is the diesel generator engine, and WT_i is the waiting time of the i-th vessel.

$$CO_2 \; emissions[ton] = Fuel \; oil \; consumption \times C_F \tag{11}$$

where C_F is the carbon factor by fuel type [35], as presented in Table 5.

- Terminal Berth Schedule Optimizer

Berth scheduling involves determining the vessel arrival schedule (BT_i) and vessel berth placement (BP_i) for coupled optimization of vessels and berths. The result is represented as a berth plan, as shown in Figure 11. A berth plan illustrates the scheduling coordination between vessel arrival time (BT_i) and berth location (BP_i) in a two-dimensional chart encompassing times and berths. This chart presents vessel allocation within a 72-hour (3-day) period across three berth locations. In Figure 11, the gray squares represent vessel berth assignments, such as V_1 berthing at 0 and departing at 12 at Berth 1.

Berth(V_b) \ Hour(V_i)	0	4	8	12	16	20	24	28	32	36	40	44	48	52	56	60	64	68	72
Berth 1	V_1	V_1	V_1			V_2	V_2	V_2	V_2	V_2			V_3	V_3	V_3	V_3	V_3	V_3	
Berth 2		V_4	V_4	V_4	V_4	V_4	V_4		V_5	V_5	V_5			V_6	V_6	V_6	V_6	V_6	
Berth 3	V_7	V_7	V_7	V_7		V_8	V_8	V_8	V_8	V_8			V_9	V_9	V_9	V_9	V_9	V_9	

Figure 11. A sample berth plan.

In this study, the berthing plan provides a consolidated schedule for vessels and berths and serves as the final output of the digital twin's collaborative scheduler. The following steps were undertaken for berth allocation:

1. Extracting the berth plan from the terminal operating system (TOS).
2. Updating real-time ETA for vessels (BT_i).
3. Reallocating berthing schedules based on carbon emission considerations for both vessels and terminal facilities.
4. Optimizing vessel berth start time (BT_i) and berth location (BP_i) once the cost of the terminal plan cost is converged.

2.5.2. Visualization Layer Development

One of the primary objectives of the digital twin is to ensure the visibility of the port's dynamic objects. In the current port operation system, visibility of all dynamic objects is not provided to the port community members, leading to several accidents. An example of such an incident occurred at the PNC terminal in 2019, as depicted in Figure 12, underscoring the need for real-time object monitoring.

On the other hand, the digital twin offers complete visibility without any shadow zones. It visualizes the real-time movement of dynamic objects such as vessels, cranes, tugs, and containers using real-time location data. AIS (automatic identification system) data are used for vessel location, high-precision RTK (real-time kinematic) GPS data for crane and yard tractor location, and terminal operating system data for container location. UNITY software, a powerful game development platform and engine, was utilized to create the digital twin's visualization. All objects within the area were modeled using real-scale CAD data.

To bring the DT of the PNIT to life, an area of 625 km^2 was modeled, encompassing anchorages, the PNIT terminal, and the hinterland. All essential objects for digital twin development, including ships, port geometry, terminal berths, quay cranes, yard cranes, yard tractors, and truck trailers, were modeled with real-scale and real-geometry information.

Figure 12. Ship and quay crane collision in Pusan New Port [36].

The digital twin archives the complete history of the moving components of the port during the study period. It is designed to store data for two weeks from a given time and can generate a timeline within two weeks in advance every five minutes. This enables users to analyze past events or forecast future scenarios.

To realize the port's digital twin, all 36 objects were modeled during development, and Figure 13a showcases examples of 9 of the 36 objects. The object locations are converted into latitude and longitude, and they possess variables reflecting their location at specific time slots based on real CAD layers. These objects not only have physical shapes but also incorporate cost modeling capabilities to calculate port performance and carbon emissions.

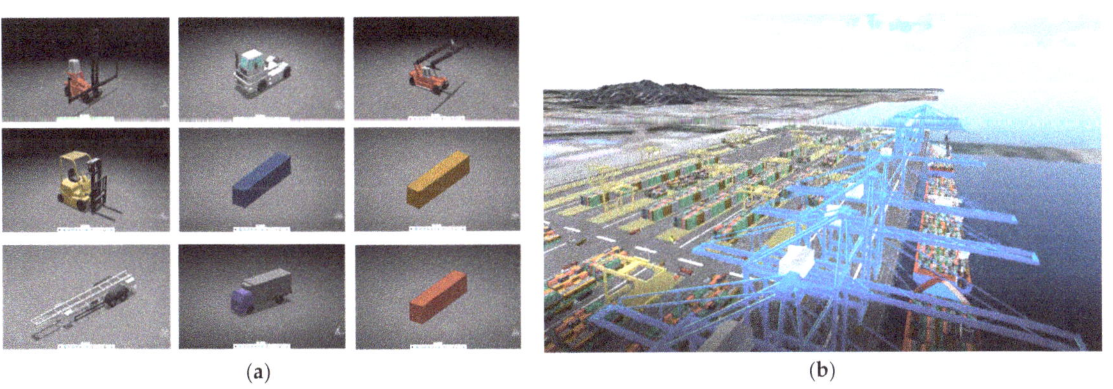

(**a**)   (**b**)

Figure 13. (**a**) Examples of digital twin objects. (**b**) Bird's eye view of the digital twin.

2.5.3. Service Layer Development

The digital twin platform includes functions that provide services for stakeholders. The prominent service functions include status monitoring, simulation of different operation schemes, and a guidance system.

- Monitoring Service

This service enables operators to monitor real-time data of the port's digital twin objects, including vessels, quay cranes, yard tractors, and tugs. Operators can track the current positions of vessels, berths, cranes, containers, yard tractors, and trailers, as depicted in Figure 14.

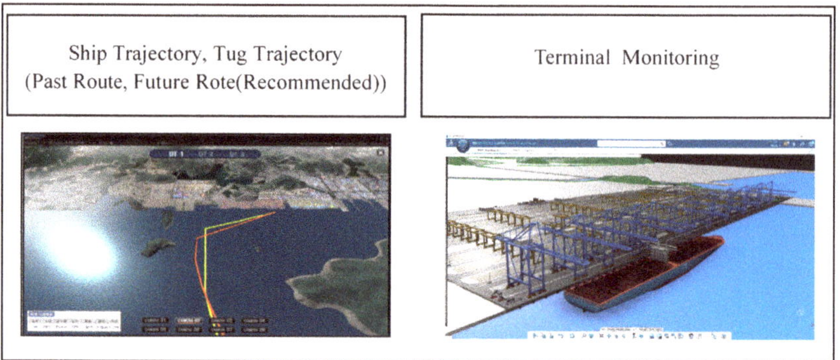

Figure 14. Graphic user interface of a monitoring service.

The monitoring service of the digital twin provides real-time status updates for vessels, including location, voyage status, and estimated time of arrival (ETA). It also offers a summary of CO_2 emissions from various objects, as shown in Figure 15.

Status		Data Connection	Vessel Name	Call Sign	ETA
Before arrival	Heading to busan	O	MSC RICC	CQIX6	2021:12:06 22:00
Before arrival	Heading to busan	O	PACIFIC TIANJIN	D5QW3	2021:11:22 22:00

Figure 15. Vessel arrival monitoring service in the DT.

- Simulation

The digital twin's primary purpose is to evaluate planned scenarios using simulations. Simulations forecasts the future performance of the digital twin and can replay past operation results. One key function is berth planning, which allows terminal operators to compare the efficiency of berth plans, as demonstrated in Figure 16. The sky-blue squares represent the original berth plan, and the yellow squares represent the reallocated berth schedule.

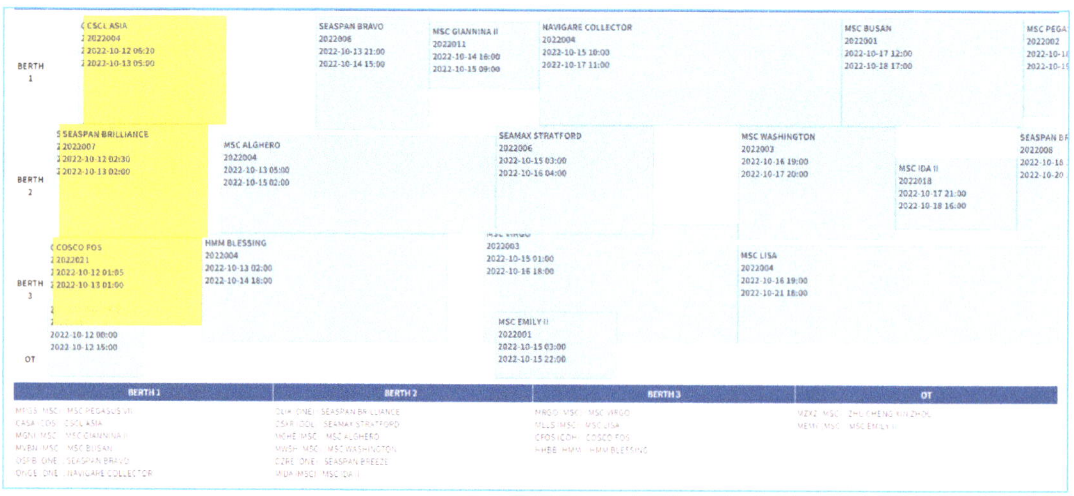

Figure 16. Berth plan using the DT simulation function.

- Guidance Service

This service offers a safe route and operational guidance for digital twin objects. It provides a recommended route for ships arriving at or departing from berths, thus enhancing safety and reducing the risk of collisions, as seen in Figure 17. The red dotted line indicates the recommended departure route for vessels, while the yellow dotted line indicates the suggested arrival route for vessels. The light green bounding box signifies the anchorage area where ships can anchor. The orange bounding box delineates the pilot embarking area, where pilots board incoming vessel, while the sky blue bounding box delineates the pilot disembarking area, where pilots disembark from departing vessels.

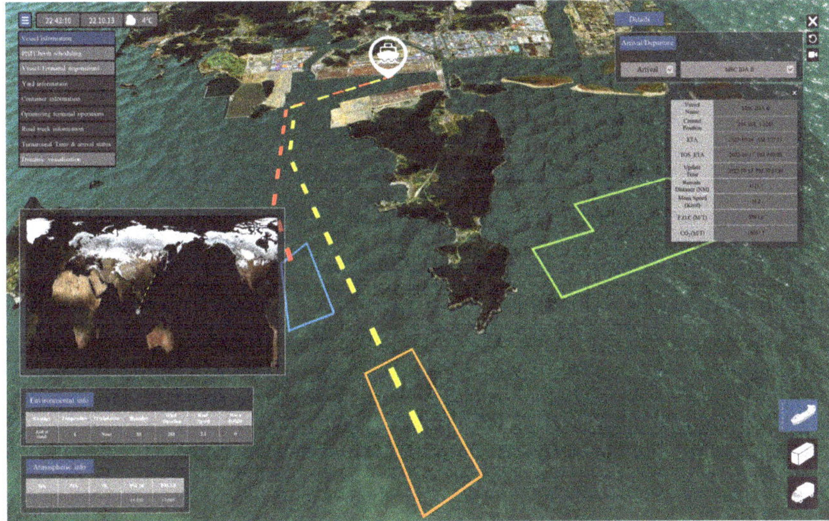

Figure 17. Guidance service.

3. Results

This section delves into the operational efficiency changes observed in the proposed DT model using collaborative scheduling in real-world cases. The real-world cases were carefully selected from the operational data from the PNIT (Pusan Newport International Terminal) spanning the month of September 2022. These analysis cases are designed to uncover temporal effects over various timeframes, including the short-term (8 h—one shift), mid-term (24 h—three shifts), and long-term (48 h—six shifts).

3.1. Ship Arrival Estimation Performance

The DT boasts a pivotal function that predicts ship arrival and departure times based on real-time location data and a sophisticated route planning algorithm, as detailed in Section 2. This ship arrival estimation function lays the foundation for collaborative scheduling between vessels and terminals. The accuracy of ship arrival predictions plays a vital role in dynamic berth planning within the DT. Prior to the development of the DT, automatic ship arrival estimation was unavailable, and thus berth planning changes were made after unexpected delays had already occurred. However, with the DT, it is now possible to proactively react to changes in berth planning based on real-time ship arrival predictions.

Of paramount significance is the precision of a vessel's ETA, as it directly impacts the quality of the berthing schedule. To estimate a ship's arrival time, this study leveraged historical data, specifically the shortest route taken by vessels from the previous port to PNIT, with reference to AIS historical data. The DT's ETA predictor provides real-time vessel arrival times, taking into account factors such as a vessel's current position (AIS), ship size, chosen route, and corresponding weather forecasts. This predictive function allows berth planners to use the DT to anticipate vessel arrival times and adjust berth plans accordingly.

Table 6 below presents a comparison between ETA errors over a three-month period obtained using the conventional method and the proposed DT-based method. The conventional method exhibited an average mean absolute error of 25.1 h, which was calculated using data recorded by ships and stored in the terminal operating system. In contrast, with the introduction of DT data, the ETA prediction accuracy significantly improved, achieving a mean absolute error of only 1.7 h, with a standard deviation of just 0.5 h. Hence, the proposed DT-based method enhanced the ETA by 95%.

Table 6. ETA error comparison.

ETA Error—3 Months	Average (h)	Standard Deviation
Without DT data (conventional method)	25.1	12.7
With DT data (proposed method)	1.7	0.5

3.2. Schedule Optimization and CO_2 Emission Quantification

To showcase the performance of the DT application, we used three simulation cases, each addressing unexpected delays at different temporal scales. These scenarios encompassed short-term (Section 3.2.1), mid-term (Section 3.2.2), and long-term (Section 3.2.3) delays. Prior to the DT application, the existing independent scheduling system lacked the ability to adapt to unforeseen delays. Consequently, when vessels experienced delays and deviated from their original schedules, an entire port's operations, including berth allocation, terminal facilities, and other port activities, came to a halt until the delayed vessel arrived. In contrast, the proposed DT-based berth planner proactively reallocates berth schedules, resulting in improved operational efficiency and reduced overall operating expenses.

3.2.1. Case 1: Short-Term Delay

In Case 1, involving a short-term 8-hour delay, the existing independent scheduling system caused downtime, as illustrated in Figure 18. Without the DT platform, a short-term delay of 8 hours led to a cascade effect, where multiple vessels' schedules were sequentially delayed within the same timeframe. The upper chart in Figure 18 shows Vessel 3's 8-hour delay marked in a red box, while the lower chart displays the statuses of Vessel 3 and Vessel 4 after this delay. Vessel 3's cargo operations start was postponed from the 44th hour to the 52nd hour, and Vessel 4's cargo operations start was pushed from the 52nd hour to the 60th hour.

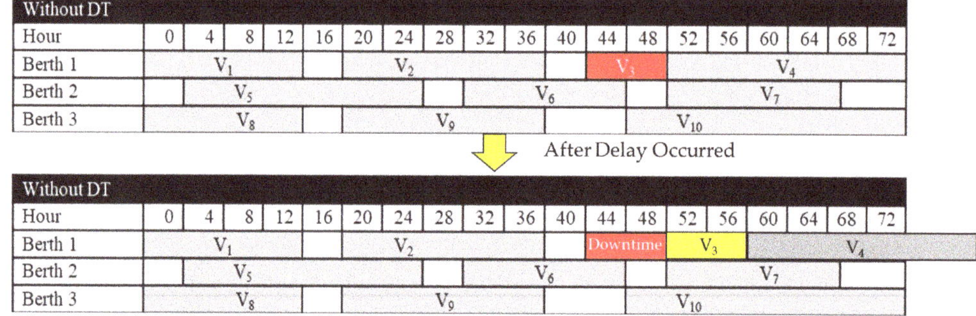

Hour		0	4	8	12	16	20	24	28	32	36	40	44	48	52	56	60	64	68	72
Berth 1		V_1					V_2						V_3				V_4			
Berth 2			V_5						V_6							V_7				
Berth 3			V_8				V_9							V_{10}						

After Delay Occurred

Hour		0	4	8	12	16	20	24	28	32	36	40	44	48	52	56	60	64	68	72
Berth 1		V_1					V_2						Downtime		V_3			V_4		
Berth 2			V_5						V_6							V_7				
Berth 3			V_8				V_9							V_{10}						

Figure 18. Separated operation without the digital twin—short-term delay case.

However, with the DT platform, Vessel 3's arrival time was adjusted simultaneously, extending from the 52nd to the 68th hour, as depicted in Figure 19. In this case, Vessel 3 sailed the same distance over the entire duration of its voyage, and the extra time gained,

16 hours, was factored into its schedule. This adjustment allowed Vessel 3 to reduce its speed, as expressed in Equation (12):

$$V_{3(reduced)} = \frac{Voyage\ distance}{Original\ voyage\ duration(V_3) + 16} \tag{12}$$

where $V_{3(reduced)}$ is the reduced vessel speed of Vessel 3.

Without DT

Hour	0	4	8	12	16	20	24	28	32	36	40	44	48	52	56	60	64	68	72
Berth 1			V_1				V_2					V_3				V_4			
Berth 2			V_5					V_6							V_7				
Berth 3			V_8				V_9						V_{10}						

With DT

Hour	0	4	8	12	16	20	24	28	32	36	40	44	48	52	56	60	64	68	72
Berth 1			V_1				V_2						V_{10}						
Berth 2			V_5						V_6						V_7				V_3
Berth 3			V_8				V_9						V_4						

Figure 19. Berthing plan—short-term delay case.

Figure 19 further demonstrates the berth plan adjustments in response to the short-term delay with and without the DT application. In the absence of the DT, the delayed schedule led to a 4-hour operation stop at Berth 1 between the 44th and 48th hour. Conversely, with the DT, the application reshuffled schedules, resulting in a mere 4-hour delay. This adjustment not only saved 4 hours in terminal berth scheduling but also reduced carbon emissions by 28.01 tons, as detailed in Table 7. Ultimately, the digital twin contributed to a significant reduction of 61.62 tons of CO_2 compared with the scenario without its use.

Table 7. Delay cost comparison for the short-term delay case.

Case	CO_2 Emissions (Tons)
Without the digital twin (conventional method)	89.63
With the digital twin (proposed method)	28.01
CO_2 saved	61.62

Figure 20 provides a visual representation of the accumulated cost savings, with the DT application significantly reducing costs compared with the scenario without the DT. The delayed scenario without the DT results in a higher cost increase due to from vessel delays.

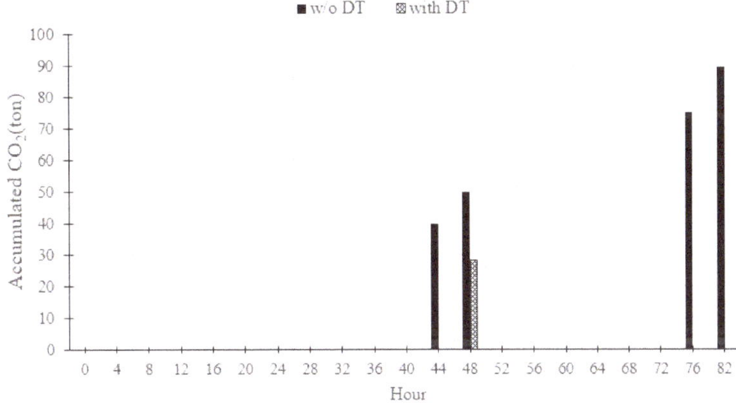

Figure 20. Accumulated delay costs—short-term delay case.

3.2.2. Case 2: Mid-Term Delay

In Case 2, involving a mid-term 28-hour delay, the independent scheduling system led to downtime, as depicted in Figure 21. Vessel 3 experienced a 28-hour delay, shifting its cargo operation start from the 44th hour to the 72nd hour.

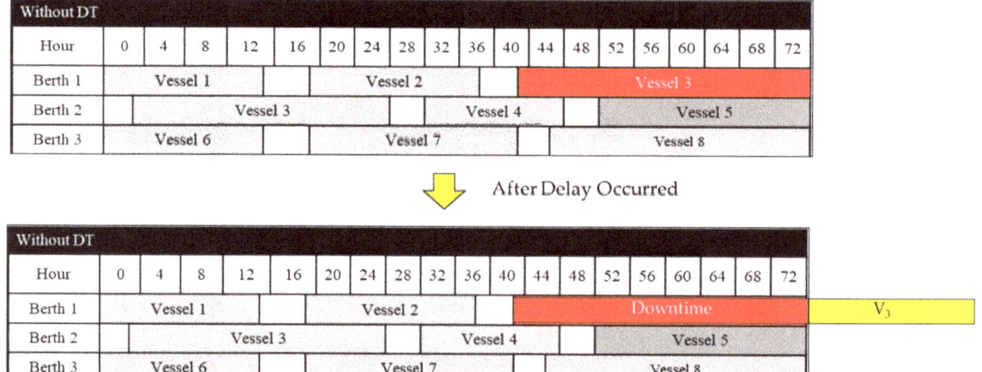

Figure 21. Separated operation without the DT—mid-term delay case.

With the DT application, Vessel 3's arrival time was adjusted from the 44th to the 68th hour, enabling a 24-hour saving achieved by slowing down the vessel's speed to meet the new required arrival time (68th hour). This speed reduction is expressed in Equation (13):

$$V_{3(reduced)} = \frac{Voyage\ distance}{Original\ voyage\ duration(V_3) + 24} \tag{13}$$

where $V_{3(reduced)}$ represents the reduced vessel speed of Vessel 3.

Figure 22 illustrates the schedule changes implemented using the DT application in the berth plan. The red-colored vessel symbolizes the delay. Without the DT, the ship and berth were both delayed by 28 hours. However, with the DT, the application adjusted the schedules for Vessel 5 and Vessel 3, resulting in Vessel 3 arriving eight hours earlier. This adjustment not only saved Vessel 3 24 hours but also ensured that other vessels expected to arrive at Berth 1 and 2 after 72 hours did not need to wait longer than originally scheduled.

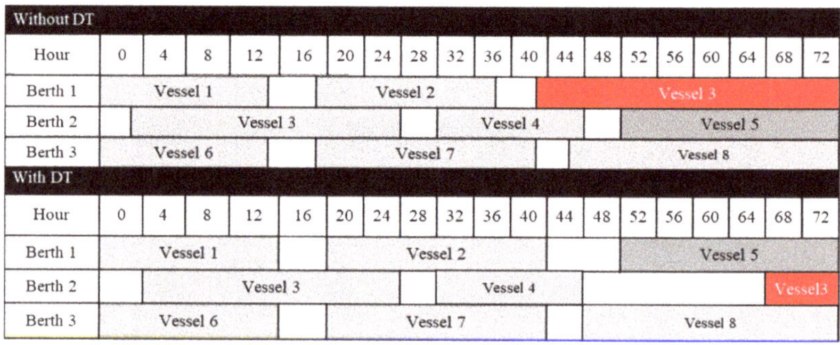

Figure 22. Berthing plan—mid-term delay case.

Table 8 and Figure 23 provide insights into the CO_2 emissions in cases with and without the DT. The DT application substantially reduced CO_2 emissions by 173.66 tons compared with the scenario without the DT. These findings highlight the substantial benefits of applying the digital twin, particularly in scenarios involving extended waiting times.

Table 8. Delay cost comparison for the mid-term delay case.

Case	CO_2 Emissions (Tons)
Without the digital twin (conventional method)	224.08
With the digital twin (proposed method)	50.42
CO_2 saved	173.66

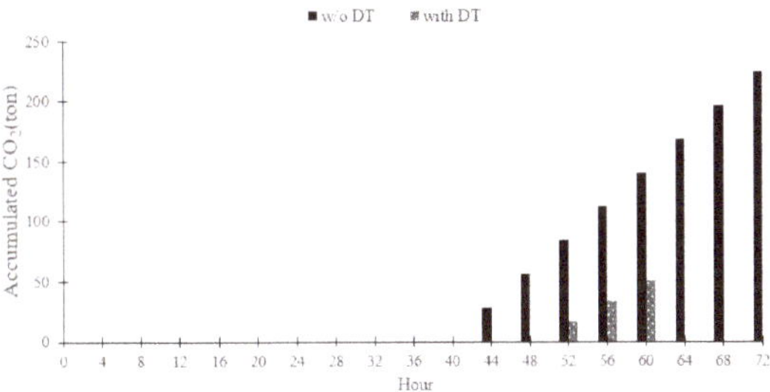

Figure 23. Accumulated delay costs—mid-term delay case.

3.2.3. Case 3: Long-Term Delay

The final case entails a 48-hour delay, as depicted in Figure 24, which showcases the schedule changes implemented using the DT application. Without the DT, Vessel 2 was delayed by 48 hours, resulting in a corresponding delay in the berth scheduling. After the DT application changed the plan, the schedules for Vessel 5, Vessel 7, Vessel 8, and Vessel 2 were modified and reassigned.

Without DT																		
Time	0	4	8	12	16	20	24	28	32	36	40	44	48	52	56	60	64	68	72
Berth 1	Vessel 1					Vessel 2 (48 hour delay)												
Berth 2		Vessel 3					Vessel 4						Vessel 5					
Berth 3	Vessel 6				Vessel 7						Vessel 8							
With DT																		
Time	0	4	8	12	16	20	24	28	32	36	40	44	48	52	56	60	64	68	72
Berth 1	Vessel 1				Vessel 7						Vessel 8							
Berth 2	Vessel 3					Vessel 4					Vessel 2							
Berth 3	Vessel 6									Vessel 5								

Figure 24. Berthing plan—long-term delay case.

Figure 25 and Table 9 illustrate the accumulated CO_2 emissions in cases with and without the DT. The DT application achieved a remarkable reduction of 313.69 tons of CO_2 emissions compared with the scenario without the DT. These results emphasize the exponential benefits of the digital twin, particularly in scenarios involving extended waiting times. These findings indicate that as the duration of vessel arrival delays increases, the benefits of the DT application become increasingly pronounced.

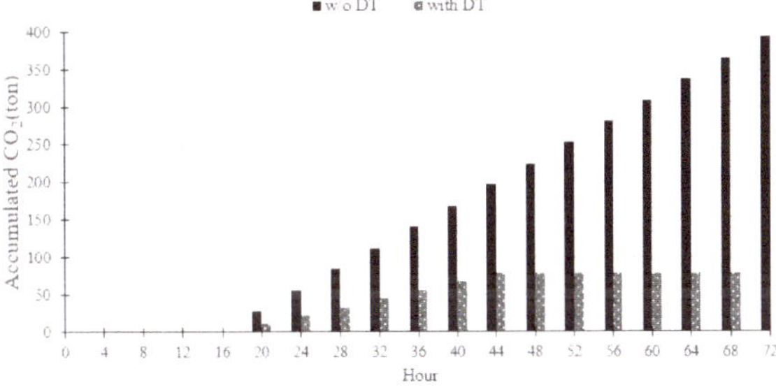

Figure 25. Accumulated delay costs—long-term delay case.

Table 9. Delay cost comparison for the long-term delay case.

Case	CO_2 Emissions (Tons)
Without the digital twin (conventional method)	392.12
With the digital twin (proposed method)	78.43
CO_2 saved	313.69

4. Discussion

In this study, we pioneered the development of a port digital twin model that caters to the diverse needs of port stakeholders, including vessels, tugs, and terminals, with the primary goal of optimizing plans and operations within container terminals. Our digital twin model proved to be capable of not only measuring but also significantly reducing carbon emissions stemming from vessels, terminal operations, and hinterland trucks, all while fostering collaborative and efficient maritime operations. The implications of this research hold profound significance for the shipping industry.

The digital twin model we created meticulously replicates the intricate decision-making processes involved in vessel arrivals and departures. Prior to our work, traditional methods for planning berth schedules, often reliant on statistical distributions of delays, struggled to predict vessel delays accurately. In stark contrast, our study introduces a novel data structure and a robust scheduling algorithm to form the backbone of our port digital twin. This innovation brings forth interactive scheduling capabilities between the port and vessels, thereby drastically enhancing our ability to predict vessel arrival times and reduce the carbon footprint associated with maritime voyages.

To validate the efficacy of our proposed digital twin model, we conducted a meticulous comparison with actual operational data from the studied terminal, focusing on September 2022. Using three compelling case studies, we demonstrated that our digital twin technology can reduce CO_2 emissions by an impressive average of 77.33% when compared with conventional independent scheduling systems. In absolute terms, this translates to an average reduction of 171.78 tons of CO_2 emissions. Consequently, the DT platform emerges as a formidable tool in the pursuit of reducing wait times for ships and port-related carbon emissions.

The performance of our model in ship arrival estimation and optimal scheduling is truly remarkable, boasting a 95% decrease in estimation errors when compared with conventional non-digital twin methods. Furthermore, our research delved into the digital twin's schedule optimization performance under variable conditions. We explored three distinctive scenarios, encompassing short-term, mid-term, and long-term delays, which allowed us to quantify the DT's schedule optimization capabilities using CO_2 reduction metrics. In the short term, we achieved a 75% reduction in CO_2 emissions, while in the mid-term, this reduction amounted to 77%. Notably, the long-term scenario exhibited an astonishing 80% reduction in CO_2 emissions when compared with legacy systems. In absolute terms, this translated to substantial reductions of 28.01 tons in the 8-hour delay case, 173.66 tons in the 24-hour delay case, and a remarkable 313.69 tons in the 48-hour delay case. This empirical validation underscores the digital twin's potential as a versatile tool for replicating current operations and optimizing schedules across port stakeholders.

Our proposed DT model, which encompasses crucial data components for vessel and berth scheduling, naturally positions itself as a potential backbone for autonomous operations within the maritime supply chain. This vision includes autonomous vessel arrivals and port operations, revolutionizing the way we envision maritime logistics and operations in the future.

In future studies, we envision using advanced optimization algorithms such as particle swarm optimization (PSO) to further enhance ship and port operational efficiency. Moreover, achieving pinpoint accuracy in predicting ship port arrival times remains a pivotal focus, and we will explore methods to enhance this accuracy further, leveraging real-time AIS data for precise ship ETA predictions. These endeavors hold immense promise for the continued advancement and transformation of the shipping industry in the digital era.

Author Contributions: Conceptualization, S.-W.K.; methodology, J.-O.E.; software, J.-H.Y. (Jeong-Hyun Yoon); validation, S.-W.K. and J.-H.Y. (Jeong-Hum Yeon); formal analysis, S.-W.K.; resources, S.-W.K.; data curation, J.-H.Y. (Jeong-Hyun Yoon) and J.-O.E.; writing—original draft preparation, S.-W.K.; writing—review and editing, S.-W.K., J.-O.E. and J.-H.Y. (Jeong-Hyun Yoon); visualization, J.-H.Y. (Jeong-Hum Yeon); supervision, S.-W.K.; funding acquisition, S.-W.K. All authors have read and agreed to the published version of the manuscript.

Funding: This research was supported by Korea Institute of Marine Science & Technology Promotion (KIMST) funded by the Ministry of Oceans and Fisheries (RS-2023-00256127).

Conflicts of Interest: The authors declare no conflict of interest.

Acronym

PNIT	Pusan Newport International Terminal
JITA	just-in-time arrival
ETA	estimated time of arrival
ETD	estimated time of departure
CTA	collaborative time of arrival
PTA	planned time of arrival
PTD	planned time of departure
RTA	required time of arrival
RTD	required time of departure
PBP	pilot boarding place
ETA PBP	estimated time of arrival at pilot boarding place
RTA PBP	required time of arrival at pilot boarding place
PTA PBP	planned time of arrival at pilot boarding place
ETC	estimated time of completion
ETS	estimated time of service
ATD	actual time of departure
ETD PBP	estimated time of departure at pilot boarding place
RTD PBP	required time of departure at pilot boarding place
ATD PBP	actual time of departure at pilot boarding place
ATA PBP	actual time of departure at pilot boarding place
YT	yard tractor
YC	yard crane
QC	quay crane
TUG	tug vessel
DT	digital twin
AIS	automatic identification system
TOS	terminal operating system
Port MIS	Port Maritime Information System
IMO	International Maritime Organization
DCSA	Digital Container Shipping Association
MRV	monitor, report, and verification
CO_2	carbon dioxide
MEPC	Marine Environment Protection Committee
PortCDM	port collaborative decision-making
RTK	real-time kinematic

Appendix A. Data for the DT Collaborative Scheduler

Table A1. Ship data for the DT collaborative scheduler (Table 3).

Data Name	Sample	Standard
Current time	2020-04-06T 08:00:00 + 02:00	ISO8601
Vessel IMO number	1801323	IMO
Vessel position, latitude	192.515, 51.9200000	ISO 6709:2008
Vessel tons	50,000	Gross ton
Previous port	KR BUS	UN location code

Table A1. *Cont.*

Data Name	Sample	Standard
Next port	SG JUR	UN location code
Facility code	PBPL	DCSA
ETA BERTH	2020-04-06T 08:00:00 + 02:00	ISO8601
ETA PBP	2020-04-06T 08:00:00 + 02:00	ISO8601
Emission	Ton	Carbon content
Berth location	Berth NR5	
Voyage type	Cargo	
Vessel type	Container	
Crew number	10	Integer
Tug usage	Yes	
Pilot	Yes	
No-go zone in port	192.515, 51.9200000	ISO15016
Wind speed	m/s	ISO15016
Wind direction	Degree	ISO15016
Wave height	M	ISO15016
Wave direction	degree	ISO15016
Current speed	m/s	ISO15016
Current direction	Degree	ISO15016
Route candidate information	192.515, 51.9200000 …	ISO 6709:2008
Optimal route information	(192.515, 51.920, speed, direction, time)	ISO 6709:2008

Table A2. Terminal data for the DT collaborative scheduler (Table 4).

Data Name	Sample	Standard
Current time	2020-04-06T 08:00:00 + 02:00	ISO8601
Vessel IMO number	1801323	IMO
Vessel tons	50,000	Gross ton
Previous port	KR BUS	UN location code
Next port	SG JUR	UN location code
Facility code	PBPL	DCSA
ETA BERTH/ ETD berth	2020-04-06T 08:00:00 + 02:00	ISO8601
ETA PBP/ETD PBP	2020-04-06T 08:00:00 + 02:00	ISO8601
Emission	Ton	Carbon content
Berth location	Berth NR5	
Operation expenditure	Korean won	
Berth air draft	50 m	
Berth depth	−25 m	
Bitt number	0~110	
Yard tractor cost	Korean won	
Quay crane capacity	30TEU/h	
Weight factor	Constant	

Table A3. Ship data for DT collaborative scheduler (Table 5).

Data Name	Sample	Standard
Current time	2020-04-06T 08:00:00 + 02:00	ISO8601
Vessel IMO number	1801323	IMO
Vessel tons	50,000	Gross ton
Previous port	KR BUS	UN location code
Next port	SG JUR	UN location code
Facility code	PBPL	DCSA
ETA BERTH/ ETD berth	2020-04-06T 08:00:00 + 02:00	ISO8601
ETA PBP/ETD PBP	2020-04-06T 08:00:00 + 02:00	ISO8601
Tug allocation result (number)	1~8	
Tug unit cost	Korean won/h	

References

1. Grieves, M. *Completing the Cycle: Using PLM Information in the Sales and Service Functions [Slides]*; SME Management Forum: Plymouth, MI, USA, 2002.
2. Grieves, M.W. Digital Twins: Past, Present, and Future. In *The Digital Twin*; Springer: Berlin/Heidelberg, Germany, 2023; pp. 97–121.
3. Deng, T.; Zhang, K.; Shen, Z.-J.M. A systematic review of a digital twin city: A new pattern of urban governance toward smart cities. *J. Manag. Sci. Eng.* **2021**, *6*, 125–134. [CrossRef]
4. Joung, T.-H.; Kang, S.-G.; Lee, J.-K.; Ahn, J. The IMO initial strategy for reducing Greenhouse Gas (GHG) emissions, and its follow-up actions towards 2050. *J. Int. Marit. Saf. Environ. Aff. Shipp.* **2020**, *4*, 1–7. [CrossRef]
5. Augustine, P. The industry use cases for the digital twin idea. In *Advances in Computers*; Elsevier: Amsterdam, The Netherlands, 2020; Volume 117, pp. 79–105.
6. Taylor, N.; Human, C.; Kruger, K.; Bekker, A.; Basson, A. Comparison of digital twin development in manufacturing and maritime domains. In *Service Oriented, Holonic and Multi-Agent Manufacturing Systems for Industry of the Future: Proceedings of SOHOMA 2019*; Springer: Berlin/Heidelberg, Germany; pp. 158–170.
7. Hofmann, W.; Branding, F. Implementation of an IoT-and cloud-based digital twin for real-time decision support in port operations. *IFAC-Pap.* **2019**, *52*, 2104–2109. [CrossRef]
8. Facilitation Committee (FAL), 47th session, 13–17 March 2023. Available online: https://www.imo.org/en/MediaCentre/MeetingSummaries/Pages/FAL-47th-session.aspx (accessed on 12 August 2023).
9. DCSA. Just-in-Time Port Call. Available online: https://dcsa.org/standards/jit-port-call/ (accessed on 12 August 2023).
10. Liu, M.; Zhou, Q.; Wang, X.; Yu, C.; Kang, M. Voyage performance evaluation based on a digital twin model. In *IOP Conference Series: Materials Science and Engineering*; IOP Publishing: Bristol, UK, 2020; p. 012027.
11. Stoumpos, S.; Theotokatos, G.; Mavrelos, C.; Boulougouris, E. Towards marine dual fuel engines digital twins—Integrated modelling of thermodynamic processes and control system functions. *J. Mar. Sci. Eng.* **2020**, *8*, 200. [CrossRef]
12. Gao, Y.; Chang, D.; Chen, C.-H.; Xu, Z. Design of digital twin applications in automated storage yard scheduling. *Adv. Eng. Inform.* **2022**, *51*, 101477. [CrossRef]
13. Wang, K.; Hu, Q.; Zhou, M.; Zun, Z.; Qian, X. Multi-aspect applications and development challenges of digital twin-driven management in global smart ports. *Case Stud. Transp. Policy* **2021**, *9*, 1298–1312. [CrossRef]
14. Wang, K.; Hu, Q.; Liu, J. Digital twin-driven approach for process management and traceability towards ship industry. *Processes* **2022**, *10*, 1083. [CrossRef]
15. Fonseca, Í.A.; Gaspar, H.M. Challenges when creating a cohesive digital twin ship: A data modelling perspective. *Ship Technol. Res.* **2021**, *68*, 70–83. [CrossRef]
16. Coraddu, A.; Oneto, L.; Baldi, F.; Cipollini, F.; Atlar, M.; Savio, S. Data-driven ship digital twin for estimating the speed loss caused by the marine fouling. *Ocean Eng.* **2019**, *186*, 106063. [CrossRef]
17. Danielsen-Haces, A. Digital Twin Development—Condition Monitoring and Simulation Comparison for the ReVolt Autonomous Model Ship. Master's Thesis, Norwegian University of Science and Technology, Trondheim, Norway, 2018.
18. Vasstein, K. A High Fidelity Digital Twin Framework for Testing Exteroceptive Perception of Autonomous Vessels. Master's Thesis, NTNU, Trondheim, Norway, 2021.
19. Raza, M.; Prokopova, H.; Huseynzade, S.; Azimi, S.; Lafond, S. Towards integrated digital-twins: An application framework for autonomous maritime surface vessel development. *J. Mar. Sci. Eng.* **2022**, *10*, 1469. [CrossRef]
20. Jahn, C.; Scheidweiler, T. Port call optimization by estimating ships' time of arrival. In Proceedings of Dynamics in Logistics: Proceedings of the 6th International Conference LDIC 2018, Bremen, Germany, 17 February 2018; Springer: Berlin/Heidelberg, Germany, 2018; pp. 172–177.
21. Cho, J.; Craig, B.; Hur, M.; Lim, G.J. A novel port call optimization framework: A case study of chemical tanker operations. *Appl. Math. Model.* **2022**, *102*, 101–114. [CrossRef]
22. Sarantakos, I.; Bowkett, A.; Allahham, A.; Sayfutdinov, T.; Murphy, A.; Pazouki, K.; Mangan, J.; Liu, G.; Chang, E.; Bougioukou, E. Digitalization for Port Decarbonization: Decarbonization of key energy processes at the Port of Tyne. *IEEE Electrif. Mag.* **2023**, *11*, 61–72. [CrossRef]
23. Alamoush, A.S.; Ölçer, A.I.; Ballini, F. Ports' role in shipping decarbonisation: A common port incentive scheme for shipping greenhouse gas emissions reduction. *Clean. Logist. Supply Chain* **2022**, *3*, 100021. [CrossRef]
24. Lu, Y.; Le, M. The integrated optimization of container terminal scheduling with uncertain factors. *Comput. Ind. Eng.* **2014**, *75*, 209–216. [CrossRef]
25. Ismail, T.; Elbeheiry, M.; Elkharbotly, A.; Afia, N.; Abdalla, K. Continuous Berth Scheduling in Port Terminals. *Int. J. Eng. Res. Technol.* **2016**, *5*, 531–535.
26. Du, Y.; Xu, Y.; Chen, Q. A feedback procedure for robust berth allocation with stochastic vessel delays. In Proceedings of the 2010 8th World Congress on Intelligent Control and Automation, Jinan, China, 7–9 July 2010; pp. 2210–2215.
27. Xiang, X.; Liu, C.; Miao, L. A bi-objective robust model for berth allocation scheduling under uncertainty. *Transp. Res. E Logist. Transp. Rev.* **2017**, *106*, 294–319. [CrossRef]
28. Park, H.J.; Cho, S.W.; Lee, C. Particle swarm optimization algorithm with time buffer insertion for robust berth scheduling. *Comput. Ind. Eng.* **2021**, *160*, 107585. [CrossRef]

29. Kim, S.; Jang, H.; Cha, Y.; Yu, H.; Lee, S.; Yu, D.; Lee, A.; Jin, E. Development of a ship route decision-making algorithm based on a real number grid method. *Appl. Ocean Res.* **2020**, *101*, 102230. [CrossRef]

30. Son, J.; Kim, D.-H.; Yun, S.-W.; Kim, H.-J.; Kim, S. The development of regional vessel traffic congestion forecasts using hybrid data from an automatic identification system and a port management information system. *J. Mar. Sci. Eng.* **2022**, *10*, 1956. [CrossRef]

31. Yoon, J.-H.; Kim, D.-H.; Yun, S.-W.; Kim, H.-J.; Kim, S. Enhancing Container Vessel Arrival Time Prediction through Past Voyage Route Modeling: A Case Study of Busan New Port. *J. Mar. Sci. Eng.* **2023**, *11*, 1234. [CrossRef]

32. Park, Y.-M.; Kim, K.H. A scheduling method for berth and quay cranes. *OR Spectr.* **2003**, *25*, 1–23. [CrossRef]

33. Kim, K.H.; Moon, K.C. Berth scheduling by simulated annealing. *Transp. Res. B Methodol.* **2003**, *37*, 541–560. [CrossRef]

34. Park, K.; Kim, K.H. Berth scheduling for container terminals by using a sub-gradient optimization technique. *J. Oper. Res. Soc.* **2002**, *53*, 1054–1062. [CrossRef]

35. Bayraktar, M.; Yuksel, O. A scenario-based assessment of the energy efficiency existing ship index (EEXI) and carbon intensity indicator (CII) regulations. *Ocean Eng.* **2023**, *278*, 114295. [CrossRef]

36. Team, P.T. *Mega-Ship Questions Raised after ONE Vessel Crashes in Busan*; Port Technology International: London, UK, 2020.

MDPI AG
Grosspeteranlage 5
4052 Basel
Switzerland
Tel.: +41 61 683 77 34

Journal of Marine Science and Engineering Editorial Office
E-mail: jmse@mdpi.com
www.mdpi.com/journal/jmse